THE VICTORIAN AMATEUR ASTRONOMER
Independent Astronomical Research in Britain
1820–1920

*To the memory of Albert Platt (1891 – 1984)
my maternal grandfather,
a late Victorian who gave me my fascination
with the history of human ingenuity*

THE VICTORIAN AMATEUR ASTRONOMER

Independent Astronomical Research in Britain
1820 – 1920

Allan Chapman, M.A., D.Phil., D.Sc., D.Univ., F.R.A.S.
Wadham College, Oxford and
Faculty of History, University of Oxford

GRACEWING

First published in 1998 by
John Wiley & Sons
in association with
Praxis Publishing, Chichester

Second Edition 2017
by
Gracewing
2 Southern Avenue
Leominster
Herefordshire HR6 0QF
United Kingdom
www.gracewing.co.uk

All rights reserved. No part of this publication may be reproduced, stored in a retrieval system, or transmitted in any form, or by any means, electronic, mechanical, photocopying, recording or otherwise, without the written permission of the publisher.

© 1998, 2017 Allan Chapman

The right of Allan Chapman to be identified as the author of this work has been asserted in accordance with the Copyright, Designs and Patents Act 1988.

978 0 85244 544 0 paperback
978 1 78182 010 0 cased edition

Original typesetting courtesy Praxis Publishing
Additional typesetting for the Second Editon by The Choir Press

Introduction to the second edition

The history of astronomy goes back as far as the science itself. Hipparchus, around 130 BC, incorporated ancient Egyptian observations into his catalogues, while Ptolemy's *Almagest*, of c. AD 150, incorporated Hipparchus's and other early observations. Then as the centuries rolled by, one finds subsequent astronomical writers incorporating the old into the new, as seen in Copernicus, Kepler, and others. John Flamsteed included a major history of astronomy in his *Historia Coelestis Britannica*, Volume 3 (1725), while Robert Grant's monumental *History of Physical Astronomy* (1852) is still regarded as a key source.

The twentieth century has also seen many distinguished scholars and scientists making fundamental contributions to the history of astronomical science, some of whom I have had the honour to know personally and to work with. These include figures of the stature of Michael Hoskin, Michael J. Crowe, Barbara Becker, Owen Gingerich, and many more besides; and the sadly late John D. North, Mary Brück, and my undergraduate contemporary Peter D. Hingley.

Yet what historians across a range of sciences have generally concentrated upon has been scientists who had an obvious professional or occupational relationship with their science, such as being professorial heads of observatories, laboratories, or research institutes.

When, however, historians looked at major figures who advanced science out of their own private resources, from a love of their subject (i.e. 'amateurs' in the original sense, from the Latin *amat*, 'he or she loves'), their circumstances would not generally be mentioned. One was left to assume that they must be 'professional' in so far as they knew what they were doing, for amateurs were mere hobbyists.

Yet two of the most admired scientists in history, whose work fundamentally changed scientific thinking, were, in the best sense of the word, 'amateurs'. For example, in his 'day job' Nicholas Copernicus was an eminent physician, an ecclesiastical lawyer, and a Polish Cathedral dignitary, whose work on a heliocentric solar system had to be fitted in around more pressing professional demands upon his time.

Likewise, no one has ever doubted the significance of Charles Darwin's contributions to biology, yet Darwin never did a day's paid work in his life, writing *The Origin of Species* and other works on the strength of a very ample private income. Indeed, the only 'paid' or 'professional' work that Darwin probably ever did was negotiate good royalties with his publishers, or make visits to his stockbroker and land agent to ensure that his assets were performing well. Even the five years during which Darwin went around the world in *HMS Beagle* were spent in the capacity of friend and companion to the Captain, and not, as it is sometimes assumed, as a paid agent of the Admiralty.

When I was a student, in 1969, it was the accepted 'mantra' that nineteenth-century British science lagged behind that of Continental Europe, where there were 'real' scientists who had publicly-funded university jobs, or were clients of the state whose talents had been noticed by a king or an archduke. But instinctive sceptic that I am when I hear independent initiative and achievement being ignored or derided, I smelt a rat. For if Britain had no real scientists in the nineteenth century, then how did one evaluate the achievements of Lord Rosse, the Herschels, the Hugginses, Mary Somerville, Warren De La Rue, William Lassell, and a host of others, none of whom got so much as a mention in the standard history of science literature that I read? The significance of their discoveries is beyond dispute, yet none of them held scientific jobs, and all were self-funded.

This led me to formulate that class of Georgian and Victorian scientist to whose achievement this book is devoted: the 'Grand Amateurs', and their more modestly-funded brothers and sisters, who became serious amateur astronomers across Great Britain and its Empire, and formed something of a parallel tradition in the U.S.A.

It was during the summer of 1988, when working on my paper on William Lassell, that the term 'Grand Amateur' first occurred to me. For some time I had been looking around for a suitable designation for that self-funded yet thoroughly 'professional' category of scientist, so perfectly epitomised by the brewer, businessman, and astronomical researcher Lassell. And it was in my 1989 paper on William Lassell (*Vistas in Astronomy* 32, pp. 341–70) that I first aired the designation 'Grand Amateur' in print.

The more I researched the world of self-funded science, from the seventeenth century onwards, the more I came to discover the sheer magnitude of its achievement, in astronomy and in other sciences as well. Across the English-speaking world self-funded scientists were everywhere: in geology, chemistry, physics, meteorology, biology, and even medicine. Scientific clergymen, for example, were widespread.

Take, for instance, the eighteenth-century Oxfordshire vicar Edward Stone, whose parishioner-based clinical trials on the efficacy of willow bark (salicylic acid) to 'break' ague fevers provided the first insights that would lead to the development of aspirin. Stone published the results of his trials in the prestigious *Philosophical Transactions* of the Royal Society in 1763. And the Revd Mr Stone was also a serious and well-equipped astronomer, who made detailed observations of the Venus transit of 1761, which he reported in his *Whole Doctrine of Parallaxes ...* (Oxford, 1763).

One finds similar circumstances at work in pre- and post-Colonial North America, with figures of the standing of Benjamin Rush, David Rittenhouse, Benjamin Franklin, and those nineteenth-century New-York-based pioneers of astronomical photography and spectroscopy, the lawyer Lewis Rutherfurd and the medical Draper family, whose achievements are discussed in the following pages.

On 29 June 2002, a new Society devoted to the history of astronomy was founded. The Society for the History of Astronomy held its inaugural meeting in Wadham College, Oxford: the very same College in which Bishop John Wilkins, Sir Christopher Wren, John Wallis, Robert Hooke, the Hon. Robert Boyle and other friends had met for their 'Experimental Club' after 1649. After the Restoration of the Monarchy in 1660, this same group of friends would re-locate to Gresham College in the City of London, to become the Royal Society. And like the early Royal Society, the S.H.A. would be composed of self-funded friends and members who

Introduction to the second edition 7

shared a common passion for the history of astronomy, and wished to advance historical research and communicate it to the wider public.

Under the early guidance of Stuart Williams, Kevin Kilburn, and the late Ken Goward and Peter Hingley, the Society grew and prospered, as it continues to do today. Peter Hingley's 'day job' as Librarian and Archivist of the Royal Astronomical Society opened up wonderful research opportunities to S.H.A. members keen to obtain access to an abundance of primary sources. And the S.H.A. and R.A.S. continue in this close working relationship. As with the early Royal Society and the early R.A.S., the S.H.A.'s members earn their daily bread through a variety of traditional and new professions, from medicine to information technology. It is their subscriptions, donations, and bequests that pay for everything, be they a student paying a subsidised rate, or a retired professional leaving a handsome bequest in their will. For just as Bishop Thomas Sprat F.R.S. styled the Royal Society in 1667, they are all '*gentlemen* [and now ladies], free and unconfin'd'. Yet independent and self-governing as the S.H.A. is, it is open and welcoming to all, its members being involved not just with research, but also with outreach to schools and to regional astronomical societies.

Society meetings are held in both northern and southern Britain, and recently, in Paris, although its hub is the Victorian gothic grandeur of the Birmingham and Midland Institute. Here, the S.H.A. holds its annual AGM in October, and it is in the B.M.I. that the Society's large and constantly growing Library of historical astronomy books is located.

Central to the research policy of the S.H.A. is a detailed regional survey of British astronomy, with regional section co-ordinators around the country. In the first, 1998, edition of the present work, I listed and looked at the biographies and achievements of a number of local astronomers who were serious in their work but not widely known: figures such as the working men astronomers John Jones and Roger Langdon. But since 2002, S.H.A. researchers have, quite literally, located, written up, photographed, and published *dozens* more. Their work includes material on individuals, houses, ruined observatories, surviving instruments and so on, to provide a 'topographical demography' or catalogue of British Isles, Irish, and wider astronomy. The S.H.A.'s official publications, *The Antiquarian Astronomer, The Bulletin,* and *e-News*, both on paper and in electronic format, put members' work on public record to build a valuable and scholarly archive of serious 'amateur' astronomy over the centuries.

And I felt it a great honour back in 2002 when I was invited to become the S.H.A.'s Foundation Honorary President – an office which I am proud to continue to hold today. But being a self-confessed 'disaster area' as far as administration goes, who always loses paper and makes computers blow up, I restricted my role to an Honorary function, and keep well away from committees, restricting my activities to writing, lecturing, and broadcasting. My relationship with the Society continues to be very close.

The S.H.A. not only works closely with the R.A.S. Library and its current Librarian and Archivist Sian Prosser, but also with other societies and organisations such the William Herschel Society and the Herschel Museum of Astronomy, Bath, the British Astronomical Association historical section, and the numerous amateur astronomical and local history societies around the country. And as one might expect, there is often a considerable cross-over in the interests of the membership.

I am greatly indebted to Dr James Dawson, a busy hospital doctor who also serves as Librarian of the S.H.A., for giving his precious spare time and technical expertise to make it possible, at last, to see this second edition of my book through the press. For while Tom Longford of Gracewing Press had long desired to re-publish *The Victorian Amateur Astronomer*, only half of the original electronic files were still in existence. It was James Dawson who, during the summer of 2017, in conjunction with the S.H.A. Council, used his skills and resources to get the ball rolling once again. Once James had shown a way forward to producing a clear camera-ready version of the text, I set about doing all the revisions and errata – many of which had been passed on to me by S.H.A. members.

I am also deeply indebted to my wife Rachel, who took a temporary break from editing and proof-reading the text of my current book *Comets, Cosmology, and the Big Bang: From Halley to Hubble* to re-scan all 80 of my original *V.A.A.* plates. I owe a debt of gratitude, too, to Dr Cristina Neagu and Alina Nachescu, both of Christ Church Library, Oxford, for their valuable contributions regarding scanning and formatting of my pages.

And once the ball had started to roll, things moved very fast. The present volume is a testimony to all of their help and kindness. I thank them all most sincerely.

Allan Chapman

24 August 2017

Additional Acknowledgements for the Second Edition

Michael Chauvin, James Dawson, Ian Elliot, Kenneth Goward, Lee Macdonald, Charles Mollan, Sian Prosser, and the Council and members of the Society for the History of Astronomy. I would also like to express my gratitude to Tom Longford of Gracewing. Yet, as always, my heartiest thanks go to my classicist wife Rachel, 'Wife, Scholar, and Best Friend', who has undertaken so much of the editing, checking, and arranging of this second edition.

Additional material in the Second Edition

Addendum: Four astronomers brought to my notice since the 1998 edition
            Four new Journals   304

Chapter 14, references 94–101, which were missing in the original
1998 edition (including material not available in 1998)   407

Errata to the 1998 Edition   408

Table of contents

Author's preface... xi
Acknowledgements... xv
List of plates.. xvii

Part 1 The Grand Amateurs

1   Amateur astronomy in the Romantic Age............................. 3

2   Gentlemen and players: amateurs and professionals in 1840......... 13
    Greenwich, Oxford and Cambridge................................... 13
    The influence of Germany.. 18
    Other professional observatories in Britain....................... 20
        England... 20
        Durham, Liverpool and the Cape of Good Hope................... 20
        Scotland.. 23
        Ireland... 24
    British 'Grand Amateur' observatories............................. 26
    Amateur observatories in Europe and America....................... 27
    William Pearson and the Royal Astronomical Society................ 29

3   An inheritance, a wife, a benefice, or a brewery: financing fundamental
    research.. 33
    Grand Amateur priorities and interests............................ 34
    Paying for instrumental innovation: the astronomical circle....... 35
        Stephen Groombridge... 36
        John Pond... 37
        William Pearson... 37
        Richard Christopher Carrington................................ 40
    New discoveries in positional astronomy: stellar proper motions
    and binary stars.. 42

viii Table of contents

	Sir James South	44
	William Rutter Dawes	46
	George Bishop	47
	Edward Joshua Cooper of Markree	48
	The American amateurs	50

4 Sir John Herschel: a model for the independent scientist 53

5 An astronomical house-party: the Bedford–Aylesbury axis 75

6 The brotherhood of the big reflecting telescope 93
   William Parsons, Third Earl of Rosse 96
   William Lassell 100
   Lassell's collaboration with James Nasmyth 102
   Nasmyth's telescopes 104
   The swan-song of speculum 109

7 The new sciences of light: spectroscopy, photography, and the
   Grand Amateurs 113
   William Huggins 114
   Warren De La Rue 117
   The American connection 118
       Lewis Rutherfurd 118
       Henry Draper 119
   Photography and the economics of astronomical working time 121
   Joseph Norman Lockyer 122
   Lord Crawford's Dun Echt Observatory 133
   Two engineers and a master builder: McClean, Common and Roberts .. 135
   The Irish photometrists: Minchin, Monck, Wilson and Fitzgerald 139
   George Higgs: a shopkeeper spectrometrist 141
   The escalation of costs: towards professionalisation 141

8 The astronomers' gentlemen: the Grand Amateurs' professional
   assistants 145

Part 2 Poor, obscure and self-taught: astronomy and the working class

   Prologue 161

9 A penny a peep: the astronomical lecturers of the people 165

10 Astronomy and the modest master-craftsman 181
   Johnson Jex: the learned blacksmith of Letheringsett 183
   James Veitch: the ploughwright of Inchbonny 185

	John Leach: the cobbler of Frodsham........................ 188
	Thomas William Bush: the baker of Nottingham 200
11	**The day-labourer astronomer**................................ 205
	John Robertson: the railway porter of Coupar Angus............. 206
	John Jones: the slate counter of Bryngwyn Bach................. 209
	Roger Langdon: a village Station Master astronomer........... 213
	Astronomers in isolation................................... 217

Part 3 The rise of the leisured enthusiast

Prologue... 221

12	**A goodly pursuit for a Godly mind: Thomas Webb and his influence** 225
	Thomas William Webb... 225
	The amateur astronomer's telescope........................ 228

| 13 | **That clubbable passion: the amateur astronomical society** .... 243 |

| 14 | **Now ladies as well as gentlemen**........................... 273 |

| 15 | **Conclusion and postscript: the amateur astronomer into the twentieth century**........................... 295 |

Notes and references 305
Index 409

Author's preface

The purpose of this book is to examine the contributions made to astronomy by those persons who were not paid professionally to do so. For within the nineteenth century, and certainly before 1880, virtually no fundamental research in astrophysics, cosmology, or planetary astronomy was performed in the British Isles that was not devised, paid for, and executed by private individuals. But in addition to those people who were involved in cutting-edge research, there were many hundreds who simply enjoyed looking at the sky with whatever instruments they could afford, trying to re-observe the objects first discovered and interpreted by their richer co-workers, and helping to spread astronomical knowledge into contemporary society. Between them, these two groups constituted the amateur astronomical community.

Nowadays, however, the word 'amateur' often has a pejorative connotation – a connotation which developed, I would suggest, only after the growing professionalisation of so much of what went on in our national intellectual life after about 1880. This led to that unfortunate dichotomy in which institutionally-funded expertise was supposed to set the standard to which 'amateurs' merely aspired.

But like its nineteenth-century users, I define 'amateur' in the Latinate sense of the word: *amat*, 'one who loves'; and by extension, one can apply the term with equal cogency to those men and women who still pursue astronomy as a serious hobby today. In our own day and age, however, not even the most committed and well-equipped amateur astronomer would pretend that he or she stood at the research frontier of the science. With institutions like NASA pioneering planetary research, and with academically-managed instruments like the Jodrell Bank, Isaac Newton, and Hubble Space Telescopes in operation, it would be absurd to make such claims for modern amateurs. But 150 years ago it was very different, for then all of the big instruments, and the expertise to use them properly, were in private hands.

To give a definition of the word 'amateur', however, only preempts a definition of 'professional' within the same context. Central to this definition, I believe, is the concept of pursuing a particular activity for money – a criterion invariably employed in nineteenth-century dictionaries – and the use of that calling to earn one's living. To be a 'professional' in any sphere of life in 1850 did not merely presuppose that one was competent and thorough in one's business, but that the business formed a vital adjunct to one's pecuniary survival.

Of course, both amateurs who worked for love and professionals who observed the heavens for money quite clearly pursued careers which, by accident or by design, could win public approval. Fellowships of the Royal Society, Gold Medals, and honorary degrees were the visible marks of success in an astronomical career, and it mattered little if one's livelihood derived from the ownership of a brewery or from an academic appointment. But a professional had a defining contractual bond with astronomy which an amateur lacked, and it is in the *earning* of money, as opposed to the *spending* of the same, that the real difference between the two classes of astronomers lay in nineteenth-century Britain.

It is not my intention in this book to treat amateur astronomy taxonomically, or to draw up a list of all known British practitioners who were active between 1800 and 1920. My abiding concern, rather, is to look at amateur astronomy as a movement taking place within the context of the 'long nineteenth century'. Although I use the term 'Victorian', I do so only insofar as it epitomises an era and a set of historical circumstances in which attitudes towards education, intellectual curiosity, politics, and finance emphasised individual initiative and minimalist public provision. This is very different from the circumstances that apply today, and I do *not* wish any readers to draw the incorrect conclusion that I believe that the Victorian approach to the financing of science and education should be used as a model for the new millennium.

Concerned as I am with tracing the history of a scientific movement within nineteenth-century British society, I shall proceed with four interrelated factors in mind. Firstly, it must be understood that I am not so much writing about *astronomy* as about *astronomers*, and this naturally involves considerable biographical treatment, as one attempts to come to terms with why a particular person chose to become an astronomer. Secondly, it is my intention to relate these individuals to the social and financial worlds in which they moved; for fundamental research was costly, and presumed access to both money and leisure. Thirdly, it is my aim to examine the development of ideas in nineteenth-century astronomy, and see how particular groups – primarily, those whom I have styled 'Grand Amateurs' – set about testing or making advances in understanding the structure of the Universe. And fourthly, attempts will be made to chart the advances in instrument technology and the financing of that technology which provided the new physical evidence against which theories could be tested.

This book breaks down into three main chapter divisions, each of which looks at a particular group of self-funded astronomical practitioners. While there is some inevitable overlap between them, especially in terms of dates, it can nonetheless be argued that these three groups of amateur astronomers fall into a clear sequence of time, so that the book will follow a broad chronology.

The first part, of eight chapters, deals with those independently wealthy individuals who took upon themselves the reform and advancement of British astronomy at the highest technical and intellectual level. These are the 'Grand Amateurs'. The chapters are organised in a way that combines an examination of the social and financial fabric of the science, like Chapter 2, with specific case studies of astronomical lives, such as is found in Chapter 4. This part falls primarily within the period 1800–1880, for while it is true that Grand Amateurs like William Huggins and Isaac Roberts were active after 1880, the growing provision for physical astronomy at South Kensington, Oxford, Cambridge and Edinburgh after that date shows that the initiative for fundamental research was moving inexorably into the hands of formally-trained persons working in endowed public institutions.

In the nineteenth century, just like today, astronomy appealed to a very wide public, and in addition to those Grand Amateurs who had the means to partake in original research, there were a large number of less well-off individuals who still enjoyed active participation in the science, even if at a less technically and conceptually innovative level. The second and third parts of this book deal with them.

In the second part – Chapters 9 to 11 – I look at astronomy amongst the Victorian working classes: the poor, the obscure, and the self-taught. Because these people rarely left diaries or correspondence that has survived, aired their views in print, or left records in the archives of learned societies, our knowledge of their activities is only fragmentary. But enough survives to show us the vibrancy of their interest, such as the letters which the Frodsham shoemaker, John Leach, sent to G.B. Airy, the Astronomer Royal, in 1869. In these chapters I look at the access to astronomical knowledge that the working classes possessed, mainly from the 1840s onwards, through a number of case studies of individuals for whom relatively detailed records survive, such as Johnson Jex, John Jones, Thomas Bush and Roger Langdon. In some cases, however, only a single detailed source survives, often published after the astronomer's death by his better-off local admirers. In many respects, these poor men are the most remarkable people in the present book, for living as they did in circumstances where the daily fight for survival must have been so overwhelming, it says something about the power of the human spirit that they could devote so much time and money from their limited resources to an activity which could have had no direct relevance to their daily lives. Considering their poverty, and lack of opportunity, these people were amateurs in the noblest Latinate sense of the word.

The last part of the book – Chapters 12 to 15 – deals with the development of astronomy as a pursuit among the later Victorian and early twentieth-century middle classes. By this time, fundamental research had largely passed into professional hands, and the Grand Amateurs, with a few notable exceptions, were becoming a thing of the past. But educated professional people – lawyers, doctors, clergy, teachers, and businessmen of more modest means – came to constitute the growing body of amateur astronomers by 1920. These were the people who sometimes found the papers which filled the R.A.S. journals to be too mathematical, and began to found societies and periodicals which catered specifically for the needs of committed amateurs whose primary concern was observation rather than theory. The first big provincial astronomical societies, such as those of Leeds, Liverpool, Manchester, Newcastle, Ulster and Wales, were founded by such people, and 1890 saw the establishment of the national British Astronomical Association. Being comfortably off, but not outstandingly wealthy, the people who joined these societies created a market for good-quality yet affordable instruments with which to study the Moon, or draw the surface detail of the planets. And unlike the Royal Astronomical Society, these societies admitted women members, who began to make active contributions to the provincial societies from the 1880s onwards. This is the amateur tradition which extended into the twentieth century, and helped to provide the local personnel and accessible knowledge which was necessary for the great expansion of amateur astronomy in Britain during the space age.

Though innumerable obituaries and short articles have appeared in popular journals and newspapers over the last 150 years detailing the careers of local astronomical worthies, I am not aware of any previous attempt to write a history of British amateur astronomy as a movement. In that respect, I hope that the present study, in which such published sources have been considerably augmented by the use of available manuscript material, can add something to our understanding of the content and context of nineteenth-century astronomy. And as I do not believe that history contains any grand plans or needs a specialist jargon in which to conduct its arguments or communicate its conclusions, I hope that my approach to the historical process, depending as it does upon the relation of biography to wider social and intellectual events, will be found to contribute something of interest. For what has always most fascinated me about the study of the past, and prompted me to follow the calling of a historian rather than that of a scientist, is the idiosyncracy of human motivation. And nowhere in the history of science, perhaps, has the peculiar diversity of that motivation been more evident than in the circumstances which attracted peers of the realm, clergymen, stockbrokers, shoemakers, brewers, bakers, barristers, railway porters – and sometimes, their sisters, wives and daughters – to the serious cultivation of astronomy.

Acknowledgements

It would not have been possible to research and write this book without the help, cooperation, and encouragement of many persons and institutions. Those to whom I am indebted are listed below.

INSTITUTIONS

I thank the Warden and Fellows of Wadham College, the Keeper and staff of the Museum of the History of Science, the staff of the Faculty of Modern History, and the staff of the Bodleian and Radcliffe Science Libraries: all in Oxford. I also thank the Librarians and staff of the Royal Society, the Royal Astronomical Society, and the British Astronomical Association. Other institutions to which I am indebted are: Armagh Observatory, Bedford Record Office, Birr Castle Demesne, Cambridge University Library, Cambridge University Observatory, University of Cork, University of Durham, University of Exeter Library, Gwent Record Office, Harwick Museum, Instituto y Observatorio de la Armada en San Fernando (Spain), National Library of Ireland, Liverpool Museums and Galleries, Manchester City Libraries, National Maritime Museum, Newcastle Literary and Philosophical Society Library, Nobles County Record Office (Minnesota), Royal Greenwich Observatory Library (Cambridge), Royal Scottish Museum, Salford City Libraries, The Science Museum (London), Sheffield Literary and Philosophical Society, Tiverton Museum, University of Wales in Bangor and Cardiff.

ASTRONOMICAL SOCIETIES

Airdrie, Aylesbury, Bedford, Cardiff, Crayford Manor House, Edinburgh, Horncastle Astronomy Weekend, Astronomy Ireland, Leeds, Lincoln, Liverpool, Loughton, Manchester, Mexborough and Swinton, Newbury, Newcastle, Norman Lockyer Observatory Society, Norwich, Reading, Salford, Shannonside, Sunderland, Tow Law Local History Group, Walsall, Webb Society, West Yorkshire, William Herschel Society (Bath).

INDIVIDUALS

C.J.R. Abbott, Rosa Alonzo, Elizabeth Amati and the descendants of Sir George Airy, William Astore, Sir Michael Atiyah, Tony Balfour, Rowland Barden, Richard Baum, Barbara Becker, Jim Bennett, Alan Bowden, Hermann and Mary Brück, Denis Buczynski, Dick Chambers, Clarice Chapman, Mary Chibnall, Michael Coates, Brian Colthorpe, Madeleine Cox, Tony Cross, Michael J. Crowe, Sandra Cummings, David W. Dewhirst, Roland Dill, John and Lou Dix, Alan Dowdell, Sheila Edwards, Ian Elliott, Jack Ells, Ray Emery, Michael Feist, John Fitzsimons, Patrick Fleckney, Robert Fox, David Gavine, Peter Gill, Gerard Gilligan, Neil Haggath, Jeff Hall, Adam Hart-Davis, Barry Hetherington, Peter Hingley, Michael A. Hoskin, Derek Howse, David Hughes, Roger Hutchins, Stephen James, Les Jepson, J.A. Jex, Kevin Johnson, Gwawr Jones, Norman Keer, Kevin Kilburn, Tony Kinder, John Lester, Sir Norman Lindop, Christopher Lord, John Lowerson, Duncan Lunan, Sir William McCrae, Joe McKie, Francis R. Maddison, Ron Maddison, Les Marsden, Michael Martin-Smith, John Mason, descendants of J.W. Meares, Paul Money, Patrick Moore, Jack Morrell, Rhys Morris, Andrew Murray, David Northrop, Michael Oates, Dame Kathleen Ollerenshaw, Colm O'Sullivan, Steffan ab Owain, Philip Pennington, Adam Perkins, Roger Pickard, Dave Powell, Francis Ring, Steve and Gillian Ringwood, Brigid Roden, Colin Ronan, the Earl and Countess of Rosse, Tony Ryan, Mary Sampson, Alan Sanderson, Gilbert Satterthwaite, Amparo Sebastián, Anne Secord, Christine and Ken Sheldon, Tony Simcock, Don Simpson, Alan Smith, Robert Smith, Pam Spence, Ron Storey, Martin Suggett, Arthur Taylor, David Todd, Michael Turbridy, David Turnbull, Gerard L'E. Turner, John Wall, Michael and Louisa Watt, Patrick Wayman, Ian Welland, Charles Hector Wheddon, Alan Whittaker, George Wilkins, Stuart Williams, Ken Willoughby, Sir Arnold Wolfendale. Particular thanks go to Bob Marriott for his advice and skills as editor and indexer, and to Clive Horwood, Chairman of Praxis, for his patience during the writing of this book.

My greatest acknowledgement, however, goes to my wife Rachel, without whose patience and organisational and word-processing skills this product of a confirmed fountain-pen user would have taken twice as long to see the light of day.

List of plates

1	The 'Shuckburgh' Equatorial by Jesse Ramsden	59
2	The 'Lee' Circle by Edward Troughton	59
3	W.H. Smyth's Bedford observatory	60
4	W.H. Smyth's Bedford observatory, with the 5.9-inch Tulley refractor	60
5	Dr John Lee observing with the 5.9-inch Tulley refractor	61
6	Hartwell House observatory	61
7	John Lee, John Herschel and W.H. Smyth, by C.P. Smyth	62
8	Sir George Biddell Airy	62
9	Sir John Frederick William Herschel	63
10	William Herschel's 48-inch reflector	63
11	Caroline Herschel's comet-sweeper	64
12	William Parsons, Third Earl of Rosse	64
13	Birr Castle	65
14	The 72-inch reflector, Birr Castle, $c.1850$	65
15	The restored 72-inch reflector, Birr Castle, 1997	66
16	Firework display at Birr Castle, 1851	66
17	William Lassell	125
18	Replica of William Lassell's 24-inch reflector	125
19	Lord Rosse's (Oxmantown's) mirror-polishing machine	126
20	William Lassell's mirror-polishing machine	126
21	William Lassell's 48-inch reflector	127
22	James Nasmyth	127
23	James Nasmyth's 20-inch reflector	128
24	James Nasmyth's observatory–residence, Patricroft	128
25	A plaster model of the lunar crater Plato, by James Nasmyth	129
26	A Troughton and Simms equatorial refractor	129
27	The Revd Edward Craig's 24-inch refractor	130
28	The Revd William Rutter Dawes	130
29	John Russell Hind	131
30	Edward Crossley's observatory–residence, Halifax	131
31	Edward Crossley's $9^{1}/_{3}$-inch Cooke refractor	132

xviii **List of plates**

32	Andrew A. Common's 36-inch reflector	132
33	Warren De La Rue's 13-inch reflector	191
34	Warren De La Rue's 3½-inch photoheliograph	191
35	The Moon, aged 17.1 days, by Warren De La Rue	192
36	Sir William Huggins	192
37	Sir Joseph Norman Lockyer	192
38	The Norman Lockyer Observatory, Sidmouth	193
39	Arrangement of a spectroscope on a bench	193
40	Multi-prism spectroscope attached to R.S. Newall's 25-inch refractor	194
41	Solar spectrum plates	194
42	Late nineteenth-century lady astronomer examining spectra plates	194
43	Isaac Roberts	195
44	A photograph by Isaac Roberts	195
45	A longitudinal section across the Galaxy, by William Herschel	196
46	The Orion nebula, by Sir John Herschel	196
47	The Orion nebula, by Lord Rosse	197
48	M51, by Sir John Herschel	197
49	M51, by Lord Rosse	198
50	M31, by Sir John Herschel	198
51	M31 and M32, photographed by Isaac Roberts	257
52	Amateur astronomers	257
53	An astronomy lesson	258
54	6-inch reflector by James Veitch	258
55	Mr Tregent and his street telescope demonstration	258
56	Roger Langdon	259
57	Silverton Station, Devon, c.1890	259
58	13-inch reflector by Thomas Bush	260
59	24-inch reflector by Thomas Bush	260
60	John Jones and his 'Jumbo' reflector	260
61	Charles Paton	261
62	Caroline Lucretia Herschel	261
63	Elizabeth Brown	261
64	Florence Taylor of Leeds	262
65	The Revd Thomas William Webb	262
66	9¼-inch With–Berthon reflector	262
67	The Revd E.L. Berthon's 'Romsey' observatory	263
68	The Temple Observatory, Rugby School	263
69	The University of Cork Observatory	264
70	Advertisement for astronomical lectures in Leeds, 1861	264
71	Leeds Astronomical Society's Ross refractor	264
72	Liverpool Astronomical Society crest	325
73	Astronomical Society of Wales crest	325
74	Dan Jones at the eyepiece of the Cardiff Observatory telescope	325
75	J. Alun Lloyd at his telescope, c.1905	326
76	John Willoughby Meares	326

77	Total solar eclipse of 22 January 1898, by J.W. Meares	327
78	H.J. Townsend	327
79	Washington Teasdale	328
80	Saturn, by H.J. Townsend	328

Part 1

The Grand Amateurs

1

Amateur astronomy in the Romantic Age

In May 1832, George Biddell Airy, who was soon to become Astronomer Royal, completed a long and detailed *Report* for the British Association for the Advancement of Science (B.A.A.S.) on the contemporary state of astronomy at home and abroad [1]. The *Report* is a fascinating document, for while it tells us virtually nothing about the wider social context in which astronomy was practised, it is penetrating in its analysis of the contemporary strengths and weaknesses of astronomy as an intellectual discipline. Astronomy, along with mechanics, was by far the most mature of the sciences insofar as observation was not only performed with the most sophisticated of instruments, but also because the observations could be directly connected by mathematics to verifiable theories, to formulate predictive laws of nature. The physical and the intellectual thereby produced a single unit, to form one of the crowning achievements of the human mind.

Lying at the heart of this process was the accumulation of exact observational data from the heavens, as right ascension and declination coordinates were built up decade by decade. To Airy's mathematical mind, real astronomy was concerned with meridional measurement, and he praised the Royal Observatory, Greenwich, for having built up the longest run of such measurements in the history of the science, for it was from such observations that further laws could be extracted. Yet Greenwich, like the rest of British astronomical institutions in Airy's opinion, was vastly better at accumulating accurate observations than it was at doing anything useful with them. Unless one reduced this ocean of raw observations, astronomy would get nowhere, and this was one of the leading defects of British astronomy across the board [2].

It was to Continental Europe that Airy looked for astronomers who were also active in reducing observations so that they could be made useful, and he lamented that even observations by illustrious British astronomers often remained unusable until a foreigner reduced them, as Friedrich Wilhelm Bessel did with the 70-odd-year-old Greenwich observations of James Bradley [3]. Airy hammered this point home again in a letter to William Vernon Harcourt, F.R.S., the scientific Canon of York Minster and luminary of the B.A.A.S. in September 1832, when he epitomised so many British astronomers as being 'mere observers', who fell in intellectual character below the 'mere chemical experimentalist', only to emphasise once again that it was to Continental Europe that we should look for guidance on how to go about doing astronomy as an intellectual discipline [4]. On the other hand, Airy was not of the opinion that Continental methods of *financing* astronomy were necessarily the ones that should be adopted in Britain.

Airy saw British astronomical instruments as still the best in the world (a decision which he came to from a masterly command of physical mechanics, for he had not an atom of xenophobia in his being), although he recognised that the French and the Germans used positional observations better than his fellow countrymen. This was because French analytical mathematics led the world, and the Germans used these techniques to process their own observations (and those of illustrious deceased English astronomers), and published tables from them. Though Airy does not say so in so many words, it is clear that English astronomers did too much observing and not enough calculating. And much of this stemmed from the fact that too few high-level mathematicians were involved in British astronomy, over and against the multiplicity of people who made observations, either with their own or with institutional instruments.

Surprisingly, Airy's *Report* is almost wholly silent on those branches of astronomy where Britain was undisputedly strong and where fundamental ground had been broken. Very little space was devoted to the work of Sir John Herschel and Sir James South in the study of binary and triple stars, while only two and a half lines were given to the work of the Herschels on nebulae and the 'construction of the heavens' [5]. This was all the more surprising, as Sir John Herschel was a friend of Airy. Nothing whatsoever is said about studies of the solar, lunar and planetary surfaces, of variable and coloured stars, or of the physical examination of comets, and even German astronomers who had led the way in these fields, such as Johann Hieronymus Schröter and Heinrich Wilhelm Matthias Olbers, are scarcely alluded to.

On the eve of the Victorian age, therefore, Airy's concept of serious astronomy is surprisingly narrow, and concerned almost entirely with the application of mathematical techniques to the study of solar, planetary, asteroidal, and cometary motion. Upon becoming Astronomer Royal in 1835, he applied his vast zeal to these theoretical and utilitarian programmes at Greenwich, to reduce and publish all observations, and make their mathematical fruits available to the state, the navy, and the world of science. One might say that Airy's vision of astronomy was thoroughgoingly professional. Because a very high level of mathematical ability was necessary to address many of the outstanding problems in celestial mechanics, he realised that those without a specialist – or, more pointedly, a *Cambridge* – education could do little that was of use. Indeed, this assumption of Cambridge excellence was brought home to W. Vernon Harcourt in November 1831 by the young Scotsman James David Forbes on his hearing that Airy was going to write the B.A.A.S. *Report* on the state of British astronomy, for 'It at once secures all the aid of Cambridge talents, and after that the Association *must* stand and be heard of far and wide.' [6]

Because of this perspective, however, it might be assumed that Airy failed to appreciate the enormous importance of and input into the science by that class of astronomers who by 1855 would be working on the most interesting and most thought-provoking departments of the science: the amateurs, or rather, the Grand Amateurs, who worked at their own expense and in their own time. Some of these men did possess high-level mathematical qualifications before moving sideways into lucrative professions which enabled them to retire early, and set up fine private observatories. Others were excellent observers, who made invaluable contributions to those non-mathematical branches of the science, such as the study of planetary surfaces, which did not really concern Airy in his *Report*. Yet between them they constituted a formidable power in the astronomical land, and to lump in their achievements alongside the professionals' simply because they were well-equipped and capable within their fields is to distort the overall picture of nineteenth-century British astronomy.

In fact, Airy, the quintessential professional astronomer, always held this breed in the highest respect, and half a century after writing his *Report*, and on the verge of his own retirement as Astronomer Royal at the age of 80, he made his position clear. In the midst of the public debate about the government's duty to fund pure scientific research in 1881, Airy stated 'that successful researches have in nearly every instance originated with private persons, or with persons whose positions were so nearly private ... that the investigators acted under private influence, without incurring the dangers attending connection with the State.' The Royal Observatory, he emphasised, had been created 'for a definite utilitarian purpose' which it still followed, whereas it had been 'private persons' who had footed the bill for 'investigations of undefined character, and, at best, of doubtful utility' [7]. In short, it had been the Grand Amateurs who had spearheaded fundamental research in astronomy and its related sciences.

Considering the role which the amateur came to play in nineteenth-century British astronomy and its related sciences, not to mention organisations like the Royal Astronomical Society (R.A.S.), it is surprising that more scholarly attention has not been devoted to them as a social group. Yet by far the greater part of the scholarly work in the history of the physical sciences which has been published over the last few decades has been preoccupied with the evolution of the professional scientist working within publicly- or commercially-funded institutions [8]. For some reason, the amateur in the natural history sciences has fared distinctly better, with figures like Darwin and the clerical geologists being the subjects of serious studies [9]. Yet the amateur physical scientist, where treated at all, generally emerges in one of two guises: as a homely dabbler who works by a species of fortunate empiricism; or else as an inspired proto-professional, who simply lacked the status and resources to do the job properly.

In nineteenth-century Britain, however, there were individuals who were neither simple empiricists nor wished to become salaried professors, and it is surprising that Airy's *Report* gives them so little credit, for they were all around him. On the other hand, it is more than likely that men such as Stephen Groombridge or Captain William Henry Smyth (who was promoted to Admiral in 1853, and is referred to hereafter as 'Admiral') in the recent generation of astronomers, or William Lassell or James Nasmyth in the rising generation, had no self-consciousness as *professional* scientists of any sort, thinking of themselves rather as independent gentlemen who *loved* astronomy [10].

In many historical studies, British science is not infrequently (and generally adversely) compared with that of Germany, especially after 1815, where a burgeoning state university system, influenced by France's École Polytechnique, was beginning to generate Ph.D. scientists and create an assumed model for science elsewhere [11]. This Franco-German model is seen most clearly in the rapid development of German research chemistry, which undoubtedly benefited from massive state investment during the post-Napoleonic decades when the German states, under a Prussian banner, were busily consolidating their new role as the intellectual arbiters of Europe. In this new way of thinking, science was becoming a centralised, directed activity, which demanded a ticket of entry for all those who aspired to be taken seriously – usually a Ph.D. – and which moved inexorably towards professionalisation.

British science, however, worked very differently, and did so for reasons that were social and political rather than scientific. As the victors of Trafalgar and Waterloo, and inhabitants of the only European country not to have been ravaged by a foreign invader for over 800 years, the early-nineteenth-century British felt no need to redefine their self-image through education or science. Independent chaos was clearly more efficient than centralised bureaucracy when it came to national survival, and as the nation that waxed lyrical about its kinship with Bacon, Newton, Locke, and Watt, it saw no reason to act any differently in matters of science and invention [12].

When scholars make adverse comparisons between the Royal Society and the French or Berlin Academies, or the École Polytechnique in the post-Napoleonic era, or take Charles Babbage too much at his word when he rails against British science policy and patronage, they must not forget that they are dealing with an approach to science that is embedded within a very different cultural tradition from that of Germany or France. As a nation obsessed with low taxation, with politicians and even monarchs who could not escape the whiplash of the popular press, Britain was a country which did not see public science policy as pre-eminent amongst its perceived aspirations. What was pre-eminent was political and economic reform, as Europe's most secure and powerful middle class evolved those strategies whereby it would acquire more power, to exercise yet more control upon an aristocracy and a monarchy which was already remarkably circumscribed by post-Treaty of Vienna European standards. Britain's middle classes were concerned with personal freedom, the widening of the franchise, the right to buy and to sell unhindered, and to pay as little to the government in taxation as the secure defence of the realm would let them get away with. Private wealth created private independence, and independence in its turn created the wherewithal to undertake great enterprises in industry, the arts, and, of course, the sciences. The formal professionalisation of science was no more of a national priority, therefore, than the professionalisation of poetry, for talents were seen as flourishing to their fullest in a situation of minimalist official involvement [13].

The lurking fear that the receipt of public money somehow compromised one's independence of action was very much in the minds of those people who were active in the early B.A.A.S. and were attempting to give British science a more coherent basis. In 1831, for instance, Sir David Brewster wrote to W. Vernon Harcourt to remind him that the Civil List pensions already being paid to nine distinguished British academics were for life, and were not conditional upon their future loyalty to government or party. But the fear expressed by men of science for their continuing intellectual freedom is perhaps better understood when one remembers that the patronage of government – in the unreformed Civil Service, the Episcopal and Judicial Benches, and the armed forces – was traditionally associated with a loyalty to one's Whig or Tory paymasters [14].

When attempting to evaluate the factors that lay behind the presence of amateurs in astronomy, and in the other sciences, one must also consider the particularly English concept of a gentleman, and the sought-after mystique which surrounded that status. Though a landed ancestry undoubtedly conferred gentility in early Victorian Britain, as it had in Elizabethan times, social status was much more open-ended and 'market-driven' than it was in Prussia, Russia, Spain, or the Austro-Hungarian Empire. In practical terms, it was money that made an English gentleman, for the conspicuous and generally accepted public badge of gentility was the ability to be the master of one's own time, live comfortably, and have the resources with which to enter public life, or to partake in expensive hobbies. A brewer, a lawyer, a stock-market speculator, a matrimonial gold-digger, or a successful industrialist could enjoy the perquisites of gentility in the fluid social world of 1840. Such men could choose to keep stables of race-horses, collect Old Masters, or build giant telescopes with which to resolve the Orion Nebula, depending upon how the quirks of fancy and temperament took them [15].

And when such rich, independent, and often self-made men took up serious scientific pursuits, they found no obstructions. A cultured but not necessarily university-educated brewer found no difficulty in moving into the existing scientific establishment, to become a Medallist, Council Member, or even President of one of the great metropolitan learned societies if his achievements merited such rewards. Indeed, such men were deemed worthy of especial respect, for they were clearly willing in their particular ways to endow civilised society with the fruits of their hard-earned personal fortunes, by undertaking long and expensive researches at their own expense. And if a man had enjoyed the advantages of a first-class university education before he decided to devote his time and money to astronomy, then so much the better.

This, therefore, was the scientific amateur. No mere tinkerer or hobbyist, but an independent personage who was willing to use his own fortune to advance knowledge. And not only was he an intellectually admirable type, he was politically and economically praiseworthy as well, for without making demands upon the taxpayer he showed that what in some countries could only be paid for by an absolute monarch could be done in England on the strength of an individual's private means [16].

Nowhere was this honoured status made more explicit than in 1849, when Sir John Herschel, President of the R.A.S., presented the Gold Medal of the Society to William Lassell, F.R.S., and styled Lassell as belonging 'to that class of observers who have created their own instrumental means – who have felt their own wants, and supplied them in their own way.' Sir John Herschel, who, in spite of his scientific eminence, took his first paid job at the age of 58 – as Master of the Mint – recognised 'a sort of hereditary fellow-feeling with Mr. Lassell', since he also saw himself as an amateur astronomer [17]. In fact, Sir John saw the status of the amateur scientist as being one of the noblest of callings, where the superior pleasures of the intellect provided an enriching occupation for educated leisure, and the researcher worked for love, and not for profit. Indeed, this was to be an amateur in the fullest Latinate sense of the word: an 'ardent lover of astronomy' rather than a self-conscious careerist.

Exactly the same point was made eight years later when the professional astronomer Norman Robert Pogson wrote to Admiral Smyth, and emphasised the 'distinction [which] is drawn between amateurs who devote their time as a labour of scientific love, and those who are paid for their work, and only successfully fulfil their duty' [18].

At first sight, however, astronomy might seem a peculiar subject upon which a person of independent leisure should choose to expend his energies and fortune. Compared with the other great amateur-directed sciences of the late Georgian and early Victorian periods, it appeared to have many disadvantages. Unlike botany and geology, astronomy did not as a practical pursuit easily engage with the prevailing culture of romanticism, and its concomitant ingredients of high mathematics and mechanism seemed to remove it from the realms of feeling. Astronomy was a singularly unsensuous science insofar as it was restricted to one avenue of contact – vision; and even that could only be explored through a complex optical technology which emphasised the physical isolation between the observer and the observed.

Diverse as the strands of romantic culture were, they were all very much imbued with the idea that the human senses had the capacity to make a powerful, direct, and polychromatic impression upon the imagination. Flowers could be handled, sniffed, examined, and each change of colour noted. Fossils could be chipped free from their rocky matrices, examined, passed around, and articulated with their adjacent bones to form the shape of an extinct and exotic beast. Astronomical bodies, on the other hand, could only be seen from one position, and with varying visibility. And even when one beheld the grandeur of the Milky Way, or looked at Sir John Herschel's published drawings of nebulae (for one could not see the actual nebulae themselves without a large telescope and a long-practised eye), the observer could scarcely avoid a sensation of disembodiment from what was being observed [19].

Botany and geology, moreover, were highly social sciences. Whether the devotees be a group of ladies and gentlemen on a country-house weekend, or a society of Manchester mill-workers heading for the Pennine slopes on a Sunday afternoon [20], one could do botany as part of a congenial social gathering. The same could also be said for geology, as Canon William Buckland, Oxford University's first Regius Reader in Geology, fired the imaginations of clerical enthusiasts, ladies, and friends to dig for bones and build up personal collections [21]. What is more, botany, and especially geology, fitted in beautifully with the romantic love of the Lake District and Scottish Highlands. Members of the leisured classes spent entire summers tracing the impact of Noah's Flood upon the glens, comparing fossil ferns with their living counterparts, moving from one castle or country house to another, and reading Wordsworth as they went.

These sciences were not only sensuous and social; they were also performed in summer, and during the hours of daylight. The combination of warmth, light, and pleasure, allied with the potential for awe and heightened religious and romantic feeling, made them ideal scientific hobbies. And although botany and geology were best performed on fair-weather days, they also offered possibilities for round-the-year satisfaction, as one could spend the winter arranging, classifying, and studying the finds of the previous summer.

And one might also say that in an age when political power was moving inexorably into middle- though not yet working-class hands, botany and geology were democratic sciences. There were no complex equations to master, and no need for technology beyond a magnifying glass and a geological hammer, so that all able-bodied people with some leisure time at their disposal could feel that these disciplines were accessible to them. They could also be pursued with as much or as little rigour as one chose. One might collect local flowers or minerals as a semi-casual elevating pastime, or one might get bitten by the bug of original research, like the landed barrister Sir Charles Lyell. Indeed, Lyell, who abandoned his legal practice to redirect geology from a foundation of original research which extended from Naples to Inverness, and his protégé Charles Darwin, who never had a paid job in his life, became the inspired leaders of the natural history sciences [22].

Amidst these conveniences of warmth, light, technical simplicity and companionship, astronomy might seem at a disadvantage for a leisured pursuit. To be a serious astronomer, one needed to master a substantial amount of abstruse physical and mathematical theory, face the prospect of spending progressively large sums of money as one exhausted each new level of instrumental power, and work at the most unsocial of hours, in the bitterest seasons of the year, standing alone at the eyepiece without moving for half the night. Although on the face of it astronomy seemed to link into the romantic spirit with its curiosity about the vastness of creation, its working procedures embodied the very antithesis of romance: big, complex pieces of precision engineering, an insistence upon measurement which excluded all avenues for sympathy or feeling, and a prevailing ethos which abhorred colourful speculation. From every angle, it seemed, astronomy was big, hard, cold, and impersonal. Where, therefore, lay its immense appeal to all classes of people, from aristocrats to unskilled industrial workers?

One very important aspect of this appeal lay in astronomy's clear logical basis, which had defined its status as a solid science back in antiquity, and the intellectual authority of which had only been enhanced since the days of Copernicus. In spite of its technical complexity and nocturnal awkwardness, and while much within the romantic ethos openly rebelled against the universal laws of Newton, astronomy possessed an undeniable power to enchant, and especially to captivate a cultured population that still read Bacon and Locke, as well as Wordsworth and Shelley.

Part of this enchantment was religious in a Christian sense, as English writers from John Wilkins in the mid-seventeenth century down to William Whewell's

Bridgewater Treatise of 1833 had placed astronomy as the cornerstone of the natural theological argument from design [23]. Simply to look at the heavens inspired one with God's glory, but to pry ever deeper into the depths of space, and to partake in original discovery and uncover further components of the divine design, exerted an appeal which went beyond that which most modern scientists would expect to get from their research.

Though the greater part of this religious awe would have been related to a relatively orthodox form of Christian expression, we must not forget that romanticism itself could take on the shades of a religion. Neither William Wordsworth nor Thomas Hardy were orthodox Christians by any means, yet both wrote of the power of the sky, and of recent astronomical discoveries, to enthral the imagination in a way which partook of the religious [24]. And one suspects that when astronomers whose lives seemed to be without much formal Christian expression, such as James Nasmyth, spoke of the glory of the celestial creation, they were using the language of romantic metaphor or of classical paganism rather than that of the Apostles' Creed. Indeed, the very act of astronomical observation could take on an aura of reverence which was not necessarily associated with an orthodox religious creed. Just as Tycho Brahe wore his ambassadorial regalia when presenting himself before the Court of Urania, on the observing platforms of Uraniborg, so Admiral W.H. Smyth, in his *Cycle of Celestial Objects* (1844), encapsulates the power of the same pagan goddess when speaking of the night sky. To a man with a classical education, it was all too easy to fall back upon the metaphors of pagan reverence, even if he took the sacrament in church on Sunday morning.

Though romanticism and classicism constituted much of the prevailing high culture of the early nineteenth century, we must remember that industry, science and invention also played a major part and were themselves imbued with images of sensory delight and human triumphalism. The Royal Institution not only epitomised this 'useful-*cum*-curiosity' culture to polite West End audiences, but in the persons of Humphry Davy and Michael Faraday unintentionally allied science first with high romanticism and then with the creed of a Christian intellectual [25].

A cultural expression that used sophisticated instruments and demanded enormous patience need not, therefore, be unappealing to the age, if it could also deliver the goods of awe, heroism, and beauty. And no astronomer captured this combination of qualities better than Sir John Herschel, who not only built his own instruments (as he reminds us in his 'Address' to Lassell), but drew conclusions across a whole range of physical sciences, including natural history, which bedazzled the reading public for half a century and showed astronomy to be a science of immense romantic power. Indeed, even Airy himself, that apparent epitome of hard-edged professional astronomy, was deeply susceptible to the broad culture of romanticism, especially literature and landscape. As a young Cambridge don conducting reading parties in the Lake District in the mid-1820s, he sometimes called in upon Wordsworth after his introduction in 1825, and committed many of the works of the English Romantic writers, along with vast tracts of the Classics, so firmly into that formidable memory of his that he could still recite them at length in his late eighties. Likewise, he returned on numerous occasions over the next sixty-odd years to the area around Windermere and Cat Bells, where English Romanticism had first been conceived [26].

Unless they are concerned almost exclusively with the Sun, however, astronomers are obliged to work at night. Considering the potential conviviality of botany and geology, one might expect, therefore, that astronomy had no social appeal. Yet this would be an incorrect assumption, for astronomers were often highly clubbable individuals. The Royal Astronomical Society was born out of a Club of wealthy amateur astronomers in 1820, while the Fellows of the Royal Society had enjoyed Club dinners together going back to Edmond Halley almost a century earlier. The Bedford–Aylesbury axis of early Victorian astronomers and related scientists enjoyed merry social gatherings in Dr Lee's mansion, Hartwell House; and from 1859 onwards, and especially after 1880, many British cities formed astronomical societies, which had social as well as purely scientific purposes [27].

Although the immediate acquisition of astronomical data may not have seemed so potentially social as that of the amateur daylight sciences of botany and geology, the pooling together and assessment of that data could be just as social. And as regards the seeming inconvenience of long night watches at the telescope, we must remember that astronomy is not just a technical activity, but also a potentially meditative one. The romantic and religious culture of the early nineteenth century is suffused with the quest for the tranquil, as well as for the social; for the calm of night as well as for the brightness of spring mornings. And nowhere are all of these strands more aptly drawn together than under a glorious midnight sky, with an instrument of great power. For it was here that a solitary Victorian Prometheus was permitted by his Creator to glimpse fresh details of His Creation, with an awe untainted by impudence, and with a reward that did not entail destruction.

The above, I would suggest, explains part of the imagery with which Victorian amateur astronomers saw themselves, what drew them to the science, and how they related to the wider educated culture of their day. Such an astronomer was no dabbler. Nor was he a frustrated proto-professional whom no government would fund. He was rather a product of peace, prosperity and optimism, with an enormous social confidence in what the unfettered private individual could achieve when working in relationship with a group of like-minded friends. And so armed, in the words of Sir John Herschel, he could compete with the mighty instruments provided by 'the Emperor of Russia and the Citizens of Boston [who] have endowed the observatories of Pulkowa and the Western Cambridge [Harvard, Massachusetts]' [28]. And what quest could be more innately romantic in itself than harnessing the inventive ingenuity of the age to fathom the 'length, breadth, depth, and profundity' of the Universe, expressed as the adventure of an inspired private individual who was answerable for his thoughts to no earthly power greater than himself? [29]

2

Gentlemen and players: amateurs and professionals in 1840

From the Renaissance until well into the second half of the nineteenth century, the development of astronomy, along with most of the other sciences in Britain, lay rooted in the amateur tradition. Most of the original Fellows of the Royal Society in the 1660s were proud to be styled 'gentlemen, free and unconfined' [1], who conducted their researches on the strength of independent means, and even men who were in receipt of academic salaries to teach astronomy and its related sciences were not generally expected to conduct original research, as opposed to passing on a received tradition. When they did conduct research, as did John Greaves, or the young Sir Christopher Wren, who both held Oxford University's Savilian Professorship of Astronomy over the middle decades of the seventeenth century, or the Astronomy Professors of Gresham College, London, they invariably did so with instruments which were their own property, and on their own initiative. The unique private collection of astronomical instruments used by Greaves for his own teaching and research in Oxford later became the property of the University, and is now preserved in the Museum of the History of Science, but it would not have been necessary for his teaching of the classical syllabus [2].

GREENWICH, OXFORD AND CAMBRIDGE

Even when an 'Astronomical Observator' was needed to direct King Charles II's royal foundation on the top of Greenwich Hill in the spring of 1675, it was a self-taught private gentleman without a formal university training who was appointed. For John Flamsteed, the 28-year-old son of a Derby brewer, was reckoned by the Royal Society to be the most competent practical astronomer in Britain. What was more, Flamsteed's official status as Astronomer Royal was ambivalent, for while the job carried with it an empty building and a modest and erratically-paid salary, it was up to Flamsteed to find his own instruments and cover most of his costs. Four years before his death in 1719, Flamsteed reckoned that his Royal Astronomical appointment had cost him over £2,000 of his own money, which was well over

£200,000 in the money of the 1990s. It could be plausibly argued, therefore, that Flamsteed was not so much a professional astronomer as an astronomical benefactor and patron who gave more to the science in material terms than he ever received from it [3].

Flamsteed's successors at Greenwich – Edmond Halley and the Revd Dr James Bradley – were never really adequately paid, and it was virtually assumed that an Astronomer Royal would also be a gentleman of independent means. It was not until Dr Bradley had already been Astronomer Royal for ten years, in 1752, that the salary assumed a serious professional dimension for the period, when he received £350 per annum, before tax. Yet Bradley combined his Greenwich appointment with the Savilian Professorship in Oxford (like Halley before him), not to mention an interest in properties in the Minchinhampton district of Gloucestershire which he probably acquired through marriage [4]. His successor, the Revd Dr Nevil Maskelyne, received the same amount, and when 48 years later in 1810 a petition was made to the Chancellor of the Exchequer, Spencer Perceval, for a salary increase, it was dismissed on the grounds that Dr Maskelyne was a well-to-do gentleman as well as being Astronomer Royal (he owned property in Wiltshire and was the absentee incumbent of the parish of North Runcton, in Norfolk), and therefore did not need more money from the public purse [5].

If one wished to do serious astronomical work in the eighteenth-century universities, the same concepts about funding applied. The zenith sector which Bradley used in his discovery of the nutation and the aberration of light while Savilian Professor at Oxford was his own property, as were the telescopes, micrometers, clocks, and the equatorial sector which he used to determine the Newtonian orbital criteria of comets, although some of these instruments were subsequently sold to the Royal Observatory [6]. Similarly, when the Revd Dr Thomas Hornsby was doing his original stellar cartographic, planetary, and Venus transit work at Oxford between 1758 and 1772, he did so with costly instruments which he had commissioned from John Bird and other craftsmen out of his own pocket. Even Hornsby's set of physics teaching apparatus, used for his professorial courses, was his own property, and was valued at £375 in 1790 [7]. Only after the funding of Oxford's Radcliffe Observatory in 1771 did Hornsby work with astronomical instruments that were not of his own providing [8]. The same state of affairs generally prevailed in Cambridge, until the founding of the University Observatory in 1823.

The concept of an adequate professional salary that was not supplemented by teaching or clerical or other perquisites did not really exist in British astronomy until 1835, when George Biddell Airy was appointed Astronomer Royal. In the businesslike negotiations which he conducted with Lord Auckland (soon to become First Lord of the Admiralty) between 1832 and 1834, Airy reminded his Lordship that unlike his Greenwich predecessors he had no private means, and required a substantial increase in the going rate of £600 per annum if he was going to come to Greenwich [9]. Airy fought for and got £800 per annum, while a Civil List pension of £300, which he had bestowed upon his wife Richarda (later Lady) Airy, gave the Astronomer Royal the total income of £1,100 a year from the public purse [10]. Yet even this was not enough to warrant the rank of knighthood which the government offered to him around the same time. As Airy reminded Spring Rice, the Chancellor of the Exchequer in 1835, he did not possess the 'considerable fortune' or independence of means which such an honour presupposed to command respect in the world, and he did not wish the 'light consideration' [11] in society that a relatively poor man who sported a high title received. Indeed, it would not be until 1872, and three offers later, that Airy felt willing to accept a Knight Commander of the Bath from Queen Victoria [12].

Adequately-paid astronomical jobs were very few and far between in Victorian Britain, and unlike today, when academic professionals dominate the science intellectually and administratively, the astronomical world was very much one of gentlemen and players. Although the Astronomer Royal, his Cambridge-educated Chief Assistant, and the Directors of the University Observatories were obviously gentlemen in manners, education and attitudes, they were nonetheless obliged to work for their livings. A former domestic servant like Abram Robertson who rose up through a Christ Church scholarship to become an F.R.S. and Oxford's Professorial Radcliffe Observer, or G.B. Airy, who climbed the scholarship ladder of Trinity College, Cambridge, must, in that rank-conscious age, have been aware that they were men who worked for their bread rather than enjoying a liberal income from an independent source [13].

This subtle sense of position is evident in the correspondence of G B. Airy. As the son of a self-made Excise Officer of considerable rank who was suddenly dismissed from the Service for irregularity at the age of 63 – when George was 12 – the Astronomer Royal was to display a lifelong concern with reminding his official superiors of his diligence, efficiency and honesty [14]. Indeed, it is very likely that the concern with bureaucratic punctiliousness, which is often thought to have been such a hallmark of Airy's professional career, stemmed from this insecurity. When applying for his first holiday leave from the Admiralty in 1836, for instance, he wrote to Sir Charles Wood, requesting a ruling upon his absence; the Secretary to the Admiralty, however, seemed a little surprised at the detailed nature of the Astronomer Royal's enquiry, and reminded him that 'you are considered as a Commander in Chief', who can act with discretion [15]. And when his son Wilfred had gone up to Trinity, Cambridge, in 1855, he was quick to point out the necessity of the young undergraduate living well within his father's modest means, for Airy was 'going to an expence almost greater than I can afford' in sending him there. Giving his son the benefit of his own strict undergraduate economies of thirty-five years earlier, he advised against accepting invitations to wine parties, which would oblige Wilfred to return them. He was to work for a First and keep out of debt: 'The booksellers will always allow 10 p.c. for immediate payment.' [16] This sense of the Airy family's modest position on their £1,100 a year with relation to their grander friends and colleagues is even captured by Richarda Airy, the Astronomer Royal's wife. In correspondence that passed between this Derbyshire clergyman's daughter and her friend Lady Margaret Herschel, wife of Sir John, one occasionally catches the perceived differences of rank that existed between the independent Herschel and working Airy astronomical families. In September 1844, for example, Mrs Airy sincerely thanked Lady Herschel for entertaining the 11-year-old Miss Elizabeth Airy at their Kent mansion, and enabling the young lady to mix in a rank of society which was above that in which the Airys usually moved [17]. Indeed, a constant caution regarding matters of money and status is never very far below the surface with Airy, whenever one looks beyond the specifically technical aspects of his business. It is not without significance that he was not elected F.R.S. until the age of 34, by which time he already had a reputation of international standing and the distinguished Directorship of the Cambridge University Observatory behind him. He certainly did not lack the merits, or powerful friends who were Fellows of the Royal Society, such as his Trinity colleague, Richard Sheepshanks, though we must not forget that Sheepshanks was a wealthy private gentleman with legal, clerical, and Halifax woollen money behind him, as well as being a Cambridge mathematics don. But in 1835, at the age of 34, Airy became Astronomer Royal, a position in which he could not avoid expending the rather costly entry fee into the foremost scientific club in the land. As soon as he was elected, Airy joined a group of Grand Amateurs from a variety of sciences, to take an immediate seat on the Council. And then again, in 1844, he was on the verge of turning down an honorary D.C.L. degree from Oxford, when the Vice-Chancellor Dr Wynter graciously agreed to waive the customary 6-guinea (£6.30) fee that was paid by an honorand upon receiving his degree [18].

As Astronomer Royal, Airy was by far the most famous and the best-paid, at least in straightforward salary terms, of all the professional astronomical observatory directors in Britain. By the late 1840s there were around ten professional observatories in Britain, in addition to Greenwich, and almost all of them were directed by salaried men, most of whom were university graduates [19].

In one important respect, however, I would argue that Airy had a very different perception of his social position from most of the other salaried observatory directors in Britain, and that might best be summed up as the 'freeholder's mentality'. Quite apart from any insecurities that might have derived from his father's disgrace, Airy had none of those historically powerful associations which came from membership of one of the old learned professions, and which gave confidence to their possessors, even if a man did earn his living. Unlike those salaried observatory directors who were clergymen of the Church of England, Airy had no parson's freehold, or that automatic standing which came from the possession of an independent landed income, even if that income derived from a lifetime's freehold, rather than from one in perpetuity. A well-beneficed clergyman had his own assured social foundation, upon which other salaried appointments could be built, and this gave him an assurance when moving amongst men of greater or more ancestrally-derived wealth. But in a world where landed money conferred a *kudos* all of its own, even a substantial annual salary could well have felt a fragile possession if it was one's only income.

The oldest professional foundation after Greenwich was Oxford's Radcliffe Observatory, which was not in precise legal terms a University foundation. It was the property of the Radcliffe Trust, a wealthy academic and medical charity which administered the enormous bequest of 1714 which Dr John Radcliffe of University College, Oxford, had created. Up to 1771, when the Trust founded the most up-to-date observatory in Europe in Oxford, it had already blessed the city with a scientific and medical library and a hospital, all bearing the name Radcliffe. By a custom going back to 1771, however, the Radcliffe Observers had also been co-holders of the University's Savilian Professorship of Astronomy, so that a University-salaried don ran what was effectively a private trust's observatory. As a well-off academic pluralist and clergyman, the first Director, the Revd Dr Thomas Hornsby, had observed for forty years without a salary, simply enjoying the magnificent instruments and elegant residence which came with the honorary title of Observer [20]. But his successor in 1810 was a poorer man, the ex-domestic servant Abram Robertson, who only had his modest professorial salary to live on. The Trustees now gave the Observer £300 a year, to augment the University income, while still letting him enjoy the residence. But Robertson scarcely needed the money, for being a deeply religious and ascetic widower, he gave away much of his income to support local charities in his native Northumberland [21]. Only in 1839, when Robertson's successor Stephen Peter Rigaud died, did the situation change. The Radcliffe Trustees gave the Observership to Manuel Johnson, a 34-year-old don who had formerly been an army officer, while the University refused him the Professorship. The grounds for the refusal have never been clear, but religious considerations no doubt played a part, for Johnson was a High Churchman and friend of the Anglo-Catholic John Henry Newman at a time when the University was gripped by a Low-Church evangelical hysteria [22].

This decision did great harm to Oxford University astronomy for over fifty years, for while Johnson did work of excellence at the Radcliffe, being able to concentrate on fundamental research with no teaching duties, the new Savilian Professor, Bryan Donkin, had no observatory to work from. Only when the dynamic 62-year-old Charles Pritchard was elected Savilian Professor in 1870 did things change. This clergyman, whose entire professional life between leaving Cambridge as Fourth Wrangler and getting his Oxford chair had been spent as a schoolmaster and educator of the sons of Sir John Herschel, took the bull firmly by the horns [23]. By 1875, with the backing of Oxford's Grand Amateur astronomer, Warren De La Rue, and the geology professor John Phillips (who ran a private observatory in Oxford), he had achieved the foundation of the Oxford University Observatory. Pritchard now went on to pioneer astrometric photography and contribute to the *Carte du Ciel* from his new Observatory, still enthusiastic and dying in harness in 1893 [24].

Airy, who was Director of the newly-founded Cambridge University Observatory between 1828 and 1835, brought a stamp of professionalism straight from the start, by demanding an increase in salary from £300 to £500 a year [25]. For as a layman who had no priestly vocation, he argued, he could not enjoy the ecclesiastical perquisites that normally went along with senior university appointments. His impact upon Greenwich has already been discussed, and when he vacated the Cambridge Directorship (and Plumian Professorship of Astronomy which was held in conjunction with it) it was taken up by his protégé the Revd James Challis, and then in 1858 by the distinguished but distinctly un-careerist John Couch Adams as Lowndean Professor of Astronomy. But the impact of Airy's rigorous professionalism, and the concept of Observatory Director as a properly paid career which he brought to it, created what would eventually become a role-model for future British public and academic observatories [26]. In return for the enhanced financial standing of observatory direction, moreover, Airy commenced a programme of systematic observation, and the publication of observations, in a way that no-one had ever done before, at least in Britain.

THE INFLUENCE OF GERMANY

Airy's professionalism of operation was partly the product of his own scrupulous and disciplined personality, but it also drew inspiration from the working procedures of the new German or German-staffed observatories on the Continent. Germany's first proper 'modern' observatory had been a product of the *ancien régime*, when Duke Ernst II of Saxe-Coburg founded Seeburg, near Gotha, which was directed by the Hungarian aristocrat–astronomer, Baron Franz Xaver von Zach, during the twenty years leading up to 1806. But since the founding of the Königsberg Observatory by the King of Prussia in 1810, under the direction of the young Friedrich Wilhelm Bessel, German astronomy had proliferated [27]. Though the first significant astronomers to conduct modern astronomical researches in Germany had been 'Grand Amateurs', the political forces that lay at the heart of Germany's cultural re-birth towards the end of and after the Napoleonic wars rapidly led to the dominance of academically-trained professionals. But it had been a Lilienthal magistrate and a Bremen physician who were the effective fathers of German observational astronomy.

Johann Hieronymus Schröter, who sat on the Bench in the north German village of Lilienthal from 1781 until French troops razed the place in 1813, acquired a series of large-aperture reflecting telescopes which he used to study the Moon, Venus, and the other planets. His work earned the acknowledgement and visits of royalty, and in his private establishment some of Germany's first generation of academic astronomers cut their teeth. Karl Harding and F.W. Bessel had both worked as assistants at Schröter's Lilienthal Observatory before going on to direct observatories of their own. It had also been this Lilienthal group, moreover, which had pioneered the search for and orbital calculation of asteroids, following the setting up of von Zach's 'celestial police' in Schröter's observatory in 1799 [28].

Heinrich Wilhelm Matthias Olbers, only a few miles from Lilienthal, in Bremen, was the world's acknowledged authority on comets, when he died at the age of 82, in 1840. Olbers was a medical doctor who specialised in diseases of the eye, and on the third storey of his house in Sandstrasse, Bremen, he used his private fortune to acquire a collection of fine achromatic refractors and micrometers with which to observe. When he died, his library, which was the most complete cometographic library in the world, was purchased by the Czar of Russia, as part of the Imperial endowment of the new Observatory of Pulkowa, near St. Petersburg [29].

It is not my intention, however, to imply that after these illustrious beginnings German amateur astronomy simply faded away in the wake of the well-funded and equipped professionals. While the professionals, it is true, took over 'big' astronomy in Germany, such as the production of those excellent published star catalogues which Airy so admired in 1832, the nineteenth century witnessed a crop of discoveries in observational astronomy by German amateurs. It was Karle Hencke, a Post Master of Driessen, whose fifteen years of careful searching led to the discovery of the fifth asteroid, Astraea, in 1845 [30], and Johann Heinrich Mädler, whose work with the Berlin banker Wilhelm Beer produced one of the first 'modern' charts of the Moon [31]. It had also been the apothecary Heinrich Schwabe who recognised the 11-year sunspot cycle [32]. Yet amateur astronomy was different in Germany, and never directed the cutting edge in fundamental research and instrument innovation as it did in Britain.

By the 1840s Germany had ten major observatories that were of royal, aristocratic, or academic foundation [33]. Five of them were founded, or else were substantially re-founded, between 1810 and 1825, while three of these – Göttingen, Munich, and Königsberg – had come into existence between 1810 and 1813. Also within the Germanic cultural and technical orbit, the Austro-Hungarian Empire had five significant public observatories by the 1840s, many of them of eighteenth-century foundation, while the Russian Empire topped the list with thirteen. Seven of these institutions came into being between the foundation of Dorpat in Estonia in 1808 and Moscow in 1825, while Russia's *pièce de resistance* was achieved with the magnificent Pulkowa in 1839 [34]. The names of the great German craftsmen Georg Friedrich von Reichenbach, Adolf Repsold, Joseph Fraunhofer and others are prominent in the instrument lists of these institutions, and in 1839 Pulkowa boasted the largest object glass in the world, with the 15 inch-aperture Merz refractor. German directors were also head-hunted by the Czars wherever possible, with Friedrich G.W. Struve first to Dorpat and then to Pulkowa. Friedrich's son Otto, and several subsequent generations of Struves came to form a Germano-Russian dynasty at Pulkowa down to the early twentieth century [35].

France had surprisingly few major observatories before the latter part of the nineteenth century, and it was not for nothing that it was from Berlin that Urbain Jean Joseph Le Verrier sought assistance in the search for his calculated planet, Neptune, in 1846 [36]. The rather moribund state of French observational astronomy contrasted markedly with France's brilliance in mathematics. France's *forte* from the seventeenth to the nineteenth centuries lay in celestial mechanics, and it was not until after 1850 that French observational astronomy began to expand in its range, with the establishment of the Meudon Observatory in 1876, and the organisation of the *Carte du Ciel* in Paris in 1887.

Switzerland, Italy, Spain, and Portugal had a cluster of public observatories between them, most of which were fundamentally re-equipped in the post-Napoleonic era, and the familiar German names, Georg Merz, Traugott Lebrecht Ertel, Adolf Repsold, and so on, ring out among their major instruments. The best-preserved expatriate Germanic observatory of this period with which I am familiar is the Old Royal Observatory in Madrid, originally founded in 1790 and re-equipped in 1853 with a beautiful Repsold meridian circle, electric chronograph, and clocks [37].

What is clear from the above, however, is that astronomy in Continental Europe was thoroughly professionalised in its major and certainly its mathematical branches by the late 1820s. Its superb German equipment, and the time of its Ph.D.-trained directors and staff, were largely bought out of the proceeds of taxation, as Emperors, Czars, Archdukes, and Kings from Madrid to Moscow rebuilt as much of the security of the pre-Napoleonic *ancien régime* as they could manage. Yet in spite of the political reaction that lay at the heart of this astronomical munificence, the results in hard scientific terms were impressive. The fruits of this patronage were found in advances in celestial mechanics, especially when applied to asteroids, and with Johann Franz Encke to comets. Fundamental developments in the understanding of binary and triple star systems, the study of stellar proper motions, and the measurement of some of the first stellar parallaxes were likewise the products of this Germanic innovation; indeed, all of those branches of astronomy where painstaking celestial measurement followed up by sophisticated mathematical analysis was needed. Before G.B. Airy had even won his historic First Wranglership in the Cambridge Tripos Examinations of 1822, Bessel, Struve, Gauss and other astronomers in the German orbit had established set techniques for the use of their instruments, and special 'skeleton forms' for the recording and precise analysis of their data, while their regular published results provided a model for how astronomy should be conducted [38].

OTHER PROFESSIONAL OBSERVATORIES IN BRITAIN

England
Against this apparent abundance of resources on the Continent, English astronomy at first sight seemed to be going off at half-cock. Not only were many of the principal actors in the drama private gentlemen, but even those academic observatories that existed seemed inefficient or lacking up-to-date instruments compared with Berlin or Königsberg, let alone Pulkowa. Oxford's Radcliffe Observatory, for instance, was still using its original 1772 mural quadrants down to 1836 – which would be equivalent to a present-day professional observatory failing to use computers or CCD imaging techniques [39]. Outside Greenwich and Cambridge, upon which Airy flung his mighty organising presence, and Oxford's Radcliffe Observatory, where his old Chief Assistant, the Revd Robert Main, became Observer in 1860, what constituted the establishment of the nation's remaining professional institutions?

There was the small observatory at Kew, near Richmond, which had been King George III's private observatory, where that scientifically-minded monarch had employed Dr Demainbray to show him the heavens, before a third and final bout of insanity had caused the unfortunate constitutional ruler to be incarcerated in Windsor Castle. By 1830 its astronomical activities had effectively ceased, though it became the foremost magnetical and meteorological station in Britain, until Airy largely usurped that role for Greenwich in 1839 [40].

Durham, Liverpool, and the Cape of Good Hope
Britain's newest academic observatory foundation in the mid-century was that of Durham University. Its genesis dates from January 1839, when the Revd Dr Thomas

Hussey, a clerical Grand Amateur of Hayes, Kent, wrote to the Bishop of Durham offering the sale of his private observatory to the new University. The matter was promptly taken up by the Revd Dr Temple Chevallier, a former Cambridge don who already held the joint Professorships of Mathematics and Hebrew in the University [41]. By June 1840, Chevallier had raised a sum of over £1,480 by subscription. The purchase and carriage of Dr Hussey's instruments, which included a fine Fraunhofer refractor, had consumed £900, while £580 had already been disbursed for buildings, though a similar sum was still wanting to complete them. In 1841, Chevallier was given the Professorship of Astronomy at Durham, which he added to his existing chairs of Mathematics and Hebrew, and retained all three for the next thirty years [42].

Yet it is probably true to say that Temple Chevallier was an astronomer by accumulation as much as by vocation, for his real goal was Durham University's Divinity Chair, which he never succeeded in getting, but for which he occasionally deputised. His Cambridge career, it is true, had received an obviously mathematical launch when he became Second Wrangler in 1817, but as a Fellow of Catherine Hall (later St. Catherine's College) and Hulsean Lecturer in Divinity, his bent was as much theological as practical astronomical. On accepting the Durham Mathematics Chair in 1835, moreover, he made it clear to the Bishop that he also wanted a priestly cure of souls in the diocese, and did not wish simply to be an ordained mathematician. He became the incumbent of Esh, just outside Durham, where his Protestant susceptibilities were immediately outraged by the presence of Ushaw, the Jesuit College, 150 men strong, within his parish [43]. Chevallier, who was just as staunch a Roman Catholic-hater as his ordained colleague at the Armagh Observatory, the Revd Dr Thomas Romney Robinson, lived in a state of semi-civil war with much of his parish for many years, in spite of the Jesuits being supportive of his observatory after 1840, and the interest which they showed in it. But in spite of any congruence which he might have had with the Jesuits on points of astronomy, Temple Chevallier railed, along with his clerical friend Dr Corrie of Cambridge, about the treachery of the 'God-denying Act', or Catholic Emancipation Act, of 1829, which granted civil and political rights to British Catholics [44]. As far as Chevallier was concerned, any Ushaw College sympathy with Anglican Durham University astronomy was just another example of Jesuitical cunning.

Indeed, it is hard to see where Chevallier found time for any long-term astronomical research, considering the teaching duties of his three Chairs, his ecclesiastical living, sustained sectarian spleen, and the subsequent Church and University preferments which came his way: Registrar of Durham University, Rural Dean of Durham, and then Residentiary Canon of the Cathedral! [45] Yet this exquisitely Barchesterian clerical and academical pluralist was genuinely active in the promotion of astronomy as well as Anglican Christianity. By June 1847, the Senate Minute Book records that Chevallier had been granted permission to take the University's Northumberland Refractor to his parish at Esh for the long vacation so that he could observe more conveniently, and in 1851 he was corresponding with Airy about the setting up of a refractor 'in the German manner' [46].

But where Chevallier was truly innovative at Durham was in his staffing of the Observatory by Assistants who did the routine work of observation and computation in his absence. The bane of British observatories in the Victorian period (as will be discussed in Chapter 8), be they public, academic, or private, was their dependence upon an astronomical proletariat of non-graduate assistants who were virtually drilled in the performance of specific skills, and who could rarely ever hope one day to direct their own professional observatories. Chevallier introduced a policy into Durham which was closer in spirit to that of the great German observatories, where the generality of assistants were gifted young university graduates, doing the slog of observation and computation for a year or two before passing on to more senior posts elsewhere. Of the ten young men who were Chevallier's assistants between 1841 and 1872, six were Durham graduates, usually in mathematics or engineering, while two more were Cambridge M.A.s. Only two of the ten did not possess a university training, and one of these, William Ellis, achieved sufficient scientific distinction in later life to be elected F.R.S. [47].

Temple Chevallier occupies an unusual position in the astronomical world of his day: a clergyman and theologian by instinct, and an able and ambitious man whom one felt was always on the lookout for any power vacuum that he might fill to advantage, be it theological, linguistic, administrative, or astronomical. Even by 1841, the Durham University Calendar showed that he was making £700 a year from his three not very well endowed chairs, while after the Reform Act of 1840, most English cathedral canons had their annual revenues fixed at £500. In 1837, the parish of Esh was worth only £65 a year, though it was later raised to £200 by the Ecclesiastical Commissioners [48]. Even before one reckons in his further stipends as Rural Dean, University Registrar, and other perquisites of office, however, one can see that Chevallier was enjoying an income of at least £1,400 a year, not to mention any free residences that might have been thrown in. He was certainly the best-paid part-time professional astronomer in the British Isles.

It was not easy to cross the divide that separated the non-graduate 'professional assistant' astronomer (to whom Chapter 8 will be devoted) and the salaried directors of professional observatories. Charles Piazzi Smyth was one of the few to have done it when he made the leap from being Maclear's Assistant at the Cape to the ill-paid grandeur of Astronomer Royal for Scotland, though he was the son of one Grand Amateur and the pupil of an ex-Grand Amateur turned professional, and was clearly a product of that gilded astronomical world of large apertures and comfortable pockets [49]. Of more humble scientific ancestry was John Hartnup. He had worked as an Assistant at Lord Wrottesley's private observatory in Blackheath (prior to its removal to Staffordshire), as a Supernumerary Computer at Greenwich, and as the salaried Assistant Secretary to the R.A.S. before becoming Director of the Liverpool Observatory in November 1843. This newly-founded Observatory was owned and operated by the Liverpool Dock and Harbour Board, and intended as a chronometer-regulating station; in 1849 it obtained a fine 8½-inch equatorial refractor, and then in 1866 migrated to an improved site on the south bank of the Mersey, at Bidston Hill [50]. Hartnup had obviously made a good impression on the influential astronomers whom he had served up to 1843, for it was the Liverpool Grand Amateur, William Lassell, himself an adviser to the Liverpool Observatory Board, who was instrumental in Hartnup's appointment. Indeed, at the meeting of the Observatory Committee held on 26 August 1843, letters were presented by Lord Wrottesley, Francis Baily, President of the R.A.S., Captain (later Admiral) Smyth, William Lassell and others 'strongly recommending Mr. Hartnup for Appointment to the Office', and all backed up by commendation from William Rutter Dawes and Airy [51]. Although the work was relatively humdrum, being concerned in the main with chronometer regulation, the industrious Hartnup soon became one of the world's leading authorities on chronometer management, and made several innovations towards improving the thermal homogeneity of these vital maritime instruments, not to mention work on double stars and comets. In 1843 he received £200 a year and a house, rising to £300 in 1851 and to £500 in 1875. His son, John Hartnup junior, was taken on as Assistant in 1862, at £200 a year, and succeeded his father as Director in 1885. John Hartnup senior, therefore, was one of the relatively few Victorian astronomers, along with John Russell Hind, to have made the transition from paid assistant to gentleman of science without the advantages of a university degree, holy orders, or an independent fortune to ease him on his way [52].

As an imperial power, Britain had been setting up observatories around its territories since 1821, when a small observatory was established at the Cape of Good Hope under a young Cambridge graduate, the Revd Fearon Fallows [53]. And the year following, the governor of New South Wales, Sir Thomas Brisbane, founded a private observatory at Paramatta, which he subsequently gave to the government to be maintained as an official Australian observatory, under the direction of James Dunlop [54]. Other small observatories were created in India by the East India Company, at Madras and elsewhere. Most of these southern hemisphere or tropical observatories had the job of improving southern star catalogues, and on a more prosaic level acting as geodetic, magnetic, or chronometer regulating stations [55]. Yet the two major British southern observers before the expansion of the Cape Observatory in the 1840s had been amateurs – Edmond Halley in the 1670s and Sir John Herschel in the 1830s – who had been working in their own capacity as private gentlemen, although observers from other European countries such as the Frenchman Nicholas Louis de Lacaille, had done good work in the southern hemisphere at the behest of home governments or other official bodies [56].

Scotland

The Observatory in Edinburgh, which had started as a subscription foundation in 1811, had been elevated to royal rank following King George IV's state visit to Edinburgh in 1822, and after its move to the present and spectacular Calton Hill site. In 1834, it became the official seat of the Astronomer Royal for Scotland, who was also to be the holder of the Regius Astronomical Professorship in the University [57]. Professor Thomas Henderson did excellent work there, though its most famous Victorian incumbent was Charles Piazzi Smyth, the son of the English Grand Amateur astronomer, Admiral William Henry Smyth of Bedford. Piazzi, who was only 27 when he became Scotland's Professorial Astronomer Royal in 1846, was not even a university graduate, although he had received an excellent practical training both from his father and from his father's former medical and amateur astronomer friend, Sir Thomas Maclear, with whom Piazzi worked when Maclear became

Astronomer Royal at the Cape of Good Hope Observatory [58]. But in spite of its royal designation, the Edinburgh Observatory was dogged by under-funding and experienced difficulty in finding high-calibre staff. Piazzi's salary was only £300 per annum (the same sum as Richarda Airy's Civil List Pension, and considerably less than the salary of Airy's Chief Assistant at Greenwich who received £350 rising to £470), although Piazzi probably enjoyed some family money as well [59]. Even so, there was never enough official funding to keep the instrumentation properly up to date, or pay for adequate assistant staff, and Piazzi was often obliged to borrow specialist instruments from his Grand Amateur friends (such as James Nasmyth) when he had need for them [60].

In 1876, a Government Commission of Enquiry, chaired by the Scottish Grand Amateur astronomer, Lord Lindsay, soon to be the twenty-sixth Earl of Crawford and Balcarres, reported upon the under-resourced and run-down state of the Edinburgh Royal Observatory, though nothing positive was done. Following the retirement of Piazzi Smyth in 1888, however, Lord Crawford took bold and munificent steps in an attempt to preserve Scotland's Royal Observatory [61]. In the same year he donated the private physical observatory which he had established on his estate at Dun Echt, Aberdeenshire, with the intention that not only should Scotland's foremost observatory retain its royal title, but that it should rival the universally-admired Pulkowa [62]. A national observatory was thereby re-invigorated and effectively re-founded by an influential amateur, and the present Royal Observatory, Edinburgh, along with its Crawford Library, is the direct descendant of this benefaction to the Scottish nation. (Lord Crawford's work will be discussed in more detail in Chapter 7.)

In 1754, the Syndics of Glasgow University instructed their London agent, William Ruat, to purchase a collection of astronomical instruments for the establishment of a small observatory. Orders for the standard right ascension and declination measuring instruments were placed with the leading London makers. There was a 2½-foot radius quadrant by John Bird at £100, a 4-foot transit by Sisson at £22 (Bird was asking £30 for the same thing), a fine regulator clock by Shelton for 60 guineas (£63), and 35 guineas (£36.75) for a 2-foot focal length reflecting telescope from the expatriate Edinburgh optician, James Short [63]. It was not until after 1840, however, that a more up-to-date observatory was founded in the city, when a public subscription (as was a common practice in America) made it possible to order a 42-inch meridian circle by the ubiquitous T.L. Ertel of Munich, and orders for a 9-inch aperture equatorial and a 6-inch refractor were eventually placed. Under Dr Robert Grant – perhaps better known today as a Victorian historian of astronomy – the Observatory published a meridian catalogue of 6,415 stars in 1883, though like so many other British official establishments, it was both cash-starved and kept largely chained to relatively routine observational work [64].

Ireland
Ireland's official astronomical establishment did rather better than Scotland's in many ways, due to university and archiepiscopal benevolence. Trinity College Dublin's Observatory at Dunsink had been founded in 1785, and the Andrews

Professor of Astronomy (who held the title Astronomer Royal for Ireland) worked there [65]. The advanced yet somewhat flawed design of the 8-foot diameter reversible meridian circle by Jesse Ramsden gave the Observatory a technical edge in many ways, yet fundamental errors caused many of Professor John Brinkley's observations made up to 1827 to be of little value [66]. But in 1863, the Observatory acquired the already well-travelled 11¾-inch aperture object glass by Robert Aglaé Cauchoix, which had been re-mounted in a new telescope by 1868 [67]. This lens, which had possessed the biggest clear aperture in the world when it was acquired new by the irascible English Grand Amateur Sir James South in 1829, was used by Sir Robert Stawell Ball at Dublin after 1874, as Ball ascended the professional astronomical ladder after working with Lord Rosse (he was actually a tutor to the Earl's sons), to eventually succeed John Couch Adams as Director of the Cambridge University Observatory [68].

Armagh was the Primate of Ireland's observatory, which had been founded by Archbishop Robinson in 1791. Its undisputed Victorian luminary, however, was the Revd Dr Thomas Romney Robinson, whose reign as Director spanned some 59 years, from 1823 to 1882. In spite of the shared name, Thomas Romney Robinson was not related to the Archbishop, being the son of an English artist who had settled in Belfast [69]. But in its original form the Armagh Observatory had not been adequately funded, and not until Archbishop Robinson's successor, Lord John Beresford, sumptuously re-endowed it with the finest instruments after 1827, did it emerge as a major observatory, as Jim Bennett has shown in his excellent history [70]. It is impossible, however, to reckon the scientific significance of Victorian Armagh separately from the driving force of Thomas Romney Robinson. The excellent transit circle by Thomas Jones, the telescopes by Thomas Grubb, and other instruments were put to constant and thoroughgoing use by Robinson, and his catalogue of 5,345 stars, published in 1859, assured the Observatory's reputation [71].

This Anglo-Irish clerical astronomer, who held in turn the Protestant Episcopal Church of Ireland livings of Carrickmacross and Enniskillen in conjunction with his Armagh Directorship, was one of the most formidable and energetic ordained scientists of the age. On intimate terms with Grand Amateurs such as Lord Rosse, Edward Joshua Cooper of the Markree Observatory, and Sir James South, he had a social confidence and ease which in the Victorian age came not from having a salaried job, but from a landed freehold: in his case, those parsonical freeholds which brought with them an influential place in society [72]. Ultra-High Tory as he was, Robinson, like Temple Chevallier, condemned the liberalising Catholic Emancipation Act of 1829, and correctly saw that in a country where the majority of the population was not Protestant, it would make tithes and church rents harder to collect. Robinson was not only an active astronomer, but an enthusiastic promoter of scientific knowledge, especially in Ireland. A brilliant orator and wit, he regularly bedazzled the annual meetings of the British Association for the Advancement of Science with his own unique combination of rigour and flamboyance, not to mention his popularity as a speaker at Tory loyalist dinners [73].

Yet in 1869 Armagh suffered a setback which was unique to the vicissitudes of the Irish situation. In that year, Gladstone's Bill for the dis-establishment of the Irish

Episcopal Church became law, with potentially disastrous consequences for an observatory which was largely financed by church rents. Romney Robinson stated that the Bill 'confiscated at least half of its income', and forced him into negotiations for compensation, which was eventually re-invested in railway and other assets [74]. At the height of the dis-establishment furore, however, Robinson gave an assessment of the significance of his observatory which reveals an attitude closer to that of a Grand Amateur than a salaried Director, for Armagh 'is entitled to rank in every respect (except wealth) with the first rate observatories... [for it]... is in fact a National one without costing the Nation a shilling' [75].

Forming in many ways a link between the more explicitly public and the private observatories was that established at Queen's College, Cork, in 1878. Although this Observatory, with an 8-inch refractor on a new type of equatorial, 4-inch siderostat, and transit circle, all by Ireland's own internationally illustrious instrument-making firm, Sir Howard Grubb of Dublin (who had succeeded his father Thomas), was intended to be part of an academical foundation, under the direction of the Professor of Natural Philosophy, it was the private benefaction of a Cork gentleman, William Horatio Crawford [76]. Like so many people who did serious astronomy in the nineteenth century, even in the once-removed capacity of patron, W.H. Crawford came from a family of brewers and merchants. These émigrés from Scotland nonetheless became liberal patrons of their adopted city of Cork. Although the Cork Academy was his primary benefaction, Crawford also saw fit to endow the institution with an astronomical observatory of the finest quality, housed within a stone building which was almost indistinguishable, when its transit roof slits were closed, from a medieval Irish church. Indeed, it was one of the most architecturally idiosyncratic observatories ever to have been built, combining as it did science, art, and religion with history [77].

BRITISH 'GRAND AMATEUR' OBSERVATORIES

When Jean Louis Emil Dreyer (who was then the Director of the Armagh Observatory, after the long reign of Romney Robinson) came to write the article 'Observatory' in *Encyclopaedia Britannica* (8th edition, 1884) he drew up a list of 26 private observatories that had operated in Britain over the previous 100 years [78]. And these were important foundations, which had contributed significantly to original research. Fourteen of these observatories (including the establishments of the two Herschels) no longer existed in 1884, ten were still in full operation, and two were the property of major public schools – Rugby and Stonyhurst – where professional schoolmasters still found time to do research of importance with school property.

Dreyer's list, moreover, was by no means complete, for he had failed to pick up many of the 24 British private observatories listed in the *Edinburgh Encyclopaedia* in 1830 [79] or some of those mentioned by John Weale in 1851 [80], or again the 48 listed in the *Astronomical Register* in 1866 [81]. Dreyer also failed to include the sumptuous private observatory of Sir George Shuckburgh, at Shuckburgh Hall,

Warwickshire, which its owner had equipped with an equatorial instrument of revolutionary design by Jesse Ramsden in 1793, though it is true that the Shuckburgh results did not come up to the original expectations that Sir George's 1793 paper had appeared to promise to the Royal Society [82]. Dreyer, furthermore, made no mention of the fine Troughton and Simms-equipped private observatory which the brewer, Samuel Charles Whitbread, set up at Cardington, Bedfordshire, in 1851, at which useful double and variable star work was performed [83]. The magnificently-equipped observatories with their large-aperture reflectors and substantial refractors which James Nasmyth established at Patricroft, Manchester, between 1840 and 1856, and then at Penshurst, Kent, from 1856 to 1890 were also left out of Dreyer's count [84], along with the West Hendon House Observatory, with its salaried assistants, employed by the Sunderland banker–astronomer Thomas William Backhouse [85]. As Dreyer was interested in functional observatories that added to scientific knowledge, one can forgive him for not including abortive failures in his list, such as the Revd Mr Edward Craig's giant 24-inch aperture refractor which was suspended from a tower in Wandsworth, South London, in the 1850s and which supposedly wasted around £6,000 [86], or the numerous private observatories mentioned in the Journals of Dr John Lee (of which more will be said in Chapter 5).

Yet when the working, failed, and lost Georgian and Victorian observatories are reckoned together, one sees that over 40 significant private institutions were founded in Britain in the 90 years before 1884, not to mention an unreckoned number of smaller ones. The capital expenditure was enormous, extending from relatively modest outfits that were probably up and running for a few hundred pounds to the princely £12,000 that it allegedly cost the Third Earl of Rosse to build the 72-inch reflector at Birr Castle, and was only possible in a country where (in spite of vast differences between the conditions of the rich and the poor) there was a considerable surplus of cash, as well as educated leisure and intellectual motivation [87].

By the 1860s, moreover, the British amateur tradition had begun to manifest itself in the countries of the Empire, to enrich and extend the work of the official observatories. John Tebbutt's observatory, at Windsor, New South Wales, which its owner had equipped with a good 4½-inch Cooke refractor and other instruments, was one of the first southern hemisphere observatories to specialise to a large extent in cometary work [88].

AMATEUR OBSERVATORIES IN EUROPE AND AMERICA

No other European country came near to rivalling the numerous substantial private observatories of Britain. According to Dreyer's list of 1884, Germany had about half a dozen private observatories, including the defunct ones of Schröter and Olbers, although it is true that he failed to list some of the smaller, yet significant, private observatories of German selenographers, such as Wilhelm Beer (where Johann H. Mädler did most of the work), or Wilhelm Lohrmann of Dresden [89]. And of course Dreyer makes no mention of modestly-equipped amateurs like the

French country doctor Edmond Modeste Lescarbault, who corresponded with Le Verrier in 1859 about his supposed sighting of an intra-Mercurial planet [90].

The only other country to experience a remotely similar ascendancy of high-level amateur astronomy outside the British Imperial sphere was the United States of America. When Airy wrote his *Report* in 1832, he pointed out that, to his knowledge, there was no significant observatory of any kind in America. Within 60 years, however, there were at least 22, most of which were in some sense privately owned. While it is true that America had two naval observatories by the 1870s, in Washington and Annapolis, and a handful of Grand Amateur establishments, such as Henry Draper's state-of-the-art photographic telescopes in New York, there were also two classes of private institution which were particularly American [91].

One of these was the observatory built by public subscription within a particular city, while the other was the Liberal Arts College observatory. Boston led the way with both forms of institution. The Harvard College Observatory came into being in 1839 when William Cranch Bond, a Rochester Massachusetts businessman, donated his private collection of instruments to Harvard. Bond's donation was quickly augmented by a public subscription and munificent donations from the businessman David Sears, which raised the substantial sum of $25,000, equivalent to £5,000 sterling [92]. This money was used to purchase a 15-inch refractor from Georg Merz of Munich, which upon its completion a few years later was one of a pair of the two largest object glasses that had ever been figured to date. The other twin 15-inch lens had gone to Czar Nicholas II's observatory in Pulkowa, of which Wilhelm Struve was the handsomely-salaried Director [93]. Bond, however, retained his amateur status by being the honorary Director of the Harvard Observatory, at least until a separate subscription had been raised to give him a salary after 1846. And almost on the heels of the Bostonians, the citizens of Cincinnati ordered an 11¼-inch Merz refractor in 1842 [94].

Both classes of observatory – public subscription and liberal arts collegiate – were accessible to non-professional scientists, especially subscribers and students, although most enjoyed the services of salaried specialists whose published researches brought lustre to their patrons. As part of the rounded education of an American gentleman, and after the founding of the Vassar College Observatory, of an American lady, most liberal arts colleges set up astronomical observatories of some sort, many of which came to be equipped with Clark refractors [95]. The science professors in charge of these observatories, like English public schoolmasters at Rugby and Stonyhurst, were employed to teach rather than to conduct original research, which produced a class of what might be called semi-amateur (or semi-professional) astronomers in American society.

The presence of these amateur, Grand Amateur, and semi-amateur astronomers in America was rooted in the libertarian political and economic traditions of that country. As in Britain, there was low taxation and, in spite of a Civil War, a mistrust of centralising influences; and as a society which was becoming increasingly prosperous and enfranchised as the century wore on, America preferred to see social and intellectual initiatives in private rather than in official hands. And as in Britain, and perhaps even more so, private individuals could generate enormous fortunes which they might use to endow scientific or educational enterprises to which the public had some assumed right of access. Indeed, the growth of American astronomy, and its relation not only to independently-financed personnel but to the enormous industrial fortunes that patronised it after 1870, would be a subject worthy of further scholarly attention, though this is beyond the scope of this present study [96].

WILLIAM PEARSON AND THE ROYAL ASTRONOMICAL SOCIETY

But the Grand Amateur tradition undoubtedly had its roots and its power-base in Britain, and nowhere was this emphasis on independent initiative more clear than in the founding of the Royal Astronomical Society.

On 12 January 1820, the Revd Dr William Pearson, Proprietor and Head Master of the fashionable East Sheen School, Richmond, London, and subsequently Rector of South Kilworth, Leicestershire, called together thirteen astronomical friends for a dinner at the Freemasons' Tavern in Lincoln's Inn Fields, London. The fourteen gentlemen included a stockbroker, a West Indies merchant, an oriental linguist, a surgeon, a couple of non-resident Cambridge dons of independent means, a Colonel in the British Army, and one or two people involved in non-university teaching. The only individuals who currently held scientific professorships in that company were Olinthus Gregory, Professor of Mathematics at the Royal Military Academy, Woolwich, and the visiting Peter Slawinski, of the University of Vilna in western Russia. The Astronomer Royal, John Pond, and the Oxford and Cambridge Professors were conspicuous by their absence, and while one might excuse them on the grounds of their loyalty to the Royal Society, it cannot be denied that the group of fourteen who resolved upon the reform of British astronomy from what they saw as its relative neglect by the Royal Society were private *lovers* of the science rather than paid professionals [97].

What concerned Dr Pearson and his friends was the apparent sluggishness into which British astronomy had fallen, and the way in which it was seeming to lag behind what other countries were doing in terms of technical excellence. No one captured this state of malaise better than John Herschel, when he remarked that 'The chilling torpor of routine had begun to spread itself over all those branches of science which wanted the excitement of experimental research' [98]. Of course, no one was denying that his father Sir William Herschel's cosmological researches were of the highest order of significance, though the nation which had led the way in the founding of the Greenwich Observatory 150 years before was now sliding behind. When in 1832 Airy summed up British astronomy as being concerned too much with routine observation and not enough with the rigorous mathematical interpretation of those observations, he was in part repeating the cry of Dr Pearson and his friends of twelve years before. But if things were starting to get better by 1832, it had been very largely through the pressures generated within the Astronomical Society of London, and not any particular sustained action on the part of the state.

On 22 January 1820 the financier–Secretary of the new Astronomical Society, Francis Baily, had already prepared, with John Herschel and others, a manifesto wherein its research objectives were to be published. Rather characteristically, this statement of the hoped-for new beginning of British astronomy was addressed not from an academical institution but from Baily's rooms in Gray's Inn, and specified his postal address as the Stock Exchange [99]. Rather optimistically – as one might think for a society of private individuals – these astronomers proposed a wholesale re-cataloguing of the sky down to the 'most minute objects visible in good astronomical telescopes'. It was their intention 'to disclose the most important secrets of the system of the universe', to detect stars of particular interest (such as binaries and triple stars) which appear the same as ordinary stars unless subjected to 'the test of careful and repeated observations' [100].

The new Society should have met for its first formal meeting on 29 January 1820, but the members postponed until 8 February, as a mark of respect to the memory of King George III, who had just died. It grew rapidly, and by March had enlisted into its ranks the instrument-maker Edward Troughton, amongst others. Sir William Herschel was prevailed upon to accept the Presidency in the spring of 1820, though the ailing 82-year-old ex-musician and pioneer cosmologist only became so on the understanding that he did not have to travel up to town to attend meetings [101]. The first active President, Henry Thomas Colebrooke, was a 'second-career' mathematician and astronomer, for as the distinguished Victorian orientalist Max Müller reminds us, his enduring reputation lay in being the 'Founder and father of true Sanskrit scholarship in Europe' [102]. When one looks at virtually all the executive officers of the Society for the first few decades, indeed, one finds a veritable roll-call of British Grand Amateurs, whose first professions had been elsewhere. There was Baily the stockbroker, South the surgeon, Groombridge the City merchant, Colebrooke the orientalist, Lee the lawyer, Pearson the parson–schoolmaster, Colby the army colonel, and Brinkley the Anglo-Irish bishop. But it was to be from this private body, which obtained its charter to become the Royal Astronomical Society in 1831, that so much of British astronomical energy was to come in the nineteenth century and which John Weale characterised in 1851 as the most efficient learned society of the age [103]. It not only set about achieving its manifesto goals described above, but instinctively did so by acting not as an organising agency, but as a clearing-house for private individuals who liked to work on their own, in their own time, and in their own way. Rather than a place from which they received instructions, the Society's rooms in Somerset House were to be a place where the Fellows would meet each other, present results, argue, and go off to dine. Corporately, however, the Society could form a powerful pressure group for British astronomy, and with the redoubtable Baily as Secretary and President, it was able to persuade the government, through the British Association for the Advancement of Science in the 1830s, to find the necessary funds to reduce and render useful the horrendous backlog of raw Greenwich observations [104]. In short, a private society took the initiative in making a government department work properly, in a way which would have been hard to imagine in France, Germany, or Russia.

The picture which thus emerges of British astronomy in the nineteenth century is very much one in which the initiative of private individuals took a clear lead against the provision of the state or corporation, and where the allocation of resources within the science – be they resources of equipment or of personal time – did not generally lie with salaried personnel. Official astronomy was conservative and intellectually circumscribed in character, and kept very much to useful functions such as the supply of accurate time-keeping and navigational data to the century's foremost maritime power. But any ventures into more intellectually exciting waters, leading to the disclosure of 'the most important secrets of the universe', as the R.A.S. manifesto put it, were matters for the private sector. A figure like G.B. Airy, who was not ashamed to tell Cambridge Syndics and Lords of the Treasury that he had no means of support beyond his salary, was very much a one-off, though he undoubtedly advanced the standing of the professional scientist in Britain. Yet for many years after 1835 it would still be assumed that even a man who was in charge of a publicly-funded observatory would have clerical, academical, or private strings to his astronomical bow. In astronomy, just as in county cricket, it was still assumed that the leaders were gentlemen who did it for love, and who were aided by professional players who needed their scholarship-earned degrees to negotiate the best incomes that they could.

3

An inheritance, a wife, a benefice, or a brewery: financing fundamental research

Conducting astronomical research of a quality that could compete with Königsberg or Pulkowa was not cheap. The instrumentation which it demanded became ever more costly with each new innovation in optics or circle-dividing, while the protracted, meticulous observation assumed that the astronomer was abundantly endowed with time. This was not an activity for people who had to be up early each morning, or whose daily occupations were time-consuming. If one was not so fortunate as to inherit money, then one had to make it from a liberal profession, a managed business, or from a dowry, if one intended to take part in high-level astronomical research.

    The inherent danger in Grand Amateur astronomy, however, was its tendency to duplication, as individual astronomers pursued their own fascinations, and in this connection, the R.A.S. and the Royal Society played key roles. If scientific research did not have an official central directing agency, then it needed forums and clubs where active persons could come together, learn who was doing what, exchange ideas, and augment and publish the growing body of data. Astronomy, after all, was not the only science to feel this organising need in a research world dominated by Grand Amateurs in a variety of disciplines. The founding of the botanical Linnean Society (1788), the Geological Society (1808), and that profusion of metropolitan learned societies that took off after 1830, many of which were to establish their standing by the acquisition of Royal Charters, was symptomatic of the tendency. Those annual scientific jamborees, the meetings of the British Association for the Advancement of Science, which started in 1831, only added to this impetus of individual researchers, their numbers swelled by the growing throng of an interested scientific laity, now coming together to discuss the state of the art in their respective sciences, in a holiday context of dinners, balls, and entertainments that followed the delivery of the learned papers [1]. Yet even those members of the early British Association who looked admiringly towards the more centrally directed sciences of continental Europe could put the wind up the staunchly independent Sir John Herschel, who warned William Whewell in September 1831 that any attempt to centrally manage scientific creativity could only result in 'an overwhelming mediocrity' [2]. To Herschel, the very Britishness of British science, and thereby, the crop of discoveries that it had produced over the two preceding centuries, stemmed from the independent character of its practitioners.

The R.A.S. provided the rallying point for British astronomy in many ways, but nowhere more obviously than in its political implications. Unlike a princely academy, its members were not nominated from on high, but came together as a voluntary group of equal partners. Like an Oxford or a Cambridge College, but unlike the Universities of Berlin or Paris, it was a members-directed body, and like the 'gentlemen free' of the Royal Society, it consisted of *Fellows*. This very term, and its Latin root-word, *socii*, as used in Oxford and Cambridge foundations, implied a self-governing society or club of voluntary participants as opposed to a hierarchical academy. The reigning monarch might be welcomed as a gracious, title-bestowing patron, but was otherwise expected to keep out of the society's internal affairs.

It was the rapidly increasing cost of the instruments that were necessary to do serious astronomy, as well as the availability of leisured personnel who could use them, which set the stage for the Grand Amateur as a scientific type. And as the instruments themselves were only a response to a particular set of questions being asked by scientists, it is important to look at what aspects of astronomy attracted the Grand Amateurs.

GRAND AMATEUR PRIORITIES AND INTERESTS

Virtually every Victorian amateur astronomer that I have encountered had a love of beautiful and delicate mechanisms. One senses that – like so many high-level amateur astronomers of the present day – they were people who enjoyed owning and handling fine technology. This meant, therefore, that first and foremost they were observers, rather than mathematicians or theorists. Indeed, Airy's complaint of 1832 – that Britain produced too many observations which never got reduced – is entirely in harmony with this circumstance [3]. To Airy's administrative cast of mind, there was no purpose in accumulating observations if one did not aim at reduction and publication, yet I suspect that he never really understood the motivations of the amateurs who surrounded him. And very importantly, as a temperamental wedge between him and most Fellows of the R.A.S., he took no joy whatsoever from the manipulative procedures of observation, and relegated all observational work at Greenwich to the assistant staff. Indeed, in this environment of Grand Amateurs who spent fortunes upon their private observatories, one wonders how many apoplectic seizures his disdain might have occasioned, as when he wrote in 1847: 'The lowest of all employments in the [Greenwich] Observatory is mere observation. No intellect and very little skill are required for it. An idiot with a few days' practice, may observe very well' [4]. Airy's opinion of the astronomical observer, however, could well have deteriorated over the sixteen years leading up to 1847, for in 1831 he had characterised the 'mere observer' somewhat more kindly as a man who was 'a compound of a watchmaker and a banker's clerk' [5].

At the beginning of the nineteenth century, the staple work of the observatory was in many ways the same as it had been in the days of Tycho Brahe's Uraniborg: the making of accurate star maps, and the constant monitoring of solar, lunar, and planetary positions to form tables. Why the same work still went on, and would continue until the early twentieth century, derived from the constant improvement that was taking place in the quality of instruments. Between Tycho's mural quadrant of 1582 and Edward Troughton's circle for Stephen Groombridge in 1806, the accuracy by which astronomers could measure celestial positions had improved by a factor of 300, from a single arc minute to sometimes less than a quarter of an arc second [6].

Astronomers had long since realised that a continual increase in the accuracy of the measured positions of the stars and planets could be made to yield up more and more of the 'secrets' of the Universe. Flamsteed's screw-edged mural arc of 1690 with its telescopic sights at Greenwich had laid up that vast treasury of observations which would go a long way to substantiating Newtonian theory with respect to the orbits of the Moon and comets. The zenith sector which George Graham built for James Bradley's private observatory at Wanstead, Essex, with its half-second of arc accuracy, had just failed to detect the searched-for stellar parallax, although it enabled its owner to discover the aberration of light and the nutation from a tiny systematic motion of the star γ Draconis by 1748 [7]. By 1810, in fact, the constant monitoring of celestial positions with increasingly refined instruments had led not only to the discoveries mentioned above, but also to the discovery of the proper motions of many stars, of the correct shape of the lunar orbit, the discovery of Uranus and its orbital characteristics, of the first four asteroids, and of the existence of several dozen binary stars. These were triumphs indeed, where a refined mechanical and optical technology elegantly complemented the predictions of Newtonian theory [8].

PAYING FOR INSTRUMENTAL INNOVATION: THE ASTRONOMICAL CIRCLE

To push back the frontiers of measurement further, however, one needed new technologies based upon the newly-developed astronomical circle, which were brought about after 1780 by Jesse Ramsden and John and Edward Troughton, along with achromatic object glasses of high resolution. And such instruments did not come cheap. In 1810, both the Greenwich and Radcliffe Observatories were still working with large-radius and rather antique quadrants that were not reliable for measurements much more refined than a couple of arc seconds or more. Yet if one wanted to measure down to 0.5 or 0.25 arc seconds one needed the superior homogeneity of the circle, as against the quadrant. In 1810, the only professional observatories in Britain that had circles were Dunsink and Armagh in Ireland, and these had been purchased by private or ecclesiastical benefactions [9]. The others were in observatories in Europe, where royal or princely patrons had recognised the superior advantages of the circle. Palermo, in Sicily, had led the way in 1789 with a

Ramsden circle, but in reality it had been the English Grand Amateurs who first commissioned circles from Ramsden and Troughton, and put them through their paces [10]. Sir George Shuckburgh, Gavin Lowe and many others had them, while astronomically-minded diplomats like the Count von Brühl were purchasing and singing the praises of Troughton's circles across Europe [11]. One influential astronomical circle by Edward Troughton was that built at the 'cost price' of £120 for his amateur astronomer friend, Gavin Lowe of Islington, in 1793. This beautiful and innovative instrument subsequently passed through the hands of the Honourable Charles Greville in 1807, and then on to the clerical amateur, the Revd Lewis Evans. The landed lawyer, Dr John Lee, then purchased it from Evans, after which Dr Lee presented it to the R.A.S. Indeed, the changing ownership through which the 'Lee Circle' passed over forty years was very much of a social slice of the Grand Amateur world of the day: a private gentleman, a son of a nobleman, a well-beneficed cleric, and a rich lawyer! [12]

Stephen Groombridge

In the early years of the nineteenth century, when British institutional astronomy was in the doldrums, and prior to the founding of the R.A.S. in 1820, the initiative lay with private individuals, who were both doing the serious work, and stimulating reform. Stephen Groombridge had spent a lifetime amassing a handsome fortune in the City of London, first from a linen drapery business, and then as a West Indies merchant. He had worked for some years from a private observatory at Goudhurst in Kent, but in June 1806 he set up an instrument of revolutionary design at his new residence–observatory at 6 Eliot Place, Blackheath, London [13]. This was a 'transit circle' by the ubiquitous Edward Troughton, which embodied the superior homogeneity of division of the astronomical circle combined with a full 360 degrees of meridian rotation upon its stone piers, so that it incorporated the functions of a normal transit instrument as well as a circle. By dispensing with the need for a separate transit instrument, the 4-foot 'Groombridge Circle' made it possible, if necessary, to take both right ascensions and declinations in a single observation, thereby reducing the time required to measure a star's position. As a composite instrument, it inevitably posed problems of engineering design, and in 1806 only an amateur had the resources with which to commission such experimental innovation [14].

Working in Blackheath, in a location so close to the Royal Observatory that his northern meridian mark seems to have been cut into the south side of the wall surrounding Greenwich Park, the 52-year-old Groombridge set about the construction of a catalogue of circumpolar stars down to magnitude 8.5 within 38° of the northern pole (circumpolar for Blackheath, at 52° north) – an area of the sky poorly covered in existing catalogues [15]. By 1817 this meticulous amateur had used his instrument to record some 50,000 individual star measurements – many of them, of course, being checks on previous observations. No doubt this labour was eased by the transit room being next door to Groombridge's dining-room, so that even in the middle of a meal he could excuse himself for a minute or two, to 'take' a familiar star which he knew would be approaching the northern meridian line [16].

Groombridge fully realised that observations without reduction were of little value, and set about the grind of computation, though he was only half way through his task when he was incapacitated by a stroke in 1827. The rest of the reduction was done at public cost after Groombridge's death in March 1832, the catalogue finally appearing, under the supervision of Airy, in 1838. The Astronomer Royal, indeed, was to describe it as 'one of the greatest which the long-delayed leisure of a private individual has ever produced', and praised the originality and quality of the Troughton circle which Groombridge had commissioned and with which he had worked [17]. It tells us something about the way in which human life expectancy was perceived in 1838, when Airy described the *Catalogue* as being the product of 'advanced age', for all of the 50,000 observations had been taken between Groombridge's fifty-second and sixty-third years. On the other hand, Groombridge had not completed the reduction of these observations by the time of his death at the age of 78, and this went on to involve the efforts first of Henry Taylor, and then G.B. Airy, to see them through the press [18].

John Pond

Another pioneer of the astronomical circle in the last years of the eighteenth century was John Pond. This son of a wealthy London merchant had enjoyed a Cambridge education and a training in practical astronomy from William Wales, who had been one of Captain Cook's scientists, before travelling on the Continent. Having commissioned an especially fine circle from Troughton, Pond began to work from his private observatory at Westbury, Wiltshire [19]. He embarked with his 'Westbury Circle' upon an analysis of the star positions contained in Maskelyne's Royal Observatory catalogues, made with the fifty-odd-year-old Bird quadrant of 8 feet radius, and compared them with the observations made with his own circle. In the same year that Groombridge began work at his new Blackheath Observatory in 1806, Pond published his results, displaying the errors of the Greenwich quadrants to the Royal Society and the Greenwich Board of Visitors, and emphasising the obvious superiority of the new circles. A private amateur had shown up the failings of the outdated instruments of a national observatory [20].

And Pond's standing as an astronomer was such that orders were quickly placed for a circle for Greenwich, with Pond acting as technical midwife to the entire venture. Moreover, when the 79-year-old Dr Maskelyne died in 1811, before the new 6-foot circle by Troughton was finished, John Pond was offered the post of Astronomer Royal, thereby maintaining the Grand Amateur traditions of that office, which would not be decisively broken until the appointment of Pond's successor, Airy, in 1835 [21].

William Pearson

Groombridge and Pond were by no means the only significant figures to pioneer the circle in high-level amateur observatories. That founding father of the R.A.S., Dr William Pearson, was equally significant, though he is perhaps remembered less as an observer than as an astronomical writer and organiser. The source of Pearson's early and sustained acquisition of liberal financial means, however, is less clear-cut than are the earned and inherited fortunes of Groombridge and Pond. William Pearson started his days on a small farm in Cumberland; and at the Hawkshead Grammar School, where he had been a contemporary of Wordsworth, he was apparently earmarked as an 'uncouth' and 'quondam clown'. But this seemingly unpromising lad – an archetypical late developer – enjoyed a rapid ascent after finding his gifts as a school teacher. From local schools in Cumberland, he became Under-Master at the Free Grammar School in Lincoln, and in the same year, 1793, matriculated at Clare College, Cambridge, although he never took a Cambridge degree. In Lincoln, Pearson also got himself ordained, so that his academical and clerical posts brought him in an income of £71 per annum. It was all the more surprising, therefore, that this relatively poor man should pay 50 guineas (£55) to become a Proprietor of the Royal Institution in 1800, and in 1805, purchase an estate worth £1,000 in the Lake District [22].

In addition to his £71 a year received from school and church salaries, however, it is very likely that Pearson was delivering subscription lectures in astronomy in Lincoln by the 1790s. It was also at this time that he began to design, commission, and possibly build with his own hands, large planetary orreries, which were elaborate teaching machines aimed not at children but at fee-paying adult audiences. Indeed, it was during these years that Pearson laid the foundation of that mastery of practical mechanics, the improved design of orreries, and astronomical instrumentation for which he became so famous in later life [23].

At an unspecified date in the mid-1790s, he married Miss Frances Lowe. This lady has recently been described as the sister of a Mr Lowe of Islington, who was an amateur astronomer, and one wonders whether she was related to Gavin Lowe (c. 1743–1815), also of Islington, who had purchased the £120 circle from Troughton in 1793. If this had been the case, one suspects that Pearson might have been acquiring money along with his bride, whose dowry could have provided the foundation for his later fortune and distinction, though I have found no solid evidence for the connection [24].

But in 1810, Pearson had the wherewithal to purchase an estate near Richmond (for which in 1806 a purchaser had paid £12,000), which he turned into East Sheen School, a successful and distinctly fashionable preparatory establishment [25]. The transmogrified 'quondam clown', now with those unassailable badges of Georgian gentility, Holy Orders, land, money, and a Commission of the Peace, and soon to acquire an honorary LL.D. degree, was excellently placed to enter the world of Grand Amateur astronomy. With a private observatory at Sheen and an even finer one at his rectory at South Kilworth, Leicestershire, he started on serious observational work after 1815.

William Pearson bought instruments from the leading makers of the day, as well as rescuing and refurbishing for South Kilworth the famous rotating dome – one of the first astronomical domes ever built – designed by John Smeaton for Alexander Aubert's private observatory in 1788 [26]. A man of natural liberality, he clearly belonged to that class of astronomers who loved to collect instruments. In 1812, for instance, he resolved something of an official deadlock in London when the export of a large circle by Edward Troughton, destined for the Imperial Observatory, St.

Petersburg, had been cancelled following the sacking of the Moscow Observatory by Napoleon's troops. Pearson paid 500 guineas (£525) for the instrument, and commissioned Tulley to supply a 3¼-inch object glass [27]. He diverted another débâcle eleven years later, by taking off the hands of the R.A.S. Council a new object glass by Tulley, at a cost of £200, and clearing over £20 in additional charges [28].

In 1829 William Pearson published a work which demonstrated his mastery of practical astronomy, and his recognition of the significance of the developments in instrumentation which had taken place over the last forty-odd years. His sumptuous *An Introduction to Practical Astronomy* (London, 1824, 1829) in two large quarto volumes was an authoritative *vade mecum* of the astronomy of this period, combined with a detailed history of the development of many instruments in the eighteenth century. Of the instruments from the decades immediately before 1829, he invariably spoke from first-hand knowledge, as a user, or an owner, of virtually every class of instrument from new designs of micrometer to Troughton circles and Fraunhofer's German-mounted refractors. The design histories, ownership patterns, correction, and usage of these instruments, most of which were depicted in a separate volume of engravings of almost photographic accuracy, bring home the fastidiousness of procedure and patience which were an essential component of manual astronomical measurement. As well as the volume dealing with instruments and their uses, Pearson had prepared for the use of working astronomers a volume of *Practical Astronomy* consisting of 544 pages of tables [29].

In addition to his contributions as an astronomer and astronomical patron, Dr William Pearson is also significant as an indicator of the relative flexibility of English society in the late Georgian period, for he died a very wealthy man. Only in a society which lacked social ceilings based upon birth or lines of descent, indeed, could a peasant boy from the Lake District, who was not the obedient dependant of a great patron, have achieved the independent distinction which he enjoyed.

By the time of Pearson's death in 1847, however, the observing of the meridian with large circles was clearly moving out of amateur hands. After a succession of Grand Amateurs had pioneered and paid for the development of the astronomical circle, and taken the circle crusade into the heart of the professional camp when Pond became Astronomer Royal, it was the professional observatories which now confirmed meridian astronomy as their own, leaving the amateurs free to embark upon fresh researches. Indeed, Admiral W.H. Smyth said as much when advising amateur astronomers how to go about equipping a serious private observatory, in 1844. By this date the reformed professional observatories of Greenwich, Oxford, Cambridge, Dunsink and Armagh, not to mention Königsberg and elsewhere on the Continent, had the task firmly in hand, and even private observers who continued to do meridian work generally tended to do it secondarily to other researches [30].

One of these had been Lord John Wrottesley. In his younger days, while still practising at the London Bar, he maintained a meridian establishment at his house in that very astronomical London suburb, Blackheath. But by the time that he came to inherit his title in 1841, and moved his observatory on to his ancestral estates in Staffordshire, double-star and related branches of the science, which demanded large equatorially-mounted refractors, began to predominate [31].

Richard Christopher Carrington

One of the last serious Grand Amateur meridian observers was Richard Christopher Carrington. Born in 1826, and educated at Trinity College, Cambridge, this son of a wealthy Brentford, Middlesex, brewer spent three years as Temple Chevallier's Assistant at Durham in the early 1850s, to gain practical experience in observatory work. Finding the Durham University Observatory's equipment to be inadequate, however, he resolved to set up a superior establishment of his own, at Redhill in Surrey [32]. By 1853 his residence–observatory was operational, and at a time when professional astronomers like Airy were doing their best to prevent railway tracks from being laid close to their observatories, Carrington pointed out that the trains passing on the nearby Brighton line occasioned no sensible disturbance to his instruments. He even included the signal gantry in the published engraving of the Redhill Observatory, to show its propinquity [33].

The 'Preface' to Carrington's *Redhill Catalogue* (1857) gives a wonderful insight into how a wealthy young Grand Amateur, whose inherited money made him 'an unfettered man', went about founding an observatory. There is an almost opulent tone to his description. The delighted Carrington tells how he commissioned a $4\frac{1}{2}$-inch equatorial refractor and a superb transit circle with a 5-inch object glass from Troughton and Simms, and a regulator clock from Henry Appleton, and then made all the critical adjustments at his new establishment [34]. Assisted by his friend, Mr George Harvey Simmonds, who received a salary of £126 per annum and had rooms adjacent to the Observatory [35], it was Carrington's intention of cataloguing the circumpolar stars, down to the tenth magnitude, using the zone-sweeping technique that had been pioneered in Germany. Though defects were later shown to exist in the catalogue, it won Carrington the R.A.S. Gold Medal in 1859 [36].

On the other hand, one is sometimes puzzled by Carrington's stupidity as a scientist. For instance, he claimed to own the most exact clock in the world, which he had installed in a deep tunnel beneath his new observatory at Churt, so as to maintain its thermal homogeneity. But the escapement was fitted within a mahogany case with an airtight glass door. When, after finally obtaining a vacuum seal on the door, he shattered three plate-glass case doors, costing £3 apiece, as a result of air-pressure differences, he wrote a letter to the *Monthly Notices* of the R.A.S. asking advice on how to maintain a successfully evacuated clock case. The elderly Thomas Romney Robinson of Armagh told him that the air-pressure problem could be avoided if the pendulum was made to swing in a cylindrical glass case.

But Carrington's enduring contributions came not from celestial cartography but from his secondary interest in solar astronomy. Inspired by the discovery of the German schoolmaster–astronomer Heinrich Schwabe, that there was a pattern in sunspot maxima and minima, Carrington began to make physical observations of the Sun, both at Redhill and to a lesser extent at his new observatory at Churt, to which he transferred his establishment around 1865 [37]. It was also coming to be suspected by many scientists during the 1850s – Sir John Herschel, Edward Sabine and others, that the Sun was probably a magnetic body. Then, on 1 September 1859, Carrington observed a brilliant flare, or energy outburst on the solar surface which affected magnetic instruments and the commercial telegraphs, followed by a brilliant auroral display, to provide the first clear indication that the Sun was indeed a magnetic body and was capable of inducing electric currents in terrestrial circuits [38]. And a few years later, Carrington discovered the 'law' whereby sunspot groups tend to drift from higher to equatorial solar latitudes during the 11 years of a solar cycle. Carrington began to record each daily appearance of the solar surface, in an extension of the techniques developed by Schwabe, so that even the life histories of small individual spots could be used to develop a wider analytical model for the Sun's rotation in different zones of latitude. He also began to number the individual axial rotations of the Sun, beginning on 9 November 1853, so that there could be a definitive archive of physical data to which scientists could relate specific pieces of phenomena in the future. And much of this was performed around the sunspot minimum of 1855, as his solar record encompassed the period 1853–1861 [39].

Yet one of Carrington's unintentional contributions to the social dimension of Grand Amateur astronomy was his reversal of that tendency whereby a strategic marriage enabled a modestly well-off man to become Grand. Already rich, the young Carrington was hopelessly smitten by an illiterate enchantress who supposedly caught his eye in Regent Street. But the beautiful Rosa Helen Rodway (or Jeffries, as was still her legal name), who soon became Mrs Carrington, was a lady with a very dubious past. She had a follower, an ex-trooper in the Dragoons who now looked after horses in a circus, whom she seems to have convinced Carrington was her brother, so that he was received at Observatory House at Redhill The so-called William Rodway turned out to have been Rosa's regular lover and common-law husband, and now that she had married a gentleman, he probably saw her as fair game for blackmail and extortion. For some years, William Rodway continued to haunt Rosa, until an incident in 1872 when he stabbed her after one of his drunken visits to the Carrington house. Rodway was arrested and imprisoned, and several embarrassing cats subsequently leapt out of the bag in court [40]. Then three years later, in November 1875, Rosa was found dead in bed, probably as a result of taking too much chloral hydrate for her insomnia. Three weeks later, Carrington himself was found dead, from what was possibly an overdose of chloral.

Carrington's career was the most tragic of all of the Grand Amateur astronomers, and one only wishes that his obvious gifts of intellectual and financial resources had been combined with less personal stupidity and better human judgement. One suspects that had he been required by circumstances to partake more consistently in the hurly-burly of life, like most of the Grand Amateurs, who in one way or another made their own fortunes, he would perhaps have achieved more. And perhaps he would have lived past 49, which, after all, was but an age when most of his colleagues in the R.A.S. were just finding their scientific feet.

NEW DISCOVERIES IN POSITIONAL ASTRONOMY: STELLAR PROPER MOTIONS AND BINARY STARS

Yet why, one might ask, was all of this money and time being expended by wealthy men to constantly measure, and remeasure, all the stars as they came to the meridian? Quite apart from the cathartic effect which it might have had in calming and prolonging the lives of once over-stressed professional men – rather like chess, or fishing – the constant revision promised rich intellectual rewards, as mentioned above. Ever since Edmond Halley had discovered the proper motions of the 'fixed' stars in 1718, and William Herschel had shown, over sixty years later, that the solar system itself was hurtling through space towards the general direction of Hercules, it was clear that the positional relationships of the stars were undergoing constant change [41]. By remeasuring every star, down to the dimmest accessible magnitudes, at regular intervals, one might be able to construct a three-dimensional geometrical model of the Universe. The stars which displayed the biggest proper motions were most likely to be the stars closest to us, in the same way as when we walk through a forest, the closest trees display the most pronounced line-of-sight displacements.

Hundreds of stellar proper motions had been monitored by 1840, and it was by then clear, though rather surprising, that some of the biggest of them were displayed by relatively dim stars. This was surprising insofar as William Herschel had worked on the assumption that all stars were more or less equal in size and in intrinsic brightness, so that the brightest stars should have displayed the biggest motions because they were supposedly the closest to us [42]. Only by sweeping the meridian decade after decade, however, could that vast internationally-harvested archive of proper motions have been built up, to make astronomers realise that the Universe was inhabited by a diverse and very heterogeneous population of stars. Groombridge himself had found such a star – or more correctly, such a star had become apparent from the reduction and analysis of his thousands of observations. Groombridge 1830, a seventh magnitude star in the constellation of Ursa Major, was a veritable 'flying star', with an annual proper motion of 7 arc seconds. It was identified from Groombridge's catalogue by Friedrich Wilhelm Augustus Argelander at Bonn in 1842 [43].

And it was the stars with the biggest proper motions, rather than the randomly bright stars that happened to pass through the zenith, which were now being thought of as the likeliest candidates to display annual parallaxes. By 1838, a large enough body of proper motions, and a sufficiently high standard of instrumentation, had become available to enable two professional astronomers working independently of each other to confirm two separate stellar parallaxes. Friedrich Wilhelm Bessel at Königsberg and Thomas Henderson working on observations made previously at the Cape of Good Hope finally extracted the parallaxes of the visually dim yet large proper motion stars, 61 Cygni and α Centauri respectively, to extend those measuring rods into the stellar Universe which had spurred astronomers ever since Galileo had aspired to find a physical proof of the Earth's motion around the Sun. Sweeping the meridian, therefore, was far more than a peaceful pursuit for gentlemen of regular habits who loved peace, quiet, and elegant playthings. It promised to be one of the chief keys whereby the 'secrets of the Universe' could be unlocked [44].

The branch of astronomy which was coming to stand at the intellectual and technical cutting-edge of the science by the 1820s was the study of double and triple stars. The scientific world had first been alerted to their presence in 1802, when William Herschel had published a list of close pairs of stars the components of which had shown a central rotation over a period of twenty years [45]. The 'binaries' promised rich rewards, for if the orbits of the component stars could be measured with sufficient accuracy as they moved about each other, the gravitational characteristics of the system could be demonstrated. This was a mighty spur, indeed, for if the inverse square law could be demonstrated in the depths of the stellar Universe, then Newtonian gravitation would be shown to be universal *de facto*.

Yet binary work needed a wholly different astronomical technology from that required for meridian-sweeping for proper motions. Instead of meticulously divided circles, large-aperture, equatorially-mounted refractors were now almost *de rigueur*. Perfect point-images of tenth magnitude stars, free from colour distortion, were essential; and this shifted the emphasis in the astronomical hardware race away from circle-dividing to high-resolution object glasses. An achromatic lens, of 6 or 7 inches diameter and of 10 feet focal length, had to be set upon a precision-engineered equatorial mount and moved by a clock drive, so that the astronomer could perform his delicate eyepiece micrometer operations upon a totally stable star image without any need to track it manually [46]. Such instruments were extremely costly, and if one wished to run in the binary race, then it was important to have an abundance of money, especially if one wanted to maintain a meridian establishment as well.

There were two important reasons for this cost. One was the high price of the optical-quality flint glass, the best of which came from Switzerland and Munich. Joseph Fraunhofer had revolutionised achromatic lens making in the years after 1806, and while he and Pierre Louis Guinand had perfected the art of making glass blanks and lenses of hitherto unsurpassed aperture and clarity, the whole technology was hedged around with trade secrets. To make matters worse, the British government placed stiff duties on the importation of European optical glass, which infuriated many English astronomers, and sometimes led them to smuggle glass blanks in their luggage when returning from the Continent, to be worked up into lenses by London craftsmen [47].

The second reason for the exorbitant cost of double star work was the need to pioneer and pay for new stable designs for the equatorial mounting of the telescopes. Fraunhofer's 'German mount' had been a technical triumph in its day, and he used it on all the fine refractors, some of almost 10 inches aperture, which he supplied to astronomers around the world up to his death in 1826. Between 1820 and 1840, all of the leading English opticians and instrument makers – George Dollond, Charles Tulley, Edward Troughton and William Simms, and others – experimented with equatorial mounts, and one of the most widely favoured designs was the 'English mount', with its long polar axis, which had been developed by the Sissons in the mid-eighteenth century.

Sir James South
The leading Grand Amateur at the start of the binary star era was Sir James South: the man, indeed, who had done so much for British astronomy and private observatory founding long before it had become as 'fashionable' as it was to be by the early Victorian era [48]. A surgeon by training and half-brother to the surgeon John Flint South, he had studied under the Royal Sargeant-Surgeon, Sir Astley Cooper, at the Royal College of Surgeons, but abandoned a promising medical career upon his marriage to Miss Charlotte Ellis in 1816. His new wife, who was the daughter and heiress of Joseph Ellis of Lambeth, brought an independent fortune to South in the days before the law prevented the transfer of a woman's fortune to her husband upon marriage [49]. Mr James South (for he was not knighted until 1830) set about the establishment of a magnificent private observatory containing what Francis Baily called 'a princely collection of instruments' [50] in the grounds of his house, in Blackman Street, Southwark, near the Borough hospitals of Guy's and St Thomas's where once he had worked with knife and saw in this pre-anaesthetic era of surgery. The growing sprawl of Southwark led him to move to still-leafy Chelsea in 1826, where his Campden Hill Observatory was generally regarded as the finest private establishment in Britain, and much better equipped than Greenwich for double star work [51].

At Blackman Street, South undertook the Herculean task of measuring the elements of 380 binary and triple star systems, assisted by the young John Herschel, who was the son of South's idol, Sir William. South was the observer and Herschel the mathematician, and when their massive catalogue was published in 1824, it took up an entire volume of *Philosophical Transactions*, and covered 412 pages [52]. South received the Royal Society's prestigious Copley Medal in 1826. And one of Sir James' many boasts in later life, though one that we have no need to doubt under the circumstances, is that it had been he who had taught the brilliant Cambridge mathematician the practical skills of working in an observatory [53]. He provided the same educative service for the young G.B. Airy, who was also a regular visitor to Blackman Street in the mid-1820s, and it is sad that South later fell out with both of his distinguished protégés, who could have remembered him so warmly had things turned out differently [54].

Like many a rich enthusiast, Sir James South was an assiduous collector of astronomical instruments, though he was not above the sharp practice of trying to re-sell some of them at grossly inflated prices if opportunity arose. Troughton's business partner and successor, William Simms, loudly protested when Pond's old Westbury Circle, which South had bought for £50, was re-sold by him to an unsuspecting fellow medical man in 1833 for £400 [55]. South had also acquired the famous Groombridge Circle, although it was still in his collection in 1851 [56].

In 1829 South pulled off the *coup*, while in Paris, of purchasing the world's largest achromatic object glass – the famous $11\frac{3}{4}$-inch aperture Cauchoix lens – for an alleged £1,000: a sum, indeed, that was equivalent to twenty months' salary for Pond, the Astronomer Royal [57]. Yet the acquisition of this glass was to signal the beginning of the end of South's career as a leading British astronomer, and commenced that descent into litigation, bitterness, and mental derangement so clearly described by Dr Michael Hoskin [58]. South commissioned his friend, Edward Troughton, who was now in the last five years of his long life and already in partnership with William Simms, to mount the lens in a scaled-up version of the very successful English mount into which the venerable craftsman and F.R.S. had incorporated an earlier, and smaller, lens for South. Troughton's instincts as an engineer told him that a mere scaling-up would not be enough for the heavy Cauchoix lens and tube, but South was adamant, and so the craftsman did his best to satisfy the client. When the finished instrument was delivered in 1832, South claimed that he could detect tremors – α Lyrae 'danced Merrily' – which would naturally make it unreliable for delicate astrometric work, and in spite of Troughton's prior *caveats*, refused to pay the £1,235 bill for mounting the lens [59].

A lawsuit then commenced which exposed the R.A.S. and its most eminent Fellows to a glare of embarrassing publicity. The Revd Richard Sheepshanks – barrister, mathematics don, and Secretary of the R.A.S. – encouraged Troughton and Simms in their suit against South, who also happened to be the first Royal Charter President of the Society, though a man whom Sheepshanks openly despised as a mathematical ignoramus who did 'not know a sine from a cosine' [60]. Sheepshanks, and his Trinity College chum, G.B. Airy, provided powerful expert witness to back Troughton's and Simms' claim, while their old College rival, Sir Charles Babbage, backed South. In 1838 the court eventually found in Troughton's and Simms' favour, and the enraged Sir James South was forced to pay compensation and costs, so that he ended up £8,000 out of pocket [61].

The great double star observer now vented his spleen by smashing up the Troughton and Simms mount which had cost him so dearly (though sparing the Cauchoix lens) and eventually, in 1842, publishing a poster announcing the sale of the broken pieces 'To Shy-cock Toy Makers – Smoke Jack Makers – Mock Coin Dealers ... and to such Fellows of the Royal Astronomical Society ... and its Botchings cobbled up By their Assistants MR. AIRY AND THE REVD. R. SHEEPSHANKS ...' This, and other scrap sales, netted a miserable £11 12s 9d for South, to offset against his £8,000 loss [62].

Sir James seems to have shown signs of mental derangement in the publishing of this notice, accompanied as it was with an engraving of all the smashed-up fragments strewn across his garden, awaiting Sheepshanks, Airy, or any other rubbish dealers who wished to take them away. And tragically, after the age of about 50, this gifted and amply endowed Grand Amateur did no more serious scientific work, in spite of having thirty-two years of potentially active life before him. Now in a permanent rage, he wrote vituperative (and often highly revealing) letters to the press, did his best to publicly harass Airy for not being an observing Astronomer Royal (while in 1846 Airy retorted that South was an 'ignorant Blackguard' no longer fit for the company of gentlemen), and provoking Sheepshanks into rages even when that distinctly uncharitable clergyman addressed the R.A.S. [63]. Only amidst the Irish calm of Birr Castle could South find any real peace, in the company of the Earl and Countess of Rosse, and Thomas Romney Robinson, who treated him like an irascible but respected old relative [64].

Though Sir James South had been the first observing Grand Amateur to set to work on the binaries, it was his protégé and co-worker, Sir John Herschel, who pioneered their mathematical analysis, and who was perhaps the second man in Europe to extract Newtonian gravitational elements for binaries, after the acknowledged primacy of Félix Savary, who had led the way when he published the computed orbital elements for ξ Ursae Majoris in 1830 [65]. Sir John went on to become a first-class observer in his own right, and continued his survey of binaries into the southern hemisphere in 1834, where he had the use of one of South's equatorial refractors in addition to his own 18-inch reflector. His work will be looked at in more detail in Chapter 4.

William Rutter Dawes
Even before litigation and rage had effectively removed South from the British binary astronomy scene, William Rutter Dawes had begun to attract attention in the scientific world. Dawes came from a family with mathematical interests. His father, as an officer in the Royal Marines, had made astronomical observations in Australia, and had been a one-time Governor of the colony in Sierra Leone and a mathematics master at Christ's Hospital. William was educated at Charterhouse, and like South, went on to walk the wards of a great London hospital, St. Bartholomew's in his case, to qualify as a doctor. In 1826 he moved to Ormskirk in Lancashire to practise medicine, and as a deeply religious man attracted to the dissenting tradition, was invited to minister to an Independent, or Congregationalist, chapel in the town [66]. With the first of a succession of great telescopes, a 3.8-inch aperture Dollond, he began serious observation, and on 23 April 1831 published the first of his double (in this case triple) star observations, of ζ Cancri, in the new R.A.S. *Monthly Notices*. By 1834 he had published observations of 121 binary and triple stars [67].

It was during his Ormskirk years, however, that Dawes met and became a lifelong friend of William Lassell, the brewer–astronomer who lived a few miles away in Liverpool. The Lassell astronomical notebooks from the early 1830s show how the two men – neither of whom had yet risen to 'Grand' status in terms of wealth or distinction – exchanged books, observed, and tested telescopes together [68]. Dawes and Lassell, moreover, were co-religionists, being members of the Congregationalist Church.

Following the death of his first wife, however, Dawes' health, which was never strong, seems to have broken down, and he left Ormskirk. Between 1839 and 1844 he worked as the salaried astronomer at George Bishop's private observatory, at South Villa, Regent's Park, London. Enjoying a relatively free hand and excellent instruments, he continued his double star and other stellar work [69]. But Dawes' career really took off, and he was launched upon the truly Grand Amateur phase of his career, following his marriage to Mrs Anne Welsby, the widow of an Ormskirk solicitor, in 1842. In 1844 Dawes and his wife took up residence at Cranbrook, Kent, and equipping his new observatory with a 2-foot Troughton and Simms transit circle and a superb $6\frac{1}{3}$-inch aperture refractor by Merz and Mähler of Munich, he set to work with a vengeance on the double stars. Over the next few years, Dawes removed his observatory to his houses at Wateringbury, near Maidstone, Kent, and Haddenham, near Aylesbury, Buckinghamshire [70]. It was at Wateringbury, in November 1850, that Dawes and his friend Lassell, who was visiting, discovered the crepe ring of Saturn, only to find that the American mails were bringing news of its co-discovery from William Cranch Bond at Harvard [71].

By 1853, Dawes' reputation as the 'eagle-eyed' observer of binaries was international, and the American telescope-maker Alvan Clark wrote from Boston, to inform him of his 7½-inch object glass with which the dim companion of 95 Ceti had been discovered [72]. At this time, it was first to Germany, followed by London and York, that the world's refractor users looked for superior object glasses – as had Bond himself in the early 1840s – and the appearance of Alvan Clark heralded the coming of age of high-quality American instrument-making. No doubt using the deceased Mr Welsby's money which had come to him by marriage, Dawes purchased the 7½-inch Clark refractor for $950, or about £195 sterling, in 1854, and found it to be better, and cheaper, than its Munich equivalents, aperture for aperture [73]. Over the next few years he bought a total of five instruments from Clark, culminating in an 8¼-inch, in 1859, which he came to compare with his 8-inch, by Thomas Cooke of York – for Cooke was fast becoming Europe's reply to Clark [74]. The 7½-inch Clark refractor, however, which Dawes always regarded in hindsight as the finest of his telescopes, later passed into the hands of the Irish double-star and planetary observer, Dr Wentworth Erck, who did excellent work with it, and then, as will be discussed in Chapter 8, the Dublin barrister and pioneer of photoelectric photometry, William Henry Stanley Monck [75].

By 1860, Dawes was probably the most skilled visual observer with a refracting telescope in the world, and the expanding firms of Clark and Cooke regarded his praise as a warranty for their object glasses. Even his fellow Kentish resident, Sir John Herschel, invited him over to test and comment upon the eyepieces of his late father, Sir William – a true mark of his standing as a practical astronomer and authority on telescopic optics [76].

George Bishop
During these years in the early 1840s between Dawes' abandonment of his work in Ormskirk and independence following his marriage to Mrs Welsby, he worked for George Bishop at the South Villa Observatory in Regent's Park. Bishop was a remarkable figure in many ways. A wine-maker and retailer whose products were said to account for one half of the British wine excise, Bishop was an astronomical enthusiast. In 1830, at the age of 45, he joined the R.A.S., where 'the amount and stability of his fortune by that time permitted the indulgence of tastes hitherto held in abeyance.' He learned mathematics as a grown man, came to grips with Laplace's *Mécanique Céleste* at the age of 50, and in 1836 founded the South Villa Observatory, in Regent's Park, equipping it with an excellent 7-inch Dollond refractor and other instruments [77].

But Bishop was not really an observer himself. More correctly, he was a Grand Patron of astronomy, for being 'determined that this Observatory shall do something', he employed several talented men over the next twenty-odd years down to his death in 1861 [78]. These included Dawes, Norman Robert Pogson, and Albert Marth, but the one whose reputation was very much founded at South Villa was John Russell Hind, who later became Superintendent of the *Nautical Almanac*, while still continuing to work with Bishop. But these employed astronomers will be dealt with in their own right in Chapter 8, which looks at the paid men who assisted the Grand Amateurs.

Bishop's South Villa Observatory was engaged in all the main research areas of the day: double stars, planetary motions, proper motions, and 'deep-sky' work. But it was in the areas of cometary and asteroidal astronomy that it perhaps won its greatest distinction. Both Pogson and Marth discovered asteroids, while John Russell Hind crowned South Villa with the discovery of two comets and ten asteroids, not to mention the first published sighting of Neptune to be made in Britain, on 30 September 1846 [79].

In the world of early Victorian astronomy, however, George Bishop was a force to be reckoned with, as an astronomical entrepreneur who used his money to advance double star and asteroid astronomy, by finding jobs for scientists who needed to earn their livings. As a man who also had a lively grasp of what he was paying for, even if he had little personal skill as an observer, one cannot be too hard on him for the pleasure which he took in being elected F.R.S. and receiving the Testimonial of the R.A.S. After all, in his view, he owned the instruments, paid the wages, and gave his observers a chance to distinguish themselves.

He was no doubt proud when in 1848 twelve luminaries, including Airy, John Herschel, and William Henry Smyth, signed his Royal Society election certificate commending him as a 'Promoter of Astronomy', and founder of a research observatory which even by that early date had led to 'the development of the solar system' by the discovery of three comets and the asteroids Iris and Flora [80]. Even so, one senses the tension which existed about the proprietorship of an astronomical observation in an observatory like Bishop's, when, nineteen years after leaving South Villa, the irate Dawes wrote to the editor of the *Astronomical Register* to correct a statement by the Revd W.A. Darby. In an astronomical publication, Darby had taken Bishop's name at face value when placed alongside a South Villa observation, whereas Dawes made it clear that, to his knowledge, Bishop had logged only one set of observations – of 8 Lacertae as truly his own. 'Dawes' or 'Hind' should really be placed alongside the rest, he emphasised [81].

Edward Joshua Cooper of Markree

In spite of its glaring disparities between rich and poor, and simmering political unrest, nineteenth-century Ireland produced more than its share of contributions to Grand Amateur astronomy within the British Isles. While the contributions of Lord Rosse, William Wilson, and others will be discussed in later chapters, it was Edward Joshua Cooper of Markree Castle, County Sligo, who was one of the earliest and most significant contributors to the tradition.

Educated at Eton and Christ Church, Oxford, though Dublin-born, Cooper epitomised in many ways the 'Englishry' that ruled Ireland, for the family descended from a Nottinghamshire soldier of fortune, who in the seventeenth century had acquired extensive lands in County Sligo. He came into full possession of his Markree properties in 1837, when 39 years of age, and as a well-travelled gentleman of abundant means, resolved to devote his fortune and energies to astronomy [82]. His passion was comets, though the study of binary stars and those other branches of high-precision astrographic work that occupied his English contemporaries also fascinated him. The ground was laid in 1831, when he snapped up Robert Aglaé Cauchoix's latest lens, with a diameter of 13.3 inches and a focal length of 25 feet [83]. This currently largest object glass in the world, which stole a lead upon Cauchoix's previous 11¾-inch masterpiece which had gone to Sir James South in 1829, was first mounted upon a large wooden altazimuth stand, though in 1834 Thomas Grubb of Dublin produced his first, daring essay in wrought iron and stone, when he remounted the lens in the equatorial plane. Cooper augmented the Cauchoix refractor with a 5-foot transit instrument, then a 7-inch aperture transit circle by Traugott Lebrecht Ertel of Munich, to produce what was to be described in 1851 as 'undoubtably the most richly furnished of private observatories' [84].

Between 1831 and 1842 Edward Cooper worked alone at Markree, until he engaged the services of Andrew Graham at the salary of £100 a year to do the daily business of the observatory (when 'the pressure of public duty' took Cooper elsewhere), and whom by 1851 he was acknowledging to be responsible for 'the great majority' of the work [85].

The year 1846 was significant for the way in which Cooper saw the research contributions of his observatory as progressing. Just before Christmas 1845 the German amateur, Karle Hencke, had discovered the first new asteroid to be found since 1807, Astraea, and this had suddenly opened up asteroid-searching as a potential new field of international research [86]. And when, in September 1846, the importance of Karl Bremiker's detailed charts of faint ecliptic stars to the Berlin Observatory's rapid identification of Neptune came to be understood, Cooper resolved to greatly extend Bremiker's surveys to cover a vastly larger band of the ecliptic [87]. Such improved charts, going down to twelfth magnitude, would greatly simplify any future searches for an 'extra-Uranian planet', as well as asteroids, by providing a definitively-mapped star field against which the presence of a small, slow-moving body could be detected. Although the Ertel Circle was no doubt useful for some of this work, Cooper was quick to point out in his Markree *Catalogue*, I, in 1851, that the secret of the success of the project stemmed from using the 13.3-inch equatorial, on a standard magnification of ×80, in conjunction with a 'square-bar' micrometer in the comet eyepiece, to give a uniform star field in all parts of the sky [88].

The prodigious workload that was entailed can be ascertained in that the four published volumes of the Markree *Catalogue*, issued up to 1856, contained a total of 660,155 stars, and were devised 'on a scale which gives an area sixteen times that of the Berlin Maps' [89]. Graham also bagged an asteroid, Metis, in 1848. One senses the sheer physical arduousness of the task, to say nothing of the endurance of poor Andrew Graham, when one remembers that the Markree refractor was not domed, and was used in the open air in the path of every blast of wind which the Atlantic Ocean might throw at the coast of County Sligo [90].

What we see at Markree is very much an epitome of the Grand Amateur tradition, set within the more feudal traditions of the far west of Ireland during the period of the potato famine: an example of how established wealth carried forth a major research project that was intended to improve upon the achievements of the Berlin Observatory, by providing the very finest instruments, yet throwing much of the burden on to the shoulders of the poorly-paid Andrew Graham. One also feels that Markree brings home the essentially luxurious status of Grand Amateur astronomy, thrown into particularly sharp focus by the contemporary Irish situation. For not only did Cooper's enterprise scarcely touch the indigenous people of Ireland – for his friends Lord Rosse, Romney Robinson and others were also men of English descent, as well as his assistant Graham – but the high watermark of the British Grand Amateur enterprise coincided with that decade known in the British Isles as 'the hungry forties' [91].

THE AMERICAN AMATEURS

The emergence of American telescope-making which came with Alvan Clark was paralleled by the rise of a Grand Amateur tradition in the United States. And with the exception of some Church revenues, which were not so easily available in the United States, the American amateurs enjoyed financial resources that were often similar to those enjoyed by their British counterparts. Indeed, as we saw in Chapter 2, Grand Amateur astronomy arose in America because of similar political and economic traditions to those which prevailed in England, and which shifted social and intellectual initiatives into private rather than public hands. While it is not within the scope of the present study to deal with the American amateur tradition, which has been treated by American scholars [92], it is important to bear in mind that astronomy in both countries developed along similar lines.

William Cranch Bond was a Massachusetts businessman before receiving a salary from Harvard College seven years after becoming Director of the Observatory which he effectively founded there. The pioneer astrophotographer, Lewis Rutherfurd, both inherited and married money, while the two Drapers – John and Henry – combined medical fortunes with strategic marriages, to parallel the astrophysical work of Huggins, Lockyer, and Common, as will be discussed in Chapter 7.

But one of the driving spirits behind the American amateur tradition from the post-Civil War era to the 1910s was Sherburne Wesley Burnham. It was also Burnham who in many respects inherited the mantle of Dawes to become perhaps the last but one major observer to use visual techniques to advance double star astronomy, the last being Thomas Henry Espinall Compton Espin [93]. Inspired by Thomas William Webb's *Celestial Objects for Common Telescopes* (1859), and feeling that he had outgrown the 3.7-inch refractor which he had bought on a visit to London, the 31-year-old Burnham splashed out $800 (about £160) on an exquisite 6-inch Alvan Clark instrument in 1869 [94]. It was used to lay the foundation of that massive project which culminated, in 1906, in Burnham's celebrated *General Catalogue of Double Stars within 121° of the North Pole* [95].

Strictly speaking, though, Burnham was not a *Grand* Amateur, though he was a comfortably-off one. He earned his living as a stenographer in the Chicago lawcourts, and while he was enticed by Edward Holden to work professionally at the Lick Observatory, California, between 1888 and 1892, he returned to the Chicago courtrooms because he could make a better living there [96].

Yet Burnham was very much an unusual case. The access which he came to enjoy to the Dearborn, Yerkes, Lick, and Dartmouth College refractors was perhaps peculiarly American insofar as these instruments came from a tradition of private benefaction and public subscription observatories which could make them available to a famous amateur. It would have been hard, however, to imagine a man in Burnham's position having invited access to Cambridge's Northumberland refractor or the 12-inch Merz refractor at Greenwich [97].

In Britain, if one wanted to use a large telescope or other piece of costly equipment to follow a private line of investigation, then one had little choice other than to make it or buy it for oneself. In short, one had to be in possession of ample funds such as those deriving from an inheritance, a wife, a benefice, a brewery, or some other thriving commercial or professional concern. And in this respect, these astronomers followed firmly in the footsteps of the first Astronomer Royal, for John Flamsteed was the son and heir of a brewer, an absentee Sussex rector, and the beneficiary of a marriage, the combined proceeds of which he used to fill the empty Royal Observatory with instruments of excellence and originality of design which remained his own personal property, and were disposed of by his widow after his death in 1719.

4

Sir John Herschel: a model for the independent scientist

John Frederick William Herschel occupied a unique position within the British scientific community. As the only child of Sir William Herschel, born in 1792 when his father was 54 years old, John was brought up in a household that was devoted to astronomical and physical research. William had long since given up that professional musical career which had first brought him to England at the age of 18, and by 1792 was securely established as a physical scientist of international standing. Observatory House at Slough, which was John's birthplace and school, also accommodated his father's laboratory and mirror-making workshops, while in the garden stood the two largest telescopes that the world had ever seen: the 18-inch aperture reflector of 1783, and the giant '40-foot' of 48 inches aperture, built in 1789. John's only regular childhood playmate seems to have been his middle-aged spinster aunt, Caroline Lucretia, who delighted in the clever little boy's company between her own discoveries of comets and error-analyses of Flamsteed's *Catalogue*. But he also formed warm and enduring bonds with his Baldwin cousins, Mary and Sophia, although the Slough house had to be kept quiet in the daytime when William had spent a night observing the nebulae [1].

John grew up bilingual in German, fluent in French, familiar with other European languages, and could impress visiting bishops with his command of Latin and Greek. The young Herschel's linguistic capabilities developed in tandem with a love of mathematics, and a flair for practical things. When an old lady, aunt Caroline recalled how little John had been fascinated by tools, and was always asking the craftsmen employed by his father to make telescopes, to show him how to work wood and metal [2].

But if John Herschel's intellectual mentors were his father and his aunt, the eventual source of his gentlemanly independence was his mother. Mary Pitt (*née* Baldwin) was the 30-odd-year-old widow of a wealthy London merchant when she married William Herschel in 1788, and without the money which she brought into the family it is unlikely that her husband's subsequent career, and that of their only child, would have developed in the way that they did.

It would be incorrect, however, to underestimate William's own means, for since establishing himself as a young musician in England, in the early 1760s, he had grown prosperous in his own right. As he recorded in his 1784 'autobiography', 'My situation proved a very profitable one, as I soon fell into all the Public business of the Concerts ... [3]' Even by 1771, his surviving accounts show that he was making almost £400 per annum from his musical enterprises in Bath, which was a sum greater than the official income of the Astronomer Royal, or of a very comfortably beneficed parson. The popular belief that William Herschel was a poor musician who made his own telescopes because he could not afford to buy them has little foundation beyond the mythology of science [4].

But when William discovered the planet Uranus, in March 1781, he became the recipient of a £200 per annum Royal Pension, which enabled him to give up music and enjoy almost total freedom to devote his time to astronomy. The cut in income that his new astronomical research time produced, however, was more than made up by the growing demand for telescopes that he received from patrons at home and abroad. A 7-foot reflector like the one with which he discovered Uranus, for instance, could be bought for 100 guineas (£105), while the large instruments which he supplied to the King of Spain and Prince of Canino cost £3,150 and £2,310 respectively [5]. Though Herschel figured the optics for these instruments, the greater part of the other work was done by those carpenters and brass turners employed around Observatory House whose tools and trades had so captivated the infant John.

Though I have found few contemporary financial documents from which Sir William's annual income can be estimated, apart from a short statement of dividends preserved in the Museum of the History of Science, Oxford [6], he nonetheless bequeathed over £30,000 when he died in 1822, in addition to personal effects, including a 'chariot and horses', valued at under £6,000. Of this estate, Sir William left £2,000 to his brother Dietrich, and invested capital sums to purchase annuities of £100 each for sister Caroline and brother Johann Alexander. As much of his money seems to have been invested in 3 per cent annuities, it is clear that very substantial premiums would have been needed to purchase the two annual incomes for Caroline and Johann. There were also several other gifts of £20 to various friends and relatives. But copyhold landed property in Slough and in Buckinghamshire and £25,000 in 3 per cent reduced annuities went to John, which would have given him an independent income of around £750 a year. None other than 'my good friend' Charles Babbage, who was Sir William's great admirer and his son's Cambridge friend, was one of the executors for the will [7]. The unspecified residue of the will went to Lady Mary Herschel, who may well have enjoyed a separate income in her own right, from the Pitt or Baldwin estates. Then, after Lady Mary's death in January 1832, John possessed the wherewithal to undertake his five-year expedition to the Cape between 1833 and 1838, and purchase the mansion 'Collingwood' and its estate at Hawkhurst, Kent, upon his return.

Although Sir William was a Royal Pensioner, who received liberal funding from the King to build his 40-foot telescope, he nonetheless acquired that independent liberality of means and freedom of time which planted him securely within the Grand Amateur tradition, rather than being a salaried professional with a specific job to do.

Even so, the £750 or so per annum which he bequeathed to his son would have been inadequate to maintain Sir John in the opulent gentlemanly style in which he lived down to 1871 without an additional large sum of money which could be securely invested for income. And as Sir John married Margaret Brodie Stewart, the daughter of an Edinburgh clergyman and scholar with abundant progeny to provide for, that money did not come from marriage [8]. In this respect, therefore, it is hard to avoid the conclusion that it was Lady Mary Herschel's money which ultimately enabled her son to assume his role as a model of independent scientific action.

John Herschel's formative years had been lonely, seriously intellectual, and, with the exception of visits from his Baldwin cousins Mary and Sophia, largely spent in the company of people who were old enough to be his grandparents. Music, it seems, had been one of the few diversions of his childhood, for like all the Herschels, he had a gift for singing and playing instruments. John spent a few months at Eton in 1800, but its barbarous régime caused his parents to withdraw him, to continue a private education at home [9]. A child less innately gifted, or naturally affectionate, could have been seriously damaged by this upbringing, but his obvious closeness to his parents and aunt, who though now elderly were still unremittingly active, gave him a secure intellectual and emotional foundation. Even so, one always feels that John Herschel was old before his time. He never really lost his shyness, while the nervous, bronchitic and rheumatic diseases that plagued him from his twenties onwards suggest that he lacked something of that indomitable toughness possessed by his septuagenarian and octogenarian parents and almost centenarian aunt [10]. Even so, in living to the age of 79, he enjoyed an almost double helping of years in an age where 40 was still an average life expectancy [11].

Going up to St. John's College, Cambridge, at the age of 17 he moved for the first time among his own age group, swept up every prize for which he was eligible, distinguished himself as a Latinist and Grecian, and became Senior Wrangler in the Mathematics Tripos of 1813. Yet even before winning the star First of his undergraduate generation, he had urged the reform of mathematical teaching at Cambridge, by pioneering the use of the new Leibnizian notation against the more antiquated Newtonian form. He moved straight into a College Fellowship at St. John's, and at 21 had already been elected F.R.S. for his work on optics [12].

But once a Fellow of St. John's, Herschel rapidly tired of 'pupillizing' the innumerate 'blockheads' of the student body [13]. Then at the age of 22 he took himself off to Lincoln's Inn to train for the Bar, retaining his Fellowship *in absentia*. But he soon threw it up in boredom on the excuse that the dusty chambers aggravated his always latent respiratory problems. Herschel's cough, wheeze, or weak chest are often referred to in the letters of his friends, and one wonders how far a nervous asthmatic condition may have lain at their root [14].

No doubt because he was a young man of mark, with a fortune to look forward to, he did not wish to dwell in his father's shadow, and seems to have avoided astronomy as a career. By the age of 24 he had given up the long-term prospect of a don's life, along with that of a barrister, and later claimed that his real fascination had settled upon optics. The behaviour of light in complex lens systems had already occupied his attention for several years, as he brought his instinctive experimental and mathematical gifts to bear upon the achromatic telescope lens [15]. But in August 1816, he joined his parents for a summer holiday at Dawlish, in Devon, where it was made clear to him that his 78-year-old father needed assistance to complete his great work on the stellar system, and as an obedient son, he willingly complied [16]. John's obedience to his astronomical destiny was made clear when he reminded his wife, many years later '... I have been *dallying* with the stars. *Light* was my first love! In an evil hour I quitted her for those brute & heavy bodies which tumbling thro' ether, startle her from her deep recesses and drive her trembling and sensitive into our view' [17]. And as one might imagine from the metaphors implicit in these sentences, John Herschel was a scientist in whom the culture of romanticism ran deep.

By the autumn of 1816, John had clearly joined the family business of fathoming the 'length, depth, breadth, and profundity of the Universe'. He packed up the contents of his rooms in St. John's, moved his books to Slough, and said *au revoir* to his university friends. Sir William set about the task of teaching him to rebuild the 18-inch reflector, and by 1820 John had figured a fine speculum for it, though it gave its best definition when stopped down to 12 inches [18]. He was made to assure his father, however, that he would never try to rebuild the great 40-foot telescope, which still stood in their Slough garden, because its performance had never been up to expectations even when new. It stayed there until 1839, when John photographed the instrument with his newly-invented 'light-drawing' process immediately prior to its dismantling [19].

At 24, therefore, this professionally-educated scientist of already international standing had reverted to type and become a Grand Amateur. And while one might argue that John Herschel was less of an astronomer by vocation than by inheritance, once he had accepted his destiny, he pursued it with the energy characteristic of a Herschel. He seems to have learned the basic skills involved in cosmological and big reflecting telescope astronomy from their fountain-head, his father; though, as we saw in the last chapter, it was Sir William's great admirer, Sir James South, who later claimed that he taught John how to use refracting telescopes, and employ normal observatory procedures, at his own Blackman Street Observatory in Southwark [20].

By the time that he was 40, in 1832, the newly-knighted Sir John Herschel had come to occupy a unique place in science. He was not just an astronomer, but a physical scientist of extraordinary range, as well as an organiser of science, an international correspondent, and a philosopher. In a world of increasingly specialised researchers, he was not merely a polymath, but a man looked up to as an authority on most of the sciences of the day, from geology and natural history through chemistry, physics, optics and mechanics, to astronomy, cosmology and instrumentation. He was a *vade mecum* of rational knowledge, to whom German professors and American missionaries communicated their exotic findings and solicited authoritative pronouncements [21]. His *Preliminary Discourse* inspired the young Charles Darwin towards a Grand Amateur career in natural history [22], while John Herschel was the big name whom the New York journalist Richard Adams Locke invoked as a way of trying to give credibility to his bogus announcement of the discovery of Moon-men in 1835 [23]. John Herschel was also a great encourager of those people who wished to take up or write about science. He was one of Mary Somerville's warmest supporters, reading through the manuscripts of her books prior to their going to press, and encouraging her to publish her researches [24].

Yet in spite of his fame and authority within the scientific community, Herschel was instinctively wary of those forces which were trying to move British science along a German or French professionalising direction, even when these moves were championed by old friends like Charles Babbage. For Herschel's conception of science was essentially 'virtuosic', and in spite of his broad cosmopolitan culture, deeply British. He saw his cultural roots as extending not from the selective largesse of autocratic monarchs (in spite of his father's pension) or Napoleonic-style bureaucracies, but from the 'gentlemen free, and unconfin'd' tradition of the Bacon-inspired early Royal Society [25]. Creativity in science, as in art or literature, sprang from independence of action, for as he reminded William Whewell when his support was being enlisted for the new British Association in 1831, no committee could 'chalk out districts for individuals or combined diligence to explore.' This was especially relevant to the English intellectual tradition (like the political), 'where freedom of action and independence of thought are highly prized and so energetically asserted on all occasions', otherwise 'an overwhelming mediocrity' would result. 'Perfect spontaneous freedom of thought is the essence of scientific progress', he firmly asserted [26].

As one might realise, Herschel's preferred scientific *milieu* was not the centrally funded research institute, but the free fellowship basis of a learned society. His involvement with astronomical and scientific societies was of enormous importance to the organisation of science in nineteenth-century Britain. It had been the energy of the 28-year-old John Herschel, conjoined with his older Grand Amateur colleagues Francis Baily and William Pearson which, as we saw in Chapter 2, led to the foundation of the R.A.S. in 1820. For at that time, according to Herschel, observational astronomy (but not, of course, cosmology) was almost in its 'last gasp', and needed the 'excitement of experimental research' to lift it out of 'the chilling torpor of routine' into which it had fallen [27]. And in 1830, the active scientific Fellowship of the Royal Society itself put up John Herschel as a candidate for the Presidency in the hotly disputed election of that year which strove to rouse the Society from its tendencies towards lethargy and jobbery. He failed to get elected President, as against the Royal Duke of Sussex, but his strong support from the nation's leading scientists clearly demonstrated his standing. Herschel was enticed, reluctantly at first, into the newly-formed British Association for the Advancement of Science in 1832, after expressing his fears to Whewell before gracing it with his Presidency, and was clearly the name to get on any new Fellowship or Patrons' list, if a fledgling body hoped to become established [28].

In September 1858, when attending the British Association meeting in Leeds, he addressed a large audience on the subject of 'Sensorial Vision' at that city's Philosophical and Literary Society. The following year, a group of local astronomers in Leeds formed Britain's first enduring provincial amateur astronomical society, and inevitably it was to Sir John, and his friend Airy, the Astronomer Royal, that the organisers looked for patrons whose names would lend lustre to the organisation. In spite of the enormous claims that the international world of learning made upon this technically independent gentleman's time, not to mention his health, John Herschel continued to be warmly supportive of the Leeds Astronomical Society, as will be discussed in more detail in Chapter 13 [29].

The first independent astronomical research with which Sir John Herschel became involved, after completing the original nebula charts for his father, concerned the binary stars. Like most interesting objects in the stellar Universe in 1820, the binaries were yet another discovery of Sir William. And by 1802, he had come to realise that certain very close pairs of stars that he had been observing for twenty years displayed a rotation about each other. Now these first two dozen binaries, and the hundreds that would soon follow, held out a most alluring prospect to astronomers. For if (as was mentioned in Chapter 3) the components in a binary system could be precisely measured with a micrometer it might be possible to extract its gravitational elements. This was one of the exhilarating challenges that Herschel realised British astronomy needed to shake it free from its 'chilling torpor of routine', for the analysis of the binaries demanded instruments of the highest optical quality, observers of great skill, and mathematicians capable of handling the most sophisticated analytical techniques.

The best-equipped British astronomical establishment for handling the observational side of the problem in 1822 was Sir James South's observatory in Blackman Street, Southwark, South London, with its equatorial refractors by Tulley, Huddart, and Troughton [30]. Although the discovery and initial measurement of the binaries had been done with Sir William Herschel's 18-inch reflector of 20 feet focal length, this department of astronomical research quickly became a refractor preserve, largely because of the equatorial mounts and clock drives which were such vital adjuncts to this work [31].

The observation and measurement of binary systems was an extremely complex branch of astronomy. Most, as the name indicates, consisted of two stars only. Yet the astronomer had to ascertain what was the plane of rotation vis-à-vis our line of sight from Earth, where the common point of rotation of the system was, and what was its period. And if the system consisted of three or four stars, such as ε Lyrae, then the whole business became even more complex. Once the delicate physical data started to come in from the telescope, then the mathematician was faced with the formidable task of trying to interpret the whole, to quantify the masses of the respective elements and their gravitational influence upon each other. Patiently sweeping the meridian was child's play by comparison, and double-star studies, like cosmology, gave astronomy 'the excitement of experimental research' [32].

Sufficient data was now becoming available from astronomers in Britain, Germany, Russia and France to see if the observed characteristics of these stars really did fit Newtonian criteria, so that the action of gravity could be demonstrated in the stellar Universe. Indeed, by 1825, η Coronae Borealis, which Sir William Herschel had first observed forty years before, had completed a full rotation, while his son was urging a more careful observation of ξ Ursae Majoris, to see if an elliptical orbit could be verified. Then in 1830 Félix Savary successfully demonstrated the Newtonian elements of ξ Ursae Majoris, while soon afterwards, Johann

Plate 1. The 'Shuckburgh' Equatorial by Jesse Ramsden, *Phil. Trans* **83** (1793) 128.

Plate 2. The 'Lee' Circle by Edward Troughton, 1793, for taking meridian transits in Right Ascension and Declination: W.H. Smyth, *Cycle of Celestial Objects* **1** (London, 1844) 333.

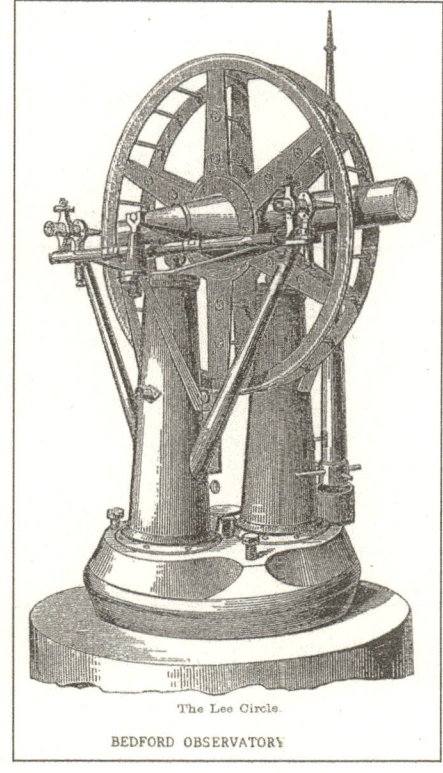

Plate 3. 'A Half-pay Officer's Work-shop, not 1,000 miles from London, 1829.' Captain W.H. Smyth's Bedford observatory, prior to the acquisition of his 5.9-inch lens and dome. Lee, 'Album' I, M.H.S. Oxford Gunther 9, 60.

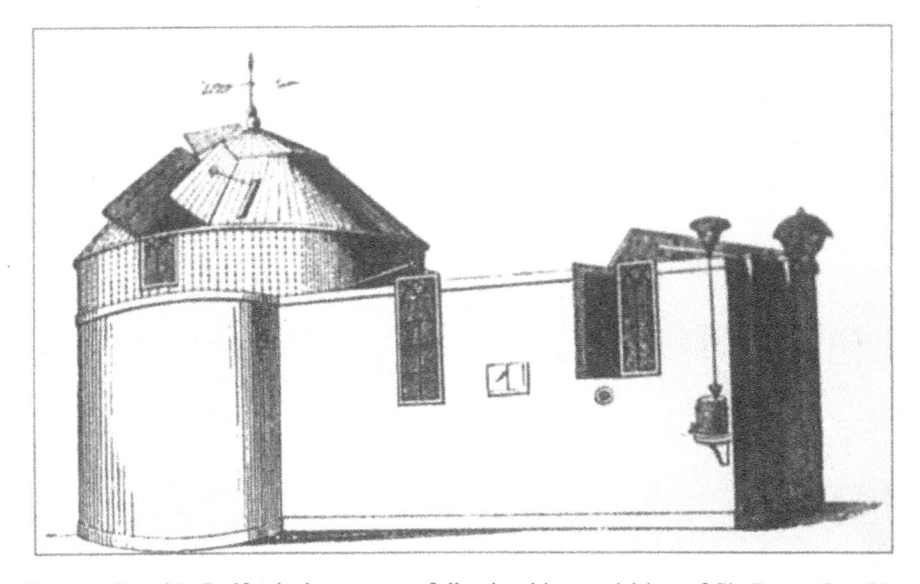

Plate 4. Smyth's Bedford observatory, following his acquisition of Sir James South's 5.9-inch Tulley refractor. The refractor is accommodated inside a truncated 'dome' which was added on to the original transit observatory depicted in plate 3. Smyth's observatory became the prototype for a whole generation of substantial amateur observatories, in which an equatorial, dome-mounted refractor was combined with a transit instrument and computing room, accommodated in an adjacent lateral building: Smyth, *Cycle* **1** (pl.2) 327.

Plate 5. Dr John Lee observing with the 5.9-inch Tulley refractor, probably after he had purchased it from his friend Smyth, and remounted it in the Hartwell House observatory in the early 1840s. This influential, English-mounted telescope is preserved in the store of the Science Museum, South Kensington, London: Smyth, *Cycle* **1** (pl.2) 338.

Plate 6. Hartwell House observatory. The figure operating the portable 'Varley-type' stand mounted refractor is probably Norman Robert Pogson. W.H. Smyth, *Cycle of Celestial Objects ... continued at the Hartwell Observatory to 1859* (1860) 129. This is the same telescope, re-mounted, as shown in Plate 3.

Plate 7. 'A Three-Fold Adieu or a Tri-partite Good Morning', 9 October 1850. Small watercolour by Charles Piazzi Smyth, showing Dr Lee, in dome-shaped observing cap, with characteristic Roman nose; Admiral W.H. Smyth (artist's father) to the right. The central figure with 'Slough' on his hat resembles Sir John Herschel, though it is unclear as to his whereabouts on 9 October 1850. The transcription of his MS Diary, in the Royal Society Library, records that he went to France on 26 August 1850, and his Diary does not begin again until 11–12 November 1850. He may of course have been back in England by 9 October, but without a Diary entry. By 1850, moreover, Herschel had not lived at Slough for a decade, but the painting may have symbolised a long-term friendship between the three astronomers more than it did a specific historical event. The date could well have been that of execution of the painting. Lee, 'Album' II, M.H.S. Oxford Gunther 10, 147.

Plate 8. Sir George Biddell Airy (1801–1892). The *Nature* portrait, *Autobiography of Sir George Airy*, ed. W. Airy (C.U.P., 1896), frontispiece.

Plate 9. Sir John Frederick William Herschel. Mid-nineteenth century engraving after Pickersgill's portrait, reproduced from Sir R.S. Ball, *Great Astronomers* (London, 1906) 253.

Plate 10. William Herschel's 48-inch aperture, 40-foot focal length reflector of 1789. Reproduced from J.N. Lockyer, *Stargazing: Past and Present* (London, 1878) 294.

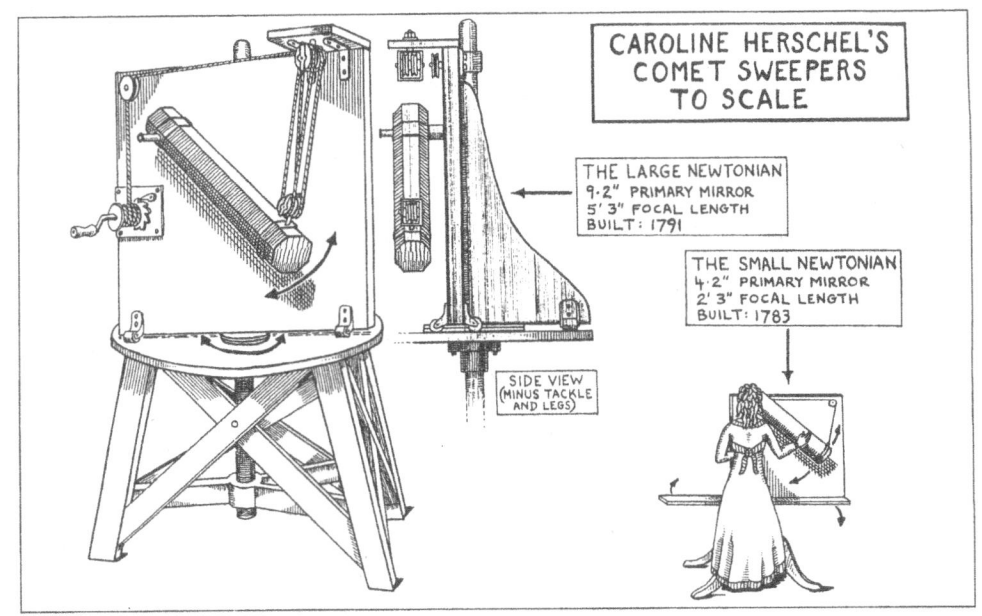

Plate 11. Caroline Herschel's comet-sweeper. Les Jepson's reconstruction of the instrument, and its mode of use, from surviving sketches. By kind permission of Les Jepson.

Plate 12. William Parsons, Third Earl of Rosse (1800–1867). Engraving of *c*.1855, reproduced in Ball, *Great Astronomers* (pl.9) 273.

Plate 13. Birr Castle. Ball, *Great Astronomers* (pl.9) 274.

Plate 14. The 72-inch aperture, 52-foot focus reflector, Birr Castle. Engraving of *c*.1850, reproduced from R.S. Ball, *The Story of the Heavens* (London, 1897) 16.

Plate 15. The Birr Castle 72-inch restored, but still awaiting its optics, summer 1997. Courtesy of the Earl of Rosse.

Plate 16. Firework display at Birr Castle, 3 February 1851, provided by the Third Earl as a local entertainment, and at which he lit the fireworks. The 72-inch telescope was about 200 yards away. *The Illustrated London News*, 15 February 1851, 137.

Encke, in Berlin, did the same for the star 70 Ophiuchi [33]. Rather characteristically, considering the social complexion of British binary star astronomy in 1832, John Herschel published an 'elegant graphical construction and numerical calculation' of several stellar orbits, which possessed the advantage of being 'well adapted for the amateur' [34].

It has already been pointed out that binary star astronomy was almost exclusively conducted with equatorially-mounted refracting telescopes. The singular exception was the Herschels. Not only did Sir William discover and successfully describe the first few dozen binaries with his 18-inch aperture 20-foot focal length reflecting telescope, but his son carried on their collection and measurement with his own 18¼-inch in the 1820s. In 1826, in the *Memoirs* of the R.A.S., John Herschel published his own great catalogue of 321 double and triple stars, and increased the list up to 1,000 by 1828, and 2,000 by 1833. After completing his southern hemisphere catalogue for publication in 1847, his reflector-harvest of binary and triple stars for both hemispheres reached 5,449 [35]. Not for nearly a century would another amateur, the Revd Thomas Henry Espinall Compton Espin, complete the next northern hemisphere search for double stars using the same reflecting telescope throughout. And by this time, the silver-on-glass mirror, and the adaptation of the equatorial mount to the reflecting telescope, had greatly eased the problem [36].

Many of John Herschel's observations of binaries made with the 20-foot telescope, however, were intended to comprise a taxonomic catalogue and general description of these stars, rather than a set of definitive measurements of their positions. When he did provide critical measurements of binaries, he used an equatorial refractor of 7 feet focal length, the clock-driven mount and micrometer of which allowed a much greater level of accuracy than the altazimuth-mounted 20-foot reflector.

Also in 1826, John published an account of the working procedures of his 18-inch telescope of 20 feet focal length which considerably amplified his father's description of his own working procedures. To minimise the amount of movement which it was necessary to apply to the great wooden tube and its altazimuth mount, both Herschels 'swept' selected zones (or 'gages') of the sky in 2-degree vertical steps. The tube would be locked to the appropriate zonal elevation, usually facing the meridian, where a handle could rack it up and down within the 2-degree step. Bells set at the upper and lower extremities of this step would tell the observer when to change direction. By slowly raising and lowering the tube within its pre-selected 2-degree step, as the sky rotated by and presented an endless stream of new objects to view, one could sweep large zones of the sky with the minimum of physical effort. And once a given zone had been thoroughly swept of nebulae, binaries, and other interesting objects, one repositioned the tube to a slightly different elevation and began to sweep out a new zone [37]. With patience, over the course of time one could sweep clean the entire sky in a series of interlocking zones, in the way that inspired a nursery-rhyme variant describing the work of Sir William and his son, which John's Cambridge friend, Adam Sedgwick, recalled many decades later:

'Oh Herschel! Oh Herschel! where do you fly?
To sweep the cobwebs out of the sky' [38].

The cobwebs, of course, were the nebulae, and like the binaries they were part of the Herschel family inheritance.

Because he lacked an assistant, such as his father had possessed in Caroline, John Herschel informs us that he had to devise one-man working procedures, which inevitably slowed things down. The biggest drawback inherent in this solo method of observing, however, came from the constant impairment of his dark-adapted vision, as he was forced to leave the eyepiece and use a candle to write down an observation – a serious irritant if one were observing faint nebulae. Yet nowhere does Sir John Herschel bring home to us the sheer danger of using the 20-foot instrument in the dark in the way that the visiting Maria Edgeworth did in 1831:

> 'I must remark that all the time we were seeing we were 18 feet aloft on a little stage about 8 feet by 3 with a slight iron rod rail on three sides but quite open to fall in front ... Herschel runs up and down the ladder like a cat (because I would not say a monkey) ...' [39]

With such flimsy safety measures, and an absence of light, so as to preserve one's dark-adapted vision, one can only marvel at the low accident rates sustained with Herschel-mounted reflecting telescopes. One of the few potentially serious accidents which occurred was when Sir David Brewster fell off the observing gallery of John Ramage's home-made Herschel-type reflecting telescope at Aberdeen. In spite of falling off when the tube was at its full elevation of around 20 feet, Brewster suffered no serious harm [40]. On one occasion at Slough, however, Caroline Herschel tripped on the snow-covered grass in pitch darkness, and impaled her leg upon a big iron hook that was used to adjust her brother's telescope: an incident about which more will be said in Chapter 14 [41].

Of all the objects that were encountered by the Herschels – father, aunt, and son – in their sweeps of the heavens with large reflecting telescopes, none were more enigmatic than the nebulae. They appeared to come in a wide variety of shapes, sizes and luminosities, and unlike the binaries, they possessed no obvious mathematical characteristics. The Herschel approach to nebulae was essentially taxonomic, and conformed in many ways to the Baconian criteria for the collection of data. Nebulae were classified by two generations of Herschels, in both hemispheres, along lines that were similar to those employed by contemporary botanists to classify plants: by shape, colour or luminosity, and with relationship to similar specimens. Nebulae were planetary, elliptical, diffuse, pea-green, with or without bright spots, and were found scattered individually among the stars, 'condensing' out of the Milky Way, or in great clusters, such as those in Virgo or the Magellanic Clouds [42].

The pressing question concerned their composition and structure. Were they dense conglomerations of individual stars gravitationally collapsing into each other – like the Hercules Cluster – or were they made of glowing, fluffy 'chevalures', or 'luminous fluid', like the nebula in Orion? Sir William had built the 40-foot telescope with its 48-inch mirror in the hope of resolving the question, but had still failed through a lack of sufficient 'space penetrating power'. John continued to wrestle with the family problem, and while he never worked with a mirror which exceeded

18 inches in diameter, he undertook several fundamental investigations over the space of the twenty years up to 1838 [43].

One of the most significant of these was his meticulous surveys of the Orion Nebula, the first of which he published in 1826. It was by far the most detailed analysis of a single deep-sky object which anyone had performed up to that date, and was based on dozens of drawings of particular zones or sections of the nebula that were brought together in the finished map. Sir John Herschel produced his detailed survey so that it might act as a benchmark alongside which future astronomers could compare their drawings, in the hope of settling the vexed question about whether change was discernible in nebulous bodies. If changes of structure or luminosity could be seen and measured over time, then it implied that the object was relatively close to us, and modest in size (by cosmological standards). On the other hand, if the nebula appeared changeless over a century or so, it was likely to be so remote and vast in size that even our finest telescopes could detect no alterations [44]. When he was at the Cape of Good Hope, a decade later, Sir John made a new survey of the Orion Nebula, as the object was much higher up, and better placed in the sky than it ever could be in England. Yet once again no changes were detected [45].

Like his father before him, Sir John Herschel saw the nebulae as holding out a key for the determination of the size of the Universe and the objects within it. Was there a single 'galaxy', containing all the stars, nebulae, and star clusters that was broadly analogous to the Milky Way, or were there physical systems beyond it: objects which would later come to be characterised as 'island universes'? If an object as large in its 2½-degree extent as the Andromeda Nebula, for instance, was made up of separate stars, none of which could be discerned individually, then how incredibly remote must it be? For by definition, it must be more distant than the stars which formed the dense clouds of the Milky Way, which could nonetheless be resolved telescopically [46].

Those objects which Sir William had called 'planetary nebulae' were even more puzzling, for while they clearly existed in the depths of the stellar Universe, they appeared as distinct, planet-like disks, with a slight hiatus of light at the centre. The planetary nebula M97, near β Ursae Majoris, was 2′ 40″ arc in diameter and appeared to glow like a faintly illuminated 'hollow spherical shell', or glass sphere. Herschel calculated that if M97 were to be placed at the distance of the recently parallax-measured (and hence cosmologically close) star 61 Cygni, it would be at least seven times bigger than the entire orbit of Neptune. If it, and the rest of the nebulae, were not made of packed stars, moreover, what could be the source of their incandescence? [47]

Back in the 1790s, Sir William had suggested that there might be incandescent sources in the stellar Universe that were made up of something other than shining stars. These were his 'luminous chevalures', or flimsy hairy structures which glowed like phosphorescence on the sea or will-o'-the-wisps above ponds on still nights. In those days, however, scientists lacked both the physical knowledge and the vocabulary to be more exact when trying to describe vague light sources. And things were not much more advanced in the 1830s, when phosphorescence was still not understood, and no-one had any idea of how a gas might behave in the vacuum conditions of space [48].

But with his mathematician's perspective, and awareness of the criteria that were necessary for gravitational explanations to work, Sir John was not easy with light sources that lacked physical mass constants. Instead, he suggested that the nebulae might be made of 'particulate matter', each particle of which was a discrete physical object in its own right, and could have gravitational relationships with similar particles. Yet unlike his father, Sir John was extremely cautious about indulging in speculations about the physical characteristics of the stuff that filled the Universe, though as he had so few hard facts to go on, he could scarcely avoid them if he hoped to provide some sort of explanation for what he could see through his telescope.

Just as microscopic particles of dust in a room could reflect sunshine to produce apparent shafts of light, so the nebulae could, perhaps, be composed of similar clouds of particles locked within a gravitational system. And what about the Zodiacal Light, which in those unpolluted days could even be seen from Slough on spring evenings after sunset? Was not this band of light always in the plane of the Solar System, thereby suggesting that the great gravitational field within which the planets and asteroids moved as gross objects was also occupied by particle streams of microscopic ones which reflected the sunlight in a dark sky? Cometary tails could well be made up of similar, gravitationally-sensitive optical agents [49]. In this way, Sir John Herschel tried, in an age still innocent of particle physics and Clerk Maxwell's equations, to explain nebulosity in the light of existing physical knowledge.

By 1833, with the publication of his great catalogue of nebulous objects [50], John Herschel realised that nothing fundamental could now be learned from the skies of the northern hemisphere with his 20-foot telescope. It logically followed from his father's and his own work that the great deep-space surveys of the Herschel inheritance should now be continued into the southern hemisphere, to 'gage' the entire heavens. Following the death of his mother in 1832, John now felt that he possessed the freedom to leave Europe for a protracted period, and in the autumn of 1833 he sailed aboard the *Mountstewart Elphinstone*, Indiaman, for Cape Town.

This was, indeed, the stately passage of a wealthy man, with Lady Herschel, their growing family, a retinue of servants, and a great deal of baggage. The three 18-inch mirrors that were necessary for keeping the 20-foot tube and its superstructure in permanent operation, along with an equatorial refractor and circle, with clocks, chemical apparatus, and tools, were stowed in the hold. This was a gentleman's household and a scientific institution on the move, and the fare for the two-month voyage was £500, which was slightly more than the entire annual salary of the Director of the Cambridge University Observatory [51].

Upon arrival at the Cape in January 1834, Sir John took out a lease for £225 per annum on the Feldhausen estate, but sensing that the lessees were having financial difficulties, bought it freehold for £3,000. One wonders how far the Baldwin and Pitt monies that he no doubt received following his mother's death facilitated this, for it was clearly not possible on the strength of the income which he received from his father's 3 per cent annuities without raiding the capital sum [52].

It was Herschel's years at the Cape between January 1834 and March 1838 which set the rest of his astronomical career into context. Not only did he become the first, and perhaps only, astronomer to scrutinise the entire deep sky in both hemispheres with the same telescope, but he drew many significant conclusions. His reason for going to the Cape, of course, was to 'sweep' the nebulae out of the southern skies, and in this respect alone he found several important differences from the skies of the north. Whereas in the northern hemisphere the nebulae appeared to be heavily concentrated in the constellations of Leo, Leo Minor, Ursa Major, Coma Berenices, Canes Venatici, and most of all, Virgo, in the southern hemisphere 'a much greater uniformity of distribution prevails' [53]. Southern nebulae seemed much less aggregated into particular sky zones, with the spectacular exception of those puzzling objects, the Magellanic Clouds, which contained no less than 1,145 nebulous objects between them. But if the nebulae, on average, tended to aggregate most thickly in the regions of the Galactic poles, the great star clusters and planetary nebulae by contrast were most commonly to be found around 90 degrees away, on the fringes of the Milky Way, or Galactic equator. By 1847, when he had prepared his total sky survey for publication, John Herschel recognised that nebulous and stellar matter were probably different in terms of the physical systems which they constituted [54].

He was also struck by the large number of planetary nebulae which seemed to populate the southern sky – mainly along the Galactic fringes – as opposed to the skies of the north. Writing from the Cape to John's cousin, Mary Baldwin, on 9 April 1834, only a few weeks after their arrival, Lady Margaret Herschel said that the South African expedition had already been 'richly rewarded' with discoveries in the southern skies. 'Herschel', as Lady Herschel always called her husband, had already discovered two new planetary nebulae, one being 'of a fine blue colour', whereas only half a dozen such objects were known in the entire northern skies, and most of them had first been seen by Sir William [55]. By 1835, however, the planetary nebulae were appearing so frequently in the 20-foot telescope that she became almost blasé in reporting them [56]. Between 1834 and 1838, Sir John discovered a couple of dozen of these objects at the Cape, and, as mentioned above, the large angular size and luminosity of the planetary nebulae baffled him. They would continue to baffle astronomers for over thirty years, until William Huggins used the spectroscope to determine their gaseous composition. We now know that each of the strange glowing 'shells' that Herschel saw is the gradually expanding remnant of gas blown off by a star [57].

It was in South Africa that Sir John's definitive idea of the 'Construction of the Heavens' took shape in his mind. He realised that the Sun was not at the centre of the Universe (a concept that also lay implicit in his father's work on the motion of the Solar System in space), and that the stars seemed to be arranged in a great flat ring [58]. But the size of that stellar Universe continued to grow with each new ingenious interpretation of his results. In 1836, for instance, he started to use his 'astrometer', which was a pioneering photometric technique whereby he tried to obtain an absolute luminosity for any given star that could be calibrated with relation to the brightness of the Sun. The method told him that Vega must be forty times brighter than the Sun, whereas Arcturus was two hundred times so [59].

These experiments in trying to determine what we would now call the absolute magnitudes of stars gave the first insights into the nature of stellar populations, and contradicted Sir William's guiding cosmological principle that all stars were probably of the same generic size in the same way that all men or mature oak trees are of the same approximate size. Sir John now came to realise that bright stars were not only bright because they were closer to us than dim stars, but because they could be much bigger and more luminous than their neighbours, thereby invalidating the precept that distance and luminosity were related [60].

And if the stellar Universe was so vast and heterogeneously populated, what was its relation to the nebular Universe? The apparent size and remoteness of the planetary nebulae, the irresolvability of the 'luminous fluid' nebulae, and their distribution throughout space, posed far more conundrums than could be answered with the physical and conceptual technology of 1840. And even when, a few years later, Lord Rosse used his new 72-inch telescope to detect the spiral structure of the nebula Messier 51, the wider problem of the 'Construction of the Heavens' seemed as puzzling as ever, for even that great instrument could scarcely resolve more than a dozen out of over 5,000 known nebulae [61].

When Sir John Herschel arrived back in England in May 1838, in time for Queen Victoria's coronation and a baronetcy to add to his Hanoverian Guelphic knighthood of 1831, his observing days were effectively over. By the age of 46, the age at which his father had been getting into his stride as an observational astronomer, Sir John had seen everything that was worth seeing in both hemispheres, in an observing career that had spanned just over twenty-one years. Surprisingly, he showed no personal interest in building bigger mirrors or in attempts to resolve the nebulae with new technologies, and one senses that he felt that his own observing work had gone as far as he wished to take it. Now it was up to his younger contemporaries – Lord Rosse, William Lassell, and James Nasmyth, who were also much more compulsive observers then he was – to carry it further.

Not only had Sir John's reputation as a scientific authority increased by the time of his return to England, but his physical circumstances had also changed. Several more children had been born to John and Margaret Herschel in South Africa, and Observatory House at Slough was becoming cramped. During his four and a half years abroad, moreover, the new Great Western Railway had reached Slough, on its way to the west, and the greater accessibility of London threatened his cherished privacy. On the other hand, it did make trips up to town quicker and easier when he wanted to go. On 10 November 1838, for instance, he travelled up behind the 'Aeolus' locomotive (using a travel vocabulary still redolent of stagecoaching), at a speed that sometimes exceeded 40 mph. He was in his club, the Athenaeum, on Pall Mall, only 61 minutes after leaving Slough, and that included contending with traffic from Paddington [62]. Indeed, trains were clearly a novelty that fascinated the returned Herschel, and he recorded many railway incidents in his Diary. On 29 November, so he confided to his Diary, his train was slowed down to no more than 6 or 8 mph because of head gales, while in January 1839 a train was de-railed and a horse beheaded when a fox-hunt charged across the tracks [63].

But it was in pursuit of more space and privacy that he purchased Collingwood House at Hawkhurst, Kent, in August 1839. It was in the depths of the countryside, and even after the railways had spread their tentacles across Britain, it would still be necessary for his coachman to drive the dozen or so miles to Staplehurst station before Sir John could catch a train up to London. Collingwood cost £10,500, and required a further £1,400 spending on improvements, but the mansion and its grounds were simply paid for by liquidating stock in the City [64]. Once again, we encounter those financial facts which enable us to account for the availability of the resources which made Herschel's independent astronomical career possible.

When one looks through Herschel's diaries, it is possible to calculate that the Cape expedition cost Sir John about £6,237 in ordinary running costs, not including the two £500 fares, or the purchase of Feldhausen [65]. Though he made £500 profit when he sold Feldhausen in 1839 – a sum which no doubt went towards the purchase of the Hawkhurst estate – Sir John seems, estimating from the sums mentioned in the Cape correspondence and his Diaries of the period, to have spent around £14,000 in fresh capital transactions between 1833 and 1839. And by far the greater part of this money went, directly or indirectly, to facilitate his astronomical enterprises [66].

The volumes of the Diary which Sir John kept following his return from South Africa also make it possible to follow a Grand Amateur astronomer about his daily business. The 20-foot telescope was never used at Hawkhurst, although he regularly wheeled out one of his father's 7-foot telescopes to observe interesting objects, such as Mauvais' Comet of 1844. By far the greater part of references to his own astronomical observations in the period after 1839, however, are records of the naked-eye glories of sunsets, or of moonlight on the Kentish fields [67].

Beset as he was with a host of people, from journalists to visiting foreign dignitaries, who wished to see and talk with him, not to mention those new photographic researches which came to fascinate him after 1839, he now needed the space and time to bring his astronomical work together [68]. And it says something about the demands which the world came to place upon him that it was not until 1847, some nine years after his return to Britain, that the Cape *Observations* were finally reduced and published.

Ill-health also figures increasingly in the Diaries after 1838. Rheumatism, sciatica, bronchitis, and an operation on his knee for varicose veins in July 1855 clearly indicate that he was less vigorous at 60 than his father had been – although Sir John still continued to father children down to the age of 63, by which date Lady Herschel herself was in her mid-40s, with a family of twelve children to look after [69].

The traumatic event of the post-Cape period, however, was his acceptance of the Mastership of the Royal Mint, between the years 1850 and 1855 [70]. Exactly why he took on this, his only properly salaried job in seventy-nine years, is not clear, and one suspects that the example of his Cantabrigian predecessor, Sir Isaac Newton, was a spur. But a man more naturally averse to administrative routines and committees than the 58-year-old Sir John Herschel would have been hard to find, in spite of the rigorous sense of duty with which he presided over the reform of the nation's currency. For while Sir John possessed an enormous capacity for sustained hard work, be it at the telescope or at the computing desk, it had to be his *own* work, and form part of his broader creative enterprise. In his working procedures, he was an archetypal loner, and while he could present and compare his scientific work in the forum context of a learned society meeting, he was constitutionally at odds with that kind of team work and team management that was essential to the efficient operation of a body like the Royal Mint. Like all of the Grand Amateurs, he was a virtuoso, and virtuosos, as a class, rarely make good 'company' men.

By 1855, the reform of the currency, and the administrative reform of the Mint itself, had driven him to nervous and physical exhaustion. The many photographs and portraits of Sir John bear witness to the effects of the Mint years upon his countenance, while in February 1855, he confided: 'Found I have myself diminished in height full half an inch' [71].

Yet Herschel's Diaries and correspondence clearly indicate that his scientific vigour, and enjoyment of the company of old and new astronomical friends, remained undiminished, especially after he escaped from the Mint. He greatly admired William Lassell, the Liverpool brewer, enjoyed the company of Smyth and Rosse, and sealed the peace around the discovery of Neptune by bringing Adams and Le Verrier together at a country-house meeting in his Kent mansion in 1847 [72]. His fellow-Kentish Grand Amateur astronomer, William Rutter Dawes, visited Collingwood, while Herschel in turn visited the Grand Amateur James Nasmyth, who had come to live in the county. Sir John and his daughter Isabella were visiting Nasmyth's mansion 'Hammerfield' at Penshurst, Kent, in May 1864, and it was at Nasmyth's dinner-table that Herschel recorded being introduced to the High Sheriff of Kent [73]. But not all of his relations with the Grand Amateur community were so felicitous. By 1846 he was eventually forced to break with his old mentor, Sir James South, after another of that unfortunate individual's furious letters to *The Times* made Herschel confide to his Diary that South now 'proves himself an *unsafe companion*' whom prudent men should avoid – a sentiment already expressed by Herschel's friend Airy [74].

By this time, Sir John Herschel was the undisputed *Grand Seigneur* of British astronomy, a catch for any intellectual gathering, and an inspiration to all of those men who resolved to take up the big reflecting telescope where he and his father had left off. As the grandest of the Grand Amateurs in terms of the devotion and thoroughness of his commitment, he epitomised the private scientist in all his strengths and weaknesses.

5

An astronomical house-party: the Bedford–Aylesbury axis

Between the late 1820s and the early 1860s the stretch of country between Aylesbury and Bedford became the site of considerable astronomical activity. Its leaders were a pair of curiously matched friends. Dr John Lee was the proprietor of two landed estates, upon which Hartwell House, Aylesbury, was his seat [1]. He was a local magistrate, a barrister, an advocate in the Court of Doctors' Commons in London, and an eccentric polymath: a man of stern aspect and punctilious habits, who was described by his obituarist as inflexible in his opinions once formed, though he never came to a hasty decision. John Lee was said to be scrupulously honest in his patrician severity [2]. His friend was William Henry Smyth, an ex-Napoleonic officer who retired from the Royal Navy with the rank of post-Captain in 1825, at the age of 37, and later rose by seniority after 1853 to the rank of Admiral. Smyth was a jovial man who loved his friends, convivial gatherings, and astronomy, and coming to live in Bedford soon after his retirement from active service, established an excellent observatory at the back of his house. After the end of hostilities in the Mediterranean, where he was stationed, Smyth became involved in hydrographic and cartographic surveys for the Admiralty, and was therefore a skilled practical astronomer who was thoroughly accustomed to the use of precision instruments [3].

It is not clear how these two gentlemen met and discovered their mutual interest, though Smyth was later to mention that when his friend was passing through Bedford to his Colworth estates to attend the assizes as a county magistrate, he would sometimes stay as his guest [4]. On these occasions, Dr Lee took a delight in the Admiral's observatory, and in the manipulation of the instruments which it contained. These sessions with Smyth in Bedford, probably around 1826, re-awakened Lee's interest in 'the tangible advantages of practical astronomy followed by physical theory, which formed a prominent part of my college reading' at St. John's College, Cambridge, over twenty years before [5].

John Lee now set about the building of his first observatory. By 12 July 1827, the architect and inventor William Cubitt had submitted a 'Design for a small Observatory', and within a short space of time some sort of observatory was operational at Hartwell [6]. The astronomical house-party had begun, as Hartwell

House became the focal point for a circle of amateur astronomers in the district and beyond. These included figures such as Samuel Charles Whitbread, brewer, of nearby Cardington, Thomas Maclear, surgeon, of Bedford and Biggleswade, Thomas Dell of Aylesbury, the Revd Joseph B. Reade, Vicar of Stone, and a widening penumbra of individuals who enjoyed the scientific and learned gatherings that started to take place in John Lee's mansion, as is testified by the surviving visitors' books [7].

The years around 1830 were significant in the evolution of the amateur astronomical observatory, and one sees this taking place in the foundations of Smyth, Lee and Whitbread in particular. These were men who did both meridian transit and double star work, and developed the architectural features in their observatories that were necessary for both. One suspects that Admiral Smyth – the technical leader of the Aylesbury pack – was in turn influenced by the East Sheen and South Kilworth Observatories of William Pearson. Pearson, after all, was not only a great collector and user of instruments, but his large and detailed articles in Rees's *Cyclopaedia* (1819), along with his *Practical Astronomy* (1829), had provided definitive treatments of how serious astronomical observatories were set up and operated [8]. Likewise, Sir David Brewster's articles in the part-published *Edinburgh Encyclopaedia* (1830) gave detailed accounts of observatory operations, and in many ways Brewster's and Pearson's works led the way as far as high-level exposition was concerned [9]. Both Smyth and Lee were members of the Astronomical Society when it obtained its Royal Charter in 1831, and as the surviving Club dinner lists make clear, would have sat at the same table as Pearson and Baily at regular intervals, although Pearson was a less frequent diner [10].

The Bedford Observatory that John Lee would have visited in the late 1820s, and which he replicated as his first observatory at Hartwell, was probably the one depicted in a beautiful ink-drawing in the Hartwell 'Album'. It shows an elegant rectangular building, not much larger than a modern domestic garage, and entitled 'A Half-pay Officer's Work-shop, not 1,000 miles from London, 1829'. Set in a garden, with a gentleman (probably Smyth), ladies and children enjoying the pleasant setting, one sees an observatory with wall and roof slits for the accommodation of fixed meridian instruments, and a computing room for calculation [11].

In 1844 Smyth described his observatory as containing an 'excellent' 3¼-inch aperture transit instrument by Thomas Jones of London, along with the original 1793 Troughton circle built for Gavin Lowe, bought fourth-hand by Lee and loaned to him, though it is less clear exactly what was there in 1829 [12]. Conspicuous by its absence in the 1829 drawing, however, is any brick drum and dome arrangement which would have been necessary for the accommodation of a large refracting telescope. But Smyth possessed a fine refractor in the late 1820s, probably the instrument of 5 feet focus, which he spoke of testing and aligning, and which was very similar to the portable instrument shown in the drawing which could be brought out of the main building and set up on a tripod [13].

In October 1829, when Sir James South carried his 11¾-inch Cauchoix lens away from Paris in triumph, he wrote to his friend Admiral Smyth offering him first refusal on the telescope which the world's largest refractor was now intended to replace at South's Campden Hill Observatory [14]. This was a lens by the leading English optician, Charles Tulley, with a clear aperture of 5.9 inches and almost 9 feet focal length – a large lens by the standards of the day. Smyth bought it for £220. His Bedford Observatory now had its circular drum, surmounted by a truncated cone 'dome' which turned upon wooden balls in a channel, added as an extension on to the end of the existing building. For this is how he depicted it in his *Cycle of Celestial Objects* in 1844, though the other details of the building, including the Gothic tracery windows, are consistent with the drawing of 1829 [15].

Smyth now had one of the most powerful and versatile refractors in Britain, but in reality it would remain, along with Cooper's Markree telescope, one of the largest working instruments for some years to come, for as we have seen, South never had the 11¾-inch Cauchoix successfully mounted [16]. The 5.9-inch Tulley lens was set up on an 'English' equatorial mount by Dollond, with a clock drive of particular regularity, devised in accordance with the principles laid down for such mechanisms by none other than Sheepshanks himself. It was an ideal instrument for double-star work, and its harvest of observations was published in Smyth's substantial two-volume *Cycle* fifteen years later.

One can guess that the added potential which the 5.9-inch Tulley refractor brought to Bedford Observatory further whetted the astronomical appetite of Dr Lee, for as Smyth recorded twenty years later in 1851, 'scarcely had the good Doctor conquered the difficulty of watching the stars across the wires while transiting, than he yearned for more power.' Lee's Cambridge education had 'instilled the physical theory of astronomy into his mind, and practice brought the conviction of its, so to say, tangible advantages' [17]. This set in motion the construction of a larger observatory at Hartwell House, which was added as an extension to the existing fabric of the mansion.

The new observatory, with its transit rooms, computing rooms, and drum and dome, connected very conveniently with the library in Hartwell House. It suited John Lee's whole cultural perspective insofar as he could walk straight from his opulent collections of books, Egyptological museum, and cabinets of rarities, without even needing to venture outdoors. But he was not a Grand Amateur in quite the same way that Smyth, Pearson or South were. Astronomy was never his consuming passion, but rather an elevating and fascinating activity amidst many other activities, and as a man who enjoyed the advantages of great landed wealth, it delighted him to host and assist a range of scientific and archaeological friends [18]. On numerous occasions in the surviving Scrapbooks and Albums of Hartwell House, one senses an underlying playfulness which is not always easy to reconcile with that stiff rigour so clearly pointed out by his obituarist.

It is possible, however, that the clear-minded and forceful Queen's Counsel became stiffer as the years passed, for Dr Lee in his late forties seems to have been a man of great charm who made Hartwell a truly merry place to visit. When the piers were being laid for the new transit room in 1831, for instance, Lee got each of his visitors to lay a brick, and gave them a certificate to commemorate the event [19]. It seems that the new observatory was expected to be slow in building, however, for on

18 December 1829, Smyth had opened a wager-book about how long it would take Lee, with his multifarious occupations, to finish the observatory [20]. Lee accepted the wager, but for fun, as he was against gambling. But as late as 25 April 1833, the visiting Royal Society dignitary, Davies Gilbert, got his certificate for laying a brick [21]. It is likely, however, that this long-drawn-out completion of the observatory came from the addition of the new enlarged dome, which became operational in 1839 [22].

John Lee is a fascinating subject for any biographer, for he was a figure of curious contrasts. Combined with the old Roman virtues and concern for hard work and efficiency (in spite of his independent wealth, he never abandoned his legal practice in London, even when in his eighties), Lee was an active social reformer. A model landlord at a time when agrarian unrest was threatening the stability of rural England, he built good cottages for his tenants, founded schools, and took a serious interest in the physical and spiritual well-being of the peasantry. Unlike most gentlemen in his circumstances, Dr Lee disliked gambling and was a staunch teetotaller, for he saw both gambling and alcohol as sources of great social mischief. The Buckinghamshire gentry must have been mystified by his firm opposition to fox-hunting and field sports, both of which he regarded as injurious to agriculture. But perhaps it was his political beliefs which seemed most *avant garde* to contemporaries, for he opposed war and used the great parks around Hartwell House to host festivals of 'Peace and Brotherhood' as well as teetotal rallies. John Lee was an eccentric in the grandest English tradition, and seemed even more so by being a highly successful man of affairs who believed that science and reason were essential in the creation of a better and a more peaceful world [23]. He even broke the social rules when at the age of 50 he married Cecelia Rutter, a woman of 'humble station but of excellent character and good disposition' [24] – so he told Sir Thomas Maclear – and who, judging from her comments in the Hartwell Albums, was not expert in the handling of a pen. But as far as one can interpret from the surprisingly shadowy presence of Cecelia in the Hartwell manuscripts, she and Lee, unlike Carrington and his 'lower class' wife, seem to have enjoyed a happy marriage. And eighteen months after Cecelia died, on 1 April 1854, the 72-year-old Lee, whose energy was still unabated, married Louisa Catherine Heath, who was almost forty years her husband's junior. There were no children from either marriage [25].

Like the great proprietor that he was, Lee saw himself as a patron. His purchase of instruments from the leading London makers was one aspect of this patronage, and when he bought the Troughton circle mentioned above, as well as the Beaufoy Circle by William Cary for £210, he handed them over to his friend Smyth so that they would be properly used [26]. One possible act of assistance to his friend was his purchase of the 5.9-inch Tulley refractor with all of its fittings in 1839. This was when Smyth and his family went to live in Cardiff to superintend the construction of the Marquis of Bute's floating dock in that city [27]. As a 'Half-pay Officer' on the navy lists, Smyth would not have been in receipt of a large income from the Admiralty. Nor would he have been retired in the modern sense, for as a half-pay post-Captain creeping up the lists to become a shore-bound Admiral, he was officially eligible for a fresh command, though he never went back to sea. Whether

Smyth had access to any substantial independent wealth, or whether he had suffered financial losses during that age of volatile money-markets, is unclear. But following the completion of his work in Cardiff, Smyth took up residence at St. John's Lodge, a house that was adjacent to Hartwell House, where he used his old instruments, now remounted in Lee's observatory [28].

Though he was born in London in 1788, William Henry Smyth was the son of an American 'loyalist' and descendant of Captain John Smith, who came to Britain following the Revolution, and any land which the family possessed in New Jersey seems to have been lost. It is said that Smyth ran away to sea as a lad, when he climbed up the side of a docked West Indiaman, subsequently joining the Royal Navy and serving under Lord Rodney and other commanders on his way up. It is very likely that he made prize money during the later Napoleonic era, for he had served with distinction in several actions, commanding a gunboat off the coast of Spain when he was 22, and a small flotilla in the Straits of Messina; but as he was not commissioned Commander until October 1815, some four months after Waterloo had finally brought peace in Europe, it is improbable that he would have enjoyed the fat prize money pickings that had made so many frigate captains rich during the War [29].

At this time, Smyth was serving with the British Squadron stationed off the Kingdom of Sicily, and like Lord Nelson some twenty years before, he fell in love with a beautiful and accomplished English lady, Miss Anarella Warington, whose father (like Sir William Hamilton before him) was English Consul to the King of Naples and Sicily. Fortunately, this naval marriage was the source of no scandal, and William and Anarella enjoyed a long and happy life together. It is not known whether Anarella brought a significant dowry to Smyth, but it is unlikely that a senior British diplomat occupying such a strategically sensitive post was poor. Anarella continued to live in the Kingdom of Naples and Sicily while William, in the peacetime Navy, did survey work in the Mediterranean, and several of their children were born there. In Sicily, Smyth came to make the acquaintance of and developed a profound respect for his astronomical mentor, Father Guiseppe Piazzi, the astronomer who had discovered the first asteroid, Ceres, at Palermo in 1801, and it was in homage to him that he named his Neapolitan-born son Charles Piazzi Smyth. The elderly Roman Catholic priest–astronomer also became godfather to the child of these staunchly Protestant parents [30].

Admiral Smyth was a dedicated observer, and from 1830, when his 5.9-inch telescope had been acquired from Sir James South, he came to concentrate on double, variable, and coloured stars. In his *Cycle of Celestial Objects* of 1844, with its *Bedford Catalogue* as volume 2, Smyth laid out his wealth of experience in the hope that this book would inspire others to come to astronomy. It is all too easy for people today to be unaware how few books of guidance were available to the early Victorian amateur astronomer, especially if he or she did not belong to the comfortably well-off and well-connected classes. John Herschel's *Treatise on Astronomy* (1833), in Lardner's *Cabinet Cyclopaedia* series had aspired to teach the broader discipline to the classes of people who read *The Times*, but it contained little practical help when it came to setting up and productively using modest instruments. Smyth hoped to fill this need, and his *Cycle* might be considered as the first real book for the amateur. As he reminded his readers in the 1859 continuation of the *Cycle*, using an obvious armed-forces terminology, 'Besides the professional regulars of the astronomical corps, the volunteer amateurs – a body unconnected with any national establishment – advanced and did good work: so much so, indeed, as to reap one-third of the unexampled harvest that followed [31].' The good work to which he referred here is that relating to the discovery of asteroids.

Yet under no circumstances could one regard Smyth's *Cycle*, and its supplements, as a beginners' book in the way that we often think of beginners' literature today. For one thing, it assumed a familiarity on the part of the reader with at least some Latin, Greek, and modern European languages. It also assumed that one had received a sufficient background in mathematics at school or university to be able to understand what positional and binary astronomy was all about. Though the vitriolic Sheepshanks had accused the medically-educated South of not knowing a sine from a cosine [32], this was clearly an exaggeration. No one who undertook double star observations and reductions of the order that earned South the Royal Society's Copley Medal in 1826 could really have been an innumerate. Quite simply, early Victorian astronomers were by definition educated people, and Smyth assumed that his readers would have come to astronomy with a good general education already behind them. Equally important, he would have assumed that they had leisure. They may not have been potential members of the Grand bracket of amateurs, but anyone who could afford to pay for two hefty octavo volumes, and then spend several weeks reading them, was likely to be cultured, leisured, and affluent.

The first volume of the *Cycle* contains an abundance of good advice from the pragmatic old sailor. Amateurs should keep away from the grind of meridian astronomy, which the professional establishments already did to an excellent standard [33]. A potential observer should get stuck in, even if his equipment was limited, for it served no purpose to put things off until one had an apparently perfect observatory. He provided drawings for stable and easily improvised mounts, such as the simple yet perfectly serviceable equatorial mount which could be devised for a 5-foot refractor of 3½ inches aperture by the village carpenter. It was also wise, if one wanted to do useful work as opposed to merely peeping at the heavens, to concentrate on a particular class of astronomy, such as variable or double stars [34].

The *Cycle* provides a wealth of information about the equipping of observatories and an evaluation of instruments, and it is here, and in Smyth's *Aedes Hartwellanae* (1851) and its continuations, that we have detailed accounts of the observatories at Bedford and Hartwell House. But it is in the second volume of the 1844 *Cycle*, known as the *Bedford Catalogue*, that we see the celestial 'harvest' of which Smyth spoke. Here is a detailed description and analysis, occupying some 543 pages, of hundreds of binary, variable, and coloured stars in 70 constellations. No one can read these pages without appreciating the thousands of hours of sheer hard observing, undertaken purely for the 'love' of astronomy, that lay behind Smyth's book. Like all of his brethren in the pre-photographic days of astronomy, Admiral Smyth possessed a minutely detailed knowledge of the sky, extending down into the telescopic magnitudes [35].

The intellectual component of hard reading and thinking that lay behind the hard observing is also transparently clear. Smyth's work on the complex binary star γ Virginis, for instance, is elegantly demonstrated along with his construction of its orbit. The analysis of this individual star system alone takes up thirty pages of text in the *Aedes* [36]. And on 3 May 1845, Smyth took great pleasure in showing this star to Dr and Mrs Lee [37]. Likewise, his treatment of coloured stars is conducted with similar thoroughness. But Smyth does not only present his own work: he also cites that of others, such as Benedict Sestini in Rome. He next enters into detailed discussion about the prevailing theories of coloured light: is colour the product of particles of light hitting the eye at slightly different velocities, as Newton proposed, or is it a product of a wave motion, as argued from the experiments of Young and Fresnel? [38] Smyth's writings bring across the formidable breadth and depth of his astronomical culture: a culture which he would have shared with his Grand Amateur friends. It was, moreover, a culture which only a man of leisure could hope to acquire.

In January 1879, however, nearly fourteen years after Smyth's death, a young English astronomer, Herbert Sadler, succeeded in getting a paper published in the *Monthly Notices* of the R.A.S. which fiercely attacked the honesty of Admiral Smyth and the reliability of his *Bedford Catalogue* [39]. Sadler asserted that in certain cases Smyth had simply lifted observations of double stars from the catalogues of John Herschel and printed them as his own. The R.A.S. was embarrassed, not only because Smyth had received the Society's prestigious Gold Medal for his work in 1846, but because Sadler's paper had slipped through the referee net. An examination of the evidence, however, showed that Sadler had been highly partial, to say the least, for 30 of the 34 double stars to which he drew attention had been published with cautions of their accuracy by Smyth himself, though Sadler had failed to mention Smyth's *caveats* [40]. Then in 1879–80 Sherburne Wesley Burnham, with access to the 18½-inch refractor at the Dearborn Observatory, Chicago, re-measured Smyth's pairs, while Edward Ball Knobel examined Smyth's original records. Burnham found that for some classes of star Smyth seemed to have *estimated*, with his practised eye, the position (or east from due north) angles of many stars, rather than measuring them; while Knobel found errors in Smyth's reductions, and hitherto unnoticed numerical printing errors in the text of the *Bedford Catalogue* [41]. Smyth's honesty was entirely vindicated, although there was a great deal of re-checking to be done before the 1881 edition of the *Cycle* and its *Catalogue* came out. But one of the things which this incident brings home to us is the enormous increase in acceptable accuracy standards that had taken place between 1840 and 1880, and the inadmissibility of the most practised of estimates in a major catalogue by the end of the nineteenth century.

On 11 May 1865, so the *Astronomical Register* reported, the Lunar Committee of the British Association for the Advancement of Science met at John Lee's legal chambers in Doctors' Commons, London, to examine the previous six years' progress in the Association's attempt to map the Moon's surface to a higher level of detail and accuracy than that achieved by the privately-funded German project of Beer and Mädler thirty years previously [42]. Although this new selenographical initiative had come, as Roger Hutchins has shown, from the Oxford geologist and Grand Amateur astronomer, Professor John Phillips, in 1852, Lee was variously involved as a patron of astronomy and as a recent (1861–1863) President of the R.A.S. [43].

As with Warren De La Rue, James Nasmyth and several others in Britain and abroad, John Phillips' primary astronomical interests lay not so much in positional astronomy as in the physical structure of the Moon and planets. (He was, amongst other things, the first scientist to construct a globe of the planet Mars.) Phillips proposed the setting up of a team of observers who, working with 6-inch aperture refractors, would produce a collated lunar chart 200 inches in diameter [44]. John Lee was additionally involved in the project, moreover, because of his possession of Smyth's former 5.9-inch refractor, which was now in the Hartwell Observatory, and in 1863 gave free access to this instrument to William Radcliff Birt who, since 1847, had been part of Lee's scientific circle.

Birt's presence in the Hartwell circle, like that of Thomas Dell, further indicates that Lee's scientific friendships were not restricted to Grand Amateurs and eminent professionals, for Birt had made his own way in the world of science, as a protégé of Sir John Herschel, a meteorologist, lecturer, writer, and recipient of small B.A.A.S. research grants. Unfortunately, the B.A.A.S. lunar map was never completed, and following Birt's death in 1882 the project was abandoned by the Association [45].

In some respects, however, the failure of the B.A.A.S. selenographical project highlights one of the fundamental weaknesses of that style of Grand Amateur astronomy characterised by John Lee and by many of his friends. For while bodies like the R.A.S. and the Hartwell connection served an excellent function in drawing independent people together for the discussion and exchange of ideas in their several branches of research, the very nature of the informal social relationship presupposed by such gatherings made it very difficult to collate effectively an extensive, detailed and sustained piece of teamwork such as was required for the production of the B.A.A.S. lunar map. In many ways, one could argue that the selenographical project only brought into focus one of the major tensions implicit within the Victorian scientific community, as the B.A.A.S. sought to direct scientific research along planned lines, whereas the independent amateurs were constitutionally inclined to doing their own thing. Excellent lunar cartographical work has been completed by private individuals doing whole-Moon maps, and by organised bodies of institutionally-based professionals working in harmony, but never really by voluntary teams of amateurs working together [46]; and one is reminded of Sir John Herschel's *caveat* (see p.57) that independent, self-funded individuals do not like working within areas 'chalked out' for them by a central organising committee.

An early member of the Hartwell astronomical connection, who nonetheless did relinquish his amateur status to become a distinguished professional, was Thomas Maclear. He was a medical man, trained at Guy's and St. Thomas's hospitals, in London, who worked as a house surgeon in the Bedford Infirmary, and practised at Biggleswade and thereabouts. In 1828 he joined the Astronomical Society, and one suspects that it was his Bedford neighbour, Smyth, who directed his early interests. As a Bedford doctor, Maclear was comfortably off, but not especially rich, in the way that Lee, South and Pearson were. This relative modesty of means is made clear in the description of his Biggleswade observatory to the R.A.S. in 1833, where the opening sentence states: 'I have been obliged to suit my astronomical instruments to economy and convenience, rather than to my wishes' [47].

Yet this modest establishment, which measured only sixteen feet from transit room to dome chamber, was by no means cheap. 'The total expense of this little room was fifty pounds', says Maclear, whose tone and context implies that he was speaking only of the new 8-foot wooden dome, which rotated on eight wooden rollers, and not the transit room to which it was attached [48]. In 1833 Maclear's 'modest' observatory extension had cost a sum which many a curate, country schoolmaster or city clerk would have regarded as almost a year's salary. When one adds to this sum the cost of building the transit room with its moving roof shutters – perhaps another £50 – and equipping the small buildings with a regulator clock, a transit instrument, and a small refractor under the dome, one begins to get some idea of the cost of serious astronomy, even to a man who was constrained by reasons of 'economy' [49].

Soon after bringing his Biggleswade observatory into operation, Maclear gave up medicine to receive the King's Warrant and become Director of the Cape Observatory, with the title of Astronomer Royal at the Cape: a title originally given fourteen years earlier to the Revd Fearon Fallows, who had died of fever in 1831. It says something about the standing of Maclear as an astronomer that he could move from being an amateur to taking independent charge of the largest government establishment in the southern hemisphere. In 1860 Queen Victoria rewarded him with a knighthood [50].

Rather than continuing him at the Bedford School, Admiral Smyth now sent his son Piazzi, who was 16, to join his friend Maclear at the Cape in 1835. Here, Piazzi Smyth learned the astronomer's trade – while enjoying an extraordinarily generous official salary of £250 per annum – and returned eleven years later to take up the post of Edinburgh University's professorial Astronomer Royal for Scotland [51]. Long absences, no doubt, came as second nature to naval and diplomatic families, although it is clear that a close correspondence was maintained not only by Smyth, but also by Lee and others, with South Africa. Indeed, even before Maclear could scarce have got under way, in August 1833, Lee was asking him to send various snakes and scorpions preserved in spirits, along with a 'Hottentot's Petticoat' for the Hartwell House museum [52], while Mrs Richarda Airy corresponded for many years with Mrs Maclear [53].

It was also at this time, in the autumn of 1833, that Sir John Herschel and his family sailed for what would be a five-year stay at the Cape, and it is possible that Richarda Airy's correspondence with Mrs Maclear was consolidated through her prior friendship with Lady Herschel. The Cape letters of Margaret Herschel to the Greenwich Astronomer Royal's wife certainly discuss their social and professional relations with the Maclears. Indeed, in the social world of the Grand Amateurs, many of the wives and families of the astronomers formed enduring friendships with each other, as the Hartwell gatherings, visits to the Herschel mansion in Kent, and convivial soirées at British Association meetings fostered familiar relations [54]. This was, after all, the same bracket of English society as that about which Jane Austen and Anthony Trollope wrote, with the added ingredient of astronomy.

For a century before Samuel Charles Whitbread became part of the Hartwell connection of astronomers, the family had been prominent in Bedfordshire. Samuel's grandfather, also named Samuel, had established a successful brewing business which still trades today, while wise landed investments and good marriages had made the Whitbreads a major force in the county. The astronomer's father had been a prominent MP in the Foxite reforming party in the 1790s, while the astronomer's mother was a daughter of the Prime Minister, Lord Grey. With service as an MP in his own right, and the High Sheriffdom of Bedfordshire behind him, Samuel Whitbread seems to have begun the serious cultivation of science in the 1830s when he was about 40 [55]. He was, therefore, a social equal of Dr Lee, but as the latter had no children or brothers with whom to divide his property, Lee may have possessed the edge in terms of independence, if not of lineage.

At some time in the late 1840s, Samuel Whitbread acquired an excellent refracting telescope originally built by Troughton and Simms for the Revd Samuel King of Chesham. It had a clear aperture of $4\frac{1}{8}$ inches and a focal length of 5 feet, and was set upon a mount that 'resembles those made by Fraunhofer'. German mounts were not common in England at this time, and it is highly likely that Whitbread's refractor was very similar to the German-mounted equatorial described by William Simms in his pamphlet of 1852 [56].

Whitbread had the instrument at his seat at Cardington, near Bedford, though it might have been the same telescope as the one with which he had mentioned observing details within lunar craters at his house in Eaton Place, London [57]. In addition to the equatorial refractor, he possessed a 12-inch transit circle, also by Troughton and Simms, which he first set up in an observatory in the kitchen garden at Cardington; however, on 31 October 1851, he recorded in his observing Diary that he had 'completely finished my new Observatory and Transit Room in every detail. I removed every article from the Observatory in the Kitchen Garden ...' [58]

The observing Diary which Samuel Whitbread kept from 11 August 1850 to 16 April 1852 brings his fascination with astronomy to life. In John Herschel's sense of the word 'amateur', here was a *lover* of astronomy, though not one who was pursuing any specific line of research beyond observing the planets, and doing some double star measurements. Whitbread was now a well-to-do gentleman in his mid-fifties, who enjoyed taking transits, observing the double stars in Smyth's *Bedford Catalogue*, and examining the Moon. Astronomy to Whitbread, moreover, was a social activity, and the Diary contains numerous references to the proud owner of the new Cardington Observatory showing off the heavens to his friends, some of whom he brought up from London, while he still kept a lesser instrument at his Eaton Place house, to use when in town. This social life naturally connected him to the scientific dining circuit. On 9 September 1851, for instance, he dined in London with five astronomical friends, including three clerical associates of John Lee, and James Glaisher from the Greenwich Observatory, 'and at 10 o'clock went to Mr De La Rue's at Canonbury. It was a cloudless night, but damp with an easterly wind, and the atmosphere by no means good' [59]. It forms an evocative picture, as the six well-fed and well-watered friends took a carriage to the London suburb of Canonbury in the hope of a good night's observing with De La Rue's reflector [60]. This same instrument, moreover, through which Whitbread had positively drooled at the Hoop Nebula in Lyra three weeks before, had a famous 13-inch diameter speculum made for De La Rue by the Manchester ironmaster-astronomer James Nasmyth, whose activities will be discussed in Chapter 6 [61].

Yet Samuel Whitbread realised, like John Lee and most of the other Grand Amateurs, that, in the words of John Weale, 'an observatory without establishment must become at length either an incumbrance or a plaything to its owner' [62]. By an 'establishment', of course, he meant financial provision for a paid assistant to undertake regular and useful observations with the expensive instruments, and Mr John McLarin was 'engaged for conducting the observations' with the prospect of 'good and useful work' [63] being achieved at Cardington.

In addition to astronomy, Whitbread was interested in meteorology. Many amateur and professional scientists had made meteorological records ever since adequate instruments had been invented in the seventeenth century, though few of these records had been properly calibrated or analysed to extract the law-like criteria which it was generally assumed must lie at the heart of weather patterns. A weather register was maintained by Lee at Hartwell from 1829, and after 1838 a Meteorological Department had been founded at Greenwich to create a national record of readings made both at the Royal Observatory and elsewhere in the British Isles [64].

A British Meteorological Society had been formed at Hartwell on 3 April 1850, and Whitbread was elected Chairman [65]. The purpose of this Society, according to Smyth in the 1860 addition to the *Cycle*, was to record and collate meteorological observations made by the Society members, and by other meteorologists with whom they were in correspondence [66]. Samuel Horton was engaged as Dr Lee's meteorological assistant at Hartwell [67]. In addition to helping to build up a body of data from which general laws might be extracted, it was hoped that, in this pre-bacterial age, a relationship might be found to exist between weather patterns and cholera epidemics, and for this reason Horton was required to supply material to the Registrar General of Births, Deaths and Marriages. And in addition to any useful medical connection, meteorology related to astronomy in its concern with the passage of light through air, water, and glass, and in the quantification of the laws of atmospheric refraction, without which one could not do critical astrometric work.

And as with astronomical research, the foundations of scientific meteorology were laid by amateurs. A network of private meteorological stations was set up across Britain, and equipped with instruments carrying uniform calibrations, so that country gentlemen, clergy, municipal officials, and others, could keep regular records. Using the rapid and reliable communications made possible by the railways and the Penny Post, local weather observers posted off their weekly readings to Greenwich on Sundays, wherefrom their collated results were published in the newspapers [68]. Although this volunteer meteorological data collection system was already up and running by the time that the Meteorological Society was formed at Hartwell House in 1850, it depended on the Greenwich Assistant who superintended the Meteorological Department, James Glaisher, F.R.S., for the collation, interpretation, and publication of its data. But it was from the Hartwell society that the Royal Meteorological Society was born, of which Glaisher later became the driving force [69].

I would suggest that why this exercise in volunteer teamwork succeeded so well over so many years, whereas the selenographical project failed, was because of the much greater simplicity of meteorological monitoring. Once a local scientist was set up with properly calibrated instruments, their daily reading became a simple routine task. Selenography, on the other hand, not only required much more complex and costly instruments, which the observer had to provide for himself, but the zonal mapping of the lunar surface, under a constantly changing illumination, involved a high degree of personal interpretation, especially if one did not use photography.

The Hartwell House gatherings brought together a wide range of people with astronomical, meteorological, and other scientific interests. Most were amateurs in the sense defined by Pogson in his letter to Smyth as those 'who devote their time as a labour of scientific love' as distinct from 'those who are paid for their work, and only successfully fulfil their duty' [70]. Yet in the person of James Glaisher one had a salaried professional who saw science both as a livelihood and as a personal passion. In 1859 Glaisher received £290 per annum as the Superintendent of the Meteorological Department at the Royal Observatory. Yet not only did he dine with Whitbread and visit De La Rue as a personal friend, but he and his family were regular attenders at Hartwell, especially in the 1850s. Glaisher was present at the Meteorological Society's foundation meeting, in April 1850, and appears thereafter as a fairly regular visitor in the Albums which recorded the comings and goings of visitors.

In 1854, moreover, the Glaisher family seems to have enjoyed Hartwell for much of the summer, and Cecilia Glaisher signed the Album on 5 August to inform Dr Lee that 'After a prolonged visit of some weeks, we are this day about to take our departure of Hartwell ... That he [Dr Lee] may be spared to continue for years to come, as he has long been the centre of a large and intellectual circle' [71]. It is likely on this occasion that Cecilia and her family were guests in their own right for at least some of the weeks, for James would have had duties at Greenwich.

So much at home were the Glaishers at Hartwell, and so attached to their friends there, that when a son was born in 1848, he was christened James Whitbread Lee Glaisher. This child grew up a prodigy, and at the age of 8 wrote a Latin letter to Dr Lee from the Glaisher home in Lewisham on 29 September 1856, to express his delight with the place, signing himself 'Jacobus W. Lee Glaisher' [72].

The presence of the obviously non-Grand professional Glaisher and his household in the familiar conviviality of Hartwell House, Cardington, and Eaton Place, also tells us a great deal about Airy's relations with his own assistant staff at Greenwich. The received opinion of Airy as an Astronomer Royal who ruled his observatory with a rod of iron and kept his 'drudges' in their places simply will not fit this evidence. Glaisher, who was elected F.R.S. in 1849, clearly moved on easy terms with the most eminent figures of the scientific world, and made more recorded visits to Hartwell than did his superior. One suspects, indeed, that Airy himself was the one who might have felt self-conscious in such gatherings – the family fireside, not the country-house gathering, was Airy's favourite place of relaxation – for on 22 December 1854, Richarda Airy signed herself into Hartwell 'for the first time alone on the occasion of the marriage of Miss Ellen Philadelphia Smyth', one of the Admiral's daughters, Airy himself being detained on 'public business' [73]. According to his diary for this period, the Astronomer Royal had been spending several busy days in Cambridge. He was also a migraine sufferer, and as Richarda's private letters elsewhere make clear, a hectic social gathering could sometimes give him a headache that lasted for days [74].

A succession of people passed through Hartwell over the years. William Rutter and Anne Dawes (the former Mrs Welsby) were there in July 1849 [75], Sir David and Miss Brewster, down from Edinburgh, 'honored Hartwell' in October 1851 [76], while William and Maria Lassell were there in August 1853 [77]. Lee had also loaned his 24-inch Cary transit circle to Lassell at one stage 'as an aid to his mighty equatoreal reflector' [78]. The German astronomers, Albert Marth and George Rümker, both of whom came to work in England, visited in 1854, and Rümker and his wife were there again *en route* for Germany from Durham in July 1856 [79].

The coming of the railway to Aylesbury about 1840 greatly facilitated travel up from town. In September 1848, Caroline Mary Smyth, another of the Admiral's accomplished daughters, mentions coming up 'by the newly-arranged and expeditious train', while her brother Piazzi, who had been at the Cape with Maclear during the decade of the railway mania, was impressed by the tunnels and great cuttings through which he passed on the way up from Euston [80]. And when Mr J.F. Cole travelled up in October 1852, he told Lee that he had left London on the 2.45, and arrived at 6.00 p.m. His train must have been severely delayed, for first-class trains were doing over 30 miles per hour by that date, and Aylesbury was only 40 miles from London [81].

And when the guests were at Hartwell, a wide range of astronomical delights awaited them. When Mr Cole, mentioned above, was there in 1852, he was especially impressed with his views of Saturn, and saw the faint companion of the star α Lyrae [82]. Ladies, it appeared, had an equal rank with gentlemen in the observatory, and Anne Camps, who substantiated Mr Cole's observations of Saturn on 12 October 1852, also observed and timed a transit of the Andromeda Nebula across the meridian, 'which to me appeared like a blue spider rapidly crawling across the lines' in the telescope eyepiece [83].

A few years earlier, before dawn on a summer's morning in July 1845, John Lee, his wife Cecelia, and four friends were observing Venus rising in the north-east, near Perseus, when they believed that they saw the planet encircled by a tenuous ring, through the 5.9-inch refractor. They wondered whether a special train should be sent to London to inform Airy and the President of the R.A.S. of this singular appearance! [84]

Novelty, and the demonstration of ingenious physical phenomena, were always admired at Hartwell, and in August 1846, Professor Baden Powell, father of the First Scout and husband of Smyth's daughter Henrietta, was visiting from Oxford, and used one of Lee's prisms and the multiple reflections made possible by the drawing-room mirrors to produce a solar spectrum in which the Fraunhofer lines D, E and F were visible [85].

It is clear that John Lee regarded his observatory as a place where friends were shown the heavens, took transits, admired the planets, and enjoyed the ingenuity of the instruments, as well as a place where serious work was done. And when the weather was bad, games of a thoughtful cast were played, such as on 30 January 1857, when Lee and his friends picked six books which, at the expense of all others, they could not do without: a sort of literary Desert Island Discs! [86] And when guests were late for breakfast at 9.00 a.m., or dinner at 6.00 p.m., a set of charitable fines were imposed, with sixpence apiece to the local school, the Church poor-box, and the Infirmary [87]. A strong incentive to be in time for meals in 1829, though the Infirmary fine was added in 1846.

The above fines, like many other of the rituals of Hartwell, were encapsulated in hymns and poems, for verses, anagrams, puns, and sketches were all part of the entertainments of the House. Eight verses were written in July 1829 about the 'Quarrell between the Great Bear and Casseopiae' [88], and soon after the visiting instrument-maker Thomas Jones had set the levels for the new transit instrument in August 1831, someone drew a sketch in the Album of a skeleton looking through a telescope, accompanied by the inscription 'Sic Transit Glorious Monday' [89].

Some delightful sketches and paintings were contributed by the returned and grown-up Charles Piazzi Smyth, including a beautiful watercolour self-portrait in 1848 [90]. Two years later, when visiting from Edinburgh, Piazzi went on to paint another picture: a detailed cartoon of his father, Dr Lee, and Sir John Herschel shaking hands after a night's observing, entitled 'The Three-Fold Adieu, or the Tri-Partite Good Morning' [91]. The three friends stand at the top of the grand staircase at Hartwell, wearing their embroidered observing caps, and carrying dark lanterns. The humour which underlies these and other pictures and verses in the Albums gives one a vivid impression of the gaiety and good fellowship between the astronomers, their wives, and families, that were also part and parcel of this 'large and intellectual circle' at Hartwell House.

As the responsible owner of a capital observatory, Dr Lee, like Whitbread, recognised that a full-time assistant was needed to do the day-to-day observations. In January 1838, James Epps was appointed and began a course of meridian observations which occupied 29 quarto pages when they were published in 1851 [92]. Epps had formerly been a private teacher of mathematics in the Commercial Road, London, and his surviving trade-card proclaims that not only did he teach mathematics 'to Youth, but to Gentlemen upon their entering the University', but that he also possessed a large library and collection of teaching instruments [93]. He had also served as the salaried Assistant Secretary of the R.A.S. for eight years. It is probable, however, that the elderly Epps, after a lifetime of freelance teaching, welcomed a regular salary, a tied residence and a relatively protected existence at Hartwell, but, alas, he did not live long to enjoy it [94]. He died suddenly at the age of 63 in August 1839, though John Lee gave a 'liberal pension' to his widow. It would appear that Samuel Horton, the

Hartwell meteorological assistant, also did astronomical work, for in January 1858 he observed a bright glowing spot within the dark body of the crescent Moon, and published his findings accompanied by a drawing [95].

It is clear that Admiral Smyth himself did a great deal of observing at Hartwell – in an unpaid capacity, of course – and using his own former 5.9-inch equatorial, though in 1858 the 70-year-old sailor 'bade adieu to my voluntary labours in practical astronomy' [96]. He recommended the appointment of Norman Pogson from the Radcliffe Observatory, Oxford, to carry on the work, where one of Pogson's duties was to observe Smyth's binary 'γ Virginis [which] has been observed so long with the Hartwell Equatoreal and wire-micrometer ...' [97]. Smyth included a not exactly flattering portrait of a rather droopy-looking figure, whom one presumes was Pogson, posing with one of Smyth's wooden equatorials, outside the Hartwell Observatory [98]. Pogson stayed at Hartwell only until 1860, and his and other professional assistants' careers will be discussed in Chapter 8.

In addition to Smyth, Whitbread, Glaisher, and the more prominent attenders at the Hartwell astronomical house-parties, there was a cluster of less famous local astronomers who, judging from their recorded comings and goings, helped to make up John Lee's body of scientific friends and correspondents. Indeed, John Weale, in his observatory survey of 1851, drew attention to the number of astronomical observatories that existed in the area around Aylesbury. Among them were the above-mentioned ones of Thomas Dell of Aylesbury, and the Revd Joseph B. Reade, Vicar of Stone, near Aylesbury. John Reade was clearly a well-beneficed clergyman, whose observatory was an 'elegant Grecian building' [99]. It contained a 4½-inch aperture transit instrument and a 7½-inch refractor of 12 feet focal length, intended for an equatorial mount upon a tower: an establishment, indeed, that was more powerful than that at Hartwell House. Thomas Dell was much more modest in his circumstances, characterised as a 'true astronomer' who had managed to acquire a good transit instrument, a clock, and a portable refractor by Tulley, although he could only afford to mount them in a simple wooden building with a sliding roof covered with stretched canvas [100]. In March 1843 Dell, whose little observatory also provided Aylesbury with its time service, was telling Lee how he and his friend Mr Blake were observing a new comet which had appeared in Orion [101]. During the summer of 1844 Dell and Reade were observing another comet, Melhaps (?), which was brighter than Mauvais', and which they followed through the skies with a small 'comet-seeker' telescope, backed up with an equatorial [102]. In April 1852 Thomas Dell was further informing Lee of a 'column of light, of a fine Rose colour' [103] seen in the north-west after sunset, and suggesting that it was the tail of a comet. From the measurements which they supply, it is clear that both Dell and Reade, in spite of disparity of wealth, possessed well-equipped observatories of their own, although Dell's 'true zeal for science [was shown] under difficulties of position and circumstances' [104]. Another observatory owner in Aylesbury was Joseph Turnbull, whose name, like that of Reade, Dell and others, runs through the Hartwell Albums. In January 1858, amongst other things, he sent Dr Lee a detailed account of the solar surface, and a watercolour painting thereof [105].

William Smyth's and John Lee's reputations as astronomical correspondents and clearing-houses for information are also indicated by the people in other parts of Britain who wrote to them. Captain Charles Shea (or Shay) wrote to Smyth in August 1860 to report his sunspot observations made in London [106]. He was wondering whether the sunspots occupied the same positions on the solar disk as the solar flares seen by Airy and the expedition which had gone to observe the total eclipse of 18 July 1860 in Spain. This was (as we saw in Chapter 3) a subject upon which Richard C. Carrington was also working. Captain Shea, moreover, had an astronomical friend – a Mr Charles Howell of Hove, with whom he had observed a lunar eclipse some time before [107]. In 1859 Mr Dillwyn Thomas was sending his drawings of nebulae to Smyth [108], while Captain Jacob and Mr Isaac Fletcher, at Clifton and Crossbarrow Collieries near Workington in Westmoreland, were forwarding sketches of Saturn made with an equatorial refractor of 9½ inches aperture built by Thomas Cooke of York [109].

Even Lee's meteorological assistant, Samuel Horton, received correspondence addressed to him personally, as when one Joseph Conter (the spelling of whose name is uncertain in the manuscript) wrote to him suggesting that 'the Comet' (Donati's) was perhaps responsible for the extremely dry summer of 1859 [110]. Yet part of Horton's job, especially if he was supplying data to the Registrar General, was to collect and organise meteorological observations from a given area, in accordance with the system managed by Glaisher [111].

The surviving Albums and Scrapbooks of Hartwell House, preserved in the Museum of the History of Science, Oxford, reveal a whole world of Victorian amateur astronomy. Not only does one find Grand Amateurs doing front-rank science, but also the 'enthusiasts' who loved observing, and wanted someone to whom to report their work. The correspondence that came to Smyth, probably as a result of the widespread circulation of his *Cycle of Celestial Objects*, is truly remarkable, and gives one a glimpse of how many small private observatories with proud and comfortably-off owners there were in England by 1860, for astronomy was now 'fashionable' [112]. Lee and Smyth were also careful preservers of handbills and newspaper cuttings pertaining to astronomy. These include a wide miscellany of documents, encompassing the itemised sales of observatory contents in auctions, attempts to establish public subscription observatories in various parts of Britain – such as those in Nottingham and Liverpool – popular lectures, and efforts, usually unsuccessful, to found amateur astronomical societies, like the proposed 'Uranian Society' of 1839 [113]. Donati's Comet of 1858 produced its own crop of preserved cuttings, including those from *Punch* and the *Illustrated London News* [114]. Many of these cuttings, moreover, came from parts of Britain well removed from the London–Bedford–Aylesbury axis, and one assumes that Smyth and Lee had friends around England who sent curious items from the local newspapers on to them.

Of the leading figures in the Lee connection, such as Lee himself, Smyth, Whitbread, Reade, Dell and the local Aylesbury and Bedford amateurs, whose names and activities run through the Hartwell Albums, only Smyth worked consistently over many years at the business of astronomical research. But the significance of Lee and his friends is not to be reckoned simply in terms of original discovery, but rather in terms of patronage, encouragement and correspondence. To Lee, astronomy was just one subject within the domain of good learning which a humane and polished gentleman should cultivate – along with his passions for Egyptology, archaeology, collecting, numismatics, and trying to make the world a better place. But what the Hartwell connection and its documented remains tell us most of all is the degree to which an interest in astronomy permeated the educated classes of early Victorian England, and how it could be enjoyed with friends in the setting of a great country house.

6

The brotherhood of the big reflecting telescope

One of the leading contributions which the Herschels – both father and son – had made to astronomy lay in demonstrating the research possibilities of large aperture reflecting telescopes. And like parallel innovations in the improved graduation of instrument scales, micrometers, achromatic object glasses, stable equatorial mounts, and clock drives, the big reflecting telescope had appeared in response to an intellectual problem that could only be addressed after an improvement in technology had taken place. While it is true that Sir John Herschel was firmly within that group of men which might be called the brotherhood of the big reflecting telescope, it was nonetheless to his father, Sir William, that the Victorian observers turned as its founder. For while Sir John had been a skilled constructor and user of large reflectors, he had done nothing to take the technology beyond where his father left off, and did his own great sky-sweeps in both hemispheres with an 18-inch aperture instrument of 20 feet focal length which was little different from the one which his father had completed by 1783 [1].

The intellectual goal which the big reflecting telescope was ideally designed to address was the fathoming of the shape and structure of the stellar and nebular universe. For though a speculum metal mirror, even when freshly polished, could rarely be more than about 70 per cent reflective, it was still possible to make bigger optical surfaces, and hence to catch more light, in reflecting metal than it was in achromatic glasses. To make further advances in cosmology, therefore, it was obvious that the researcher was going to have to effect significant improvements upon the image quality and handling properties of William Herschel's 48-inch aperture instrument of 40 feet focal length of 1789 [2].

Yet cosmology was not the only area in which the big reflector had already proven its worth. It had been with an excellent 7-foot focus instrument of just over 6 inches aperture that William Herschel had discovered Uranus in 1781, and with his 20-foot instrument that he had discovered its satellites Oberon and Titania in 1787 [3]. The mirror telescope had enormous potential as a powerful planetary instrument, where its lack of chromatic aberration made it ideal for the detection of colour changes on the surfaces of Mars, Jupiter, and Saturn, not to mention its ability as an outer Solar System satellite detector.

Relatively little commercially-inspired innovation went into producing the big-aperture reflecting telescope before the 1860s. For once James Short and a handful of mid-eighteenth-century craftsmen working in London had taken up the designs of David Gregory and Sir Isaac Newton, and turned them into replicable commercial artefacts incorporating 2- or 3-inch diameter mirrors with which to view one's estate or enjoy views of the lunar craters, the optical development virtually stopped. And while several English and French craftsmen had experimented with equatorial mounts, the vast majority of commercially-manufactured reflecting telescopes were intended for use upon altazimuth mounts [4].

The development of the speculum mirror reflecting telescope as an instrument of fundamental research in astronomy lay entirely in the hands of private individuals who pioneered the necessary optical and mechanical technologies as adjuncts to their own astronomical interests. It is true that Sir William Herschel made a considerable amount of money out of the sale of telescopes, and James Nasmyth might have sold one or two 13-inch mirrors to friends (such as the one that went to Warren De La Rue), but all of these sales were secondary to the maker's normal line of business. They were virtually all the by-products of Grand Amateurs to whom the commercial sale of telescopes was not a primary concern [5].

The big reflecting telescope, indeed, was very much an amateur's instrument. Because such instruments were the products of specialist interests, for users whose lines of astronomical enquiry put light-grasp above all things, large reflectors lacked the wide usage or sales potential that was essential to attract most professional instrument makers. In consequence, almost every major reflecting telescope maker between 1775 and 1865 was also the same telescope's user. As late as 1877, that pioneer and son of a pioneer of commercial large-aperture reflecting telescope manufacture, Sir Howard Grubb, made the remark that 'Reflectors very seldom do good work except in the hands of their makers' [6].

One of the outstanding exceptions to this rule had been the German amateur Johann Hieronymus Schröter between 1790 and 1812, whose observatory at Lilienthal, near Bremen, contained large reflectors by Johann Schrader and William Herschel, including one of 20 inches aperture by Schrader with which important lunar work was performed. But Schröter was the exception that proved the rule [7].

Building and maintaining large reflecting telescopes was as costly in money and in manpower as maintaining racehorses. For these instruments, in spite of their physical awkwardness, with wooden tubes like factory chimneys set in a forest of rigging, were the fleet-footed beasts of Georgian and early Victorian astronomy, that could run farther into the depths of space than any refractor of the age. Not only were they costly to construct, but their mirrors needed regular repolishing, the woodwork and cordage required constant maintenance, and their precipitously-placed observing galleries were fraught with danger to their users. They were far too much trouble for active professional observatories, where it was important that the same instrument could be used efficiently by different shifts of observers, without the need for endless preliminaries, adjustments, or broken necks [8]. Like fox-hunters or Stradivarius 'cellos, big reflecting telescopes were owner–user creatures.

The construction of these instruments, moreover, was an arcane skill until William Lassell published a detailed account of the metallurgical casting, figuring, and mounting of his great equatorial instrument in 1849 [9]. No books were available to give an aspiring constructor any consistent guidance, apart from the Revd John Edwards' account of how to cast a small speculum, which was issued as a Supplement to the 1787 *Nautical Almanac* and never reprinted [10]. Unless, like William Lassell, one was willing to copy out this 30-odd-year-old article in longhand from a borrowed copy, one did not even know how to start [11]. William Herschel, we must not forget, never published an account of his mirror-making techniques, though he did publish detailed descriptions of the wood and cordage structures in which the mirrors were mounted [12]. It must have been extremely frustrating, therefore, for an astronomer who had read William Herschel's numerous papers describing the multifarious nebulae and star clusters visible in deep space through his telescope, yet who had no guidance as to how to acquire or build such an instrument for himself. As Sir William had tantalisingly remarked, the nebulous objects which he swept out of the sky and classified could not be seen through 'common telescopes' [13]. The brotherhood of the big reflecting telescope, therefore, was a brotherhood in which intellectual curiosity and experimental doggedness aspired to overcome all odds. And such a level of doggedness, sustained over decades in many cases, demanded the means and the leisure of a Grand Amateur [14].

By 1820, however, several people had succeeded in casting and figuring large mirrors. James Short himself, who had died in 1768, even before William Herschel had begun his own experiments, had supplied an 18-inch aperture Gregorian to the Duke of Marlborough's private observatory at Blenheim Palace, Oxfordshire, an instrument which still survives in the Museum of the History of Science, Oxford [15]; while the early-nineteenth-century telescope enthusiast and collector, William Kitchener, was told by the London optician Charles Tulley that he had completed a good 15-inch Cassegrain in 1802 [16]. James Veitch of Inchbonny, in Scotland, was making modest aperture reflectors for himself and for select customers (Veitch will be discussed in more detail in Chapter 10), and around 1817 the Aberdeen tradesman, John Ramage, had built a 15-inch mirror of 20 feet focus, which had been mounted after the altazimuth fashion of Herschel, and off which the visiting Sir David Brewster fell [17].

But none of these instruments (with the exception of Herschel's) seem to have been designed with any particular research goals in view. Nor do they seem to have made much contribution to the technical development of the large reflecting telescope, insofar as their optical performances were inferior to the Herschel instruments. When the Astronomical Society tested Ramage's instrument at Greenwich in 1825, for instance, it was only found to be satisfactory when using its lower-power eyepiece [18].

Three men, born in 1800, 1799 and 1808 respectively, were the first astronomers not to bear the name of Herschel to take up the cosmological and big reflector gauntlet where Sir William had thrown it down. They were an Irish Earl, a Liverpool brewer, and a Manchester iron-master.

WILLIAM PARSONS, THIRD EARL OF ROSSE

William Parsons, or Lord Oxmantown as he was called prior to inheriting his full title and becoming the Third Earl of Rosse in 1841, had admired the work of Sir William Herschel since at least his undergraduate days at Trinity College, Dublin, and at Magdalen College, Oxford. He became a good mathematician at Oxford, took a first-class honours degree, and no doubt benefited from contact with Dr Charles Daubeny, F.R.S., Magdalen's distinguished science don [19]. Returning home to Birr Castle in central Ireland, he took over some disused buildings on the estate, and began to experiment with making telescopes. Abandoning attempts to make fluid lenses, he turned to reflectors, but 'The task was evidently a very difficult one, as the late Sir W. Herschel had apparently almost exhausted the subject', or so he wrote in 1840 [20].

Lord Rosse was an academically trained mathematician with the practical instincts of an engineer, and his goal was to solve the problems which Sir William Herschel could not. Were the nebulae composed of luminous 'chevalures' of glowing, perhaps gaseous, material, or were they conglomerations of individual stars, like the Hercules cluster? And why did the star clusters and nebulae tend to occur most frequently in different regions of the sky? As Herschel saw it in 1785, and Rosse still saw it fifty years later, these questions could only be answered by increasing the light grasp and resolving power of telescopes. The prevailing theory in 1835 favoured the idea that the nebulae were 'particulate', and made up of vast numbers of individually radiating stars or other particles, for each increase in telescopic power certainly brought more stars into view; however, the proof must lie with yet larger-aperture telescopes [21].

Between Herschel's greatest achievement – the disappointing 40-foot instrument of 1789 – and his own day, moreover, Lord Rosse realised that a major revolution had taken place in metallurgical knowledge and in precision engineering, so that it seemed likely that a bigger and more powerful telescope than Herschel's could be built. Yet the Birr estates were of only moderate profitability, and one cannot underestimate the significance of Lord Oxmantown's marriage to Mary Wilmer Field in 1836 to his development of the big reflecting telescope. This Yorkshire heiress (indeed, Lord Rosse himself had been born in York) brought valuable properties in the Bradford area, at Heaton and Shipley, into the Rosse family, and one suspects that without the revenue which they produced, the 72-inch might never have been built. All of the evidence suggests, furthermore, that the Countess was firmly behind her husband's scientific work. She became a distinguished photographer and user of scientific instruments in her own right, and one of the many learned wives within the Victorian astronomical community [22].

Speculum metal, however, is an awkward optical medium. It is hard to cast properly, cracks easily when cooling, is brittle to work, and extremely heavy to mount in a telescope. Some of Rosse's early experiments at Birr, where he came to build up a highly skilled workforce under the direction of a local blacksmith named William Coghlan [23], were directed at making mirrors lighter in weight and easier to handle. Indeed, a journalist writing for the *King's County Chronicle* in 1857 emphasised the Hibernian patriotism implicit in the entire Birr telescope-making enterprise, where 'it was not only Irish genius which directed, but Irish diligence and skill which executed the task'. Indeed, 'Common Irish labourers working under his Lordship's eye, were found quite adequate to accomplish all, where the nicest precision of mathematical exactness was required at every point' [24].

In his paper of 1840, Lord Rosse described how he and his workforce made a 15-inch diameter mirror by soldering eight segmented wedges of speculum metal onto a light, ribbed brass base. The resulting mirror took a good figure, and gave excellent views of the Moon at 600× magnification. His lightweight segmented mirrors were increased from a 15-inch to a 24-inch and then on to a 36-inch diameter, all of which performed well, though large segmented mirrors were always susceptible to thermal disequilibrium between the segments. By 1840, however, he succeeded in casting a solid mirror of 36 inches diameter, which made Lord Rosse's reflector the largest and most powerful telescope then in use in the world [25].

Not only was Lord Rosse able to get his alloys to a relatively high degree of purity by 1840, but his use of a turf-fired annealing oven made it possible to control the heat loss, and hence reduce the likelihood of a freshly-cast speculum cracking. It must also have been an enormous asset to Lord Rosse not only to have the resources of space and a picked workforce from the estate at his disposal, but to have a limitless source of heat from the turf of the adjacent Bog of Allen. The melting and casting of specula could therefore be very largely achieved with home-ground resources.

For the next few years, the 36-inch, mounted on the broad lawn in front of Birr Castle, would be referred to in correspondence as Lord Rosse's 'Great Telescope'. News of its size and potential had reverberated around Europe, as when, on 11 November 1843, Mary Somerville, who was then living abroad with her elderly husband, wrote from Rome asking if his Lordship had yet resolved the nebulae, or made advances in understanding the geological formation of the lunar surface [26]. Rosse wrote back to this eminent mathematician telling her that while he expected that 'with sufficient optical power the nebulae would all be reduced to clusters' of stars, it was still too early to draw firm conclusions. He also reminded her of a fact of astronomical life, especially as it applied to the misty climate of central Ireland: 'You recollect Herschel said that it was a good observing year, in which there were 100 hours fit for observing, and of the average of our hours I have not been employed above 30' [27]. Lord Rosse later recorded that on winter nights, the local skies were best for nebula observation before 11 p.m., presumably before the mist rose up from the damp ground [28].

These estimates of observing time, by Herschel and Rosse, bring home to us what little opportunity for big-aperture astronomy was available in the British Isles, and the level of sacrifice in terms of money and resources which the Grand Amateurs were willing to make in pursuance of their researches. Indeed, only a private pocket could undertake such a fearfully expensive and irritatingly time-consuming task.

Towards the end of his letter to Mary Somerville, however, Lord Rosse says that his 72-inch diameter mirror of 52 feet focal length 'is nearly finished, and I hope it will effect something in astronomy' [29]. In fact, the 6-foot speculum, containing one part tin and four parts copper, and weighing four tons, had originally been cast on 13 April 1842, using a cast-iron mould of novel design aimed at creating uniform heat loss, and slowly cooled in an annealing oven. Sadly, this first mirror was accidentally broken, though a second casting was successful. Before Lord Rosse had obtained the pair of 72-inch mirrors which were necessary to keep a speculum instrument in constant readiness for use, however, he had to make a total of *five* castings. The other three either cracked or were found to have porous, unworkable surfaces [30].

In the same way that Lord Rosse employed annealing ovens and similar new metallurgical techniques to cast his mirrors, he became the first scientist to use a steam-powered machine, which imparted sixteen strokes per minute, to figure and polish them [31]. And by 13 February 1845, the 72-inch mirror of the 'Leviathan' of Parsonstown, as Birr was called in Victorian times, was producing spectacular views of the heavens. On that night, Lord Rosse, his great friend and mentor Thomas Romney Robinson, and Sir James South obtained the first images. Soon afterwards, Jupiter appeared as large and as bright as a 'coach lamp' placed within the tube, and several double stars were seen, brilliant and enlarged, as they crossed the meridian [32]. Immediately upon the instrument's formal commissioning, in April 1845, it revealed the spiral structure of the nebula M51, and within a further five years had detected structure in over a dozen other nebulae. Although it could still not resolve the great majority of nebulae, such as that in Orion, it strongly implied that stellar, or particulate, and not glowing misty 'shining fluid' structures lay at the heart of the nebulae. Within a few weeks of April 1845 it seemed that the Leviathan had answered the great question for which it had been built, and what followed was further substantiation, rather than fresh fundamental discovery [33].

The Rosse reflector became a Victorian icon of the power of science and technology to resolve doubt. And most significantly, it was a piece of private property, built at a cost of £12,000, notwithstanding all the in-house resources that were available on the Birr estates, by the wealthiest of the Grand Amateur astronomers [34]. For in spite of his mathematical education and Presidency of the Royal Society between 1848 and 1854, Lord Rosse's hereditary calling was that of an Anglo-Irish statesman. Lord Rosse took his public duties very seriously, both in Ireland and as the Irish Representative Peer in the House of Lords at Westminster. And almost as soon as the first nebula had been resolved in 1845, the potato famine struck Ireland. Rosse now found himself involved in relief operations, attempting to find, and often pay for, public works projects with which to provide wages for the unemployed, and trying to explain the Byzantine system of Irish land tenure to a London Parliament which still insisted upon trying to raise taxes from the destitute [35].

Birr Castle became one of the astronomical centres of Britain, if not of Europe, after 1845. Like Hartwell House, it became a focus for scientific gatherings, where the leading figures of British astronomy, along with visitors from abroad, came to use the Leviathan and be delighted by it. Thomas Romney Robinson and Sir James South were the regulars, while the archives preserved in Birr Castle give a taste of the range and frequency of visitors that came, especially after the completion of the railway from Dublin around 1850. But one excellent example of how the wonders of the Birr instrument reverberated through the astronomical community, including the families of the astronomers themselves, is to be found in the letter written by Richarda Airy to her friend Lady Herschel in October 1848. George Airy himself had just written to his wife from Ireland, where he had 'delayed his departure from Parsonstown for two or three days in hopes of getting a peep at the moon, but the weather was thoroughly spiteful.' Even so, Airy saw Saturn as 'a perfect blaze', observed the Milky Way, and enough of other things to be very impressed with the great 72-inch telescope. Although the mirror 'performed beautifully when directed to the zenith, [it] gave quite a distorted image when tilted to a lower part of the heavens', as Airy told his wife, though he was certain that Lord Rosse would overcome these difficulties [36]. This level of sustained technical discussion between two astronomers' wives, one married to a professional and the other to a Grand Amateur, only confirms the degree of awareness which these women had of their husbands' work.

The British Association invariably had outings to and entertainments at Birr whenever it met in Ireland. The Association came in 1843 and in 1857, and on the latter occasion Dr Daubeny from Oxford called the assembled company to drink toasts to Lord and Lady Rosse, as well as to the rising Lord Oxmantown, the future Fourth Earl of Rosse, who was showing himself to possess considerable scientific gifts [37]. And not only did Lord Rosse provide fine entertainments for visiting ladies and gentlemen of science, but as a good landlord who took his ancestral occupation seriously, he provided them for his tenants and local citizens as well. Early in 1851, when serving as President of the Royal Society, he put on a local celebration culminating in a firework display in the grounds of the Castle. The fireworks alone cost £400, and his Lordship lit them personally. As a journalist reported, 'It is highly interesting to see a man of Lord Rosse's capacity blending the pleasant with the useful – amusing his friends, and getting up that amusement, expending so much money among his people' [38].

As one might expect, Lord Rosse corresponded with many astronomers and engineers, especially when designing the 72-inch. In 1840, Romney Robinson was writing about mirror-polishing tools from Armagh (where he had a 15-inch aperture Cassegrain reflector by Thomas Grubb), while in 1841, when the plans for the 72-inch were getting under way, Robinson was suggesting methods of preserving wood in a damp climate, as well as advising his old pupil about employing Fairbairns of Manchester to undertake the large iron castings [39]. Robinson also passed on news about Professor John Pringle Nichol of Glasgow, who had acquired an old 24-inch mirror by Ramage, in the hope of resolving the nebulae, but now 'seems in despair at what he hears you have done' [40]. Another astronomically-minded peer, Lord Adare(?), moreover, requested Romney Robinson to pass on to Rosse his suggestion that he 'would use the Herschelian form' for the 72-inch instrument, though Robinson counters 'I hope that you will not' [41].

But Lord Rosse also corresponded with, and greatly valued the opinions of, the two other contemporaries whom I consider in particular to have constituted the brotherhood of the big reflecting telescope. They were the Liverpool brewer, William Lassell, and the Scottish–Manchester iron-master, James Nasmyth. I speak of them as a 'brotherhood' because all three men shared a common intellectual commitment to those branches of astronomy best done with big mirrors, spent a great deal of time and money in bringing the large speculum metal reflector to its peak of perfection through a variety of intermediary instruments, and then used their developed instruments to produce significant results. The exchange of letters and ideas went three ways, thereby indicating a common awareness of a mutual set of problems.

WILLIAM LASSELL

William Lassell was born in Bolton, Lancashire, in June 1799, and then educated at the Rochdale Dissenting Academy [42]. No doubt aided by his family, which had extensive commercial connections in Liverpool, he became a partner in a brewery at the age of 25, and had by then already mastered the art of producing good speculum mirrors of 6 or 8 inches diameter. He had gained this knowledge from John Edwards' article of 1787, which was mentioned above [43]. At this time, when Liverpool was the world's fastest growing industrial port, the first maritime rail-head, and the centre of innumerable works of civil engineering, there was an unquenchable thirst to slake, and a shrewd brewer could scarcely avoid becoming rich. Lassell tended his business carefully, as he often reminded his scientific correspondents [44], for it was the goose which laid the golden eggs which made possible the development and construction of three superb reflecting telescopes to his own original specifications.

In the 1820s, when Lassell was finding his feet as a businessman, an amateur telescope maker, and an astronomer, he also enjoyed the social dimension of his science, and partook in astronomical entertainments that were similar to, if less grand than, those of Hartwell House. Lassell's surviving journals record several such events, along with the names and observing locations of his Liverpool friends. There was Mr Roskell the watch manufacturer in Church Street, Mr Davis, Mr Bywater, and a Mr Findlow in nearby Bootle. There was also a lady member of this group of scientific friends, a Miss Harrison, also of Bootle, and after 1826, William Rutter Dawes, who lived a few miles away in Ormskirk. In addition, there were Alfred and Joseph King, the sons of a Liverpool teacher of navigation, and it seems to have been through them that he met, and married, their sister Maria, though we do not know if Maria was an astronomer in her own right. These friends often observed the heavens together, testing their Dollond and Tulley refractors against Lassell's early reflectors [45].

As early as 1821, when still only 22 years old, Lassell had cast and figured the mirrors for a 7½-inch Gregorian reflector, and by 1833 had produced a 9-inch mirror of 112 inches focal length which possessed a truly exquisite figure. By testing it on the star α Lyrae, and constantly adjusting the figure, as one can do with a speculum, he finally became satisfied with the stellar image in all the zones of the mirror [46]. Some years later, he loaned it to his 'eagle-eyed' friend Dawes for critical testing on double stars at Ormskirk, and for comparison alongside Dawes' Dollond refractor. By 1833, Lassell had clearly acquired a mastery of the art of mirror-making, and pointed out that commercial reflecting telescope makers often 'hurt' their specula by over-polishing to a high lustre, whereas what was necessary for serious astronomy was perfect evenness of the optical surface [47].

About 1837, Lassell's business had prospered sufficiently to enable him to move out of central Liverpool – where he had observed from the top of his house in Milton Street – to his rural mansion 'Starfield' in the fashionable West Derby district of Liverpool. With money and large grounds at his disposal, Lassell built an observatory. He now pioneered a development which was to be fundamental in the future of the reflecting telescope [48]. He took his 9-inch speculum and reset it in a balanced equatorial mount working at the Newtonian focus. The 9-inch mirror was placed in a tube and fork mount made entirely of cast iron. The polar axis was a large cone, also made in cast iron, rotating upon its point. The upturned base of this cone carried the cast-iron fork and box arrangement through which the telescope tube passed. Lassell recognised that in a Newtonian optical system it was essential that the eyepiece should be accessible without too many gymnastics, if the observer was going to be able to work for any length of time. In an altazimuth mount such as those used by the Herschels, this problem never arose, for the eyepiece only moved in a vertical plane with the tube. But with an equatorial it was very different, for a fixed Newtonian eyepiece could present itself at the most awkward angles, depending on the region of the sky to which the instrument was pointing.

Lassell solved this problem by making it possible to rotate the entire tube, mirror and eyepiece system around the optical axis, by making the tube pass through the iron box set between the mounting forks [49]. As the tube rotated inside this box upon a set of steel rollers, it was easy for Lassell, when working fifteen feet up in the air, to turn the tube around within the box and work at an eyepiece which was always upright in front of his face. The whole mounting was further balanced and adjusted with gears and wheel-bearings, so as to ensure the even tracking of objects across the sky. It was a world apart from the battleship-rigged altazimuth reflectors of the Herschels, Ramage, and Lord Rosse, for it was now possible to follow a dim object over a long period of time with a reflecting telescope, and work within the comfort of a dome. It is hardly surprising that Lassell's equatorially-mounted 9-inch reflector was the subject of a major paper to the R.A.S. in 1842 [50].

On the other hand, we must not forget that Lassell was not the first person to mount a reflecting telescope in the equatorial plane, nor to use iron mounts. It would be interesting to know how far he might have been influenced by the work of Thomas Grubb of Dublin, who in the mid-1830s constructed large, iron equatorial mounts for long telescopes. In particular, there was Grubb's new mount for the Markree Observatory's 25-foot refractor, and the 15-inch aperture Cassegrain reflector which Grubb had recently completed for Thomas Romney Robinson at the Armagh Observatory [51]. Yet the Cassegrain focus of the Armagh reflector enabled Grubb to mount it like a refractor, in a simplified version of a German mount, for its optical arrangement, with its eyepiece always at the bottom centre of the tube, contained none of the peculiarities of position inherent in an equatorially-mounted Newtonian. But where Grubb's Armagh instrument was truly innovative (as will be discussed presently) was in its incorporation of the first three-point astatic suspension system within the mirror cell [52]. Yet as far as the equatorial mount for reflecting telescopes was concerned, Lassell's was the more portentous instrument, for it was designed and used as a research tool in its own right, rather than as a visual adjunct to an otherwise meridian-based observatory, as was the case at Armagh.

LASSELL'S COLLABORATION WITH JAMES NASMYTH

Around the time that he was perfecting his balanced iron equatorial, William Lassell made the acquaintance in 1840 of James Nasmyth, a 32-year-old Scottish ironmaster and engineer whose factory was at Patricroft, Manchester. James was the son of the distinguished Edinburgh artist Alexander Nasmyth, and had been articled to Henry Maudslay before beginning to reap the reward for his recent invention of that fundamental Victorian machine tool, the steam hammer [53]. This man, who came to build steam engines for a living, had a life-long love affair with fire and molten metal. His indulgent father had allowed the boy to cast brass in a home-made furnace at the family house in York Place, Edinburgh, while young James had been shown the Moon and planets through his father's Dollond refractor, and enjoyed the guidance of Sir David Brewster, who was a visitor to the York Place house. While still only in his teens, Nasmyth had started to cast and figure speculum mirrors for telescopes, and developed the 'chill' method of casting onto an iron plate, which gave a firmer crystalline structure to the metal than a sand mould [54].

With a foundry and engineering works at his command, it was only natural that the adult Nasmyth's astronomical interests should incline him to the reflecting telescope. By 1840 he was able to produce excellent mirrors of 10 and 13 inches diameter, such as the one which he supplied to Warren De La Rue in 1842 [55], and was experimenting with powered machines for these operations. When Lassell saw one of Nasmyth's early mirrors, he claimed that it 'made my mouth water' [56], and the two men began a friendship and a collaboration which lasted until Lassell's death in 1880. Nasmyth made numerous mirrors, and in his first letter to the R.A.S. in 1844 he enclosed a sketch which depicts a figure (looking more like Lassell than himself) using his 12-inch Newtonian set upon a portable mount [57].

When his 9-inch iron equatorial had proved its worth, both optically and mechanically, around 1843, Lassell decided to upgrade the design to an instrument with a 24-inch mirror of 20 feet focal length. While Lassell made it quite clear in the published account of this instrument in 1849 that it was his own design and idea, he paid tribute to Nasmyth for his contributions to the figuring and polishing machine which had been built to produce the exquisite specula [58].

When designing the 24-inch, Lassell visited Birr Castle, and inspected the polishing machines for the 36-inch and possibly the 72-inch mirrors [59]. Lord Rosse was free with his information and entertainment, and he and Lassell began a correspondence which spanned many years, and which was no doubt backed up by personal meetings at the R.A.S., the Royal Society (of which Lassell became a Fellow in 1849 and Royal Medallist in 1858), and perhaps in Liverpool. Yet when Lassell built a replica of the Rosse polishing machine in Liverpool, he found that he could not get results which compared in perfection with his hand-made metals [60].

It was probably at this juncture that Nasmyth's help was sought, and by 1845 a polishing machine of seminal importance had been built. Unlike the Rosse machine, that of Lassell and Nasmyth showed the hallmarks of a professional engineer. No wooden parts were used, and no swaying, lateral motions. Instead, the new machine was in heavy cast iron, in a vertical design, like a drilling machine, and secured to a strong wall. Its motions were governed by a series of iron cams and gear races, so that when the steam power was applied through a driving belt, the mirror virtually ground itself [61]. The significance of this machine to the design and construction of large precision mirrors was noted by the Revd Richard Sheepshanks, as an officer of the R.A.S., when he had 250 additional offprints of Lassell's descriptive paper in the Society's 1849 *Memoirs* run off, to secure a wider distribution to scientists and scientific institutions at home and abroad [62]. It was to become the prototype of most subsequent optical figuring machines.

Lassell's 24-inch telescope of 1845 became the world's first large reflector to have full motions in the equatorial plane, and to be possible for a single individual to use [63]. The fastidious Lassell considered it to be as nearly perfect, in its optical and mechanical operations, as he could reasonably expect it to be [64]. On 3 October 1846, using the coordinates, obtained from *The Times*, for the recently discovered 'Le Verrier's Planet' (the planet Neptune, which had been so diligently searched for in Cambridge and Berlin), Lassell's 24-inch immediately picked it up as a disk. A week later, on 10 October, the instrument's light grasp was so great that he discovered the planet's largest satellite, Triton [65]. As the pair of 15-inch aperture Merz refractors in Harvard and St. Petersburg picked up Triton independently soon afterwards, John Herschel could rightly say in 1848 that a Liverpool amateur had beaten a track which American and Russian professional observatories soon followed [66].

Lassell's dominant interest in astronomy was the planet Saturn, its rings and surface markings. He co-discovered the crepe ring with W.R. Dawes and William Cranch Bond in 1850 (using Dawes' refractor, while in Kent), along with the Saturnian satellite Hyperion (also with Bond), and Uranus' satellites Ariel and Umbriel, as well as Neptune's Triton [67]. Unlike John Herschel and Lord Rosse, Lassell was not really an observer of nebulae, though in 1854 he published the results of a detailed study of that perennial object of fascination, the Orion Nebula. Inevitably, he was unable to resolve this complex gaseous object with the 24-inch, but he did discover a further star at its heart, in the Trapezium [68]. He was cautious, however, not to offer an opinion upon the nebula's possible structure, for Lassell was no speculator. He was willing to describe it as resembling several layers of pea-green tinged cotton wool, but that was all [69]. His only other published work on nebulous objects was the paper which appeared in the *Memoirs* of the R.A S. in 1866. It described the visible characteristics and positions of 600 new nebulae observed with his final and greatest speculum mirror telescope, the 48-inch equatorial of 40 feet focal length. Yet the very secondary character of the observation of these objects within Lassell's planetary priorities is brought out when he designated them as the work of his assistant, Albert Marth, made when 'the objects I especially wished to observe were not to be seen at all' [70]. One suspects that Lassell's pragmatic cast of mind felt that so prodigious an increase in optical power was necessary before any firm conclusion could be reached about the nebulae, that it was not worth speculating upon the internal organisation of what appeared as indistinct smudges of light. The Solar System was much more rewarding.

Lassell's main studies of Saturn, his work on the Orion Nebula, Marth's work on 600 new nebulae, and many other observations, had been made during Lassell's two sojourns on the island of Malta in 1852–3 and 1861–5. On the first occasion, he had taken his 24-inch away from the indifferent skies of Liverpool, in the hope of 'observing the larger and more distant planets' to greater advantage [71]. His second sojourn was made with his great 48-inch instrument, with the same object in view [72]. One also suspects that he hoped that Malta, which is 7 degrees closer to the equator than Harvard, and 18 degrees nearer than Liverpool, would give him a latitude advantage over Bond for objects in the plane of the ecliptic. Lassell's Malta expeditions, totalling a residence with his wife, family, and household of around six years upon the island, bring home to us his serious astronomical commitment. And while we do not know the precise sums which he expended upon his great telescope, some remarks made by James Nasmyth to Lord Rosse in 1853 indicate that at least £3,000 would be required if the government went ahead in commissioning a Lassell-type 48-inch reflector for use in the southern hemisphere [73]. Yet when Lassell requested a letter of introduction from Lord Rosse (along with another from Sir John Herschel) to the Governor of Malta in 1852, he made his self-financing status crystal clear when he stated that 'I want no pecuniary aid from any quarter' [74].

Perhaps his independent means and the elegant lifestyle of the Lassell family gave the wrong impression, for he was complaining to the R.A.S. by December 1852, after a few months in Malta, that 'People here think I am a gentleman of fortune, who has brought a telescope here "for his amusement". How different from the fact: I call it real labour ...' [75]

NASMYTH'S TELESCOPES

Like Lassell, Nasmyth was a reflecting-telescope user who showed no especial concern with the nebulae. Indeed, this lack of concern with faint astronomical objects was brought home around 1848, when he was developing his triple-mirror 'comfortable telescope', an arrangement which tends to be known today as a 'Nasmyth focus' instrument. In this optical configuration, a train of two secondary mirrors directs the light of the primary mirror through a hollow trunnion upon which the telescope is balanced, thereby enabling the observer to direct the instrument to any part of the sky without leaving his seat [76]. All that he needed to do was to operate a pair of wheels which controlled the geared altazimuth mount. Considering the low reflectivity of speculum metal, however, this meant that the observer's eye received only 37½ per cent of the light that originally went down the tube [77]. But as Nasmyth's concerns were the solar and lunar surfaces, he could afford to be profligate with his light.

It is easy to see the connection between Nasmyth's interest in fire and molten metal and his astronomical concerns. From the late 1830s, he had been surveying and drawing the formations on the Moon's surface, and had developed an explanation for lunar geology based upon internal heat and vulcanism that was in accordance with the non-meteoritic interpretations for the lunar surface that were prevalent at the time. The Moon's low gravity and lack of blanketing atmosphere had produced the stark and uneroded features that we see today, as hot ejecta from below formed mountains and craters. He saw parallels between lunar formations and the slag and scoria that one finds forming on the top of a crucible of molten iron. In 1850 he regaled the Edinburgh meeting of the British Association with his researches, and seems to have caused something of a sensation [78].

In addition to the Moon, Nasmyth studied the solar surface. He tried to devise an instrument to reveal those prominences which had been seen during the total eclipses visible at Turin in 1842 and Sweden in 1851, having one of his devices tested by Airy, but he was not successful [79]. Nasmyth also examined the interior structures of sunspots at high magnifications, and in 1860 made his most significant discovery, that of the 'willow leaf' granulation of the solar surface [80]. Like his friend Lassell, he was a planetary observer, and during the Martian opposition of 1863 was examining the planet's surface under high magnifications. He sketched continental zones of red and green along with the polar caps, and compared the colour sensitivity of his large reflectors and his 8-inch Cooke refractor [81]. Nasmyth saw no 'canals' on Mars.

By 1849 Nasmyth had developed a 'comfortable telescope' with a mirror that was 20 inches in diameter: an instrument which is still preserved in the store of the Science Museum, London. Nasmyth also built a model of this telescope which was offered for display at the Great Exhibition of 1851 (though it is not clear whether it was accepted), and the following year he was writing to John Williams, Secretary of the R.A.S., about its return [82].

But one of the most significant innovations in reflecting telescope design during the 1830s and 1840s was the development of the astatic mirror support system. One of the great problems with reflecting telescopes, especially those with heavy speculum mirrors, was the distortion of the image as the instrument moved from horizon to zenith, due to the mirror sagging under its own weight. The astatic levers that brought pressure to bear upon the back of the mirror as it was elevated seem to have been one of the very few innovations in speculum reflecting telescope design that were the products of a professional instrument maker. The 15-inch aperture Cassegrain reflector which Thomas Grubb of Dublin built for Thomas Romney Robinson's Armagh Observatory in the mid-1830s seems to have been the first instrument to have such a mirror support system [83]. It divided the back of the mirror into a series of three-point locations, each with a 'tree' of three further points branching from it, so that 3, 9, 27 or more (depending on the size of the mirror) metal pressure pads could be brought to bear on the back of the speculum as the elevation of the tube changed. Though Grubb's Armagh reflector does not survive in its entirety, its astatic mirror cell does [84].

Lord Rosse adopted the same system for the support of his own mirrors, as did Lassell and Nasmyth, though Rosse's supports do not seem to have been working properly in 1848 when Airy told his wife about the distorted image which the 72-inch telescope gave when it was brought down from the zenith [85]. Though Lassell had used a simpler pressure-lever device to maintain the equilibrium of his 9-inch mirror as the elevation of the tube changed, he developed a Grubb–Rosse system for his 24-inch of 1845. In its original form, however, this arrangement was still found to produce optical distortions, such as the supposed 'ring of Neptune' seen in 1846, though when Lassell improved his mirror support system for this instrument in 1852 – adding the further support of an outer iron band – the Neptunian 'ring' disappeared [86]. The pressure-support points of the 1852 rebuild of Lassell's telescope mirror cell are still quite clear on the back of the surviving 'B' mirror preserved in Liverpool Museum, and were used as a guide when the mirror cell of the replica Lassell telescope was built in 1996 [87].

Indeed, when one looks at the Lassell and Nasmyth instruments, developed between 1839 and 1859, when Lassell completed his 48-inch equatorial of 40 feet focal length, one sees how far the reflecting telescope had come, both optically and mechanically, especially during that brilliant decade 1839 to 1849: iron parts, large equatorial and geared altazimuth mounts, superior mirror casting techniques, annealing ovens for controlled cooling, steam-driven machines to figure and polish their optical curves, and astatic mirror supports. In comparison with John Herschel's Cape of Good Hope reflector of 1836, with its 18-inch mirror and rope and timber construction, and even with Lord Rosse's vast 'Leviathan' the reflectors of the 1850s seem to belong to another age.

It also says something about the acknowledged expertise of Lassell and Nasmyth and their close relationship with Lord Rosse that when his Lordship, as President of the Royal Society, was forming a Committee in 1852 to consider the setting up of a large reflecting telescope in the southern hemisphere, they were invited to join it. The purpose of this proposed large instrument was to commence an examination of the southern skies of a kind that the Herschel and Birr Castle telescopes had already done for the northern. During his five years at the Cape, Sir John Herschel had shown that the skies of the southern hemisphere possessed riches which his 18-inch telescope could survey, but not resolve. These included the Magellanic Clouds and the seemingly different distribution of nebulae from that of the northern hemisphere [88]. A large-aperture telescope in South Africa or Australia, built and staffed at the government's expense, was the perceived way forward for a part of the world where no Grand Amateurs resided. But as the Grand Amateurs were the innovators and custodians of the necessary technology, it was inevitable that Rosse should invite both Lassell and Nasmyth, along with the Astronomer Royal and others, to join the Southern Telescope Committee. Once again, it was to the acknowledged expertise of Grand Amateurs that officialdom looked when considering a capital astronomical project.

In the late autumn of 1852, William Lassell was enjoying the pellucid skies of Malta, so that the first Grand Amateur reflector man to respond to Lord Rosse's enquiry was James Nasmyth. On 15 December he told Lord Rosse that as he had recently lost his 'right hand man' at the Patricroft Works, he was immersed in 'commercial Engagement[s]' which demanded the whole of his time [89]. As the letter progressed, however, one can see Nasmyth becoming increasingly fascinated by the idea of a Great Southern Telescope, and by its end he was submitting a full proposal. It must be an equatorial, not an altazimuth, made of iron on an English mount, with geared parts. The extraordinary innovation which he proposed was that the observer, who would be working some 40-odd feet in the air at full elevation, should be seated within a 'snug box', which would hang down in the vertical from a pivot at the eyepiece, to protect him from the cold. One sees how highly Nasmyth recognised the need for comfort when working for hours at the eyepiece. The telescope, with its 48-inch mirror, would cost well over £3,000, he estimated, and as if to demonstrate his impartiality, he recommended Messrs Maudslay and Field (the firm in which he had served his own articles) to undertake the construction [90].

Over the next few months, Nasmyth produced several improvements in his original design. In July 1853 he was proposing the abandonment of the original English mount, and its replacement with a geared cone and fork mount. The tube, mirror, and eyepiece, moreover, would rotate on rollers within an iron box, in an arrangement which was almost identical to Lassell's 24-inch equatorial [91]. Ancillary details are also discussed, such as the use of sprung waggons running on railway tracks to convey the re-polished specula to and from the telescope, along with a drawing of a hinge system whereby the 48-inch speculum could be easily attached and detached from the tube [92]. All of Nasmyth's beautiful watercolours of these stages of the Southern Equatorial Telescope are still preserved in the archives of the Royal Society [93]. One of the most perceptive suggestions which Nasmyth made in July 1853 was the abandonment of the usual solid tube and its replacement with a skeleton tube made up of iron bars and rings; its purpose was to prevent the build-up of heat above the mirror, which would cause an air turbulence and spoil the image. With the possible exception of De La Rue's 13-inch reflector (the main mirror of which had been made by Nasmyth), I am not aware that anyone had built a skeleton tube reflector by this early date [94]. The first large instrument to incorporate one was Lassell's 48-inch telescope of 1859, which he used at Malta between 1861 and 1865 [95].

Lassell's first letter on the Southern Telescope project, dated from Valletta, Malta, on 30 December 1852, shows his mastery of what would be involved in building and running a big Newtonian reflector away from European resources. He also expressed doubts about the stability of Nasmyth's proposed 'snug box' at the eyepiece [96]. Yet what is remarkable in all of this correspondence is the willingness of Nasmyth, the articled and professional engineer, to bow to the experience of 'your Lordship and Mr. Lassell' when it came to the finer practical points of big telescope making and using, for they were 'so vastly more competent' than he was [97]. Indeed, Nasmyth regarded that it would have been 'presumptuous' of him to stress his own opinions against theirs, for Rosse and Lassell were the real big reflector astronomers. These surprisingly modest comments cast new light upon a Victorian ironmaster–scientist who is perhaps better known for his business acumen and determination to get his own way than for his willingness to take off his hat to his perceived scientific superiors. They only confirm my suspicion that Nasmyth approached astronomy and big telescope making as a hobbyist, turning down proposals for an F.R.S. in

1865, and a St. Andrews University honorary LL.D. degree, whereas Lassell and Rosse approached their scientific activities as a vocation [98]. One might also argue that the jocular tone of many of Nasmyth's letters to astronomical friends, with their puns, jokes, and sketches [99], also expresses the sheer pleasure that he got from astronomy, though they should not be allowed to mask the fact that he was still willing to expend major portions of his time and thousands of pounds of his money in the serious pursuit of the science, acquiring large-aperture Cooke refractors in addition to his own reflectors [100].

There are not many astronomical letters written by Nasmyth which do not pay some sort of respect to 'my very worthy friend, Mr. Lassell'. And nowhere does this admiration come over more clearly than in 1863, when Lassell was in Malta with his 48-inch reflector, and Nasmyth was writing to Richard Hodgson at the R.A.S. about the problems of big telescope management: 'for none that have dabbled a bit in such matters can duly appreciate the vast amount of down right hard work of head, hand and body that is involved ... to say nothing of the L.S.D. [money] part of the affair' [101]. Indeed, Nasmyth only regretted that his friend Lassell had not gone one step further, and instead of building a 48-inch equatorial reflector for his Malta expedition, had stolen an aperture lead upon Lord Rosse, and built an instrument with a 7-foot mirror on an equatorial mount [102].

It is not easy to ascertain how far the discussions and ideas for the Southern Equatorial Telescope in 1852–3 encouraged William Lassell to upgrade his celebrated 24-inch Newtonian to produce his 48-inch in 1859. When one looks at Nasmyth's engineering drawings for a proposed 48-inch Southern Telescope submitted to the Royal Society in 1853 – based as they were on a scaling-up of the Lassell 24-inch – one sees a prototype for the instrument which Lassell took to Malta in 1861–5 [103]. Indeed, the only significant alteration is Lassell's abandonment of Nasmyth's proposed 'snug box' at the eyepiece, and its replacement by a detached sentry-box tower at the Newtonian focus [104]. Yet, as was mentioned above, one major idea for the proposed Southern Telescope which Nasmyth did not include in his drawings but mentioned in correspondence was the skeleton tube [105]. It is hard to avoid the conclusion that the apparent foundering of an expensive official scheme to place a large-aperture reflector under the southern skies had indeed spurred the Liverpool brewer to grasp for the laurel and take an instrument of identical specification, complete with a skeleton tube, to $35\frac{1}{2}°$ north, where he would see the planets in the skies over Africa to better effect than anyone had ever seen them in the whole history of astronomy.

The opulence of Lassell's outfit at Malta, with its 48-inch telescope and fittings, workshop for polishing the mirrors, and elegant residence, was commented upon by Piazzi Smyth in November 1864, when the ship carrying him out to Egypt called in at the island. In particular, Piazzi was told that the open skeleton tube gave a much more steady image than a solid one, and prevented the 'twirling and twitching' of star images caused by warm air above the mirror [106].

Upon his return to England in 1865, Lassell let it be known that he was willing to donate his 48-inch telescope for public or academic use in the southern hemisphere, but sadly his offer of this excellent instrument was turned down. Perhaps it was felt to be too much of a one-man instrument for a public observatory. Instead, Sir Howard Grubb of Dublin completed the 48-inch Melbourne Telescope for £5,000 in 1874 [107]. This was the last major speculum mirror telescope ever to be built, and the first major instrument of that class not to be in the hands of a Grand Amateur. And almost from the start, the skeleton-tubed Melbourne reflector was found to be a disappointment [108].

THE SWAN-SONG OF SPECULUM

At the end of the day, however, a reflecting telescope is only as good as its mirror, and while it had become possible to give excellent optical figures to large mirrors, even by the time of William Herschel's 18-inch mirror of 1783, the problem with speculum was its tendency to tarnish. The high copper content of the alloy naturally aggravated this tendency, while the tin itself soon dulled in the presence of urban smoke or Irish mists. Part of the expense of maintaining a big reflecting telescope came from the need to remove the heavy mirror from its tube every few weeks, wheel it into the workshop, and take off a perfectly even film of metal to reveal a bright fresh layer below the oxidation.

This operation could not be performed with cloths and a cleansing agent, but needed the application of the original polishing tool, jeweller's rouge, and precision machinery, to ensure that the original optical figure of the mirror was preserved in a way that would be impossible if one simply polished it as though it was a piece of table silver. It was a labour-intensive job, and Lord Rosse later built a stretch of railway track from the mirror box at the foot of the 52-foot telescope tube, so that the 72-inch speculum could be pushed with the minimum of jogging to the steam driven polishing machine in his workshops [109]. Yet even regular polishing did not prevent it from deteriorating quite rapidly when inside the telescope, and when William Lassell and Otto Struve from St. Petersburg were Lord Rosse's guests in August 1850, Lassell commented upon the 72-inch speculum 'which appeared to me to be in a far more smeared, dewy or tarnished state than mine ever was when observing with it' [110]. It was for this reason that all speculum reflectors had to have at least a pair of identical primary mirrors – commonly referred to as 'A' and 'B' – if they were going to be ready for regular use. One of the mirrors could always be in the telescope while its companion was in the workshop being re-polished, in a ceaseless alternation [111]. This was neither a simple nor a cheap procedure if one's mirrors, like Lord Rosse's, weighed four tons apiece.

Amongst the brethren of the big reflecting telescope, it was Lassell who seems to have paid the greatest attention to the purity of the copper and tin of his alloys and the enduring reflectivity of his mirrors. He tried many experiments with speculum metal, and in addition to his manuscript notebooks, left a good published account in 1849 [112]. Some of the crucible remains, along with a few polished 'flat' mirrors which are preserved in the Science Museum store, still have a reflectivity which seems scarcely inferior to that of modern chromium plate [113].

In October 1996, one of the original pair of 24-inch mirrors – marked 'B' – was placed in the recently completed full-scale replica of Lassell's telescope, as part of the commemoration of his discovery of Neptune's satellite, Triton, on 10 October 1846. This 150-year-old speculum, while clearly dulled with the passage of time, still produced a decent image. Yet truly remarkable was the superb lustre still present on the surface of the 4-inch Newtonian finder of the 24-inch. After a century and a half, it seemed to perform almost as well as a modern aluminium-on-glass mirror, and showed the rings of Lassell's favourite planet, Saturn, with a wonderful clarity [114].

The above examination of Lassell's surviving optical surfaces makes it clear that it was no idle boast on his part when, after returning from Malta, he was able to inform Thomas Bell, on 2 July 1853, that 'My two-foot speculum I have within the last day or two replaced in the tube, in apparently as perfect a condition of lustre as well as figure, as when taken out for its journey to Malta' [115]. And this had been nine or ten months earlier, and notwithstanding exposure to salty air, damp at sea, and the abrasive Sirocco wind which blew from Africa. But as Lassell reminded Thomas Bell, this excellence of reflectivity (surviving in part as it does down to 1996) was the product of painstaking metallurgical experimentation, as well as optical figuring, for 'To ensure this property ... of the metal, the alloy must be carefully compounded by an assaying process before the casting is made.' The total weight of the speculum alloy mirror, moreover, was 449 pounds [116].

The era of the brotherhood of the big reflecting telescope, in which the participating members recognised prevailing optical frontiers and sought to transcend them with instruments that were both designed and built by themselves, lasted little more than eighty years: from the completion of Sir William Herschel's 18-inch to William Lassell's return from Malta with the 48-inch. Yet during those years, a new type of astronomy – physical observational astronomy – had been brought from a feeble infancy to a vigorous adolescence. Cosmology had been taken from the realms of speculative philosophy, and given a physical basis, while the planets and satellites of the Solar System had ceased to be interesting solely on the grounds of their motions among the constellations, and were studied as worlds in their own right.

While there were other amateur astronomers who were to use the reflecting telescope later in the nineteenth and into the twentieth centuries, conditions were changing by the 1860s. Warren De La Rue, for instance, continued to use the equatorially-mounted reflector with a 13-inch speculum mirror by Nasmyth and an early skeleton tube in his pioneering photographic work. Yet when the German optician Karl Augustus von Steinheil and the Parisian physicist Léon Foucault turned the silver-on-glass mirror into a viable technology in the late 1850s, they rang the death-knell for speculum, and these arts and sciences of casting, figuring, and mounting it that the brotherhood had brought into being [117].

Glass is a very much lighter optical material than speculum metal, can be bought in ready-made blanks without the need for any private foundry technology, and is much less awkward to work. It is less thermally sensitive than speculum, can hold its superior reflectivity for years, can be easily re-coated, and requires fewer feats of engineering to mount it, inch for inch, than metal. The reflecting telescope makers and users of the glass mirror era, therefore, needed neither the same wealth, freedom of time, nor heavy ongoing maintenance costs as their speculum predecessors, while still continuing to see further into space. If the big speculum reflecting telescope needed the same dedication as a racehorse to keep it in optimum condition, one might say that the glass mirror telescope only needed that of a sports car. For like a sports car it was less expensive to maintain, yet was capable of running faster and further down a ray of starlight than anything that had come before it.

7

The new sciences of light: spectroscopy, photography, and the Grand Amateurs

By the mid-1850s, it seemed as though cosmology had encountered obstacles which the optical technology of the age could not overcome. Lord Rosse's 72-inch instrument had resolved a dozen or so nebulous objects into star clusters, yet until the means were found whereby even bigger mirrors could be mounted upon greater equatorials, further progress seemed to be blocked for lack of light. This may well have been a reason why neither Lassell nor Nasmyth displayed any sustained research interests beyond the Solar System, for men of their pragmatic cast of mind no doubt recognised an apparent dead end when they thought they saw one. Visual observations using 48- or 72-inch mirrors would never resolve the Orion and thousands of other nebulae, for no more cosmological blood could be squeezed from speculum stone. The French philosopher Auguste Comte argued that the composition of the stars and their related matter represented a classic category of unknowable knowledge [1].

Within a decade, however, two new science-based technologies became available which broke the eye-on-metal deadlock, and allowed light both to be stored on a recording medium, and analysed chemically. While it is true that photography as an art medium had been around since 1839, and John Draper in America obtained a daguerrotype image of the Moon soon after [2], it was not until the development of the collodian sensitised glass plate in the mid-1850s that exposure times could become sufficiently short for it to be of real practical value to astronomers. But the initial breach in the cosmological barrier occurred somewhat by chance in a Heidelberg University chemical laboratory in 1859.

Towards the end of that year, the chemist Robert Wilhelm Bunsen and the physicist Gustav Robert Kirchhoff realised that they could make clear connections between the dark lines which Fraunhofer had first detected in the solar spectrum in 1814, and specific chemical elements [3]. By admitting sunlight through their spectroscope, and then incandescing pure reagents in the colourless flame of Bunsen's famous burner which was placed in the same path of sunlight, they found that bright lines replaced the dark lines in the Fraunhofer spectrum. The analytical potential of the spectroscope seemed truly enormous. For once one had classified the 'signatures' of pure chemical substances in the laboratory, one could equate their individual coloured emission lines with the dark Fraunhofer lines in the solar and perhaps other astronomical spectra, to begin the chemical analysis of astronomical bodies [4].

Kirchhoff also proposed a model which sought to explain why the solar spectrum was seen in dark Fraunhofer lines, whereas the spectra of substances burned alone and without sunlight in the flame of a Bunsen burner were coloured. The surrounding luminous atmosphere, through which the incident light and energy from the radiating solar mass below passed, acted as a 'reversing layer'. This layer absorbed energy to produce black lines. Incandescent laboratory substances, however, sent their light straight into the spectroscope without its first passing through an absorbing layer which reversed it. Light which had passed through a hot gaseous source prior to reaching us, therefore, had undergone *absorption*, whereas non-gaseous incandescence produced a continuous, coloured *emission* spectrum [5]. In the light of Kirchhoff's work, it seemed that the spectroscope could tell us not only about the chemistry of distant light-emitting bodies, but also about their solid or gaseous composition.

Although the foundations of chemical spectroscopy were laid by academic professional scientists in Germany, the early astronomical application and development of the technique lay very much in amateur hands in Britain and America, as will be discussed below.

WILLIAM HUGGINS

William Huggins was a comfortably-off amateur whose boyhood passion for science had been largely directed to microscopy. Living in the city of London, where his father owned a silk mercery business in Gracechurch Street, he had difficulty observing the skies, and found that he learned more from his microscope than from the telescope which he bought for £15. Although he received a first-class schooling in London, the necessity for him to take a more active part in the management of the family firm when he was 18 made him abandon hopes of going to Cambridge. Even so, he enjoyed the life of the metropolitan learned societies such as the Royal Astronomical and Royal Microscopical Societies, as well as the opportunity to travel on the Continent [6].

In the mid-1850s, the 30-year-old Huggins sold up the family firm and moved into a comfortable house at Tulse Hill, south London, with his elderly parents. Tulse Hill was to be Huggins' home for the remainder of his long life, while its spaciousness and clear skies enabled him to turn his attention to serious astronomy in a way that had been impossible in the City. While he was not wealthy in the way that Nasmyth, Lassell, and Lord Rosse were wealthy, he had enough money to live in decent upper-middle-class comfort without the need to go out to work. Even if he did not exactly belong to the grandest class of Grand Amateurs in terms of wealth, therefore, he was certainly independent and in full command of his own time.

Huggins had become a friend of William Rutter Dawes through the London learned societies, and in 1858 purchased from him, for £200, an 8-inch object glass by

Alvan Clark (which he had mounted by Cooke) when Dawes acquired a larger instrument [7]. After the usual round of planetary observations with the instrument, his attentions were suddenly riveted by the announcement of the findings of Bunsen and Kirchhoff in Heidelberg. It came to him, as he later claimed, like 'a spring of water in a dry and thirsty land', so full of possibilities was spectroscopy to his mind [8]. What appealed to Huggins' imagination was not so much the chemistry of the Sun, as the composition of the stars themselves. Huggins was a lateral thinker insofar as he possessed the ability to see around scientific corners, while combining that insight with a strictly experimental approach which eschewed speculation. To a 36-year-old bachelor of independent means in possession of an excellent 8-inch equatorial refractor, the spectroscopy of the stars offered an alluring prospect.

With support from his friend Professor William Allen Miller of King's College, London, Huggins built spectroscopes with two and then six prisms which could be fitted onto his 8-inch telescope. Realising, however, that no definitive work could be done on the stars until more exact spectral maps had been observed and recorded in the laboratory, Huggins prepared sets of 24 spectra for the most common laboratory elements [9]. The telescope with its spectroscope was next directed to the heavens, and by 1864 a series of laboratory and telescopic spectra observations had been published by the Royal Society [10]. It was soon clear that the bright stars which Huggins examined bore strong similarities to the Sun. Their absorption spectra showed the presence of atmospheric 'reversing layers', like that of the Sun, although the distribution of chemical elements differed markedly between individual stars.

It was in 1864, however, that Huggins made a discovery which promised to resolve the cosmological deadlock. Turning his spectroscope to one of those consummately puzzling objects, the 'Cat's Eye' planetary nebula in Draco, he found that instead of a now characteristic stellar spectrum, there were only a few single lines. While the character of these lines told Huggins that the source was gaseous, they did not seem to correspond to any known terrestrial element. He provisionally ascribed them to a new element, 'nebulium', although in the present century it was discovered that the lines were produced by ionised nitrogen and oxygen. He announced the gaseous composition of six nebulae in 1864 [11].

Within five years, the problem which had haunted cosmology ever since William Herschel had spoken of 'luminous fluid' and glowing 'chevalures' in space had been largely solved. There were clearly two kinds of nebulous object: those which broke down into star clusters when seen through powerful telescopes, and those which were composed of glowing gas.

And when Huggins had resolved these questions with his spectroscope, his talent for lateral thinking soon posed others. Could the different distributions of chemical elements within the stars tell us things about their individual physical structures or histories? Could the principle which Christian Doppler discovered in 1841 be applied to the waves of starlight, as Doppler had applied it to sound waves, to discover which stars might be approaching or receding from us? This was to lie at the heart of all subsequent determinations of radial velocities and eventually redshifts. In 1868 Huggins computed from spectra of Sirius that it was receding from the Sun at 29.4 miles per second, though he later modified the figure [12].

Because of the relative modesty of his means, however, Huggins was to some degree dependent upon the assistance of the Royal Society when it came to acquiring more powerful telescopes than his 8-inch and its early spectroscopes. In 1870 he received the value of the Oliveira Bequest from the Royal Society, which enabled him to commission a 15-inch refractor and an 18-inch Cassegrain reflector from Sir Howard Grubb of Dublin. In a way, this Cassegrain almost qualified Huggins for membership of the brotherhood of the big reflecting telescope, for it must have been one of the last capital reflecting telescopes to have a mirror made of speculum metal, following hard on Grubb's Melbourne reflector design of 1866 [13]. It was ingeniously mounted on the same axis as the refractor, in a configuration that presaged several subsequent instruments which, like it, would be used jointly for spectroscopic and photographic work. And in 1875, the 51-year-old Huggins married Margaret Lindsay Murray, who had admired Huggins' scientific achievement since her teens. But the future Lady Huggins was considerably more than another learned scientific wife, for she soon became her husband's collaborator and colleague. She became an active observer with the telescopes and spectroscopes in their garden at Tulse Hill, a skilled organiser and interpreter of spectra, and a noted astronomical obituarist. And after Sir William's death in 1910 (he had been knighted in 1897), it was Margaret who drew their surviving work together for publication down to her own death in 1916 [14].

It would be incorrect and imbalanced, however, to imply that Huggins was the only astronomer to take serious notice of work so fundamental as the spectroscope discoveries of Bunsen and Kirchhoff. The Jesuits Giovanni Battista Donati and Angelo Secchi in Italy, Joseph Norman Lockyer in England, and Lewis Rutherfurd and Henry Draper in America all became active in the 1860s, along with many professional observatories. In particular, Father Secchi, of the Jesuit Collegio Romano Observatory in Rome, had been active in stellar spectroscopic work since 1862, and was the first scientist to develop spectral sequences into which to classify stars of different types [15]. And even Greenwich had become interested by 1863, for once it could be shown that the spectra were authentic astronomical phenomena, they had to be measured and monitored just like meridian coordinates. In that same year, 1863, Airy was being challenged by Huggins about the most accurate method of observing and recording stellar spectra. It seems, however, that Airy, who at this time was no expert in this new branch of astronomy, was defended by his Greenwich Assistant James Carpenter, and the schoolmaster astronomer Charles Pritchard. In 1863 Carpenter was the Greenwich Assistant who seems to have had charge of what was gradually becoming 'physical astronomy' in the Royal Observatory, and in January 1864 Airy was sending memoranda relating to the observation of the Orion Nebula to him [16]. But such studies would always remain tangential to the main meridian work of the Royal Observatory.

It is clear, however, that by the mid-1860s many professional observatories felt it incumbent upon themselves to make a few cautious forays away from the meridian, to look at spectroscopy. With the exception of Father Secchi, the amateurs had very much stolen the lead as far as spectroscopy was concerned, and it would be some years before the professionals began to catch up, most especially after the founding of the Potsdam Astrophysical Observatory in 1874, and the appointment of Hermann Carl Vogel as Director in 1882 [17].

The impact of the spectroscope upon astronomy was immediate after 1859. Yet the realisation of its twin innovation, photography, took more time. For over a decade and a half after its invention, photography remained little more than an interesting possibility to astronomers, although by 1858 Angelo Secchi was experimenting with photographic photometry, and noticing the different images produced by the light reflected from different planets in a given exposure time [18]. But it was a Guernsey printer living in London who began to conduct some of the first sustained experiments in the use of photography to record detailed astronomical information.

WARREN DE LA RUE

In some respects, Warren De La Rue became one of the brethren of the big reflecting telescope after an inspiring contact with James Nasmyth had led him to start casting specula [19]. And after purchasing an excellent 13-inch mirror made by Nasmyth himself, De La Rue devised and built with his own hands an iron-framed skeleton tube for it, which he placed upon a modified German equatorial mount. It was this instrument, which De La Rue first set up in his garden at Canonbury, London (and later at Cranford, Middlesex) which attracted Samuel Whitbread and friends for an observing session in 1851 (as mentioned in Chapter 5). Although the 13-inch aperture of this telescope made it rather modest for Grand Amateur purposes by the 1850s, it was an extremely influential instrument in other ways. Its skeleton tube and accurate clock drive were significant, but more important was its adaptation for photographic purposes after 1852.

Using Scott Archer's wet collodian process, with guided exposures of up to 30 seconds, De La Rue was obtaining lunar photographs of unprecedented quality, and capable of substantial enlargement, by 1858 [20]. Then for some years after 1858, he began to use the 13-inch for what was probably the first explicitly photographic investigation in the history of astronomy: the search for suspected changes in the area around the lunar crater Linné.

Realising the research potential of photography, especially in the wake of the German amateur astronomer Heinrich Schwabe's discovery of the 11-year sunspot cycle, De La Rue devised his first photoheliograph, to take daily photographs of the solar surface. The photoheliograph was the first specifically designed astronomical camera, with a 3-inch object glass colour-corrected for the early blue-sensitive emulsions, and a plate holder in place of an eyepiece [21]. Although De La Rue designed this instrument, and had use of it on eclipse expeditions to Spain and in his Cranford Observatory, the photoheliograph was not his personal property, and was eventually installed at the Kew Observatory.

By the early 1860s, therefore, it had been an amateur astronomer who had shown the scientific world what photography could really do, especially in solar studies, where De La Rue had photographed and measured the elusive prominences at the total eclipse of 1860 and confirmed the depression characteristics of sunspots, and in recording lunar detail. While De La Rue was not especially concerned with spectroscopy, it soon came to be realised that if photography could be adapted to the recording of spectra, a great deal of time could be saved and extra detail recorded.

Warren De La Rue would also continue that tradition, as seen already in figures like John Pond and the founders of the R.A.S., whereby private individuals took it upon themselves to take in hand and bring up to date a moribund 'professional' institution. For De La Rue was one of the driving forces behind the founding of the Oxford University Observatory after 1870, to which he would donate his 13-inch photographic reflector. (The other Oxford driving forces would be Professor John Phillips, academic geologist and Grand Amateur astronomer, and the Revd Charles Pritchard, who, as we saw in Chapter 2, had combined Grand Amateur astronomy with the profession of headmaster before taking up his professional Directorship of the new Oxford Observatory) [22].

THE AMERICAN CONNECTION

Although this book is concerned with the work of English amateur astronomers, it has already been pointed out in Chapters 2 and 3 that the United States of America was the only other country to have a significant Grand Amateur tradition. And nowhere was the vigour and originality of that tradition more clearly seen than in the development of photography, spectroscopy, and astrophysics. Indeed, it is impossible to assess the impact of amateur scientists upon those new investigations and the development of their related technologies, without considering the work of three Americans in particular. For in addition to their other achievements, it had been the Anglo-American physician–astronomer John Draper who first successfully photographed the Moon, and his Grand Amateur son Henry who first succeeded in recording stellar spectra on photographic plates to advance the painstaking visual recording technique of Huggins, while Lewis Rutherfurd performed fundamental work in both spectroscopy and photography.

Lewis Rutherfurd

Lewis Morris Rutherfurd belonged to the patrician class of the American east coast. He trained as a lawyer, but was safely relieved of any further prospect of earning his living by marrying into the even wealthier Stuyvesant family, which owned much of the land around New York. Like many old American families of the pre-Civil War period, the Stuyvesants had astronomical interests, and it had been to the private New York City Observatory of Peter Stuyvesant that William Cranch Bond had sent some of the world's first telegraphic time signals from Harvard in 1848 [23].

In 1856, Rutherfurd set up a private observatory on the Stuyvesant estate, at what was later to become New York's Lower East Side. Photography was his original interest, and he had succeeded in taking some fine photographs of the Moon, planets, and star clusters with an 11½-inch equatorial refractor by 1865. He envisaged a photographic atlas of the entire sky.

But it was his learning of the work of Bunsen and Kirchhoff in 1861 which directed his attention to spectroscopy. By 1864 he had made spectroscopic studies of the Moon and planets, and completed a photographic survey of that part of the solar spectrum which comprised Fraunhofer's H and F lines. It was over 30 inches long, and contained three times more lines than Bunsen's and Kirchhoff's recent visually observed solar spectrum. Unfortunately it was never published [24].

Rutherfurd acquired a 13-inch aperture refractor in 1868, which could be fitted with a corrector lens to turn it into a specialised photographic instrument, and came to pioneer the use of diffraction gratings in preference to prisms for obtaining the most useful dispersions. The ruling engine which he devised enabled him to produce the best diffraction gratings of the day, with over 17,000 lines to the inch [25].

Like the Grand Amateurs of the Old World, Rutherfurd combined means, leisure, operative dexterity and that ability to see fruitful connections between emerging technologies, to become one of the first Americans to grasp the possibilities of astrophysics and astrophotography. But Rutherfurd's achievement was to be eclipsed, in many ways, by a man who was twenty-one years his junior, yet whom he would survive by a decade.

Henry Draper

Henry Draper was the son of an English physician, Dr John W. Draper, who had emigrated to Virginia shortly before Henry's birth in 1837. John Draper soon began to practise medicine in New York, took up the Chair of Chemistry at the University of New York City, and became the first scientist to photograph the Moon on a daguerrotype plate in 1840 [26]. By the time that Henry completed his medical studies at the remarkably premature age of 20, his father was one of America's leading medical and scientific men, and in command of substantial means. Because Henry was legally too young to practise medicine at the time of passing his qualifying examinations, he travelled in Europe during parts of 1857 and 1858, in preparation for joining the staff of New York's Belle Vue Hospital. And it was in Ireland that his attentions were decisively turned towards astronomy, when he was shown Lord Rosse's 72-inch telescope at Birr Castle. On returning home to New York, he made a speculum of 15 inches aperture, but was not pleased with it. His polishing machine, which was similar to that of Lord Rosse, failed to impart an even curve to the metal, and various attempts at using acids and a variety of abrasive techniques failed to make the mirror optically correct. But when his father John communicated these problems during a visit to Sir John Herschel in June 1860, he was recommended to try glass [27]. The young Draper immediately took up the suggestion, and was soon making good silver-on-glass mirrors. He made a succession of these instruments, which he employed both in visual and photographic astronomy, culminating in a 28-inch aperture silver-on-glass equatorial, built between 1867 and 1872.

The chemical skills of both Drapers were also important in perfecting a chemical technique, based upon those of European scientists, for depositing a film of perfectly even thickness upon the curved surface of a glass mirror. Henry Draper used one developed by Cimeg in 1861 for back-silvering looking-glasses, but modified so that it could surface silver a telescope mirror [28]. In the infancy of the silver-on-glass process, a real problem was experienced in trying to deposit a layer of silver that followed precisely the curve imparted to the glass. Although Draper experienced some problems in getting glass mirror blanks that were perfectly even in their composition, once such blanks could be found, they could be worked using techniques that were similar to those used to impart curves to glass lenses. By the early 1860s he was overcoming these problems and, employing Cimeg's process, which used potassium tartrate and Rochelle salt, was able to produce fine glass mirrors, with 'bright, hard and in every respect perfect [silver] films' [29]. Henry Draper thereby became one of the leading innovators of silver-on-glass reflecting telescopes, and while he (like his fellow countrymen) was neither an 'artist in speculum' nor did he work primarily with visual techniques, one can see his relationship to the brotherhood of the big reflecting telescope.

Indeed, one cannot help but notice parallels between the Drapers – father and son – and the Herschels. Both sets of men were Grand Amateurs as far as their astronomy was concerned, and while there was ample money in the Draper family, as there was in the Herschel, the Drapers nonetheless continued to be professionally engaged in medicine or other non-astronomical branches of science such as chemistry, and up to 1873 Henry Draper was Dean of the New York Medical Faculty. On the other hand, the young Henry Draper faced very different scientific circumstances from those encountered by the 24-year-old John Herschel. In 1816, the astronomy of the nebular Universe had already reached something of an interpretational *impasse* which no new foreseeable data could resolve. But in 1862, when the younger Draper was serving as a surgeon with the Union Army during the Civil War, astronomy was full of excitement. Photography, silver-on-glass mirrors, and Bunsen's and Kirchhoff's newly-published work on spectra had opened up enormous possibilities for advancing the science.

Henry Draper's early astronomical work was performed at an observatory on the family estate at Hastings-on-Hudson, outside New York City. But in 1867 he followed the well-known Grand Amateur path, firmly beaten by Sir William Herschel, Sir James South, William Rutter Dawes and his American predecessor, Lewis Rutherfurd: he married a wealthy heiress. Anna Mary Palmer, like the future Lady Huggins, was not only gracious in herself and as a hostess, but actively assisted her husband in his astronomical work. When his father-in-law Courtlandt Palmer died in 1873, Henry largely gave up his medical appointments to devote his time to astronomy and to the management of the Palmer estates. With a world-class astrophysical laboratory at their home in Madison Avenue, New York City, and a 'country' house at Dobbs' Ferry, near the Draper estate and observatory at Hastings, Henry Draper's grandeur in the amateur astronomical community was beyond doubt [30].

In 1872, using his newly-finished 28-inch silver-on-glass reflector, in conjunction with a sensitised wet collodian plate, he succeeded in photographing the spectrum of the star Vega, and developed an instrument which he came to call the spectrograph. Attempts had been made to photograph spectra before 1872 – most notably by Huggins – but Draper was the first to succeed. His success derived in part from the bright spectral image produced by the 28-inch mirror – as against the 8-inch and 15-

inch object glasses that were at Huggins' disposal – combined with an excellent clock drive which was capable of maintaining the position of the lines on the plate during an exposure [31]. Yet as in the earlier work of Lassell and Bond, one could not deny the latitude advantage. Observing from the New York latitude of 41° north, as against Huggins at 51½° north, Vega was higher up in the sky, and shining through a lower density of air and turbulence.

In the way that the spectroscope revolutionised our knowledge of the physical composition of astronomical bodies in the 1860s, photography began to do the same in our wider understanding of deep space in the 1870s. For in addition to spectrographic recording, Draper became the first astronomer to use photography to record nebulous objects when, on 30 September 1880, he finally succeeded in obtaining a good photograph of the Orion Nebula using the 11-inch aperture photographic refractor by Alvan Clark [32]. By the spring of 1881 he was obtaining improved, high-definition photographs of the Orion Nebula which showed fourteenth magnitude stars. And once again, his early photographs were very much dependent upon the excellent clock drive and guidance system of his telescope, which enabled him to track objects for well over an hour [33]. His Clark refractor and his own 28-inch mirror were both mounted on the same polar axis – in a configuration which was pioneered by Sir Howard Grubb and used by William Huggins in 1870 – so that the same precise driving clock could power both instruments, depending upon whether one was photographing nebulae or spectra. By 1880, moreover, he was using the new gelatine dry plates, instead of the less sensitive wet collodian emulsions which were unsuitable for very long exposures due to their tendency to dry out while still being exposed. Draper now realised that the photographic plate could be used to store up light through long exposure, so that if the telescope was so accurately guided that star images did not drift, one could detect objects and structures which were impossible to see visually.

In addition to the straightforward photography of star images and their corresponding spectra, Draper had come to the recognition that there were often differences between the visual appearances of individual stars and their images on a photographic plate. This recognition was to lie at the heart of photographic photometry, and the realisation that the spectrum and photographic 'colour' of a star might be an indicator of its surface temperature. In this respect, Draper's work laid the foundation of a great deal of modern stellar classification [34].

PHOTOGRAPHY AND THE ECONOMICS OF ASTRONOMICAL WORKING TIME

The photographic recording of nebulae, and in particular stellar spectra, also represented a milestone in the eventual processing and interpreting of astronomical data which would soon become vital to the professionalisation of astrophysics. Firstly, photography greatly reduced the amount of telescope time that was necessary to make a full record of a star's spectrum. What had taken Huggins or Secchi many hours or many nights, working with a micrometer at the eyepiece of the spectroscope attached to the telescope, could now be accomplished in minutes or less. This was of particular importance in professional observatories, like Potsdam, when the maximum research usage had to be obtained from a telescope. Secondly, it allowed the spectra to be measured at leisure, on a plate-measuring micrometer, by a human observer working in a comfortable position, in a heated room, during office hours. As James Nasmyth had been at pains to point out in the 1850s, 'snug' working conditions were more conducive to meticulous work than awkward and cold ones. Thirdly, the production of hundreds of spectrograph plates now generated a new type of astronomer: professionals, who never needed to go to a telescope or work after 6 p.m. if they did not wish to. Lady Margaret Huggins and Mrs Anna Draper initiated this new type of astronomy in some respects, as they worked by day on the plates which they and their husbands had exposed at the telescope by night. In the big professional observatories, however, it meant that teams of plate-measurers could spend days or weeks in the laboratory carefully interpreting what had taken minutes or hours to expose at the telescope. And as the flow of data was out-running its rate of interpretation by 1900, professional astrophysics quickly developed aspects of an industrial organisation. Plate-measuring, either for spectra, binary star motions, or celestial cartography, moreover, became the first avenue by which women could enter professional astronomy, especially in the American university observatories, like Harvard [35].

One can argue with some conviction, therefore, that the astronomical techniques first recognised and developed by amateurs made possible the economics of professional astronomy, and then led to the extinction of amateurs at the forefront of the science. Henry Draper's work had undoubtedly facilitated this trend, and one might say that his sudden death from pleurisy in the autumn of 1882 hastened it along, for Mrs. Draper established instruments and salary endowments for the Henry Draper Memorial at the Harvard Observatory to advance research in astronomical photography and spectroscopy [36]. Edward Pickering, Director of the Harvard Observatory, commenced a spectroscopic survey of the sky, along lines proposed by Draper, and Annie Jump Cannon was largely responsible for the working out of Draper's improved system of stellar classification (which was a fundamental advance on the simpler visual system of Angelo Secchi), and the publication of the Harvard Draper Catalogue in 1924 [37].

Lewis Rutherfurd, similarly, donated his great collections of instruments, plates, and spectra to Columbia University, of which he was a Trustee, in 1881, to establish academic astrophysical studies there. And in the case of Joseph Norman Lockyer in England, one can clearly trace the transition from amateur to professional astrophysicist and back again to independent amateur in the span of a single career.

JOSEPH NORMAN LOCKYER

Because he spent most of his working life as a scientific civil servant, and thirty-two years of it in charge of Britain's first professional astrophysical observatory, Joseph Norman Lockyer is often, and quite rightly, seen as belonging to the emerging class of professional astronomers. On the other hand, we must not forget that he only became Director of the Solar Physics Observatory at South Kensington (now Imperial College, London) in 1879 because of his prior distinction as an amateur astronomer. For rather like John Pond in 1810 or Charles Pritchard in 1870, Lockyer was an amateur who became a government-funded professional on the strength of what he had achieved in his own time, and at his own expense, after leaving his desk each evening at the War Office or on the Secretariat of the Devonshire Commission [38].

Like so many young astronomers in the 1860s, Lockyer was fascinated by the work of Bunsen and Kirchhoff, which he recognised as providing the key to many problems in physical science. For Lockyer was the son of a Rugby surgeon who himself had wide scientific interest, and was keen to encourage his son, while his mother Anne Norman was the daughter of a Warwickshire squire. The family seems to have been double-blessed with money and with strong intellectual interests. Joseph Norman received a thorough formal education, and though he never went to university, had the means to travel on the Continent, before entering the War Office at the age of 21 in 1857 [39].

In many important respects, Lockyer was similar in his circumstances to his older spectroscopic contemporary, William Huggins; for both men came from comfortably-off (but not 'Grand') middle-class backgrounds where youthful interests in science had been encouraged. Yet neither had received a university education, although they had enjoyed the means and the time to buy expensive instruments and follow their scientific inclinations wherever they led. On the other hand, there were fundamental differences between them. Huggins was shy, whereas Lockyer was forceful. Both, it is true, were skilled observers and experimenters with instruments, but where Huggins was cautious about drawing over-arching theories from his results, Lockyer could speculate boldly from his.

Lockyer began to observe the Sun with a spectroscope in the mid-1860s, for unlike Huggins' interests, it was the Sun, its chemistry and physics, rather than the stars themselves, which remained his abiding object of concern [40]. Since 1866, Lockyer had been considering ways by which he might observe the spectral lines of solar prominences without the need for a total eclipse. The problem which had to be overcome was the dispersion of the unobscured sunlight in the earth's atmosphere, which had foiled the earlier attempts of James Nasmyth and Piazzi Smyth to see the prominences with obscuring-disk projectors. To achieve the desired result, it was necessary to obtain a sufficiently large dispersal of the Sun's light in the Earth's atmosphere to make the prominence spectrum visible [41].

Lockyer had lacked the time during the two years following 1866 to devise an appropriate high-dispersion spectroscope, until the Indian eclipse of 18 August 1868 led the French solar physicist, Jules Janssen, to India to apply his spectroscope to the Sun's limb during totality. This quickly led to another of those simultaneous (often Anglo-French) scientific discoveries which took place in the nineteenth century [42]. After catching the spectra of the solar prominences during totality, Janssen had the idea of modifying the slit of his spectroscope and applying the instrument to the unobscured solar limb. Lockyer, quite independently, had exactly the same idea some time later, and examined the solar limb through the fine slit of the spectroscope attached to his own 6¼-inch refractor. Both men now succeeded in obtaining prominence spectra in full sunlight [43].

The congruity of the Lockyer and Janssen discovery also related to another congruity in the careers of the respective scientists, for Janssen had entered the French scientific establishment only after working as a bank clerk and a schoolteacher. Like many fundamental breakthroughs in the technology of science, spectroscopy permitted original figures who might otherwise have remained outsiders in the scientific establishment to rise to prominence [44]. This tendency was perhaps even more significant in Continental Europe, where the established sciences were already more professionalised than they were in Britain.

Both Lockyer and Janssen announced their independent discoveries in obtaining out-of-the-eclipse prominence spectra almost simultaneously in the autumn of 1868, although Janssen had obtained his results before Lockyer. The French government generously commemorated the event with a double-portrait medal [45]. And following upon the analysis of the new spectra which the prominences displayed, Lockyer identified a new yellow line which did not correspond to any known terrestrial element. He named it 'helium', from the classical Greek Sun-god Helios. It was not until 1895 that helium was found to exist on Earth, by Sir William Ramsay [46].

While Lockyer's emergence into scientific prominence took place when he was working on his own, as an amateur, he belonged to a new generation of scientists whose attitudes were less proudly independent than those of the older British Grand Amateurs, and less patrician than contemporary Americans'. Lockyer could see no reason why the state should not pay for scientific research (an attitude which he held in common with several of the younger generation of astronomers), and from the early 1870s onwards, this adroit civil servant began to move towards professionalisation within his own career. Working in the Science and Art Departments, he became responsible for what was called the *1876 Loan Exhibition* of scientific artefacts, which would form the nucleus of the Science Museum in South Kensington [47]. Then he succeeded in persuading the government to use the South Kensington site and some of the other profits which had been made out of the Great Exhibition of 1851 to establish a Solar Physics Observatory, which was formed at first around his own private collection of spectrographic equipment. Lockyer was a powerful persuader, and by 1881 had become the first properly salaried astronomer to be paid by the British government to undertake physical researches that had no practical bearing upon the well-being of the navy or the nation [48]. If the government was willing to pay salaries to the curators and librarians of the British Museum, and the new museums that were planned for the South Kensington site, why should it not pay for astronomy? Solar physics, after all, was no less meritorious as a demand upon the public purse than housing the Elgin Marbles.

Although he was an observing astronomer, Lockyer was also a theorist, willing to draw far-reaching conclusions from his evidence. One of these was his 'dissociation hypothesis', in which he tried to explain the different spectral lines that were displayed by the same chemical element in different stages of stimulation. Lockyer argued that the chemist's elements were in fact aggregations of sub-atomic materials,

Plate 17. William Lassell (1799–1880). Daguerreotype, *c.*1845, in the possession of the Liverpool Astronomical Society. The damaged original has been skilfully computer-scanned and 'cleaned up' by Michael Oates of the Manchester Astronomical Society, 1996.

Plate 18. Replica of William Lassell's 24-inch aperture, 20-foot focus reflector, 1845. This fully-operational full-size replica of Lassell's cast-iron equatorial was built between 1994 and 1996, with a grant from the Royal Insurance Company, Liverpool. It is shown at its inauguration, 10 October 1996, in the playground of St Michael's Primary School, Guion Street, Liverpool, virtually on the spot of Lassell's 'Starfield' Observatory, where the original telescope had revealed Neptune's satellite Triton on 10 October 1846. The instrument is now housed in Liverpool Museum. Photograph by Paul W. Roberts.

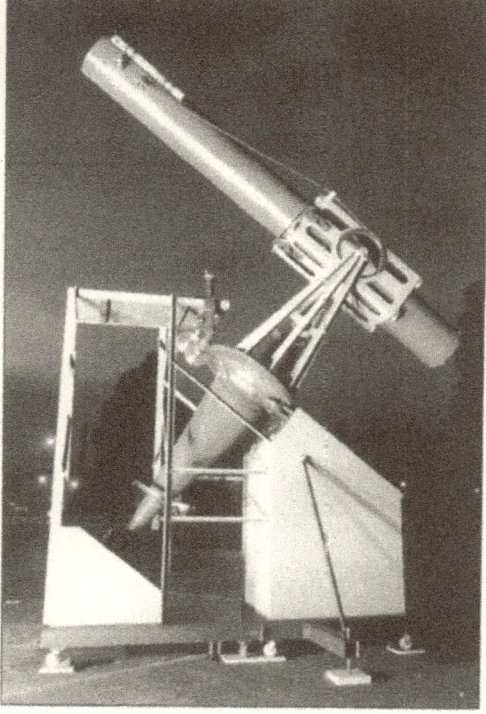

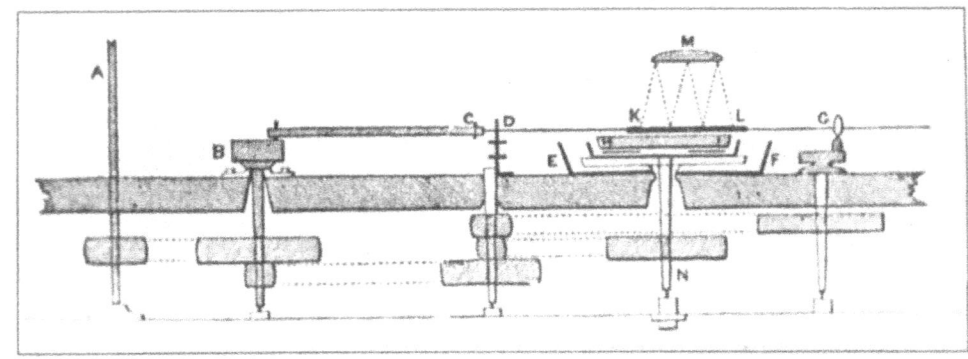

Plate 19. Lord Rosse's (Oxmantown's) mirror-polishing machine, *Phil. Trans.* **130** (1840) 515; reproduced from Lockyer, *Stargazing* (pl.10) 131. The wooden beams and laterally-moving parts, connected to a steam engine, introduced possible sources of error in the resulting optical surface.

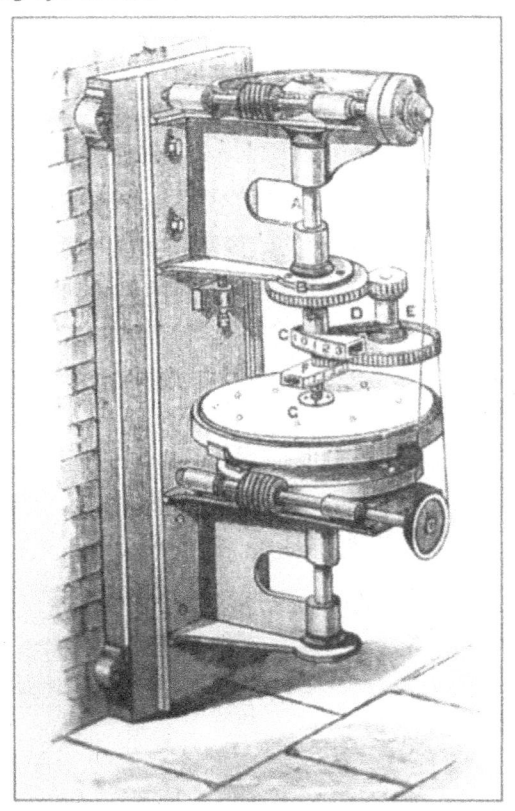

Plate 20. William Lassell's mirror-polishing machine, built by James Nasmyth, 1845. The vertical, rigid cast-iron structure, secured to a strong wall, enabled 24-inch and 48-inch mirrors of superb optical figure to be produced. W. Lassell, *Mem. R.A.S.* **18** (1849), reproduced from Lockyer, *Stargazing* (pl.10) 132.

Plate 21. William Lassell's 48-inch aperture, 40-foot focal length equatorial reflector, set up in Malta in 1861. A man working a ratchet lever powered the equatorial drive. Note the ingenious rotating 'sentry box', the four opening galleries of which enabled easy and wind-protected access to the Newtonian eyepiece. Simon Newcomb, *Popular Astronomy* (London, 1898), 132.

Plate 22. James Nasmyth (1808–1890). *James Nasmyth, Engineer. An Autobiography*, *ed.* Samuel Smiles (London, 1889), frontispiece.

Plate 23. Nasmyth's 'Comfortable Telescope', 1848, which introduced the through-the-trunnion 'Nasmyth focus', whereby an observer could direct the 20-inch aperture telescope to any part of the sky from his seat. *Nasmyth, Autobiography* (pl.22) 338.

Plate 24. Nasmyth's observatory–residence, 'Fireside', at Patricroft, north Manchester, *c*.1850. *Nasmyth, Autobiography* (pl.22) 315.

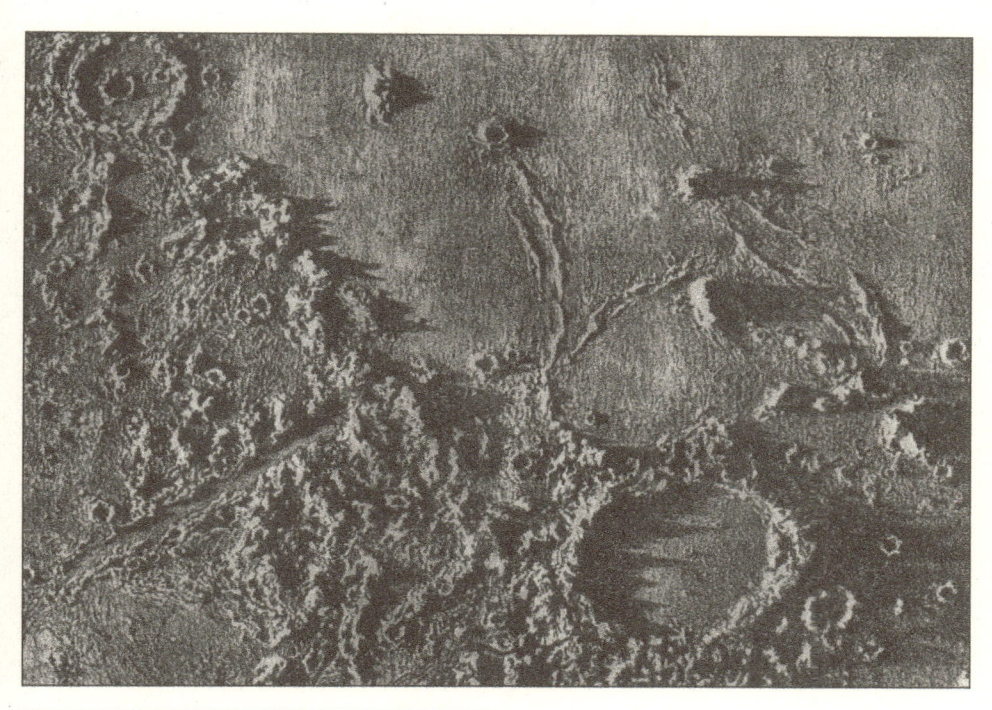

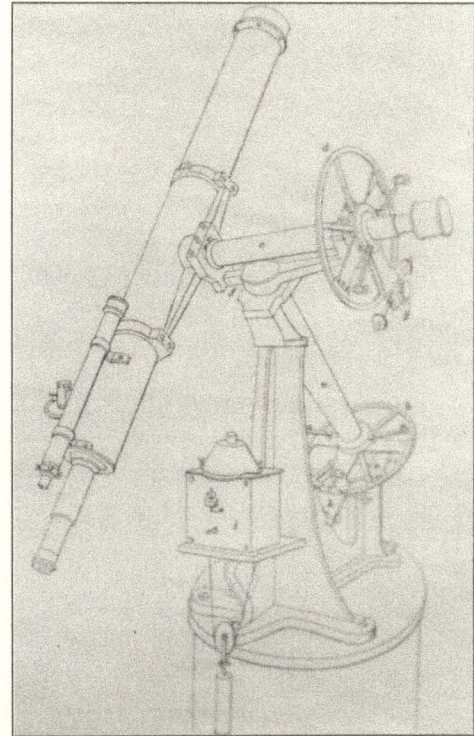

Plate 25. One of Nasmyth's detailed plaster models of the lunar crater Plato and its region. Two similar models of the lunar surface by Nasmyth, each about 1 foot square, are preserved in the Science Museum, South Kensington. Reproduced in Ball, *Story of the Heavens* (pl. 14) 81.

Plate 26. A Troughton and Simms equatorial refractor of about 4 inches aperture, on a German mount, similar to the instrument which Samuel Whitbread recorded using in the late 1840s. W. Simms, *The Achromatic Telescope, and its Various Mountings, especially The Equatorial, To which are added some hints on Private Observatories* (London, 1852) 45.

Plate 27. The Revd Edward Craig's (Vicar of Leamington) 24-inch refractor, with its 85-foot cigar-shaped iron tube, with optics by F. Slater of Euston Square, London. The instrument, which was erected on Wandsworth Common, south London, rotated around the central brick column on rails. It was an expensive failure. *The Illustrated London News*, 28 August 1852, 168.

Plate 28. The Revd William Rutter Dawes (1799–1868). Hutchinson's *Splendour of the Heavens. A Popular Authoritative Astronomy* I, *ed.* T.E.R. Phillips and W.H. Steavenson (1923) 305.

Plate 29. John Russell Hind (1823–1895). *The Illustrated London News*, 28 August 1852, 168.

Plate 30. Edward Crossley's observatory–residence, 'Bermerside', Halifax. Edward Crossley *et al.*, *A Handbook of Double Stars* (London, 1879), frontispiece.

Plate 31. Crossley's $9^{1}/_{3}$-inch Cooke refractor, at 'Bermerside'. *Handbook* (pl.30) 11.

Plate 32. Andrew A. Common's 36-inch reflector for Crossley, after its removal to Lick Observatory, California. The original skeleton tube was later modified and strengthened. Hutchinson's *Splendour of the Heavens* II (pl.28) 524.

which could emit their characteristic spectra under the right conditions. Similarly, his studies of the spectra of meteors and comets, and the examination of meteoritic fragments in the laboratory led him to argue that meteoritic material was a type of building material from which all self-luminous bodies in the Universe were formed [49]. In some respects, this controversial theory of Lockyer's has some tangential relation with Sir John Herschel's suggestions of 'particulate matter' as a fundamental cosmological building-block.

In their day, these and others of Lockyer's conclusions were seen as far-fetched, and speculations in advance of his evidence. They were in marked contrast with the interpretative caution of evidence displayed by Huggins.

In 1911, however, when the Solar Physics Observatory was to be moved to Cambridge, the 74-year-old Lockyer retired from professional astronomical research, and returned instead to his amateur roots. The well-equipped private observatory which he founded at Salcombe Regis, Sidmouth, in Devon, was to be his new scientific base until his death in 1920. In many ways, the 'Norman Lockyer Observatory', which still operates on the same site today, is one of the great monuments of the British amateur astronomical tradition, and is still run by serious astronomers involved in research and the public teaching of astronomy [50].

LORD CRAWFORD'S DUN ECHT OBSERVATORY

The move towards professionalisation, though in a form which was rather different from that of the mainstream English astronomical tradition, was reflected in the establishment and operation of the Dun Echt Observatory, on the Aberdeenshire estates of Lord James Ludovic Lindsay, twenty-sixth Earl of Crawford and Balcarres, between 1872 and 1892. Like Lords Wrottesley and Rosse before him, Lord Lindsay was an astronomical nobleman who was willing to spend liberally upon what he recognised as an intellectually adventurous branch of science [51]. Yet in many respects, Lord Lindsay's noble patronage and active participation in research is perhaps closer to that of the European courts than that of Britain, where direct action by feudally powerful figures was a more commonly encountered way of funding science. One also wonders how far Lord Lindsay's great observatory, rather like Lord Rosse's in central Ireland, was intended to draw attention to those local power bases within the British Isles where old landed interests, as opposed to *nouveau riche* commercial fortunes, still took initiatives.

Dun Echt had a 15-inch refractor by Grubb, a 40-foot focus solar telescope, an optical laboratory, siderostats and heliostats, a heliometer, and an impressive range of powerful spectroscopes. As Professor Hermann Brück reminds us, it was his Lordship's intention to found an establishment which would rival the Imperial Russian Observatory at Pulkowa [52]. Dun Echt, which in 1876 was valued at £20,000, was a Grand Amateur's observatory insofar as it was the property of a single enthusiastic individual, yet it was a professional one in the respect that it had a properly salaried Director who undertook the business of research. Unlike Parsonstown (Birr), however, this was not an observatory conceived of, built, and used at first hand by an aristocratic engineer–astronomer, who after a few years found it necessary to employ an assistant. Instead, it was more obviously an act of patronage, intended to be used on a daily basis by others.

And as a patron's observatory, it was essential that the Directors should get on well with Lord Lindsay. The first Director, the Aberdonian David (later Sir David) Gill, stayed only a few years until 1876, before going on to a major career in astronomy culminating in the post of Astronomer Royal at the Cape. Gill – or so his biographer George Forbes argues – enjoyed a relationship 'entirely devoid of friction' with Lord Lindsay, although it is clear that the two men decided to part – on amicable personal and financial terms – because Gill wished to take Dun Echt in a different research direction from that envisaged by Lord Lindsay [53]. The second, Ralph Copeland, was a Göttingen-educated Lancastrian who was to be connected with Lord Lindsay in various capacities for the rest of his career. And as in a proper professional establishment, the Directors had qualified assistant staff [54].

In addition to work in Scotland, Lord Lindsay financed and accompanied major expeditions to observe eclipses and similar phenomena in several distant parts of the world. The expedition which he sent to Mauritius to observe the 1874 transit of Venus was later described as 'perhaps the most completely equipped one ever undertaken by a private individual in the interests of astronomy' [55]. For Dun Echt, and its expeditions, accomplished pioneering work in the study of the solar corona, in photography and spectroscopy. Ralph Copeland was, moreover, one of the first scientists to obtain a spectrum for the Andromeda Nebula supernova (then thought to be a nova) in 1885, while in 1888 he detected helium – an element still only known to exist in the Sun – in the Orion Nebula [56].

The Scottish scientific patriotism that was implicit in the whole Dun Echt enterprise took on a sharper focus in 1888, when the now elderly Charles Piazzi Smyth retired from his post of Astronomer Royal for Scotland. And as was pointed out in Chapter 2, Lord Lindsay was outraged when the British government now threatened to close down Scotland's national observatory along with its standing as a royal institution. Lord Lindsay's proposed removal of the Dun Echt Observatory into Edinburgh, and Copeland's appointment as Astronomer Royal for Scotland in 1889, transformed Scottish official astronomy at a stroke. The Observatory was eventually relocated on Blackford Hill, to become the Royal Observatory, Edinburgh, while complete with his Lordship's comprehensive astronomical library, Edinburgh rapidly became one of the world's major centres for astrophysics. In terms of its far-reaching consequences, and the fresh impetus which his work and patronage gave to the growing professionalisation of astrophysics, Lord Lindsay was probably the most influential Grand Amateur astronomer of the nineteenth century. And it is not without significance that his two consecutive Dun Echt Directors came to bear the title of Astronomer Royal [57].

One cannot avoid drawing the conclusion that the technologies both for gathering and analysing photographic and spectroscopic data were of enormous importance in moving fundamental research towards those economies of time and resources which were essential for the professionalisation of the science. Yet while the new royal foundations of Europe, such as Potsdam, the millionaire graduate schools of the

American universities, and the aristocratic patronage going on within British astronomy were professionalising astrophysics, one must not lose sight of the fact that important individual Grand Amateur work was continuing to be done. In particular, three British astronomers, Frank McClean, Andrew Ainslie Common, and Isaac Roberts, who had made their money in aspects of the construction trades, were significant in the continuity of independent astrophysical research.

TWO ENGINEERS AND A MASTER BUILDER: MCCLEAN, COMMON AND ROBERTS

In some ways, one can see parallels between the careers of Frank McClean and Henry Draper. Both men were born in 1837, the sons of distinguished scientific men – McClean's father, John Robinson McClean, was a successful civil engineer, F.R.S., and Member of Parliament – as well as the recipients of first-class educations. Both men were drawn to original work in astrophysics, and by the early 1870s had ample means to enable them to devote much of their time to astronomical research. Frank McClean was born in Glasgow, and in the way that Draper had followed in his father's footsteps with regard to medicine, so McClean followed his, and became a civil engineer. But five years after his marriage in 1865, McClean gave up civil engineering to occupy himself with linguistics, collecting *objets d'art*, and astrophysics, building observatories at his country house near Tunbridge Wells, Kent, and later at Rusthall, while still retaining his London house in South Kensington. How far McClean's marriage to Ellen Greg in 1865 augmented his financial independence is not clear, but it took only eight years for him to pass from articled engineer to a partnership in his father's firm, thence to a senior post at Barrow-in-Furness docks, and into retirement [58].

McClean's scientific significance began when he started to conduct original work on the solar spectrum, and used his engineering knowledge to produce the improved 'McClean' spectroscope. But in 1895 he acquired a 13-inch aperture equatorial refractor from Sir Howard Grubb of Dublin, and with a large prism mounted in front of the object glass, began a spectroscopic survey of all those northern hemisphere stars that were brighter than magnitude 3.5. Upon the completion of this project in 1897, Sir David Gill, Astronomer Royal at the Cape, invited McClean to take his prismatic apparatus to the Cape of Good Hope, where he extended his survey into the southern hemisphere, to cover the entire heavens. This extremely important astrophysical project – completed by one man, without an assistant, using one prism, and hence possessing an all-important uniformity of results that was otherwise hard to maintain in the 1890s – won him the R.A.S. Gold Medal in 1899 [59].

And also like Draper, and many other wealthy Grand Amateurs, Frank McClean became a major patron of professional astronomical research. He gave a 24-inch photographic telescope with a twin 18-inch visual instrument, along with the necessary spectroscopic apparatus, to the Cape Observatory; and, in addition, presented his old University, Cambridge, with a munificent £12,500 to endow Isaac Newton Scholarships, and made donations to the Fitzwilliam Museum [60].

Andrew Ainslie Common was the son of a Newcastle-upon-Tyne surgeon who had won distinction for his work on cataract. While he was still an infant, however, his father died, as a result of which Andrew's uncle eventually took him into the firm of Matthew Hall and Company, sanitary engineers of London. In this profession, Common made his money, and retired from business at the age of 49, in 1890.

Common had shown astronomical interests even as a 10-year-old when his mother borrowed a telescope for him from a medical friend. Twenty-three years later, in 1874, he bought a 5½-inch aperture instrument, which came to be set up at Ealing, London, which was to be the site of his future astronomical work [61]. Moving from visual to photographic astronomy, Common began to consider reflecting telescopes, and commissioned first an 18-inch, and then a prodigious 36-inch silver-on-glass reflector from George Calver. With these instruments he became increasingly concerned with the photographic investigation of the Orion Nebula, so that he, and Henry Draper, have sometimes been thought of as engaged in a race with each other. Indeed, the congruity of Common's and Draper's enthusiasm for the research potential of photography with large-aperture instruments was brought home when they took independently simultaneous photographs of the Great Comet of 1881 on the night of 24 June 1881 [62].

Common first photographed the Orion Nebula on a gelatine dry plate in January 1880, using his 36-inch telescope at Ealing. But the result was not satisfactory, due to a defect in the drive mechanism of the equatorial mount [63]. Draper, it will be remembered, had obtained improved photographs in October 1880 and in March 1881, and then, on 17 March 1882, Common re-took the lead with an Orion photograph 'which excited the admiration of all the astronomers who had an opportunity of inspecting it' [64]. Sadly, death caused Draper to drop out of the race, at the onset of the Orion-observing winter season of 1882–3. Then on 30 January 1883, Common obtained his finest photograph of the Orion Nebula with a 37-minute exposure. It showed the superiority of the photographic plate as against the drawing when it came to nebula work, and pointed towards the astronomical technology of the future. This picture won the R.A.S. Gold Medal for Common in 1884 [65].

Both Common and Draper had realised that the accuracy of the telescope's driving mechanism was fundamental when photographing deep space objects. Although Calver had made the 36-inch mirror, Common had mounted it himself. As an engineer by profession, Andrew Common had developed a sophisticated equatorial fork mount, floating in a mercury bath, for his 36-inch at Ealing. Yet the drive clock was still not perfect, and in 1880 was still showing the stars as little lines and the Orion Nebula itself as a blur on long exposures. While he made important adjustments to the drive, his engineer's ingenuity led him to take the line of least resistance. Instead of worrying about a critically accurate guidance system for the entire telescope, Common developed a guided photographic plate holder. The cross wire in a sighting telescope attached to the plate holder could be fixed upon a chosen star, so that Common could guide the plate manually, in a way which enabled him to override any irregularities in the overall guidance system of the telescope [66].

Having proved its worth by 1885, Common sold the 36-inch to the Halifax textile manufacturer Edward Crossley for £2,500. Finding that his Pennine observatory and mansion were very poorly located climatically for the proper use of this instrument, and that he was unable to use it as effectively as his $9\frac{1}{3}$-inch Cooke refractor, Crossley presented the 36-inch to the Lick Observatory in California [67].

But as soon as he had got the 36-inch off his hands, Dr Common (who received an honorary LL.D. degree from St Andrews in 1891) put all his energies into a mirror of 5 feet aperture. Common resolved to figure the mirror himself. By September 1886 the polishing machine was finished, and glass blanks ready for working. February 1889 saw the instrument completed, although Common was to give the great glass mirror further polishes, and re-silverings, before he was pleased with it [68].

Then one night, in a situation reminiscent of the earlier reflecting telescope accidents and near-accidents alluded to by Maria Edgeworth and Sir David Brewster, Common almost fell off the observing stage when working at the Newtonian focus with the 5-foot. The incident so clearly frightened him, that he decided to redesign the entire instrument so that it would work at an oblique Cassegrain focus, although it was never a success in this arrangement [69]. The 5-foot reflector consumed many of Common's resources, and a good five years of his time. It never fulfilled the photographic potential for which it was designed, and never took photographs of the same ground-breaking significance as the 36-inch. Yet it was a pioneering instrument, built entirely within the Grand Amateur tradition, and for a few years made Ealing the home of one of the most powerful telescopes in the world.

The incentive to use large-aperture photographic instruments, with accurate guidance systems, to resolve the structures of more deep-space objects, became one of the driving concerns of astronomy within the last twenty years of the nineteenth century. Even though the gaseous composition of many nebulae could now be demonstrated spectroscopically, astronomers realised that their physical structures were still a mystery. Long-exposure photography seemed the way forward towards solving that problem. And the foremost operator in this field after 1885 was the Welshman Isaac Roberts, whose fortune, accrued in the Liverpool building trade, enabled him to take his place as a pre-eminent Grand Amateur astrophotographer.

In 1879, Roberts had acquired a 7-inch refractor by Cooke of York, which he set up in his private observatory at Maghull, a suburb of Liverpool. He then augmented his observatory with a 20-inch photographic reflector by Sir Howard Grubb of Dublin, and like Huggins and Draper before him, mounted both instruments upon the same polar axis, so that one telescope could be used to guide the other for long photographic exposures [70]. But when the supernova was discovered in the Andromeda Nebula in 1885 (of which Copeland obtained an early spectrum), key questions which had lain unresolved for decades were once more brought into prominence. Could this lenticular object with a larger, brighter centre, and extremities that faded away into space – the Andromeda Nebula – really be composed of individual stars if one single new star could suddenly outshine the rest? It re-opened the whole question about whether the nebulae in general were component parts of the Milky Way, or vastly remote 'island universes'. Photography, with its ability to store up light in long time-exposures, offered a solution [71].

Roberts had already taken several exposures of the Andromeda Nebula with the 20-inch reflector, using the 7-inch Cooke as a guiding telescope, before he began to obtain significant results. On 10 October 1887, however, his plate revealed better structural detail, which seemed to confirm what the Bonds – father William Cranch and son George Phillips – had detected with the Harvard 15-inch many years before [72]. The Andromeda Nebula seemed to consist of a nucleus, surrounded by circular arcs. And as the eye of the observer is rarely without prejudice when it comes to interpreting faint detail, it had been argued that the Andromeda Nebula was an example of a condensing system, similar to the one that Pierre Laplace had suggested decades before as an origin for our Solar System. Gravitationally controlled bands of matter would eventually condense into planets around a central sun [73].

A better plate, taken with a guided 4-hour exposure on 29 December 1888, shows what we now recognise as the left-hand spiral arms of the nebula, along with many individual stars in these arms. Yet Roberts still cautiously retained his Laplacian model of this object until he re-interpreted it from the superb 90-minute exposure plate taken on 17 October 1895, which he published in the second volume of his *Photographs of Stars, Star Clusters, and Nebulae* in 1899 [74]. Finally, by the end of the century he had come to recognise the spiral structure visible in his own photographs, and rapidly confirmed his new interpretation with superior photographs in 1899 and 1902. But these latter photographs were taken in the clearer air of Crowborough, Sussex, where Roberts had moved to in 1890 [75]. Like so many of his Grand Amateur predecessors, Isaac Roberts had moved his observatory–residence away from the original urban location where the enabling money had been made, to become an astronomical country gentleman. One wonders how closely he was conscious of following a similar path to his Liverpool predecessor, William Lassell, who in 1837, 1852, 1858 and 1861 had tried to move his astronomical establishment a few steps ahead of the smoke.

And yet the air of suburban Liverpool was not only sufficiently clear to obtain 4-hour exposures of the Andromeda Nebula in the 1880s, but also to capture the nebulosity surrounding the stars of the Pleiades, which Roberts, along with the Frenchmen, Paul and Prosper Henry, was one of the first astronomers to photograph in the mid-1880s [76]. The Pleiadean nebulosity had first been detected visually by Professor Tempel working with a 4-inch refractor in 1859, although photography, it was hoped, might solve the vexed question of whether the nebulosity was changing [77]. Photography now promised to be a tool of arbitration in cosmology, where its stored-up light was capable of detecting detail which lay beyond the visually accessible. And Isaac Roberts' resolution of the Andromeda Nebula provided a classic vindication for this usage of photography, along with the prospect of detecting any possible changes in the Pleiades or Orion Nebula.

One must also remember that Isaac Roberts followed another social trait which had developed within the ranks of the spectroscopic and photographic Grand Amateurs: in 1901 he married Dorothea Klumpke, who became his active co-worker.

It says something about the changing role of women in the late nineteenth century that whereas earlier generations of Grand Amateurs had often married wealthy women who took a keen interest in their husbands' scientific work, the post-1870 generation of amateurs often became research duos: William and Margaret Huggins, Henry and Anna Palmer Draper, and Isaac and Dorothea Klumpke Roberts. Indeed, Dorothea Klumpke Roberts, who was from a Californian family and educated in Europe, showed a development beyond the two previous co-worker wives, for she was not only a professional scientist prior to her marriage, but a senior astronomer working on the *Carte du Ciel* in Paris [78].

THE IRISH PHOTOMETRISTS: MINCHIN, MONCK, WILSON AND FITZGERALD

The accurate measurement of starlight, whereby stars of particular colours and spectral classes could be related to an absolute laboratory standard, was crucial to the development of astrophysics, and lies at the heart of the Hertzsprung–Russell diagram of stellar classification. And it was in Ireland, using Grand Amateur resources, that the foundations of photoelectric photometry were laid, in the early 1890s.

It began when the Anglo-Irishman George Minchin, who held the Chair of Applied Mechanics at the Royal Indian Engineering College, near Staines, London, began to experiment with the photoelectric properties of selenium-coated aluminium plates sealed into glass tubes with acetone. Professor Minchin had devised a suitable photoelectric cell by August 1892, and hoped to try out the apparatus at the private observatory of his Dublin lawyer–philosopher friend, William H. Stanley Monck [79]. Monck, who broke the Protestant Church of Ireland clerical tradition of his family by becoming a distinguished Dublin barrister, had set up a fine observatory in the garden of his Dublin house in Earlsfort Terrace. Its principal instrument was the same 7½-inch Alvan Clark refractor which W.R. Dawes had purchased for $950 in 1854 (see p. 47), and which had subsequently passed through the distinguished hands of Frederick Brodie, Dr Wentworth Erck, and others to become one of the most acclaimed refractors of the century [80].

Unfortunately, Minchin, Monck, and his young neighbour from next door, the future civil engineer Stephen Dixon, were foiled by bad weather from obtaining photometric results from the stars, although in the early morning of 28 August 1892 they did get measurements for Jupiter and Venus. The experiments were taken further, however, at the sumptuously-equipped astrophysical observatory of William Edward Wilson, at Daramona House, County Westmeath, directly north of Lord Rosse's County Offaly [81].

Since 1871, when still a 20-year-old student at Trinity College, Dublin, Wilson had established an observatory complete with a 12-inch Grubb refractor, on his father's estate at Daramona House, after travelling to Iran to observe a total eclipse. The Daramona establishment was augmented in 1881 with a 24-inch Grubb reflector, which he had re-mounted on an electrically-driven equatorial of advanced design. At this observatory, William Wilson began to do important work in astronomical photography, although his scientific instincts were drawn more towards experimental physics than straightforward observation. However, Wilson elegantly combined both of these traits in 1898 when he pioneered the use of time-lapse cinematography to study the changes taking place within a large sunspot [82].

Using the telescopes at Daramona, combined with an improved photometric apparatus built by the Dublin instrument-makers Yeates & Sons, Wilson and his friends obtained the first estimates of the temperature of the solar surface in 1894, while the revised figure obtained in 1901 is very similar to that still accepted today [83]. Similarly, on 11 April 1895 they had obtained the first relative temperatures for the stars Regulus, Arcturus, and η Boötis. In many of these experiments, the management of the electrical recording instrument, which had been set up in isolation in a room beneath the telescopes at Daramona, was in the hands of the distinguished Irish physicist George Francis Fitzgerald, who is better known for his work on the Fitzgerald–Lorenz Contraction, which was soon to be so important in the development of relativity theory [84].

This pioneering photometric work conducted in Ireland was also destined to sow the seeds for an overhaul of the way in which astronomers classified stars by spectral class, brightness, and assumed distance. By 1894, it was generally thought that stars displaying characteristics similar to our own Sun were those of the most physically mature type. But William Monck realised that the parallax-distances of so many stars were unreliable, and approached the problem from a different direction. While parallaxes became unreliable for distant stars, he realised that proper motions were likely to be a more trustworthy guide, for while these motions could not give a star's absolute distance, a sufficiently large statistical sample of proper motions could arrange the stars into zonal distances from the Solar System. He then found that when the proper motions, spectral and photometric characteristics of the stars were examined, the closer, Sun-type stars were duller, and the brighter, non-Sun-like stars were more distant. This reversed the way in which spectroscopists had come to interpret their results when dealing with stellar types and distributions. For Monck, after all, came close to recognising the distance–luminosity ratio of stellar distribution which was to achieve its definitive expression in the Hertzsprung–Russell diagram by 1910.

It is surprising that more scholarly attention has not been paid to these Irish scientists in the development of astrophysics. For it was a group of Trinity College Dublin graduates – consisting of two professional scientists, a lawyer, and a country gentleman – who collaborated to lay the foundations of photometric analysis and stellar distribution theory. Their work further demonstrates how, even in the 1890s, it was an easy social relationship between amateurs and professionals, rather than employment in a research institute, which still so often guided initiatives in new fields of astronomical investigation [85].

GEORGE HIGGS: A SHOPKEEPER SPECTROMETRIST

By the 1880s it was becoming virtually impossible for anyone to carry out fundamental work in astronomy, especially in its astrophysical branches, unless one had plenty of time and money. It is the more remarkable, therefore, that one of the most complete maps of the solar spectrum to be made in the nineteenth century, and published privately in 1893, was the work of a small suburban jeweller and watch-assembler.

Considering the contributions which that climatically-disadvantaged city made to serious amateur astronomy in the nineteenth century, it may not seem surprising that, as Laurie Brock has shown, George Higgs lived and worked in Liverpool [86]. During the 1880s Higgs was living in a modest terraced house on Liverpool's West Derby Road: an area which had been covered by acres of streets since its rural days in the 1840s, when William Lassell's mansion and observatory had stood only a few hundred yards away. Indeed, it is not clear how a man in the modest circumstances of George Higgs acquired his serious interest in solar spectroscopy, and obtained the 10-foot spectrograph with optics by the eminent American instrument-maker John A. Brashear, and one of the precious diffraction gratings of Professor Henry A. Rowland of Johns Hopkins University, Baltimore [87]. When we remember that Piazzi Smyth paid $100 – or about £20 – for a Rowland grating in 1883 [88], and that even before he could begin to use it profitably, George Higgs needed the Brashear optics, a powerful Rhumkorff coil and other electrical parts, along with chemical and photographic apparatus, one appreciates the dent which serious spectroscopy inevitably made in the means of a small shopkeeper. It is hardly surprising that his family seem to have resented his expensive hobby and grudged him the entire back bedroom which the spectrograph took up. It is also likely that they were aware of the enormous amount of time and money which was consumed in privately publishing *A Photographic Atlas of the normal Solar Spectrum* around 1893. The core of this work consisted of 106 individual paper photographs, printed from the original negatives, to form Higgs' great map of the solar spectrum. He was obliged to sell his *magnum opus* for 8 guineas (£8.40) a copy, and one suspects that he failed to recover his costs [89].

THE ESCALATION OF COSTS: TOWARDS PROFESSIONALISATION

The new sciences of light transformed the whole scale of astronomical research after 1860, not only by providing techniques for the analysis of starlight and the recording of nebulae, but in those unexpected ways which fundamental innovations inevitably do. Although astrophysics and astrophotography in Great Britain and America continued to be dominated by wealthy amateurs for some years further, the photographically recorded image played a decisive role in professionalising the science by facilitating, as we have seen, a more economical use of telescope time, followed by the 'industrial' processing of photographic plates by trained specialists. And perhaps this was nowhere more propitious than in the *entrée* which it gave women into professional astronomy, especially in the United States. Although the lone amateur still continued to be significant until after 1900 – as was George Higgs down to 1909 – the whole scale of serious astronomy had irrevocably changed by that date, and threatened his extinction.

Moreover, those very instruments which the Grand Amateurs had pioneered – such as big reflecting telescopes, astrophotography and photoelectric photometry – were now outstripping the resources of even well-to-do businessmen. For while an amateur like Common could afford to build a sophisticated reflecting telescope with a 5-foot mirror in 1888, the climates in which such people invariably lived rendered them of very limited value. Liverpool, Dublin and central Ireland, New York, or the London suburb of Ealing, were poor locations indeed for the siting of major new instruments by 1890. For once an astronomical institution had a budget to work within and staff to pay, it was wasteful to continue working in a location where the sky could only be accessed on occasional good nights. The mountain-top observatory which Piazzi Smyth tried out at Teneriffe in 1856 [90], which Henry Draper proposed for the Rockies [91], and upon which Edward Crossley's 36-inch was eventually mounted at Lick [92], with its clear and much more predictable atmosphere, was becoming a matter of necessity if 'industrialised' research was really going to develop. And this required that major instruments should no longer be wasted by locating them in their owner's private gardens, no matter how large and elegant those gardens might be. So it was inevitable that, when the retired steel-founder Sir Henry Bessemer set up a 40-inch reflecting telescope in the gardens of his Denmark Hill, London, mansion in the 1890s, it would remain little more than a retired gentleman's plaything [93].

By physically divorcing the individual astronomical patron from the Rocky Mountain, Andean or Pyrennean sites where big instruments could be used to the greatest advantage, however, a vital nerve in the anatomy of Grand Amateur astronomy came to be severed. And this physical separation between the parent institution and the distant observatory was only possible in an age when steamships, railways and telegraphs had made it relatively easy to control what was going on in an isolated location several thousand miles away from its home base. Indeed, this problem had been foreseen by Richard A. Proctor in his 1873 letter to the *English Mechanic*, written as part of his campaign to obtain adequate government funding for British science. Although Proctor praised the Herschels, Rosse, and Lassell, he argued that modern astronomical research demanded mountain-top locations, and 'who can reasonably be expected to devote two or three hundred thousand pounds to such an undertaking' out of their private pocket? But a major mountain-top observatory would cost the government much less than the construction of a single ironclad battleship! [94]

Yet it was the United States of America which became the undisputed beneficiary of that train of events which led to big-technology astronomy, backed by both government and commerce. For while the independent Grand Amateurs may have been dying out by 1900, it was becoming possible to find industrial magnates like John D. Hooker and the family of George Ellery Hale who were willing to finance mountain-top observatories, and were happy to let the hired professionals do the work [95]. In England, however, it was a different story. Though there were some observatories within the Empire, with the exception of the wholesale move of the Radcliffe Observatory, Oxford, to Pretoria in Africa in 1935, Great Britain lacked both the industrial magnates who were willing to found great research observatories in far-flung locations, and the home-university endowments that were necessary to make proper use of them. Not until the founding of the cloud-unencumbered radio observatories of Jodrell Bank, Manchester, and Cambridge in the late 1940s did British observational astronomy begin to regain its nineteenth-century distinction in fundamental research. But by 1950, Grand Amateurs as initiators of fundamental research were a thing of the past. Indeed, one could argue that in its spectacular success in the nineteenth century, Grand Amateur astronomy contained the seeds of its own destruction, by creating a technical and an intellectual momentum which became too big for private individuals to control and sustain.

8

The astronomers' gentlemen: the Grand Amateurs' professional assistants

When John Weale published his book describing the five public and fourteen private astronomical observatories that were operating in the vicinity of London in 1851, he made a point which said much about the Grand Amateur enterprise. For while private science had become 'fashionable' by that date, if the proprietor of an observatory did not bother to make financial provision for the employment of a salaried observer, it 'must become at length either an incumbrance or plaything to its owner' [1]. For even a gentleman of leisure often had demands placed upon his time, by virtue of his social position, which could make regular nightly observation virtually impossible. The particular astronomer with regard to whom Weale made this general point (and who was discussed in Chapter 5) was Samuel Charles Whitbread, who in addition to his passions for astronomy and meteorology, was a businessman, magistrate, and leading figure in the county of Bedfordshire. It was a wise and responsible act, therefore, for him to pay John B. McLarin to look after the daily work of his Cardington Observatory, in the same way that his similarly situated Buckinghamshire friend, Dr John Lee, had placed James Epps in charge of his [2].

While it is true that many Grand Amateurs – such as the Herschels, Admiral Smyth, and even Lord Rosse for the first few years of the Birr Observatory – seem to have no paid astronomical helpers, the appearance of the 'professional assistant' became a necessary fact of life in Victorian astronomy, especially if one had other responsibilities to attend to. Indeed, these were the men who did much of the routine work in the nation's observatories, both public and private, in much the same way that bailiffs, stewards and clerks had their functions in the running of a great man's household. On the other hand, astronomical observers often occupied an ambivalent place, betwixt servant and gentleman; for while their job demanded intelligence and a good education, their prospects were limited, especially before the 1870s. In many respects, they were the male counterpart of that other lost Victorian soul, the governess, who dined at the family table and yet was expected to hold her tongue. Yet as with governesses, the social relationship of an assistant astronomer to his employer could be coloured by a myriad of personal circumstances which could create genuinely fruitful working relationships. And in the right circumstances, being a Grand Amateur's assistant could open up wonderful prospects which could transform one's entire life.

When, for instance, the Duke of Marlborough set up a well-equipped astronomical observatory at Blenheim Palace, Oxfordshire, in the middle of the eighteenth century, it became the workplace for two gifted estate workers. Thomas Phelps had started his working life as a stable-boy, while his colleague, John Bartlett, had been a shepherd. Although I have not been able to find documents which indicate their salaries and conditions of employment, it is clear that astronomy had taken Phelps and Bartlett many rungs up the social ladder by the time that the Duke had their portraits engraved as part of a picture of the interior of his observatory. The now elderly ex-peasant boys are shown elegantly dressed and wigged, holding a discussion at the eyepiece of the Blenheim transit instrument, and visually quite indistinguishable from any pair of Fellows of the Royal Society [3].

The constant threat to the assistant astronomer's life, however, especially if it was not spent in the service of an aristocratic family, was its uncertainty, for it only needed the master to die, or switch his interests elsewhere, for the assistant to be out of a job, and often a home. What sort of people, therefore, in addition to gifted estate workers, became professional assistants?

The best initial guide that we possess into this 'below stairs' world of the Victorian observatory are those letters preserved in the papers of Sir George Airy from young men, or from their fathers, who were seeking employment on the staff at Greenwich. For an hourly-paid 'Supernumerary Computership' at the Royal Observatory could be the first step on the ladder.

Most of the candidates applying to Airy from 1836 onwards were the sons of modestly well-off London or Kentish tradesmen. For as Airy assured Algernon West, who was Gladstone's Secretary, in 1869, it was their 'pliability' or receptiveness to training which was all-important: 'For Assistants I usually begin with lads, the sons of small tradesmen ... I do not like to begin with men who are past the pliable age' [4]. These lads would often have been educated at local grammar schools, or else privately tutored by local clergy up to the age of 15 or 16. They would invariably proclaim a love of looking at the Moon and stars, and possess a facility at arithmetic, geometry and some languages. Airy liked his staff to have an acquaintance with French or German, and perhaps Latin, but he was watchful of candidates who were too well educated, and sometimes spelled this out to unsuccessful applicant [5].

It may seem strange that, with such obvious skills already in hand, these young men were not aiming for higher things in life. We must bear in mind, however, that 160 years ago, university was rarely seen as a viable option for lower-middle-class English youths (although it was more so for their Scottish counterparts), and that open and adequate scholarships were rare. If one were a Roman Catholic, a Methodist, a Dissenter, or a Jew, moreover, one was ineligible for scholarships to Anglican Oxford and Cambridge before the 1870s, while London's 'Godless' University College, which had no religious barriers, was still too expensive for families which could not afford to maintain their 16-year-old sons for another four or five years. From what we can glean of the religious backgrounds of Airy's junior assistants, therefore, it is no surprise that non-Anglicans figure prominently among them, though the evidence is too scanty to quantify them as a percentage of the total number of professional assistants employed at Greenwich, or indeed elsewhere. At Greenwich, William Rogerson (whom Airy inherited as an adult in 1835) 'is a Dissenter ... I think [a] Methodist' [6]; while the two generations of the Breen family that were on the staff in the early 1840s were Irish Roman Catholics. Andrew Graham, who worked first at Markree as Edward Joshua Cooper's Assistant, and then under John Couch Adams at Cambridge, was a 'staunch Wesleyan Methodist' [7].

As a Supernumerary Computer, a young man worked between 8 a m. and 8 p.m. each day in the Royal Observatory's Octagon Room, at a rate of between 6d (2.5p) and 10d (4p) an hour, for six days a week [8]. Here, he would be 'drilled' in the performance of specific mathematical tasks, using filled-in forms and volumes of tables, such as those needed for the reduction of the Moon's true place from meridian observations. If the lad showed an aptitude for instruments, he might – after a year or so – get himself attached to an observing department of the Observatory, such as the meridian transit circle, or the altazimuth, and through a mixture of regimented instruction, practical experience, and private study, turn himself into a precise and 'regular' observer by the age of 24 [9]. The Royal Observatory possessed a good library, which Airy constantly sought to expand and improve, and the Supernumerary Computers were encouraged to engage in private, improving study, in what Airy called the intellectual centre of the Observatory [10]. Airy generally regarded computing as young men's work, and while there were some older men, the goal of the dozen or so such people on the Observatory's staff at any one time was to step into the vacant shoes of one of the six 'Warrant Assistants', when one of them should retire or die, to obtain a permanent Civil Service post. If he was so lucky, as was the 24-year-old Edwin Dunkin, who was warranted in 1845, the young man faced an untroubled run to the grave. An established salary scale and promotion from £130 to around £240 a year was his for an official five-hour working day, plus some night duty [11]. He might also have up to six weeks' annual holiday, and a decent pension at 65. There was in addition a day's holiday each month, when there was no Moon to observe [12].

If he was not so fortunate, however, the young Computer either left astronomy altogether for commerce or a different Civil Service department, or else got a post in another observatory. Even so, after a stint at the Cambridge Observatory, or working in a Grand Amateur establishment, he often kept an eye open for senior Greenwich vacancies, for the Royal Observatory paid roughly double the wages that were obtainable elsewhere, and there was a blessed security of tenure to be had nowhere else.

In this small world where aspiring candidates inevitably exceeded the available number of decent jobs, personal contacts and a good 'character' were fundamental. High-flying originality or scientific research abilities were not the characteristics that were looked for, and nowhere were the requirements more clearly spelled out than when Airy offered the Chief Assistant's job at Greenwich to the Revd Robert Main M.A., formerly a Fellow of The Queen's College, Cambridge, in 1835: 'The employment[s] in general will be more like those of a head clerk in a bank than any other that I can compare them with: and as in a bank or manufactuary, punctuality and regularity in a routine of very dull business are quite indispensable' [13]. And if these were the conditions for the senior professional assistant astronomer in Britain, with Holy Orders, a Cambridge first-class honours degree in mathematics, and a salary of £350 per annum [14] behind him, one can imagine the qualities looked for in the more humble lads who had risen up through the 'apprenticeship' system!

Yet there was never a lack of applicants from whom to choose. Hugh Breen junior was one of that dynasty of Irish Greenwich Assistants whose Roman Catholic faith barred them from English higher education. Hugh junior was brought into Greenwich as a Computer by his Assistant father, Hugh senior. Young Hugh's brother James also worked as a Supernumerary Computer, and in June 1846 James was strongly recommended by Airy to Challis, for a temporary Assistantship at the Cambridge Observatory [15]. Hugh had got himself warranted at Greenwich in 1844, but resigned his post in 1859 to establish a school for 'young Catholic Gentlemen' at Enniskillen, the same Irish town where Thomas Romney Robinson of the Armagh Observatory was the Anglican rector. Sadly, the school failed, and following the resignation of his Warrant Assistantship, Hugh Breen was facing destitution by the summer of 1859 when one Mrs Catherine Wicks of the Isle of Man was writing to Airy on his behalf, requesting work [16]. He seems to have gone back to working as a Computer, which only emphasises the precariousness of employed astronomers once they made a bid for an autonomous existence.

Hugh's brother James returned from Cambridge Observatory to Greenwich computing, and finally left the Royal Observatory for a City clerkship in 1860 [17]. William Ellis fared better. He started as a Greenwich Computer in 1841, and moved to Durham University Observatory as Temple Chevallier's Assistant, before getting warranted at Greenwich in 1853, and many years later, becoming an F.R.S [18].

Airy was in fact sometimes criticised for the salary and employment conditions under which his Assistants – especially the Supernumerary Computers – were required to work, and in September 1856, the *Morning Chronicle* attacked him for treating them like servants [19]. Considering the highly-sought-after nature of the established Greenwich posts, however, one is forced to ask what employment conditions were like in the academic and Grand Amateur observatories. One also wonders what kind of career and social position a modestly-educated young man envisaged for himself when he resolved to become an astronomer.

Precise information about the pay and working conditions of the university and Grand Amateur professional assistants is not always easy to find, for unlike Greenwich, formal records are more patchy, as jobs seem to have been considered more *ad hoc*. In 1854, however, when Airy was borrowing assistants from Oxford, Cambridge, Durham, and from Richard C. Carrington's Redhill Observatory for his experiment to remeasure the Earth's gravitational constant at South Shields, records of official payment were made. As each man was paid for one month of his normal salary, it is possible to calculate what they would have made per annum in their employing observatories. One of the best-paid assistants involved in the South Shields project in 1854 was Norman Robert Pogson, who was borrowed from Manuel

Johnson at Oxford's Radcliffe Observatory, where he received a salary of £120 a year [20]. Temple Chevallier's Durham assistant, the German George Rümker, received £100 a year, though as it will be recalled from Chapter 2, Durham University was unique in British public observatories in the respect that it generally employed graduates on short-term contracts [21]. Rümker, indeed, was also a man of standing, and the son of the Director of the Hamburg Observatory. George Criswick, who had been one of Airy's Greenwich Supernumerary Computers, was receiving the rather miserable £70 per annum at the Cambridge Observatory in 1854 [22]. And George Harvey Simmonds, who was also styled as the valuable 'friend and assistant' of his employer, Carrington, during his work at the Redhill Observatory, seems to have been one of the best-paid assistant astronomers of the day. Carrington's correspondence with Airy in 1854 indicates that Simmonds was receiving 10 guineas a month, (£126 a year), which was a generous income for the period [23].

The provision of accommodation was obviously an important factor when employing a professional assistant, for while it was important for him to be near to the observatory where he was employed, it was an advantage to have a house of one's own. The Royal Observatory Assistants' salaries included a 'rent allowance' which grew with one's grade, so that a man could take a house of his choice in Greenwich. This was important, for it gave the Assistant a life of his own, as only the Astronomer Royal had an official residence on Greenwich Hill. One suspects that live-in astronomers would have been acutely aware of occupying the same social no-man's-land as the governess, although if it was in a childless household with less social paraphernalia, and the employer regarded his assistant as his friend, as was the case with Simmonds and Carrington, it could have been much more pleasant.

But one of the most telling insights that we possess into the financial and social dynamics of the assistant's life can be gleaned from a letter to Lord Rosse from Thomas Romney Robinson in November 1847. Lord Rosse was considering employing an assistant to look after the 72-inch (perhaps because of his own involvement with potato famine relief, and forthcoming Presidency of the Royal Society), and Robinson was advising him on the going rates for salaries. Robinson's own assistant at Armagh, Neil McNeil Edmondson, 'gets £75 per annum but then he has a house rent free and has something as a member of the [cathedral?] choir, so he is not altogether dependent on what he gets from them' [24]. Robinson went on to make it clear in his letter that Edmondson could not have lived properly on an unsupplemented salary of £75. The assistant at Dunsink Observatory, Dublin, got £100, 'and a small house and garden'. Robinson clearly respected an assistant's right to a life of his own, and firmly castigated the live-in arrangement. 'Cooper [owner of the Markree Castle Observatory, in County Sligo] I think gives his assistant £100 and he lives in the house, an objectional arrangement for more reasons than one.' This was Andrew Graham, who served Edward Joshua Cooper loyally from 1842 to 1860, without a pay rise, and did a great deal of the work at Markree. Robinson admitted that Greenwich rates of pay were higher than those of Ireland, but so was the cost of living in London [25]. Even so, the Irish observatories generally employed only one assistant apiece, so that a man lacked the chance to inherit better-paid dead men's shoes that he had at Greenwich.

The salaries at the Royal Observatory, Edinburgh, were no better. Alexander Wallace, who was Piazzi Smyth's Assistant in the 1840s, received £100 a year, with accommodation provided in the old Tower House, on Calton Hill. The second Assistant received £6 per month, in a salary structure very similar to that of the Second Assistant at the Cambridge Observatory. Piazzi Smyth was appalled by the low salaries that were the going rate for educated men doing meticulous and responsible jobs [26]. The same point was taken up on a national level thirty years later by Richard A. Proctor when he accused the British government of supporting science in the 'abstract', yet consistently failing to fund it or pay adequate salaries to its 'Science-Workers' [27].

What Robinson's and Piazzi Smyth's remarks most clearly do is open up the wider question of the self-perceived role and expectations of an assistant astronomer. When the gate porters at the Royal Observatory received 12 shillings (60p) per week, or £30 a year [28], a salary of £100 must have seemed ample. Assistant astronomers, on the other hand, generally saw themselves as 'gentlemen', and Airy often pointed out the desirability of 'gentlemanly habits' in candidates [29]. And a salary of £100–£150 a year clearly put one into the middle class in 1840, especially if it came with a house and presumed some education. Indeed, this was also the financial bracket of the City clerk, and one must remember that it is the income range to which several of Dickens' young heroes aspired; such as Nicholas Nickleby, who regarded himself as made when he landed a City job with the Cheeryble brothers at a salary of £120 a year [30]. Yet these expectations were still modest compared alongside the prospects of a man with a good university degree. Airy with his £1,100 a year, or even Robert Main with his £350 rising to £470 [31], had expectations which were so much greater than those of the men who served them.

From 1848 onwards, the Birr Castle observatory had professional assistants, starting with the two Stoney brothers – Dr Johnstone Stoney and Mr Bindon Stoney [32]. What is more, the working arrangement at Birr was clearly a fruitful one, for of the eleven men who worked with the Third and Fourth Earls of Rosse between 1848 and 1916, several went on to careers of real distinction: these included an Astronomer Royal for Scotland (Ralph Copeland), a knighted Cambridge professor (Sir Robert Stawell Ball), and a President of the R.A.S. (Jean Louis Emil Dreyer). An account even survives for the tea, coffeee, and snacks that were served to the men observing with the telescope in 1848 [33]. Although I have not been able to find out what salaries Lord Rosse paid to his assistants, the surviving correspondence shows that they were addressed and treated as gentlemen, as one might expect from the shape of their subsequent careers. Sir Robert Ball left a very warm account of his time at Birr in the mid-1860s, where he was included in the round of balls and dinner parties given at the castle and in London, and of his enduring friendship with the Fourth Earl, Dr Johnstone Stoney, and others [34]. Like Temple Chevallier's University Observatory at Durham (discussed in Chapter 2), it seems that Lord Rosse saw Birr less as a place where an able young man would spend a working life than as a prestigious springboard for him to launch himself onto a distinguished career elsewhere. On the other hand, the last Birr assistant, Dr Otto Boeddicker, who was appointed by the Fourth Earl in 1880, stayed until forced to return to Germany in 1916; his departure concluded eighty years of astronomical research at the Castle [35].

What is clear when one looks at professional assistants in Victorian Britain are the very different aspirations of the individual astronomers themselves. Many of those who aspired to Greenwich as their long home were perhaps more concerned with job security than with intellectual adventure, and those who found Warrant posts too unchallenging generally left. For some this could lead to spectacular failure and ruin, as in the case of Hugh Breen junior, though for others, it could lead to a career of eminence. Charles Todd, after a Supernumerary Computership and a stint at the Cambridge Observatory, threw up his secure new Warrant post at Greenwich after only a year in 1854. He had become an expert in the new 'whizz-kid' technology of electric telegraphy, and became supervisor of telegraphs in Adelaide. Todd's explorations in Australia occasioned the famous springs in the centre of the continent being named after his wife Alice, while his patriarchal position in Australian science eventually led to his being elected F.R.S. in 1889, and knighted in 1893 [36].

Other ambitious young men also looked to the Empire. Although Piazzi Smyth was hardly a typical Victorian assistant, he used an extremely well-paid assistantship to his father's old friend Thomas Maclear, at the Cape Observatory, to help him on his way to becoming Astronomer Royal for Scotland (as was discussed in Chapter 5). Similarly, David Gill moved on from directing Lord Lindsay's Dun Echt Observatory to become the knighted Astronomer Royal at the Cape [37]. Yet a move overseas was not always a good one, and one cannot help feeling that in the case of Norman Pogson, something went wrong. By the time that he was 18, in 1847, this Nottingham youth had drawn attention to himself as a calculator of cometary orbits. This work brought him to the attention of his older Nottingham contemporary, John Russell Hind, who brought him into Grand Amateur employment on the staff of George Bishop's South Villa Observatory, Regent's Park, which Hind superintended [38]. The gifted young man next moved on to Oxford's Radcliffe Observatory in 1852, where it will be recalled that he was receiving £120 a year in 1854 when he assisted Airy in the Harton Pit experiment and over 1859–60 took charge of Dr Lee's Observatory at Hartwell House [39]. But by this time, Pogson was clearly aiming at being something more than a scientific gentleman's gentleman – as well he might, with four asteroids and a French Académie Lalande Medal to his credit – and in 1861 he took on the superintendence of the Madras Observatory.

Yet it was in India that things seemed to go awry, in spite of the discoveries which he made there. He lacked proper assistance, and was eventually obliged to employ, without pay, his own children in that capacity. In particular, Pogson employed his gifted daughter Isis, who was nominated as an F.R.A.S. in 1886, but failed to get elected [40]. It is also clear that, now he was in charge of his own observatory, he did not like having to report back to Airy at Greenwich, and one reads between the lines of the impeccably-penned letters which he sent to the Astronomer Royal that he resented the new 1869 Code of Regulations which was being enforced upon him from London [41]. Pogson in Madras comes over as a lonely and rather embittered figure, especially after the death of his first wife. For though his second wife was the daughter of a Lieutenant-Colonel, one feels that he was never at ease in Raj society, while his heavy single-handed duties and lack of proper holidays made it impossible for him to come home. At the age of 32, in 1861, he had an impressive *curriculum vitae* and list of discoveries and honours to his credit, yet in India his career seems to have stuck fast in a situation where neither relief nor recognition would ever be forthcoming. Norman Pogson died in India at the age of 63, after thirty years of unrelieved service [42].

One extraordinary professional assistant who in reality transformed himself into a full professional scientist, however, was James Glaisher [43]. Although he spent most of his working life at Greenwich, Glaisher created his own empire within the Royal Observatory, where he became Superintendent of the Meteorological Department, and built up a national weather monitoring system which laid the foundations of modern scientific meteorology. Glaisher was certainly not a man who saw the Royal Observatory primarily as a safe scientific job, and while it is clear that Airy had great respect for the increasingly famous F.R.S. on his staff – an Assistant, moreover, who (as we saw in Chapter 5) moved on familiar social terms with Grand Amateurs like Lee and Whitbread – one feels that their relationship must never have been an easy one [44].

One of the most serious weaknessess of the British Grand Amateur tradition was its inability, as a movement, to provide a regular research training-ground for the rising generation, in the way that European state observatories and American graduate schools did. It was in this respect that Birr Castle, George Bishop's Observatory in Regent's Park, Dun Echt, and even Durham were significant, for they began to produce men who were not 'assistants' in the more limited Greenwich sense, but who perceived themselves as professional scientists. Nor is it without significance that these observatories were all involved in fundamental research, rather than routine stellar cartographic work.

Attention has already been paid to Lord Rosse's staff at Birr and the careers on to which several of them passed. One also sees how George Bishop's South Villa Observatory in Regent's Park (mentioned in Chapter 3) must also have been conducive to a growing perception of professionalisation, especially in the person of John Russell Hind, who directed the Observatory from 1844 to the 1860s. Indeed, Hind brought enormous renown to this private observatory, where he was given enough rope to discover ten asteroids and two comets in his own right, win an F.R.S. in 1851 and a medal from the French Institut [45]. It further says something for the standing of Hind's work at Regent's Park that it was to him that Dr Franz Friedrich Ernst Brünnow wrote to announce the discovery of Neptune in Berlin in September 1846 [46]. Hind was the first British astronomer to successfully identify Neptune, and he announced his results in a letter to *The Times* dated 30 September 1846, which was the day on which he had sighted the planet [47]. Although Hind, who was the son of a Nottingham lace manufacturer, had started his astronomical career as a 16-year-old Supernumerary Computer at Greenwich in 1839, one surmises that he, like Pogson, would have found a Warrant position at the Royal Observatory to be too restrictive. Yet even after being honoured with a Civil List Pension, and beating John Couch Adams to the Superintendentship of the *Nautical Almanac* in 1853, Hind still continued to work at Bishop's establishment at South Villa, Regent's Park, to add to his own discoveries, and reflect some glory on to his patron [48].

It is quite clear that by the middle and later years of the nineteenth century a different type of professional assistant was emerging: men who saw themselves as being more professional than assistant. And it is not for nothing that this change took place at the same time as the three above-mentioned private research observatories – Birr, Regent's Park, and Dun Echt – were operational. The perceived change in the social and professional status of these astronomers is also evident in the tone of their correspondence. Men like Hind, Copeland, Ball, and Gill – two of whom were knighted and three elected F.R.S. – corresponded in their own right from their observatories, and were not so much their employer's 'men' as the earlier generation of assistants had been. The manuscript autobiographical reminiscences of Ralph Copeland's daughter Fanny, preserved in the Library of the Royal Observatory, Edinburgh, bring out the easy social relations which existed between her father's family and that of the Earl [49].

Another self-perceived professional assistant was the German Albert Marth. Although educated at Berlin and Königsberg Universities, Marth spent his entire professional life in Britain, mostly in Grand Amateur establishments. He joined Hind on the staff of Bishop's South Villa Observatory in 1853 at the age of 25, when Hind was spending part of his time superintending the *Nautical Almanac*, and where Marth co-discovered (with Pogson in Oxford) the asteroid Amphitrite soon after. Indeed, asteroid discovery had become almost routine at South Villa by 1855, making it a centre of excellence among the European observatories of the day [50].

From South Villa, Marth went on to Durham to work with Temple Chevallier. Then, between 1863 and 1865, he became what I believe was William Lassell's only professional assistant, at Malta. It was at Malta that Marth carried out important nebula work with Lassell's 48-inch, which Lassell published with full accreditation to Marth in 1866 [51]. For some time after 1868, Marth was assistant at the Gateshead observatory of Robert Stirling Newall, whose great Cooke refractor with its 25-inch object glass was, however, only used to full effect on one single night between 1870 and 1885. In 1874, Marth tried, and failed, to get a job with the Fourth Earl of Rosse at Birr Castle, and in 1883 moved to Markree – where Edward Cooper's nephew, Colonel Edward Henry Cooper, had revived the family observatory – and retained this position down to his sudden death in 1898 [52].

Marth's career is significant for what it tells us about the experiences of a university-educated German expatriate in the British Grand Amateur world. In no way at all did Marth see himself as an astronomer's helper, and the directness of manner recalled in several of his obituaries probably had something to do with it. Nor was he averse to controversy. In 1860, for instance, he submitted a criticism of the Greenwich transit circle observations to the R.A.S. *Monthly Notices*. Although it was rejected, its subsequent appearance in *Astronomische Nachrichten* infuriated Airy, who did not expect employed men to bite [53]. The Astronomer Royal certainly recognised Marth's talents as an astronomer, but reminded the Fourth Earl of Rosse, when being solicited for a reference for Marth in 1874, of that earlier 'transaction' which 'gives [me] a low opinion of him'. With his strong sense of

Roman justice, Airy advised his Lordship to contact Lassell for a more informed reference, before going on to further blast Marth as 'a man who is so weak as to be made the tool of another' [54]. Airy believed that Marth had been put up to criticising Greenwich, but weakness was certainly not one of this punctilious German's character traits. What Marth's career really tells us, though, is that even talented professionals had to know how to keep their mouths shut if they lacked an independent position.

Personal relationship clearly lay at the heart of all of the dynamics which operated between a patron and his professional assistant. In the way that the blunt Teutonic honesty of Marth probably offended his potential English patrons, so the relationship between Thomas William Backhouse and his assistant, Charles Paton, was one of great warmth and humour.

Backhouse was a Sunderland banker (described as a 'Coal Owner' on his 1873 R.A.S. Election Certificate), educated at Durham University partly under Temple Chevallier in the early 1860s, and became subsequently a leading Quaker philanthropist. His mansion, West Hendon House, Sunderland, had an observatory mounted on the roof which accommodated a fine 4¼-inch equatorial refractor by Thomas Cooke of York, along with other instruments [55]. He published privately several books, with the West Hendon House *imprimatur*, together with his catalogues. Backhouse's astronomical notebooks survive in Durham University Library, and comprise over 1,350 leaves [56].

In 1883, Thomas Backhouse was trying to engage an 'astronomical secretary'. He expected the candidate to be adept in the management of astronomical, meteorological, and photographic instruments, speak German, and be a skilled draughtsman. For these skills, the successful candidate would receive between £130 and £150 per annum, with board and lodging [57]. Between 1879 and 1883, Backhouse had enjoyed the services of his fellow-Quaker Albert J. Edmonds, while between 1897 and 1920 he had Irwin Sharp, on an increased salary rising up to £180 a year [58]. Surviving evidence suggests that Backhouse enjoyed good relations with these men, who also accompanied him on eclipse expeditions, and business visits to Friends' Missions in India and elsewhere. But the assistant about whom we know most is Charles Paton, whose great-grandson, Don Simpson of Sunderland, has uncovered a great deal of archival material on Backhouse and his observatory [59].

The financial relationship between Backhouse and Paton is not clear, however, and while he may have worked as a paid assistant between the years 1883 and 1897 (for which there is no record of any other observer or 'astronomical secretary'), their friendship and cooperation in matters scientific and religious continued down to Backhouse's death in 1920.

Charles Paton was a Scotsman, born in 1849, and educated at Edinburgh University. He moved to Gateshead and Sunderland, however, to found a grocery business, and seems to have come to know Backhouse when the banker came to give him financial advice. But even if Paton did receive a stipend from Backhouse for his astronomical services, it seems that the assistant's wife, Emily Paton, objected to his observing by night and running the business by day, even when Backhouse sent round a carriage, and later a motor-car, to collect him [60]. How much the relationship between Backhouse and Paton was based on paid-for professional services, and how much on mutual interests and genuine friendship, is hard to say, although Paton clearly felt obliged to the banker for several acts of personal favour. On one occasion, when Paton's 12-year-old consumptive daughter was excluded from treatment in the Sherburn Hospital because of her age, the enraged Backhouse gave a 'substantial' donation to the Hospital to secure her admission. On the other hand, Paton was no yes-man to Backhouse, and fortunately his willingness to stand his ground and state his honest opinion was respected, and on occasions, applauded [61].

But where I would suggest that the West Hendon House astronomical enterprise differs from all of the others that have been discussed above, is in the conscious charitable motivation of its owner. Astronomy, in many respects, was Backhouse's serious hobby; it was not his life's work. His money came from banking, but one of his principal occupations seems to have been charitable work, mainly conducted through Friends' organisations at home and abroad, and over the course of his long life he gave away very large sums of money. It is also likely that his religious views on human relationships, and the responsibilities of the rich towards their dependents, strongly coloured his association with his assistants, or 'astronomical secretaries'. Although Paton does not seem to have been a Quaker himself, he did become a teetotaller, and also wrote addresses for the Meeting House [62].

Paton was an unusual figure in the astronomical assistant world, for in spite of his university education, he clearly did not see himself as a professional career scientist, in the way that Glaisher, Hind, Pogson, Ball and Gill did. On the other hand, his familiar relationship with Backhouse, and the independence which his grocery business brought, meant that he most definitely did not see himself as a superior servant either.

One of the most tragic Victorian professional assistants, however, was Gabriel Goldney, whose career began with great promise, and ended in destitution. He was the younger son of an armigerous landed family who was sent up to Lincoln College, Oxford, in 1868 at the age of 19 [63]. His entry into Oxford at this date, however, is somewhat puzzling, for Goldney claimed to be a Roman Catholic, and until the repeal of the Test Act in 1871, all Oxford students were required to be communicants of the Church of England. But Gabriel Goldney failed to complete his degree requirements, perhaps because of a Roman Catholic conversion while at Oxford, which would have required his withdrawal from Lincoln College. By the early 1870s he was working as a Supernumerary Computer at the Royal Observatory, and living at 3 Royal Hill Terrace, Greenwich. It was from that address that he wrote to Professor Pearce, who had recently succeeded Temple Chevallier at Durham University, about a job in the observatory [64].

Whatever weaknesses may have lain within Goldney's character, however, they had not become too conspicuous by 1874, when Airy gave him the warm recommendation which secured the Durham job. After eleven years at Durham, however, poor Goldney was faced with the sack for 'neglect of duty'. Exactly what that neglect consisted of is not clear, but the shaky handwriting in which the 36-year-old man wrote to Professor Pearce in July 1885 suggests that it was probably alcohol-related [65]. After leaving the observatory, Goldney began a tragic descent, truly 'falling upon evil days', in the words of his obituarist, and died in the Durham workhouse in November 1905 [66]. Yet even in this ultimate degradation for a Victorian gentleman, he was not entirely without dignity, eking out a pittance of £30 a year as live-in workhouse porter, and impressing everyone with 'his polished manners, ripe scholarship and brilliant conversational powers' [67]. Gabriel Goldney is unique amongst the ranks of Victorian professional assistants, for in the vast majority of cases the astronomical calling was a vehicle for social advancement, whereas in his the spectacular descent from Gloucestershire broad acres to the Durham workhouse only reminds us that not even talent and connections can avert tragedy if other things go seriously wrong in an individual's life.

Over the course of the nineteenth century, the professional astronomical assistant underwent a number of changes, including a maturation into full-scale professional astronomers from institutions like Birr Castle, South Villa, and Dun Echt. Quite simply, by 1870 serious astronomy had become too complex in both its intellectual and instrumental operations to be entrusted on a day-to-day basis to men who were expected to demonstrate obedience rather than personal creativity. Indeed, even Airy had long since come to recognise that the reformed Royal Observatory of the 1870s depended upon men of much higher calibre than the original set of observers whom he had inherited when he became Astronomer Royal in 1835. For even by 1846, he was describing the work of each individual Assistant by name in the annual *Report* to the Board of Visitors [68], and in 1872 admitted that the days had long passed 'when the position of the Assistants was really very humble and the education and acquirements expected were of the lowest class' [69].

In the same way that the changing technologies and observing practices that came with spectroscopy and photography opened the door whereby women first came to earn their living in professional astronomy in some American observatories, so the Royal Observatory employed its first women staff in 1890. But the appearance of women at Greenwich, unlike Harvard, was less the product of technological than social innovation, on the part of the new Astronomer Royal William Henry Mahoney Christie.

Alice Everett and Annie Scott Dill Russell occupied paradoxical positions at Greenwich, however, for while these 'Lady Computers' were on a similar footing to the young lads who entered the Royal Observatory in their late teens on the 'apprenticeship' system, they were high-flying mathematics graduates from Girton College, Cambridge [70]. Although serious efforts were made by Christie and university supporters like Sir Robert Ball to augment their derisory £4 per month salary, no progress was made, and after two or three years the women left for posts elsewhere. Alice Everett joined the staff of the Potsdam Astrophysical Observatory in 1895, although the new astrophysical technologies which were bringing women into professional astronomy in other countries were in operation at Greenwich by 1892, for Christie recommended Miss Everett as 'a skilful photographer who made herself expert in the manipulation of the new photographic equatorial' [71]. Also, in 1895, Annie Russell resigned her post to marry Edward Walter Maunder, of the Solar Department at Greenwich. Her subsequent career in independent astronomical research, and eclipse photography in particular, will be examined more thoroughly in Chapter 14.

The eminent historian of nineteenth-century astronomy, Agnes Clerke, was also offered a Greenwich Supernumerary Computership, but turned it down, as she already made a better living as a scientific author. Even so, she had long desired to have the practical experience of working in an observatory, and in 1888 Sir David and Lady Gill invited Agnes Clerke to Cape Town to spend two months working in the Cape Observatory [72].

Yet the Greenwich Lady Computers were not by any means the first women to be employed as assistants in British observatories. Anne Walker was a local Cambridge girl who had done well in her school examinations, and was taken on to the staff of the Cambridge University Observatory at the age of 15 in 1879. She was to work there for the next twenty-four years, as the Second Assistant to Andrew Graham, who had become the Assistant to John Couch Adams at the Cambridge Observatory following the initial closure of the Markree Observatory in 1864 [73]. Her salary increased from £40 a year in 1885 to £100 by 1895, which made her by far the best-paid woman in British astronomy for the period, and she earned commendation in the Cambridge Observatory *Reports* as 'a good observer and an expert calculator' [74]; however, as was the case with so many other professional assistants, she found that there were no real promotion prospects. Roger Hutchins, who has done much fundamental research on Anne Walker, has shown that she was one of a total of five women employed in the Cambridge Observatory, generally for only one or two years apiece, between 1876 and 1904 [75].

It is impossible, however, to understand the place, prospects, and possible transformation into salaried research scientists of the professional assistant astronomers in Britain without recognising the influence of the Greenwich Supernumerary Computer system. Right through the period covered in this book, and onward to the 1950s, the Supernumerary Computer remained one of the most evolutionarily versatile forms of astronomical life in Britain [76]. The Supernumerary system could produce what Charles Dickens called a 'band of astronomical clerks', [77] a succession of noted F.R.S.s – James Glaisher, William Ellis and Edwin Dunkin, amongst others – a patriarch of Australian science, the distinguished asteroid and variable star discoverer, Hind, and some of Britain's first women professional astronomers who were also university graduates.

While the professional assistant of 1830 may well have lived in a social half-world, the changes through which astronomy had passed by 1910 had edged the Grand Amateur astronomer towards extinction, as the technologies which they had first pioneered eventually became too multifarious for single private fortunes to sustain [78]. Britain's relative industrial and economic decline by the end of the nineteenth century further led to a crisis of confidence in the old *laissez-faire* virtues which had generated the fortunes that financed Grand Amateur science, and by degrees, Oxford, Cambridge, the new 'Victoria' and other British provincial universities, and the government began to pay for the increasingly costly enterprise of research out of public or corporate money. And by similar degrees, and through an increasingly accessible system of higher education, that class of person who needed to earn a living out of astronomy came to hold executive posts in academic and public scientific institutions, so that by 1910 the old servants of astronomy had effectively become its professional masters.

Part 2

Poor, obscure and self-taught: astronomy and the working class

Prologue

One of the most extraordinary features of amateur astronomy in Victorian Britain was its cultivation among the poor. Although working-class astronomers lacked the sustained leisure and financial resources to make serious contributions to fundamental research, a significant number of working people were active in observational astronomy, and the making of their own instruments. And as far as one can tell from their documented remains, the inner motivation of a railway porter towards the pursuit of astronomy was no different in its essentials from that of a wealthy barrister: they both shared a delight in the beauty and complexity of the night sky, a love of orderly procedures, a fascination with delicate mechanisms, and very often, religious awe. Yet where the sacrifice for the working man seems so prodigious lay in astronomy's long night watches; for a twelve-hour working day, from 6 a.m. to 6 p.m., for 5½ days per week, was standard in most industrial workplaces, and one wonders when a serious working-class astronomer ever found time to sleep.

Although hours were long and pockets short, the Victorian working class possessed a spectrum of intelligence and talent that was not significantly different from that of any other population group. But where the working class differed from those groups that came above it on the social scale was in the relative inability of its individual members to cultivate and express their talents in a way that had an influence beyond their immediate circle. A talented brewer or clergyman experienced little difficulty in contributing to contemporary scientific debate and winning peer-group acknowledgement, such as an F.R.S. A self-educated shoemaker or dock worker, by contrast, had no ladder by which he could ascend into a wider world, so that knowledge of his achievements rarely spread beyond the local area, unless some journalist or patron chanced to publicise his work.

When looking at Victorian society, one cannot fail to be struck by the gulf which existed between the working and the middle classes. For while it is true that energy and ingenuity could facilitate great social advancement in many branches of commerce or skilled manufacture, sometimes taking a man from the shop floor to the Aldermanic Bench, it was impossible to avoid the obstacles that lay in the path of advancement by intellectual endeavour. The founders and managers of the Mechanics' Institutions from the 1820s to the 1890s were virtually unanimous in their belief that a literate and technically-informed skilled labour force was important for industrial development, but none of the people in authority seem to have realised the importance of linking such studies with incentives for social and economic advancement for the individuals making the effort. Even science education enthusiasts like Lord Henry Brougham believed that when the skilled artisan had given up his evenings to master calculus or optics, he should remain content to work his sixty-odd-hour week for his twenty shillings wages for the rest of his life [1].

But what is truly remarkable in the light of the obvious lack of connection that existed between self-education and improving one's material circumstances is that many Victorian working people were enthusiastic to make the sacrifice of time and energy. For within a social system which gave so few opportunities for self-expression to ordinary people, one cannot help but be impressed by the alacrity with which available avenues of cultural expression were seized upon by working men and women: dissenting religion, trades union activity, musical performance and amateur science [2]. Nonetheless, it is sad that while working-class astronomers pursued their passion from motives that could offer no material reward, they had to operate on an entirely local level, for unlike their middle-class counterparts, they had no learned societies or international periodicals which could give them voice or eminence, and travel was beyond their means.

Because there were no social forums or journals to which they had easy access, and because their overall social position was one in which literary record was not necessary for daily survival, the activities of working-class astronomers are extremely patchy in their documentary remains. Our knowledge tends to derive from five broad classes of document: occasional letters from working-class astronomers that have survived in the archives of more eminent astronomers; letters to periodicals, such as the *English Mechanic*; published journalists' interviews with local autodidacts; reminiscences under an individual figure's own hand; and obituaries appearing in provincial newspapers. Consequently, evidence for the life and work of an individual working-class astronomer consists very often of a single, though perhaps substantial, written source.

Unfortunately, what is absent are the diaries, observing books, and exchanges of letters between peers which provide the historical grist when looking at the careers of the Grand Amateurs, and their successors, the genteel 'Enthusiasts'. The historian, therefore, must never lose sight of the fact that his working-class subjects only occasionally speak directly for themselves, for their activities are frequently filtered through the perceptions and priorities of the middle-class persons who write or edit the articles. And very often these articles are cast in a distinctly heroic tone, as if to pacify the fears of their polite readers by showing that not all working men are given over to drunkenness, vice, or social insurrection. Yet sanitised as some of these articles might be, their collective existence, especially when combined with occasional direct reportage from working-class astronomers themselves, makes it clear that one of the most intellectually abstruse of all the sciences still had a lively following within the ranks of the disadvantaged.

And an examination of working-class astronomy also makes it clear what a gulf existed even between the ranks of the artisans and the men who worked as paid professional astronomical assistants, such as those described in Chapter 8. For most of these assistants, it will be recalled, came from middle-class backgrounds, where schooling up to the age of 16 and a knowledge of Euclid, French, and applied mechanics were necessary qualifications to even begin paid work in an astronomical observatory. No matter how modest – in bourgeois terms – the backgrounds of these assistants might have been, they were still a world removed from that of the working man, whose first step towards acquiring astronomical knowledge inevitably lay in finding the time and the shillings necessary to gain some rudimentary instruction.

9

A penny a peep: the astronomical lecturers of the people

The nineteenth century was the golden age of the lecture both as an agent for the mass communication of ideas and as a form of entertainment. And the popularity of astronomy as a subject can be gauged from the abundance of lectures recorded in the newspapers and private journals of the period. Lecturers of eminence, such as Sir George Biddell Airy, the Astronomer Royal, or Sir John Herschel, could pack in hundreds of people who came to witness their gratuitous performances when addressing a Mechanics' Institution gathering in the wake of a British Association meeting, and less eminent professional lecturers could invariably draw paying audiences which made it well worth their time. While it is true that by no means all lectures were intended specifically for working-class audiences, it is clear from surviving reports that many were aimed at the relatively uninformed: references to the presence of costermongers or mechanics on the benches give some indication of the potential clientele. And lectures varied in quality from the gravely (and perhaps incomprehensibly) learned to those that were so pitiably bad that even their simple audiences rolled about in laughter.

In spite of his sometimes stern aspect, G.B. Airy was committed to the astronomical education of the public. As a young don at Cambridge in the 1820s his lectures had attracted large academic audiences, who were impressed by their thorough preparation, clarity, and lucid delivery [1]. These skills stood him in good stead, for on several occasions in his subsequent years of fame, Airy stood up before enormous audiences of ordinary people to lecture on astronomy. His private Diary and letters, prior to these occasions, moreover, make clear the meticulous preparation, with attention to visual aids, which he made, combined with his concern that large numbers of tickets should be made available to working people. On 24 October 1854, for instance, Airy packed 500 people into a room measuring 46 x 30 feet at the Central Hall, South Shields, Northumberland — the biggest public room in the town — for his lecture describing the measurement of the gravity constant in the nearby Harton Pit [2]. So assiduous had Airy been in expressing his wish to make the lecture accessible to local working men that Temple Chevallier, of the Durham University Observatory, commended one Mr J. Watson for a ticket.

According to Chevallier, Mr Watson, who worked in a local chemical factory, was 'a self-educated practical astronomer, who finds his latitude by a transit placed in the prime vertical, & the kind of man that I know you are always willing to encourage' [3]. The South Shields lecture, with its elaborate electromagnetic clocks and Kater pendulums all packed into the same 46 x 30 foot room, was not only reported in the Tyneside newspapers but also in the *Illustrated London News*, and parodied in *Punch*, much to Airy's amusement [4].

A few years earlier, in March 1848, Airy had complied with a wish from 'the people of Ipswich' to give a course of six astronomy lectures in the town. The lectures, delivered to 700 people, were a great success, and an account of one of them, given in a letter to his wife, provides a glimpse of the elaborate visual aids apparatus which he had prepared, with mechanisms to demonstrate the motions of the heavens, and planetary transits, although he found that 'the rollers of the moving piece of sky would squeak' [5]. But Airy faced his biggest audience in the Free Trade Hall, Manchester, in 1861, when, in spite of a weak voice, he 'riveted' 3,000 persons in a lecture on the solar eclipse of 1860 [6]. It is true that this Manchester lecture was a star performance as part of a British Association meeting, and not aimed specifically at ordinary people, although the size of this audience suggests that it was swelled by local astronomical enthusiasts.

When figures like Airy lectured, however, one must always bear in mind that the fame and high office of the individual might have helped to swell the crowd. In the small towns of provincial England, or in the industrial cities of the north, there was a lack of spectacle and entertainment, especially for the working classes, and a visit by the Astronomer Royal to the local Mechanics' Institution must have had a crowd-drawing power that was similar to that of a visit by Jenny Lind, the opera singer, or of a famous prize-fighter.

Indeed, the impact of lectures from celebrities from afar could have enduring effects upon a community. When Sir John Herschel addressed the Leeds Phil. and Lit. Society, following a meeting of the British Association in that city in 1858, it led to the formation of a local Astronomical Society which, as a deliberate act of policy, fixed a low sum for the annual members' subscription so that artisans would be attracted to join [7]. Other records survive of distinguished professors speaking to working-class audiences, as when in 1839 the academic astronomer with a genius for public lecturing, Professor John Pringle Nichol, from Glasgow University, addressed 250 workmen in their 'working jackets' at a public lecture in Salford [8]. Nichol had another triumph at the Sheffield Mechanics' Institution in 1847 [9]. Publications like the *Astronomical Register*, which probably qualifies as Britain's first journal which catered specifically for an amateur astronomical (but by no means poor) readership, carries frequent notices from the 1860s and 1870s of high-level amateurs like Thomas Gwyn Empy Elger, or the professional William Radcliff Birt, giving lectures to local self-improvement societies [10]. Several Grand and even modest amateurs are on record for their occasional lecturing activities. William Pearson, it will be recalled from Chapter 2, laid the foundations of his opulent fortune as an astronomical lecturer in Lincoln in the 1790s, while Thomas Dell and the Revd Joseph B. Reade, of the Hartwell House connection, held forth in village halls in the area around Aylesbury [11].

But as one might expect from a city which had over a million inhabitants by the start of the nineteenth century, it was London which put on the widest range of public lectures in astronomy, generally on a commercial basis, and virtually all of which tried to attract the pennies of the working man.

In the spring of what was probably 1839, a journalist writing in an unspecified newspaper cutting preserved in the Hartwell House scrapbooks reviewed at some length the current season's crop of astronomical lectures open to the public in London under the title 'The Astronomical Mania' [12]. As the anonymous journalist, moreover, speaks of 'the last Lent', and some of the lectures were being offered in theatres, one wonders whether the 'improving' lectures were deemed to be a suitable Lenten alternative to regular theatre, in the same way that in Italy sacred oratorio replaced secular opera during the forty days and nights before Easter.

The activities of eight lecturers were reviewed in the distinctly satirical 'Astronomical Mania' article. They included a Mr James Howell, a young man who worked for a guinea (£1.05) a week as a City clerk during the daytime, but who, being 'in his own opinion, a *wonderful clever fellow*', had displayed for the last two years his 'affected speaking and monkey-looking personage' to paying audiences at the Strand and Adelphi Theatres and elsewhere by night. The other seven lecturers included a Mr Williams, whose Westminster lectures often ended in fits of stammering, a Mr Henry Altham, an Independent Minister who lectured for various charities near the Tower, and Professor Dewhirst, whose range included not only astronomy but also anatomy and phrenology. The only name that possessed any broader fame was that of Dr Dionysius Lardner, the Irish author and editor, whose lectures, like those of his less famous colleague Mr Wallis, were so abstruse that they put the audience to sleep.

It was essential that a successful popular lecturer should have demonstration models, although some of the equipment on show in London that season had greater defects than the mere squeaks of Airy's moving sky device that would be displayed at Ipswich in 1848. The article refers to orreries, 'telluria' (or Earth-Moon machines), and 'transparent scenes', which were probably hand-painted lantern slides.

As public lectures there is no indication that any of the above performances were aimed specifically at artisans, but their elementary contents, price, and other remarks by the author indicate that working people probably constituted a substantial part of the attenders at many of them. For instance, when Mr Adams, resplendent in black court dress, held forth in the Italian Opera House, there were chimney sweeps and costermongers willing to pay a 'bob' – one shilling (5p) – to gain admittance, though the author suggests that the real motive of these people for attending was to avail themselves of the opulent delights of being in a great theatre for such a small sum. A similar motive was attributed to the apprentice boys and Thames watermen who came to watch James Howell's antics at the Strand Theatre. We must not forget that the West End of London also had dense pockets of working-class slums in 1840, accommodating people who worked on the Westminster river wharfs and elsewhere, while the criminal 'Rookery' of St. Giles, so beloved of Charles Dickens, and the infamous vice quarters of Soho and the Haymarket, lay to the south of Oxford Street [13].

When material survives which reports the content of many lectures aimed at popular audiences, one is often struck by their apparent abstruseness. Airy's six Ipswich lectures, which were published in book form, were devoted to astronomical dynamics and celestial mechanics. They progress from an examination of the Earth's physical characteristics and relation with the Moon to gravitation and the physics of planetary motion, and then proceed to parallax theory [14]. Nothing is said about how the telescope works, or how the surface features of the Sun, Moon and planets appear, while the nebulae do not even get a mention. One senses that no-one leaving the lecture hall would have been instructed in how to find a single constellation, or delighted with images of the rings of Saturn or the craters of the Moon. No doubt the carefully prepared models, of which Airy wrote in his letters, would have provided visual spectacle – and their celestial squeaks might have raised a few laughs – but the content of the lectures themselves looks tough going. It would be hard to imagine an audience of 700 ordinary people, many of whom might have been scarcely literate, getting a great deal out of the Astronomer Royal's discourse on celestial mechanics. But perhaps it was the sheer sense of occasion that was the real draw, combined with a pride in their possession of Airy himself, for he was a local lad.

Celestial mechanics, the Earth–Moon system, tides and eclipses, albeit demonstrated with sometimes rickety models, also seem to have been the stock-in-trade of the eight lecturers reviewed in the 1839 'Astronomical Mania' article discussed above. On the other hand, three other ingredients seem to have played an important part in successful astronomical lecturing, especially when an economic motive was present: the promotion of a cause; the topicality of an event; and what the self-perceived working man's educator, David Mackintosh, characterised as 'elevating recreation' [15].

Of all the causes which an astronomical lecturer might choose to promote the most common was religious, and it was not for nothing that one of the grounds upon which the satirical 'Astronomical Mania' article ridiculed James Howell was his courting of the religious market, being the husband of a 'regularly dipped Baptist'. And around 1850 the Welsh minister, Mr W. Rees, was lecturing in his native tongue on the wisdom of God 'as made manifest in the history of astronomical discoveries' at Dr Bennett's Chapel in Falcon Square, London [16]. It seems that the Revd Mr Rees was using his lecture to address expatriate Welsh congregations in order to raise money to pay off the debts of four Welsh chapels in Liverpool. No doubt his silver-tongued oratory, upon a subject where there was then very little published literature in the Welsh language, combined with the cultural appeal to one expatriate community to assist another, provided a galvanising force. We are reminded of the London lecturer mentioned above, Mr Altham, who was also a dissenting minister, who lectured to raise money for various charities [17]. And lecturing on astronomy for an ulterior motive was clearly present when Dr Barber took to the boards, in September 1848, and two of Admiral Smyth's daughters, Georgiana Rosetta and Caroline Mary, wrote down their individual 'Recollections from An Astronomical Lecture for promoting Emigration' [18]. This lecture, delivered in an unspecified location but probably in Aylesbury, dealt with the familiar topics of eclipses, gravitation, and the causes of day and night, and one wonders quite how Dr Barber related these themes to emigration. But as would-be emigrants usually came from the most socially disadvantaged sections of the population, one realises that the lecture was most likely aimed at a working-class audience, which the Misses Smyth no doubt graced because of their connections with both astronomy and good causes. One wonders whether Dr John Lee of Aylesbury, who was concerned with social reform issues and was Admiral Smyth's close friend, had anything to do with the lecture, which was delivered when the ravages of the Irish potato famine were causing starving people to flood into England [19].

The single cause most commonly promoted by astronomical lecturers, however, was that of Christian belief itself. Numerous lectures, recorded in newspapers or preserved in handbills, were either given or else promoted by the clergy of various denominations. And unlike the case of the Welsh chapels, many of these lectures were not connected with any particular fund-raising project, and were sometimes without admission charge. The motive was to teach ordinary people, and more pointedly 'the poor', about the glory of God's Creation, and, in an age when sceptical and foreign revolutionary information was filtering down into the ranks of the working classes, confirm the truths of Christianity through an appeal to rational knowledge. Their model was the Argument from Design, academically established in the Revd William Paley's *Natural Theology* (1802), and extended in the eight *Bridgewater Treatises* after 1830, which showed that things so complex as the human eye, or the Solar System, could not have occurred by chance, but must be the products of Providential design. The specifically astronomical development of this mode of argument owed much to the *Astronomical Discourses* of the Glasgow academic and divine, Dr Thomas Chalmers, and to Dr Thomas Dick of Dundee [20].

This willingness not only to talk about the glory of God but also to take a firm Christian apologetic stance was shown by the Anglican Revd W. Walton of Waddesdon, a village near Aylesbury, before the assembled ranks of the Aylesbury, Buckinghamshire, Young Men's Christian Society, in late January 1854. After an animadversion of the virtues of self-reliance in a town where so many people were 'without the prospect of eating or drinking', and a swipe at 'The Popish doctrine ... – Don't inquire', the lecturer came to grips not only with astronomy and geology, but with popularly perceived scientific challenges to religion. Grappling with Laplace's nebular hypothesis, the resolution of nebulae into stars by the 'Clapham-Common telescope' [21], the geological antiquity of the Earth as demonstrated from fossils, and the importance of scientific instruments to modern knowledge, he argued that science can never teach the *origins* or first causes of things. Describing phenomena was one thing, explaining them was another; and when it came to the absolutes of existence, religion had as much to say as ever it had. Indeed, the newspaper report of Mr Walton's lecture captures the enthusiasm and intellectual energy of the speaker, and though no references are made to the use of slides or demonstration equipment, one suspects that a man of his lucidity and persuasiveness did not need them. After all, he was not teaching elementary facts, but discussing surprisingly *avant garde* ideas, of the type that would soon be aired in the notorious *Essays and Reviews* of 1860, which tells us something of the sophistication of the young Christian men in his audience. Mr Walton's lecture ended in resounding applause, a prayer, and the singing of the Doxology [22].

This popular lecturing tradition in the Aylesbury area received newspaper attention again in mid-January 1861, when the Revd R.B. Burgess, also of Waddesdon, gave a lecture in the Upper School Room in nearby Steeple Claydon [23]. This lecture may not have been so intellectually scintillating as that of Mr Walton in 1854, although 'the attendance was larger than any previous occasion of a like kind – there being about 250 persons present'. It was clearly aimed at the more straightforward imparting of basic knowledge, within a Christian context, employing as it did diagrams, an orrery, and a magic lantern, operated by the lecturer's clerical colleague the Revd Mr Alfred Everitt. It is also clear that Mr Burgess, who had quite a popular following in the parish where he had previously been curate, was deliberately trying to attract the working classes to his lecture, for his exciting demonstrations were available to 'the poor to whom, in these hard times, a penny is an object', and who could come for only a penny [24]. It is interesting to note that both Walton and Burgess drew particular attention to the poverty prevailing around Aylesbury, no doubt occasioned by agricultural distress, and wished to make their discourses available to the poor.

Not all the people who lectured in a declared religious cause, of course, were necessarily clergymen, as when Mr W.H. White, M.B.S., lectured at The Rectory House, Cold Harbour Lane, Camberwell, London, in June 1842. But Mr White, who was designated as a Meteorologist and Professor of Mathematics, obviously knew how easy it was to go above the heads of one's audience, when he emphasised that not only would 'The Power and Goodness of the Creator' be demonstrated 'through Astronomical Facts', but that 'Distant Views of the Celestial Bodies ... [would be] conveyed in the most familiar language' [25]. And as an indication of the enhanced costs of these delights in London, as opposed to rural Buckinghamshire, he charged what seems to have been the metropolitan going rate of one shilling for his lecture, or about one-twentieth of a workman's weekly wage.

One also wonders whether the six astronomical lectures which the Revd Dr Cahill delivered in the Manchester Corn Exchange early in December 1853 were also intended to advance any particular cause, although if they did, the cause does not seem to have been specifically religious, judging from the 'polemics' and 'politics' which excited such disagreement. But whatever Cahill's motives, the laudatory account of his performance which appeared in the *Manchester Weekly Advertiser* only confirmed him as a man of established popular fame, for in science and astronomy, 'he stands without a rival', for 'The highest departments of science are mere playthings in his hands' [26].

The two most common types of topical astronomical event upon which popular lectures were delivered were eclipses and comets. Both were rare, dramatic occurrences, capable of striking fear into simple folk, and providing a ready theme upon which a lecturer, be he motivated by altruism or financial gain, could find ready listeners. Indeed, it was not only the poor who were often mesmerised by these events, for even the readers of *Punch* were regaled with a diet of cartoons and outlandish tales following the spectacular appearance of Donati's Comet in 1858 [27].

While it is not clear which comet was being referred to in his undated handbill (though as the previous bright comets of 1811 and 1835 were referred to in the past tense, it was probably the Great Comet of March 1843, which was believed to herald Armageddon), Jabez Inwards displayed a natural talent for advertising. His lecture, to be given at the Camden Hall, Camden Town, London, on 6 April (at around the same time as the 'Millerite' comet was expected to end the world), was arrestingly billed as 'THE COMET. IS THE WORLD IN DANGER?' [28] The lecture syllabus contained a skilful blend of sound astronomical knowledge and scaremongering, which could well have tipped a worried waverer into attending. A series of brief 'tabloid' sentences set out the topics to be covered, such as 'What are Comets?', 'Is it possible for a Comet to strike the Earth?', 'What Herschel and Newton said', and indications that subjects as diverse as astrology, gravitation, and the future fate of the Solar System would be fully and satisfactorily explained in 'The Answer'. Clearly, Mr Inwards was expecting a capacity audience, for his tickets were astonishingly cheap by metropolitan standards, being four pence (2p) each, or four persons for a shilling (5p), and children half-price. The style of advertising, the assumed cometary fear, reference to 'the [Mechanics'?] Institution', and the cheapness of the tickets, not to mention the fact that Camden Town was a poor district of London (Scrooge's starving clerk, Bob Cratchit, lived there), all conspire to the conclusion that this was a lecture aimed at working people [29]. While we have no indication that any active working-class astronomers attended the lecture, the reference to technical terms, such as 'Gravitation', 'Is there a resisting Medium?', along with references to Newton and Herschel, imply that a certain rudimentary acquaintance with astronomy was expected by Mr Inwards on the part of at least some of his listeners. There was nothing in the advertising, moreover, to suggest that the lecture aimed at promoting any religious opinions [30].

Much more restrained, and ostensibly erudite, was the lecture given by Thomas Dell in the 'upper room' of the Aylesbury Mechanics' Institution (perhaps the same 'Upper School Rooms' in which the Revd Burgess lectured) in the spring of 1860 [31]. Dell, as we have already seen in Chapter 5, was a well-known Aylesbury astronomer, and friend of Dr Lee, and his lecture on the forthcoming total eclipse of the Sun, due on 18 July 1860, was much more than an exercise in simple popularisation. Indeed, the Aylesbury mechanics were treated to a state-of-the-art discourse on solar dynamics, with references to the views of Arago, Airy, and others. Dell spoke, on the very eve of the birth of solar spectroscopy, of the Sun's gaseous photosphere (which Bunsen and Kirchhoff in fact were just beginning to analyse in Heidelberg), and the possible source of the 'rose-coloured prominences' first seen at the eclipse of 1842 and then again in 1851. These prominences were still a subject of considerable speculation amongst astronomers, and it was hoped that the forthcoming 18 July eclipse might help to resolve their nature. Although we saw that Thomas Dell was a relatively poor man himself, the *ex cathedra* standing of his astronomical utterances was no doubt strengthened by his intended inclusion in the official British expedition going to observe the July 1860 eclipse in Spain, while his remarks about the chill which accompanied totality at the eclipse at Pavia, Italy, in 1842, implied that he had witnessed it personally [32].

At the opposite end of the scale when it came to explaining an eclipse, however, was the interpretation offered by a sailor in a Shields, Tyneside, barber's shop during what was probably the total eclipse of 28 July 1851. When the people ran outside to view the eclipse, the bemused tar explained, in shipboard language, 'Oh, its only the moon, ma'am, thats broke adrift and got athwart the sun. It'll all be right by-and-by, if the old boy [God ?] only puts the helm hard over' [33].

While astronomical lectures to ordinary people might be used to advance a cause or explain a topical event, it is likely that the largest single category of lecture was aimed at 'elevating recreation'. It was certainly the one out of which the lecturer best stood to make a decent livelihood.

Many people who took it upon themselves to stimulate scientific interests in the working classes, ranging from Thomas Morrison the 'shoe-making philosopher' of Sheffield, and Edward Mills the itinerant Welsh orrery-demonstrator, [34] to the Revd Henry Solly, recognised the need to do so pleasurably and entertainingly. Solly, in fact, stressed the importance of the 'play impulse' in mankind, and how the attentions of hard-worked people were best obtained and retained by making learning enjoyable [35]. Astronomy as entertainment, however, opened the floodgates to a remarkable array of personages, that included not only Mechanics' Institution and theatre lecturers but also itinerant demonstrators, travelling showmen, and street-corner patterers. And while it is true that these figures may not have supplied regular, structured learning to serious working-class amateur astronomers, they could sometimes strike a vital spark in the mind of a beholder, as well as show members of the poorest sections of society natural wonders which they would never otherwise have been able to see.

Sometime around 1835, Mr Robert Children, a small-scale master bootmaker of Bethnal Green, London, whose business turned over only £3–£4 per week, attended a lecture on astronomy delivered by a Dr Bird. The lecture so enthused Robert Children that on arriving home he insisted upon demonstrating the Sun's axial rotation to his wife in the kitchen. He picked up a wet mop, and began to spin it, soaking the kitchen as he did so, so determined was he to communicate the wonders of astronomy to his household. His biggest drawback, however, was his illiteracy, and Children was obliged to depend upon the more established erudition of his journeyman, the son of a Wesleyan Methodist preacher with an already developed taste for astronomy and experimental science. Raising £500 from his father, Children bought a large, powerful magic lantern, capable of projecting an image twelve feet across, along with twelve smaller lanterns to show each of the zodiacal signs. Complete with a substantial set of astronomical slides, and apparently guided by his journeyman, the irrepressible Robert Children began his lecturing career.

But the enthusiastic astronomical lecturer's ignorance soon shone through, especially when it came to the pronunciation of big words. When he said 'Now, gentlemen, I shall present to your notice the *consternations*' of the sky, cat-calls and laughter broke out in the audience. Undaunted, Children persevered, and was soon able to piece together an astronomical lecture that people were willing to pay good money to hear, so that he abandoned boot-making. Aided by his spectacular visual aids, enthusiasm, and performing talent (which he must have possessed in abundance to survive his early ridicule), Robert Children acquired a fortune of £6,000 from his lecturing, and became 'a great farmer' in America.

As is the case with so many working-class astronomers, Robert Children's career is recorded in one single source. It was passed on to the journalist Henry Mayhew by the journeyman who had first instructed him, an unnamed man who, by the mid-1850s, had also abandoned boot-making, to earn £80 a year demonstrating his microscope and collection of slides, at a penny a peep in the streets of London (a figure to whom I shall subsequently refer as the 'Microscope Demonstrator') [36]. With him, and Children, and other figures interviewed by Mayhew, one encounters a world of something more than just 'elevating recreation', for their peep-shows display an abundance of genuine scientific enthusiasm, and remarkable erudition, combined with the racy repartee of the born popular entertainer. With his acute ear for inflection, and speed of shorthand reportage, Henry Mayhew was able to capture both the personalities and the verbatim patter of these working-class philosophers, which makes Mayhew's interviews major documents in the history of popular adult science education.

In addition to relating his narrative about Mr Children, the unnamed street Microscope Demonstrator regaled Mayhew with a great deal of information about his own intellectual ancestry. He was the son of 'a wonderful clever' Wesleyan minister, who died at the age of 27, as mentioned above, and was apprenticed at the age of 13 to William Knock, a kind-hearted shoemaker, and 'a profound philosopher'. Knock owned 3,000 books, had an astronomical observatory on the roof of his Spitalfields house, and lectured on many branches of science to religious and other audiences. Like the Microscope Demonstrator's father, William Knock was also a Wesleyan minister, who not only taught the Demonstrator shoemaking, but astronomy and many other branches of science, and prepared him, in turn, for the Wesleyan ministry. At the time of the interview, probably in the mid-1850s, William Knock was still alive, and aged about 70 [37]. But only after the Microscope Demonstrator's own prosperous shoemaking firm, with its weekly turnover of 50 guineas (£52.50) and £10 weekly profit, had collapsed because of a defaulted loan, did he turn to street teaching. Prior to this, however, he had become an expert teacher during fourteen years spent managing the Sabbath Schools which he had founded in Deverell Street, New Kent Road, London, and lecturing to the Young Men's Improvement Meetings [38]. Then, seeing a telescope demonstrator showing Jupiter to crowds in Regent Street one evening and making five shillings an hour in takings, he acquired his own telescope and became a street astronomer in his own right. And then his sister presented him with a valuable microscope worth £15, so that he abandoned astronomy lecturing, which was so weather-dependent, and turned to microscopy.

While both the Microscope Demonstrator and his master, William Knock, had lectured on astronomy as part of their calling as Wesleyan ministers, their activities also open up the wider question of how far a deliberate component of entertainment was mixed into the discourses of men who lectured for apparently serious religious causes. Small masters like Knock, after all, were culturally close to working-class life, and from the laughs and cat-calls that inevitably came forth at a dull or a bad lecture, they knew that liveliness was necessary in holding a popular audience's attention. Wesleyan ministers, after all, were famous for the pyrotechnics of their sermons, let alone their secular lectures [39]. And one would also like to know more about the relationship between shoemaking and popular astronomical lecturing, or whether Knock's, Children's, and the Microscope Demonstrator's connections with that trade were purely fortuitous.

Suffice it to say, however, that the more intimate ambience of the small craftsman's workshop, with its opportunity for conversation and even reading, was more conducive to intellectual exploration than the relentless haste of a cotton mill. In the hands of a generous master, such as Knock or Children or the Microscope Demonstrator's shoemaker–minister father, the pace of work could even be adjusted to suit his own intellectual enthusiasms, so that religious, political and scientific conversations at the workbench could set a young apprentice's mind on fire. Elizabeth Gaskell, the Manchester novelist, recorded how, before mechanisation overwhelmed the textile industry, the handloom weavers of Oldham, Lancashire, had thrown the shuttle while working on a problem in Newton's *Principia*, the volume of which stood open on the frame before them [40].

Mr Tregent (we are not given his Christian name) first acquired a passion for astronomy and for scientific instruments while still a boy, probably in Yorkshire [41]. He was put to tailoring at the age of 6, orphaned at 10, in 1818, and after working 'on the tramp' as a journeyman tailor, ended up in London. His former love of instruments was rekindled around 1848, however, when he and his wife were enjoying a Sunday outing to Greenwich Park, and a man showed Tregent the view along the river with his pocket telescope, and invited him home to have a look at Venus through his superior astronomical telescope [42]. Tregent was immediately hooked, and soon struck a bargain with an Islington optician to make him a telescope in return for five new suits of clothes. By October 1856, when Mayhew interviewed him, Tregent had owned a progression of telescopes, mainly by a Mr Mull, of Clerkenwell, culminating in a 4¼-inch refractor with a magnification of up to 300 diameters, at a cost of £80. With this instrument, and a smaller one which he entrusted to his son to work upon another patch of pavement, he *taught* astronomy, at a penny a peep [43].

What is so impressive about Tregent is his remarkable technical mastery of telescopes and basic astronomy. He mentioned, for instance, that 'the very fine climate' of Cincinnati allowed the observatory there to see things, such as planetary satellites, which would not be visible in London with an instrument of equivalent aperture [44]. Unfortunately, Tregent says very little about the sources of his erudition, for there are no references to classes, or Wesleyan connections, and only one author, Dr William Kitchener, is mentioned, although his knowledge of telescope prices among the various London makers, and most especially, his connoisseurship of object glasses, begs more questions than he answers. The 'Revd Mr Cragg's' (Craig's) 24-inch aperture refractor on Wandsworth Common, for instance, was a failure costing more than £2,000 for the object glass alone, and would need another £6,000 spending before it could work properly [45]; the large Ross object glass displayed at the Great Exhibition in 1851 had taken four years to finish, while 'The making of object-glasses is a dreadful and tedious labour.' So trying was it, that 'Men have been known to go and throw their heads under waggon wheels, and have them smashed, from being regularly worn out with working an object-glass, and not being able to get the convex right' [46].

Mayhew's shorthand catches the nuances of the Tregent lecturing patter very elegantly, and shows how a mixture of authorative pronouncements, build-ups of expectation before the penny 'student' got his or her peep, and a humorous banter made him so successful. For Tregent was no mere street demonstrator, but a lecturer to and teacher of the public. 'When I exhibit', said Tregent, 'I in general give a short lecture whilst they are looking through. When I am not busy I make them give me a description. ... Suppose I'm exhibiting Jupiter, and I want to draw customers, I'll say, "How many moons do you see?" They'll answer "Three on the right, and one on the left" ... Perhaps a rough standing by will say "Three moons! thats a lie! there's only one, everybody knows." Then when they hear the observer state what he sees, they'll want to have a peep' [47]. Here was an ingenious blend of commerce, entertainment, and real instruction. People were amazed to learn that Jupiter went around the Sun at 26,000 miles per hour [48], watch the moving shadow of a lunar eclipse, or see the Sun's rays illuminate the ramparts of the lunar crater Tycho when it stood on the terminator. Indeed, a poor charwoman who saw this last-mentioned spectacle almost fell off the observing steps in her proclaimed contemplation of the God who had created such wonders [49].

Rather annoyingly, Henry Mayhew does not tell us in which part of London he met Tregent, although one suspects that this professor of the people operated in the poorer districts. For it can be surmised that the man who shouted out 'My God! there's a star hit me slap in the eye!' [50] when the tripod leg slipped, or those drinkers who rolled out of public houses when they heard Tregent lecturing outside, were not the denizens of Belgravia.

In an average year, Tregent made £125 in pennies, and was professionally proud of the fact that he still retained this price even for big events, such as eclipses, when he could have charged double [51]. As he was still working as a tailor during the daytime (as a small master, one suspects, from his freedom to come and go, to Portsmouth and elsewhere), and only demonstrated the telescopes between 6 p.m. and midnight on good nights, he was clearly making a decent lower-middle-class income: the same, in fact, as a junior Warrant Assistant at Greenwich. But as suggested above, Tregent poses more questions than he answers. For instance, how did he, in his enthusiasm for conveying astronomical knowledge, and with his handsome income, compare with those other four contemporary London street telescope-demonstrators of whose existence Tregent had informed Mayhew, or with his predecessor seen by William Wordsworth holding forth in Leicester Square several decades earlier and commemorated in a poem? [52]

As Ian Inkster has shown, astronomical lectures were being delivered all around England by 1850, and astronomy was the most popular of the sciences, although not all of these lectures were intended for working-class audiences [53]. Yet in spite of the popularity of astronomy as a subject, there was a remarkable shortage of books that would have been accessible to the non-technical reader. While the works of William

Pearson and Robert Ferguson would have provided sound treatments of celestial mechanics and instrumentation to the well-off and well-educated amateur, there was very little that was cheap and simple. Working-class astronomers, be they private devotees or aspiring lecturers, had so little to go on, and to acquire 3,000 books like Knock, or the 300 possessed by Mayhew's Microscope Demonstrator in his days of prosperity [54], poses a host of questions about the educational resources that were available to the early nineteenth-century working classes. Large private libraries such as that of Mr Knock, whose shoemaking business made only £2 or £3 a week, must have been acquired through the second-hand trade, and from 'throw out' sources. Inevitably, such private libraries, like those of the Mechanics' Institutions of the same period, were largely governed by what could be picked up cheap, rather than what was really appropriate [55]. Even so, one presumes that a man like William Knock, who lectured on zoology, natural history, and electricity as well as astronomy, had been able to pick up an adequate range of books on these subjects to allow him to teach.

A small number of books aimed at a relatively elementary readership at a cheap price were available by 1800, such as The Religious Tract Society's *The Solar System* (1799). Yet it is impossible to underestimate the impact of the writings of Dr Thomas Dick, the Dundee divine, schoolmaster and scientific populariser, when assessing the availability of astronomical knowledge to working people in early Victorian Britain and America. Dick's *The Christian Philosopher* went through eight editions between 1823 and 1842, and together with his *Celestial Scenery* (1838), *Practical Astronomy* (1845) and many other works, presented a great deal of astronomical information, set within the religious context of the Argument from Design [56]. As a young man, working as a handloom weaver, Dick had acquired his astronomy and most of his other learning from second-hand books which he, like Elizabeth Gaskell's Oldham weavers referred to above, studied at the loom and at meal times. From the loom, he had gone on to study at the cheaper and more socially accessible University of Edinburgh, as part of a process of educational mobility that would have been impossible for an English artisan, so that he possessed a first-hand appreciation of the popular mind for which he wrote. The scientific writings of the mid-eighteenth-century populariser Benjamin Martin and others whom Dick had read in his youth around 1790 would also have been available on second-hand stalls in London, and no doubt entered by that route into libraries of men like William Knock [57].

And for astronomical aspirants whose understanding or grasp upon the printed word was less secure (as was probably the case with Robert Children), there were those books aimed at the juvenile market. The Revd Jeremiah Joyce's *Scientific Dialogues* (1808), Dr Ebenezer Cobham Brewer's *Guide to ... Scientific Knowledge* (1848) and Dr Mann's *Guide to the Knowledge of the Starry Heavens* (1856) employed an easy conversational, or question and answer, style of the kind pioneered around the turn of the nineteenth century in the juvenile books of Richard Lovell Edgeworth [58]. In those early Victorian juvenile works, which sold new at around 2–3 shillings (10–15p) per copy, big words were generally avoided and technical concepts explained by way of homely demonstrations. Indeed, from the juvenile authors alone, an intelligent shoemaker–lecturer could easily have assembled a simple course on astronomy and optics that would have a powerful educative effect, especially when supplemented with globes and an orrery, upon simple audiences.

It is clear, however, that the men interviewed by Mayhew were familiar with stronger intellectual meat than Brewer's or Joyce's juvenile publications. Tregent, in his discourse on eyepieces delivered to Henry Mayhew, refers to Dr William Kitchener [59], whose *Practical Observations on Telescopes* (1818) he no doubt knew, while the Microscope Demonstrator mentions Dr Dionysius Lardner, referred to above as an over-technical London lecturer, yet whose *Cabinet Cyclopaedia* succeeded in putting a great deal of authoritative scientific information before the lay public in the early 1830s [60].

Indeed, Lardner successfully induced some of the finest scientific minds of the period to write concise and elegant critiques of their sciences for his *Cabinet* series, around 1830, in volumes selling at five or six shillings (25–30p) apiece [61]. Sir John Herschel's *Treatise on Astronomy* (1833) in Lardner's *Cabinet*, a volume of 422 pages, was a milestone in the history of astronomical publication insofar as one of the leading research astronomers of the age laid the entire science bare, without the use of a single equation. The *Treatise* passed through several reprintings, while in 1847 Herschel expanded it into 661 pages in his masterly *Outlines of Astronomy*, both books being illustrated with fine-quality plates.

And once Herschel had shown the way, other distinguished professional astronomers tried their hands at writing good popular works, often based on their lecture courses. Thomas Henderson, Astronomer Royal for Scotland, brought out his *Treatise on Astronomy* (1843), and Professor John Pringle Nichol his *Architecture of the Heavens* (1843) and *System of the World* (1846), which placed the Laplacian nebular cosmology before the general reader. And of course, the pre-eminent asteroid discoverer and Director of George Bishop's private South Villa Observatory, John Russell Hind, brought out his *Introduction to Astronomy* complete with an invaluable 'Astronomical Vocabulary' of technical terms in 1863, while even Airy published his *Popular Astronomy* in 1866. Professor Ormsby McKnight Michell and Denison Olmsted in America had also entered the popularising race by the early 1850s [62]. Of course, these post-Herschel popular books were not written with the working classes in mind, costing as they did around a third of a working man's weekly wage per volume, but it is impossible to gauge their impact upon working-class ideas once copies came to be recycled via the second-hand trade.

Perhaps one reason why so many established scientists were becoming keen to popularise was because of its financial profitability. In 1876, Richard Anthony Proctor admitted that a successful populariser could make between £2,000 and £5,000 a year from his writing, while a new edition of a popular work could bring in as much money for the author as the entire annual salary '(say from £500 to £1,000)' of a university professor [63].

Considering the importance of globes, planetaria, and magic-lantern slides to popular astronomical teaching, it is hardly surprising that publications connected with their use started to appear by the 1840s. Thomas Keith's *Treatise on the* ...

Globes (revised edition 1831) was a teaching treatise, while Henderson's *A Treatise on Astronomy*, mentioned above, carried a detailed appendix devoted to planetary machines.

Yet these books could have been aimed at the grammar school and university as much as at popular teaching, although the published lecture course which accompanied a box of commercially-produced lantern slides, preserved in the Liverpool Museum, was clearly aimed at an elementary or popular teaching market. This anonymous and undated 30-page pamphlet provided short pieces of text to accompany 38 lantern slides, or 'diagrams', which in their entirety would have constituted a well-rounded elementary course in astronomy in 1851. The spherical figure of the Earth, the planets – including Neptune and the ten asteroids discovered up to 1851 – comets, eclipses, the zodiac and tides comprise the main body of the text to accompany the slides [64]. There is also some cosmological material in the course, with illustrated discussions of the Newtonian system, the Milky Way and nebulae. Interestingly enough, the nebulae are described as being composed 'of a multitude of stars, situate at an incalculable distance from us, and some of these seem to have a planetary revolution among themselves' [65]. Whether this view was influenced by the Laplacian speculations of Professor J.P. Nichol, or by the recent discovery of the spiral arms and stars of the nebula Messier 51 (the 'Whirlpool') by Lord Rosse, is not stated, although one is impressed by the up-to-date content of this elementary course. One wonders if Mr Children invested in a slightly earlier edition of slides of this kind, to use in conjunction with his impressive battery of twelve projectors.

By 1850, hand-painted slides of great detail and artistic beauty, measuring about three inches square, were readily available for astronomical demonstration, and many survive in our national science museum collections. The Museum of the History of Science in Oxford, moreover, has several 'compound' slides, where two plates of painted glass are contained within a geared wooden slide frame. By turning a handle the plates within these 'compound' slides can be made to move with relation to each other, to produce a true moving picture. One wonders how the people at the slide show responded when they saw the Earth begin to orbit a golden Sun set in a rich blue sky, or else saw the Earth's moving tidal bulge follow a silver Moon across the sky on the magic lantern screen. Beautifully-coloured, hand-painted lecture slides survived well into the photographic era.

The active transmission of recondite scientific information into the ranks of the Victorian poor cannot help but impress us. And while we have no statistics that indicate the percentage number of the population, the gender-distribution, or the occupational groups which encountered organised astronomical knowledge, it is nonetheless clear that the popular lecturers rarely lacked a willing audience. For while it cannot be pretended that anything but a fraction of the people who attended these lectures were serious amateur astronomers, one realises that with street telescope demonstrators like Mr Tregent, Wesleyan ministers like William Knock, and enthusiastic buffoons of the stamp of Robert Children harnessing science both to popular entertainment and to religious awe, astronomy struck deep into the psyche of the Victorian working class.

But nowhere, perhaps, did it strike deeper than into the mind of the loquacious and surprisingly erudite Cockney pavement artist Bozo, whom George Orwell befriended during his tramping days around 1930. For Bozo – who was born towards the end of the Victorian era – knew much more about astronomy than did the Eton-educated Orwell, and had received two letters from the Astronomer Royal acknowledging his observations of meteors. When trying to sleep on the freezing benches on the Thames Embankment, and seeing the planets shining across the river, he sometimes speculated as to whether there were destitute people living on Mars or Jupiter. While Bozo was never referred to as a street astronomical teacher by Orwell, it is clear that he was free with his knowledge, and could well have held forth to his pavement art patrons if he had the chance. Orwell, indeed, was clearly amazed at the strong intellectual life which the destitute Bozo maintained and enjoyed, and perhaps his defence of his art and of his astronomy would have struck an immediate resonance in all the working class astronomers mentioned in this chapter, for even extreme poverty 'don't need turn you into a bloody rabbit – that is, not if you set your mind to it' [66].

10

Astronomy and the modest master-craftsman

A record of the activities of several working-class amateur astronomers – both modest master craftsmen and day labourers – survives from the middle decades of the nineteenth century. While it cannot be argued with certainty that astronomy, as a pursuit of working people, was firmly in the ascendant by the 1860s, evidence suggests that it might have been increasing. We know from the remarks of John Weale in 1851, and from the commencement of the *Astronomical Register* periodical in 1863, that astronomy was coming to be cultivated on a considerable scale by the middle classes by those years, so that it is not unlikely that working-class interests moved accordingly [1]. It is certainly the case that the political and economic stability of the mid-century, following the troubled earlier decades and the 'hungry forties', gave more stable employment prospects, even if wages stayed low and hours remained long. Yet in many respects, what we know of the vigour of working-class intellectual life has surprisingly little to do with times of dearth and plenty, as people who were intellectually inclined would often continue their passions irrespective of circumstances – perhaps, indeed, as an anodyne against them. The mid-century record of working-class scientists, therefore, probably has more to do with the activities of hero-collectors like Samuel Smiles, and the occasional individuals who chanced to attract the attention of local men of influence, than improving economic circumstances, for original documents from under the hands of working people themselves are relatively scarce.

But while it is true that all Victorian working people were poor, in terms of income, free time, and social resources, there were, nonetheless, considerable gradations between individuals in the overall social group. A small master craftsman – be he a handloom weaver, a blacksmith, or a shoemaker, like the figures discussed in this chapter – was more likely to have an income that was roughly double that of a journeyman or day labourer, work from his own independent premises, and enjoy greater flexibility of working hours, so that such people might be looked upon as belonging to a working-class élite.

Yet the branch of science which had a long and proven working-class association, especially in the old handloom weaving districts of Lancashire and Yorkshire, was botany. As Dr Anne Secord has shown, there were botanical clubs active around

Manchester dating back into the eighteenth century, and frequently operating from that working-class stronghold, the public house [2]. They often had large memberships and were far more than simple gardening clubs. There would be a stress on mastering the correct Linnaean Latin taxonomy for the different specimens displayed at meetings, local experts would know where a particular species of moss or flower was to be found, and aspirant botanists would be set tests by the *cognoscenti*, such as the correct identification of individual types from a jumble of greenery. Printed herbal volumes were highly prized, and passed on as treasures, in this serious study of taxonomic botany. One important incentive was, no doubt, the practical applications of herbs at a time when the working population had no reliable access to formal medical assistance.

One looks in vain, however, for similar clubs devoted to astronomy. Perhaps the climate and the unsocial hours of astronomy had something to do with it. But the overall reasons were probably similar to those expressed by the botanist–handloom weaver, John Horsefield, in a communication to the *Manchester Guardian* in 1850. In this autobiographical recollection, Horsefield, who was born in 1792, mentioned having been taken by his father to see 'old Robert Ward, astronomer, resident in Blackley', Manchester. 'He was so fortunate as to possess a telescope, and it being a clear, starlight night, when we were at his house, we must look through it. We looked at the moon, it being then about a week old; also at Jupiter, and saw his four satellites. That was the only time I ever looked through a telescope at "the flowers of heaven", for so somebody somewhere calls the stars. He also showed us what he called a solar system [orrery], being a rude mechanical contrivance to show the positions of the planetary bodies to each other (with their orbits) with the sun in the centre.' Although in later life Horsefield built an orrery himself, 'My attentions to astronomy were confined to reading books which treated on the subject: the objects being so very distant, and instruments to observe them too expensive for my limited circumstance' [3].

It is a pity that John Horsefield did not tell us what the books were that he read, and whether they included old copies of the works of Benjamin Martin, or the more recent ones of Thomas Dick, or perhaps Sir John Herschel. At a time when even a modest astronomical telescope, according to Mr Tregent, could cost anything above £3 [4], one understands why a handloom weaver – in a highly skilled yet dying trade in the early nineteenth century – could never afford one. On the other hand, several men in ill-paid occupations did scrimp and save to buy telescopes, though one can only assume that in their cases astronomy, rather than the cheaper science of botany, was their dominant interest. Considering the prices of books and telescopes with relation to prevailing wages, however, one suspects that a love for astronomy could lead a working-class family into penury no less rapidly than a love of the ale-house!

But amateur astronomy was not without its uses, as recollected by the political radical Samuel Bamford, from the time when he was imprisoned in Lincoln gaol in 1820. It seems that the governor, who was a man of little education but 'a genius in his own way', had an active interest in astronomy, and would often spend the night looking at the heavens through his own well-mounted astronomical telescope. His unexpected vigilance in the middle of the night successfully foiled several escape attempts [5].

Yet while the above-mentioned incidents provide examples of the practice and perceptions of astronomy within certain echelons of the skilled working class, it will be instructive to look at four case studies of modest master-craftsman astronomers, two of whom were in the blacksmithing trade, the third of whom was a cobbler, and the fourth a baker.

JOHNSON JEX: THE LEARNED BLACKSMITH OF LETHERINGSETT

When Johnson Jex died, in Letheringsett, Norfolk, on 5 January 1852, at the beginning of his seventy-fourth year, his local fame was such that *The Norfolk Chronicle* newspaper carried a detailed obituary essay that was over 3,000 words long. This obituary remains the only substantial source for the life and work of a rural scientific autodidact of acknowledged genius and considerable personal pecularity [6].

Astronomy to Jex was but one of several technical occupations which absorbed his energies and filled his waking hours, for he lived unkempt and in squalor, and avoided the company of women after the death of his mother, to live 'a life of complete solitude – *a scientific anchorite*'. Yet neither his solitude nor his motivation stemmed from a religious cause, for while Jex was always superstitious about Sunday observance, his obituarist had to admit – somewhat embarrassedly – that he was neither inspired by nor took any particular interest in religion. In spite of his irreligion, however, Jex seems to have enjoyed friendly relations with several clergymen, and presented one of them, the Revd T. Munnings, of Gorget, near Dereham, with an inscribed silver watch which was entirely, like the tools of its manufacture, of Jex's own personal making.

Johnson Jex's interests in astronomy lay not so much in the visual observation of celestial bodies in their own right, as in their relationship with wider mechanical principles and with instrumentation. His driving passion throughout life seems to have been horology, from his first hanging around a local watchmaker's shop as a boy, to his independent construction of gold-cased chronometers as a middle-aged man. It was the motions of the heavens that fascinated him, and his desire was to replicate those motions with gear trains and escapements to harness them to the business of time-telling. He timed the transits of heavenly bodies using his own doorpost and a distant chimney, though there is no reference to his devising an optical transit instrument. Even so, Jex seems to have made, according to his obituarist, 'telescopes, and *metallic reflectors*, which are universally acknowledged to be extremely difficult of construction'. It is unfortunate that his obituarist did not say more about these instruments; possibly he may not have possessed a sufficiently thorough knowledge of astronomy himself to supply more detail about Jex's astronomical activities.

But it is obvious that Johnson Jex must have mastered a considerable amount of astronomical knowledge if he was capable of designing, making, and regulating his own chronometers, while the same can be said for his grasp of optics when he was making refracting and reflecting telescopes. We are told, however, that 'His knowledge of astronomy, *as of everything else*, was SELF-ACQUIRED', and while nothing is recorded of the individual books that he read, he did teach himself French at the age of 60. This was to enable him to read the works of the Continental horological writers, though he found nothing in them that he had not already found in English. One is tantalised, moreover, by his obituarist's frequent reference to Jex always thinking things out from first principles by his own powers of reason. The obituarist, whom one suspects had not got a particularly firm grasp on the principles of scientific procedures, tends to give the impression that original ideas sprang complete from Jex's brain, and only needed embodying in glass or metal to make them perfect. What the historian would like to know, however, is where Jex first learned about the isochronal properties of watch-balances, temperature corrections in metal, or the differences between parabolic and spherical optical curves, for such things predicate an extensive knowledge of contemporary mechanics which could only have been acquired from books if one lived in rural Norfolk in 1820.

It would indeed be interesting to know how Jex learned the basics of speculum mirror casting and figuring. Perhaps it was from the same article by the Revd John Edwards in the 1787 *Nautical Almanac* from which the young William Lassell obtained his first instructions around 1820, for it was still the only step-by-step instruction in print. On the other hand, Jex's easy access to the furnaces, moulds, and metal casting processes that formed part of his blacksmithing business might have enabled him to re-invent the metallurgical technique for himself. One also wonders how Jex figured and polished his glass lenses from first principles, though even this was quite within the grasp of a determined amateur, as we shall see in the lives of Thomas Cooke and Samuel Lancaster, in Chapter 12.

But what is surprising about Johnson Jex was his failure to make an impact upon the wider stage, for the extraordinary versatility of his mechanical genius had been first proclaimed to the world in 1804, when he was still only 26 years of age. For in that year, the agricultural author, Arthur Young, had included an account of Jex's inventions to date, as part of his survey of the agricultural state of Norfolk. According to Young, by 1802 Jex had already made several pieces of clock and watch work, improvements to precision horological tools such as the fusee-engine, and built an electrical machine. Young had remarked that it 'is melancholy to see such genius employed in all the work of a common blacksmith', though he predicted that he will 'ere long, move in a much higher sphere' [7]. And while Jex did indeed move from his native village of Billingford, Norfolk, where Young had met him, to set up a blacksmithing business in Letheringsett, he never left rural Norfolk, in spite of encouragement from certain London chronometer makers, and the casting of his 'fine head' when the artist Bianchi was working in Norwich [8].

Several factors seem to have prevented Johnson Jex from ever leaving rural Norfolk for the wider world. One was his natural timidity – especially his fear of catching contagious diseases, which seemed the more odd to his obituarist because of Jex's personal uncleanliness. But apart from contagion, Jex does not seem to have relished the company of strangers or of unfamiliar situations. Another reason was his avowed dislike of travel, for he rarely left Letheringsett, never visited London, nor set a foot outside Norfolk in seventy-four years. In spite of his love of machinery, Johnson Jex never saw a railway locomotive, although they were commonplace by the 1840s, and the nearest station was only twelve miles away from his home.

On the other hand, we must not underestimate among Johnson Jex's motives the likelihood that he was fulfilled and content in his rural Norfolk home. And as a man 'entirely destitute of the love of money [who] sought out truth *for its own sake*, and with no view to any personal gain' (or, one suspects, fame), he felt no urge to enter upon the wider world stage [9].

Johnson Jex is an extraordinary figure among working-class astronomers, and brings home the paradox that while many people have ambitions that outrun their talents, some people have talents that outstrip their ambitions. For one could speculate *ad nauseam* about what could have happened to a man who was able to build a chronometer escapement from first principles and similarly fabricate refracting and reflecting telescopes, if only his personal motivation had been different, but that would serve no historical purpose.

JAMES VEITCH: THE PLOUGHWRIGHT OF INCHBONNY

One wonders if there was something in the calling of a rural blacksmith which led to contentment, for James Veitch, like Johnson Jex, his exact English contemporary, was an artisan–astronomer whose talents outstripped his ambitions. Unlike Jex, however, James Veitch, of Inchbonny, Jedburgh, in the south Scottish border county of Roxburgh, was a man of established fame within his own time – especially in his native Scotland – who ranked figures like Sir Walter Scott among his personal friends, while he had been the man who taught Scotland's leading optical physicist – Sir David Brewster – how to make his first telescope. But in spite of this reputation, which grew to a very high level by the time of his death at the age of 68 in 1838, Veitch would never desert his roots as a working artisan [10].

On the other hand, James Veitch was a very Scottish figure insofar as he lived in a society which operated in many respects by different rules from those of England. Mention has already been made of the cheaper and more socially accessible education system of Scotland, compared with that of Oxford and Cambridge, and one facet of that approach to education was a much higher level of meritocracy in Scottish society. Although Veitch was not himself a university graduate – unlike his neighbouring Jedburghian, Brewster – the recognition of his talents by Brewster, Sir Walter Scott, Mary Somerville (another Jedburghian, whose work will be discussed in Chapter 14) and others meant that he had to turn down invitations to join the 'philosophers' in the Athens of the North during Edinburgh's golden age [11]. For in the way that Robert Burns became Edinburgh's 'ploughboy poet', so Veitch could have become its 'ploughwright philosopher'.

James Veitch, therefore, was not an artisan scientist who stayed in his place because of a rigid social system which locked him out, but because he chose to contribute on his own terms and in his own independent way. Like Jex, he was happy as a blacksmith who specialised in making agricultural implements, and while scarcely out of his apprenticeship, he applied his knowledge of mathematics and mechanics to the invention of a lighter yet more efficient plough. He continued to make ploughs and similar tools until he was 55, after which he devoted himself to making those artefacts which had won him his first fame in the world outside Inchbonny: astronomical telescopes and other scientific instruments. And even then, his friend Brewster could not understand why he spent so much time doing the routine woodwork and metalwork on his telescopes personally, instead of concentrating upon the optics and sub-contracting the rest, beyond the simple fact that Veitch enjoyed it [12].

It is not known how Veitch's interests in astronomy and telescopic optics were first triggered, although at some date in his teens one Mr Alexander Scott gave him a 15-inch focal length speculum which he proceeded to make up into a telescope. And by the time he was 20, in 1791, his optical knowledge had become sufficiently sure as to enable him to teach the young David Brewster how to make a telescope. Much of what we know of Veitch and his early telescopes comes from the later optical writings of Brewster [13].

The speculum metal reflecting telescope was the instrument in which he most came to specialise, and by 1818 Sir Walter Scott – with whom he corresponded on the friendliest of terms – described Veitch as being 'one of the very best makers of telescopes, and all optical and philosophical instruments, now living ... [but who] prefers working at his own business as a ploughwright' [14]. And not even Sir Walter Scott – perhaps the most famous living Scotsman in 1818 – could lure his friend from Inchbonny to Edinburgh.

Veitch made Newtonian, Cassegrain and Gregorian reflectors, calculating the optical characteristics for a particular telescope himself beforehand, and then casting, figuring, and polishing the specula in the workshops where he also made his ploughs. In 1816, at Sir David Brewster's request, Veitch undertook a series of tests wherein he compared the 'light-giving power' of the various types of reflecting telescope under different magnifications. He tested the perfection of his polished specula by their ability to resolve the sparkle of the eyes of individual birds that perched in the uppermost branches of an oak tree that stood half a mile out of Jedburgh. As one might expect, his instruments soon came to be in demand, and he supplied them to Professor Heinrich Christian Schumacher at the Altona Observatory, Germany, Lord Minto, and Mrs Mary Somerville, amongst others. A Gregorian telescope with a 5½-inch diameter speculum of 5 feet focal length cost a mere £21 from Inchbonny, whereas William Herschel's 7-foot reflectors, with their 6¼-inch mirrors, by contrast, cost 100 guineas (£105) [15].

It is unfortunate that very few of Veitch's telescopes still survive, and the items that are in the possession of the Royal Scottish Museum, Edinburgh, probably represent the biggest collection in any one place. In 1993, however, I was invited to examine a large reflecting telescope of Veitch's manufacture by a firm of antique furniture restorers who were conserving it for the private client who had purchased the instrument in Scotland. Sadly it was lacking its optics, although the rest of the wood and brasswork was in surprisingly good condition, considering the neglected state in which its present purchaser had found it. One of the brass strips which bound the tube bore the inscription 'James Veitch, Inchbonny', the letters being beaten into the metal with punches rather than engraved. The hexagonal mahogany tube would have originally carried a primary mirror some 6 inches in diameter, working at either the Cassegrain or the Gregorian focus. This tube is still thumb-screw-fitted into a hexagonal clamp of plate brass, to swivel upon a 5-foot-high stand which could well have been adapted from a picture easel, if Veitch did not make it himself. The lateral motion of the tube in the altazimuth plane would have been guided by a rack and pinion mechanism (now missing) which engaged a denticulated brass track, which still survives. Indeed, it was a stand like none other that I have seen, and could well have been a product of Veitch's own original design, or else have incorporated several items which he had conveniently to hand.

Indeed, several features of this telescope suggested the incorporation of pre-existing parts, rather than an instrument that was designed and built of a piece. One half of the hexagon which secured the 4-foot-long wooden tube, for instance, was made from solid plate brass, one-twelfth of an inch thick, whereas the matching half was made by two thinner pieces, laterally soldered together to build it up to the right thickness. The brass locking screws had edges with different decorative knurlings and different threads, and one surmises that the denticulated track could have had a previous incarnation in some other device. On the other hand, the whole collage was brought together with wonderful design effect: the product of a craftsman who possessed an artist's eye for elegance, and knew how to select items from his brass oddments box with consummate discernment. It is unfortunate that the mirrors by which he could see the eyes of distant birds are no longer present.

In addition to theoretical and applied optics, however, Veitch was an accomplished astronomical observer. He liked to claim the discovery of the Great Comet of 1811 and the recovery of Halley's Comet in 1835 [16]. Although the generally agreed credit for the first sightings of both of these comets goes to Continental astronomers, Veitch wrote to Mary Somerville, mildly chastising her for her 1836 *Quarterly Review* article in which she accredited the first 1758 sighting of Halley's Comet to Georg Palitzch ('a peasant with a small telescope'), while at the same time ignoring his own achievement of 1811 [17]. And James, it would seem, was not the only Veitch who understood astronomy, for when Mary Somerville was visiting Inchbonny, she recorded a meeting with his wife, Betty, who was 'a person of intelligence, for I remember seeing her come from the washing-tub to point out the planet Venus while it was still daylight' [18]. Indeed, Betty Veitch is the only artizan-class woman of the period that I have encountered who displayed a serious interest in or knowledge of astronomy.

James Veitch's observatory in Inchbonny also became a place of some celebrity, not only for the excellent home-made telescopes which it contained, but also as a place to meet the ploughwright–philosopher himself. It seems that if Veitch was not willing to go to Edinburgh, then some of the city's most distinguished *savants* were willing to come to him. Sir Walter Scott – who even in his days of fame still followed the law-circuits – would turn up on Veitch's doorstep to show judges and leading advocates the wonders of the heavens when they were in the nearby Jedburgh courts. Veitch sometimes became irritated by the callers who took up his working time [19].

Although James Veitch may have remained relatively poor, and was certainly self-taught, he was by no means obscure. While his fame is perhaps best understood within the peculiarly meritocratic context of Scottish intellectual culture, his refusal to give up simple manual work tells us much about the man. For while he sent his own sons to Edinburgh University, and one of them, the Revd Dr James Veitch, became a scholar and theologian of note, James senior enjoyed his rural calling, and a life in which he could combine a satisfying manual dexterity with a vigorous and creative intellectual life. It is unlikely, moreover, that, as with Johnson Jex, his reluctance to move away from his rural setting was occasioned by reclusiveness or timidity, for Veitch had a self-confidence and fondness of straight talking in his 'racy Scotch' which is hard to reconcile with shyness [20].

It would appear, though, that Scotland had something of a tradition of autodidact reflecting telescope makers and users, and astronomical expositors. For whilst James Veitch was the most prominent and best documented, he was by no means the only one. There was, for instance, his slightly older contemporary, Gideon Scott of Harwick, Roxburghshire, a carpenter, wheelwright, and agricultural implement maker, who was self-taught in astronomy. At least one of his telescopes survives in Harwick Museum, a $3^{3}/_{10}$-inch aperture Cassegrain of just under 2 feet focal length, and Scott calculated and observed a transit of Mercury, as well as many other astronomical events [21]. It was Gideon Scott who first kindled a passion for astronomy in the young Harwick schoolmaster John Pringle Nichol, who subsequently acknowledged Scott in his own years of fame as Regius Professor of Astronomy at Glasgow [22]. And mention has already been made in Chapter 6 of John Ramage, the Aberdeen tradesman who built the first large-aperture reflecting telescopes after the Herschels, which were visited by the ubiquitous Brewster and described in his *Treatise on Optics* [23].

But if James Veitch, Johnson Jex and, as far as we are aware, Gideon Scott, were content to remain relatively prosperous small-town craftsmen, all the evidence suggests that John Leach would have quitted the workbench in an instant if circumstances had permitted him.

JOHN LEACH: THE COBBLER OF FRODSHAM

The artisan astronomers discussed so far in this section were in some respects men of modest property, with workshops, extensive tool collections, and possibly hired hands to assist them. John Leach, on the other hand, seems to have been a man living in much more restricted circumstances than either Jex or Veitch, and like the Wesleyan astronomy lecturers discussed in Chapter 9, followed the poorly-paid trade of a small-scale shoemaker. Although we know nothing about John Leach's religious beliefs, the modesty of his circumstances, as either a journeyman, or more likely a small master-cobbler (his exact position is unclear), combined with the patronage sought for him, points to the extremely limited resources at his disposal. But in addition to his *prima facie* standing as a working-class astronomer, Leach is historically interesting on two other grounds. Firstly, he came to correspond directly with major professional scientists, but in a much more limited way than Veitch; and secondly, he might well have been recognised as a suitable candidate for a form of patronage that was associated with Liberal Party politics.

Towards the end of 1869, one Mr Robinson, a gentleman of Frodsham, Cheshire, wrote to William Ewart Gladstone, the reforming Liberal Prime Minister, trying to secure some form of patronage for the local shoemaker whose talents and enthusiasm had so impressed him [24]. Exactly what Mr Robinson was is not clear, but he was obviously a man of comfortable means, who spent the first few weeks of 1870 travelling in the south of France and visiting in Montpellier. One can guess that his writing to Gladstone (who was also fond of wintering in the south of France) might have had something to do with his association with the Liberal Party, at a time when the Premier was pressing ahead with a policy for social reform that was more far-reaching than any practicable policy that had preceded it [25].

According to Robinson, John Leach had detected an error in one of the mathematical books of the Cambridge scientist Isaac Todhunter, and had even corresponded with him about the same. Leach had, furthermore, taught himself French so that he could study a borrowed copy of Phillipe Gustave Doulcet de Pontécoulant's *Théorie analytique du système du monde* (1846), and had made a simple telescope for himself, although it was a 'poor affair' which his patron feared might damage his eyesight [26]. Robinson enquired to see if Gladstone's reforming government could bestow any patronage upon the shoemaker, and perhaps find him employment in some scientific establishment. Robinson's letter to Gladstone was passed on to the Astronomer Royal, and with his usual thoroughness Airy set out to investigate the merits of the claim for himself.

After a correspondence with Todhunter and others, Airy advised Gladstone that it would be unwise to remove the 39-year-old cobbler 'from the employment by which he is gaining a livelihood' [27]. Airy was also firmly opposed to the suggestion that some form of job could be found for Leach at Greenwich, for the Royal Observatory was an institution dedicated to the 'mechanical' activities of accumulating and reducing the movements of the heavens, and not to providing opportunities for the curious. Leach would find the Royal Observatory disappointing, Airy insisted, for initial entry was invariably restricted to youths, and the arrangement of duties left no room for private astronomical interests.

On the other hand, Airy was in favour of the government making a small gift to Leach, which might further his astronomical knowledge and use to his local community. After communicating via Mr Robinson, Airy was writing to Leach directly by March 1870, to ask him what he would like within the constraints of the modest sum available, which seems to have been less than £20 [28]. Leach's letters to Airy reveal a man who combined intelligence with modest self-assurance and dignity. Beautifully penned, yet scarcely punctuated (like the letters of most working people applying for servants' jobs preserved in the Airy papers), Leach's correspondence expresses a desire to own a good achromatic telescope mounted on an equatorial stand, so that he could undertake serious observations. Yet while Leach's letters are respectful, they are in no way self-abasing, and suggest that he was a man who had the measure of his own worth and a sense of what he deserved [29].

Airy suggested to Gladstone that an educative present should be given to Leach: 'a pair of globes would cost but a few pounds ... [although] A telescope even of a moderate class, would be expensive.' A 4-inch aperture achromatic refractor, with a simple stand and 'low class' fittings, would cost at least £28, and a 3-inch £15, while a good equatorial refractor could run to *hundreds* of pounds, said Airy [30]. Leach still held firm for some kind of telescope, however, and via Mr Robinson informed Airy that he had seen a device called a 'Star Finder' advertised in the catalogue of a London optician, which cost £5 15s (£5.75) [31].

One might assume that by this stage in the proceedings, and after the passage of many letters, Airy would have been willing to wash his hands of the whole affair, and leave Gladstone's Treasury underlings to make the simple purchase and forward it on to Frodsham. But Airy's didactic enthusiasm seems to have been fully aroused, and on 12 March 1870 he wrote to Messrs Horne and Thornthwaite of Newgate Street, London, for information about their 'Star Finders', requesting that details of their two models, costing £5 15s 6d (£5.77½) and £8 8s (£8.40), should be sent to Greenwich for his personal evaluation [32].

The 'Star Finder' turned out to be a simple equatorial instrument, with graduated right ascension and declination circles, and mounting a small telescope. Airy wished to satisfy himself as to its instructive properties (for even the astute Leach had suspected from the advertisement that the instrument might well be a 'catch penny') and whether it would further the cobbler's astronomical understanding. The Astronomer Royal then wrote to the manufacturer to say that while the design and general functions of the 'Star Finder' were fine in themselves, the instrument could be substantially improved by the fitting of a 1¼-inch object glass in place of the ¾-inch lens supplied with the instrument. The manufacturers dutifully modified one of their 'Star Finders' at the Astronomer Royal's request, after which it was recommended to the Treasury for purchase [33].

Airy initiated a similar correspondence of inquiry with Messrs Malby the globe-makers of London, for he insisted that the best way to obtain a thorough grounding in astronomy (for the Astronomer Royal was a geometer and applied mathematician at heart, and not an observer) was to gain a familiarity with the use of both astronomical and terrestrial globes. In his ensuing correspondence with Leach, Airy urged him to choose a pair of 12-inch diameter globes, astronomical and terrestrial respectively – the best size for teaching and the same as the Astronomer Royal's own set of globes – which were subsequently purchased along with the 'Star Finder'. According to Malby's Catalogue, their 12-inch globes were sold at three prices – £2 2s (£2.10), £3 15s (£3.75) and £8 18s (£8.90) – though it is not clear which quality of globe was purchased for Leach [34].

On 7 June 1870 the delighted John Leach wrote a fulsome letter to Airy, to acknowledge the receipt of a box containing an improved 'Star Finder', with its 1¼-inch object glass, and globes. Leach wished to thank Airy, Gladstone, and all the other gentlemen who had taken so much trouble on his behalf, for now he was truly 'set up' to do astronomy [35].

The patronage of John Leach, however, poses far more questions than the available documents can answer. Like so many working-class astronomers, he

Plate 33. Warren De La Rue's 13-inch reflector, with its then revolutionary skeleton tube. This instrument, after subsequent service in Oxford University Observatory, is preserved in a dismantled state in the store of the Museum of the History of Science, Oxford. Engraving in the possession of the author.

Plate 34. Photoheliograph by J.H. Dallmeyer, used to photograph the 1874 transit of Venus. The object glass was designed to work at photographic optical wave-lengths, and a plate-holder replaces the eyepiece: J.N. Lockyer, *Stargazing* (pl. 10), 461. This instrument was a development of Warren De La Rue's original design of 1857, which was the first specifically designed astronomical camera, with a 3½-inch-aperture object glass by Ross. Both instruments are now preserved in the Science Museum, London.

Plate 35. The Moon, aged 17.1 days, by Warren De La Rue, 12 August 1862. The photographic emulsions of 1862 made it very difficult to obtain contrast upon the terminator without washing out other parts of the image. Original print in the possession of the author.

Plate 36. Sir William Huggins (1824–1910). Hutchinson's *Splendour of the Heavens* (pl.28) 51.

Plate 37. Sir Joseph Norman Lockyer (1836–1920). Courtesy of the Norman Lockyer Observatory Society, Sidmouth, Devon.

Plate 38. The Norman Lockyer Observatory, near Sidmouth, Devon, c.1920. Courtesy of the Norman Lockyer Observatory Society, Devon.

Plate 39. The spectroscope on the bench is collimated to receive a constant beam of sunlight from the clockwork-driven equatorially-mounted heliostat mirror placed outside the laboratory. This would have produced a conveniently observable spectrum, showing black Fraunhofer (absorption) lines. When a pure chemical substance was incandesced between the carbon points of the electric arc (which was collimated to be within the incoming beam of sunlight), however, it would produce a bright emission spectrum, that would replace its appropriate black absorption lines. In this way the elements in the Periodic Table could be identified in the Sun. Lockyer, *Stargazing* (pl. 10) 481.

Plate 40. Multi-prism spectroscope attached to R.S. Newall's 25-inch Cooke refractor at Gateshead. Sadly, this spectacularly-equipped Grand Amateur observatory achieved virtually nothing. Lockyer, *Stargazing* (pl.10) 427.

Plate 41. Solar spectrum plates, comparing Kirchhoff's original manually-observed spectrum with Rutherfurd's photograph. Lockyer, *Stargazing* (pl.10) 480.

Plate 42. Late nineteenth century lady astronomer examining spectra plates on a measuring machine. Hutchinson's *Splendour of the Heavens* I (pl.28) 53.

Plate 43. Isaac Roberts (1829–1904). S. Evans, *Seryddiaeth a Seryddwyr* [*Astronomy and Astronomers*] (1923) 277.

Plate 44. Isaac Roberts's Observatory at Crowborough, Sussex. A 7-inch Cooke visual refractor was mounted upon the same axis, by Sir Howard Grubb, as a 20-inch photographic silver-on-glass reflector, also by Grubb. Isaac Roberts, *A Selection of Photographs of Stars, Star-Clusters, and Nebulae* 1 (London, 1893) pl.2.

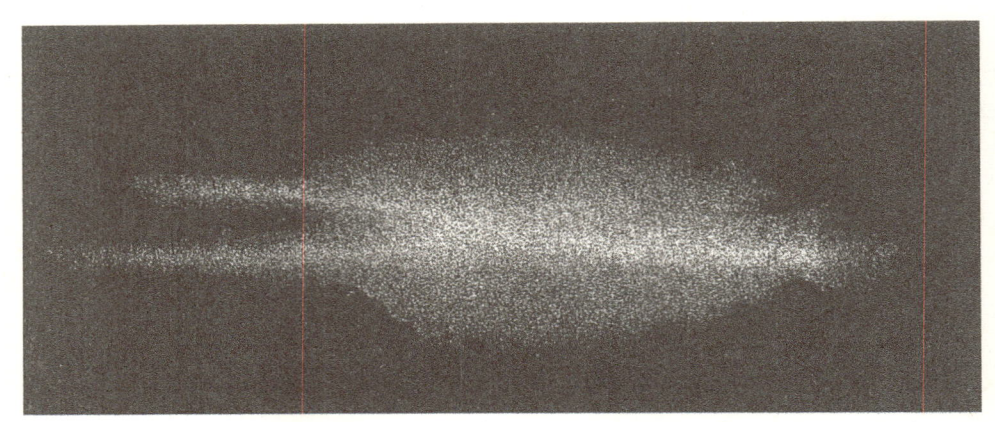

Plate 45. A longitudinal section across the Galaxy, originally published by William Herschel, *Phil. Trans.* **75** (1785) 266, re-engraved in J.P. Nichol, *Views of the Architecture of the Heavens* (Edinburgh, 1853), pl.II, p. 19. From his original 'gages' of the density of stellar distribution in the Milky Way, Herschel saw the Galaxy as roughly disk-shaped, with the Sun near the centre. The split, seen to the left, was interpreted as being due to a breakaway section of stars. We now know that this apparent split is caused by non-luminous matter which obscures the light of stars behind it. In 1785, however, Herschel had no concept of what is now known as dark matter, and interpreted these dark bands as the products of the mechanical separation of whole star-clouds.

Plate 46. The Orion Nebula, observed by Sir John Herschel, through an 18-inch-aperture reflecting telescope, and re-engraved in J.P. Nichol's *Thoughts on some important points relating to the System of the World* (Edinburgh, 1856), pl. VIII, p. 51. A detail of the central section of the nebula appears in J. Herschel's *Outlines of Astronomy* (2nd. edn, London, 1849), pl.IV, fig.1. Herschel's minutely detailed survey of the Orion Nebula was intended to serve as a benchmark against which future potential changes in its structure could be measured. We now know that the nebula is too remote for changes to be discernible within a human telescopic time-scale.

197

Plate 47. The Orion Nebula, observed by Lord Rosse, reproduced in Ball, *Story of the Heavens* (pl.14) 456. Rosse's very different and almost cubist interpretation of the Orion Nebula, compared to John Herschel's, is due to his greater, 72-inch telescope, which could detect structures unresolvable in an 18-inch telescope. Differences in ophthalmic sensitivity between the two men could also have played a part.

Plate 48. Messier 51 (the 'Whirlpool'), observed by Sir John Herschel, and re-engraved in Nichol, *System of the World* (pl.45), pl. II, p. 19.

Plate 49. Messier 51, observed by Lord Rosse, in Nichol, *System of the World* (pl.45), pl.VI, p.23. The greatly superior light-grasp of Rosse's 72-inch revealed the famous nucleus and spiral arms connecting to a satellite galaxy as one structure, whereas Herschel's 18-inch could show only a nucleus, a split ring, and a secondary galaxy as three separate structures.

Plate 50. Messier 31, the Andromeda Nebula, observed by Sir John Herschel with his 18-inch reflector. Herschel, *Outlines of Astronomy* (pl.46) pl. II, fig. 3. A similar drawing was published in Herschel's *Treatise on Astronomy* (London, 1833) pl.3. Apart from its lenticular shape and increased brightness in the centre, no structure is discernible.

remains a tantalisingly elusive figure, about whom nothing is known before and after his passing encounters with Airy, Gladstone, and Robinson. The most complete statement about the man himself comes from the 1871 Census Return, where he was described as a 40-year-old shoemaker living in Chapel Lane, Frodsham, and a native of Heywood, Lancashire. He had a 38-year-old wife named Frances, a son, and four daughters, and his neighbours in Chapel Lane included a shepherd, a retired farm worker, and several charwomen [36]. Nothing is known about the genesis of his passion for astronomy, how he cultivated it up to 1869, and to what use he put his present after 1870. Nor is anything recorded about Leach's role in the intellectual life of Frodsham, his membership of a Mechanics' Institution, or his desire to teach others. We do not even know what branches of astronomy most appealed to him, though his reading of Pontécoulant and his desire to own an equatorial refractor suggest that they extended across both the mathematical and observational branches of the science. Indeed, one suspects that it was Leach's proclaimed love of the mathematical branches of astronomy which aroused Airy's sympathies, for in his view optics and celestial mechanics constituted the foundation of the science.

Airy's initial involvement in the affair arose from his public position, while his insistence to Gladstone and others that he could not employ the shoemaker, interview him at Greenwich (even if Robinson paid for the journey), or enter into a continuing correspondence after the gift had been made, are all characteristic of the Astronomer Royal's obsession with encroachments upon his time [37]. On the other hand, it is curious that the Astronomer Royal should have been willing to spend so many hours (in 1870 Airy was paid at the approximate rate of 10 shillings (50p) per hour) corresponding with globe-makers and telescope-makers and others about the intellectual well-being of a Cheshire cobbler whom he had no intention of meeting personally. One can only assume that Airy's active interest in the Leach case derived from what Temple Chevallier in 1854 had called his desire to 'encourage' intelligent workmen [38].

Probably Robinson's own involvement stemmed from altruism and a sincere wish to help a worthy man. Not only does his initiation of the correspondence imply as much, but also his willingness to pay for Leach's visit to Greenwich and act as a go-between for Leach and Airy before each man wrote to the other direct. But it would be interesting to know Robinson's politics, and whether he or Gladstone (or even Airy for that matter, for he was a Liberal in his sympathies) saw the value of encouraging an articulate working man at a time when Gladstone – 'the people's William' – was riding high on the shoulders of the newly-enfranchised artisan class [39]. After all, it cannot have done the populist Liberal Government any harm if the pubs and Institutes of Frodsham rang with the story of a self-educated artisan being given official recognition in his studies. The more political aspects of the patronage, moreover, were implied in Airy's Diary, when he referred to a 'letter to Mr. Gladstone about his client at Frodsham' [40], although whether the client was Leach or Robinson is not clear.

Yet officialdom's treatment of John Leach was entirely characteristic of the period, at least in England, and only bears out the attitudes towards working-class intellectual advancement which I discussed in the introduction to this present section: artisan scientists should be praised as role models for their class, encouraged within acceptable bounds, but under no circumstances whatever removed from their existing social places.

THOMAS WILLIAM BUSH: THE BAKER OF NOTTINGHAM

It is curious how certain nineteenth-century towns produced or nurtured several astronomers, whereas others did not. Bedford and Liverpool, as we have seen, were perhaps the most prolific, and Nottingham too generated more than its expected statistical share. John Russell Hind and Norman Robert Pogson, who spent their lives as employed astronomers, came from Nottingham, while an offer of a public observatory worth £10,000 was made to the town by one Mr Henry Lawson in 1852, although the problem of finding a suitable site and maintenance endowment caused the donation to lapse [41]. One wonders whether the biographical article and portrait of the already famous 29-year-old Hind, which appeared in the *Illustrated London News* in August 1852 [42], spurred the 13-year-old Thomas Bush in an astronomical direction.

Bush was born at 2 Canal Street, Nottingham, in 1839, and was probably introduced to the trades of baking and greengrocery by his stepfather. He was still living, and trading in his own right, in the same 'back street of Nottingham', but at a different house, where the 'canal, and ... smoke yielding factories [meant that] ... his observations have been hindered by the state of the atmosphere in his vicinity' in 1870, when his telescope excited the attention of the national press [43].

No clear evidence survives about Thomas Bush's education, beyond a Wesleyan association, and an ordinary 'three Rs' schooling. Yet like the other figures mentioned in this section, his intelligence strove for expression, leading him to acquire a knowledge of the scriptural languages, Latin, Greek, and Hebrew, along with French [44]. But his passion was astronomy. Nothing is recorded concerning the books which he read, except Sir John Herschel's *Treatise on the Telescope* [45], but one presumes that in a city like Nottingham he would have had access to a Mechanics' Institution and perhaps public libraries.

But it was during the summer of 1870 that the hitherto obscure Thomas Bush sprang into public attention, for he had loaned his telescope to form part of the Nottingham contribution to the 'Working Men's Exhibition', being held in the Agricultural Hall at Islington, London. Among the things being exhibited that summer, which included 'doll's houses, and a steam engine in leather, cathedrals made of cork, and pictures painted by the toes', Thomas Bush's 13-inch aperture equatorially mounted reflecting telescope seems to have stolen the show [46]. For as an extraordinarily narrow-minded, utilitarian, and sarcastic journalist reported, all of these expressions of working-class creativity 'were absurd, useless, and deplorable examples of wasted time and toil', except Bush's telescope [47].

It is not known where Thomas Bush, 'a small tradesman in a back street of Nottingham' [48], acquired both the theoretical and the practical knowledge necessary to construct his excellent instrument. One would like to know, for instance, if he had seen a picture, or read an account, of Warren De La Rue's reflector, for the two telescopes were very similar. Both had 13-inch mirrors, were set upon German-type equatorial pillar-mounts (Bush's instrument was defined as having a 'Browning Equatorial'), and used the new technology of the 'rods and rings' arrangement of a skeleton tube [49]. All that we are told, however, is the obvious point 'that Mr. Bush has not acquired his practical skill without long and patient toil extending over many years'. But on 19 July 1870, one Mr S. Alex. Renshaw wrote to the editor of the *Nottingham Journal* to claim his personal acquaintance with the town's sudden celebrity, and report upon how Bush had gone about his optical work [50].

Renshaw clearly knew something about telescopes, and discussed the problems of casting metal specula and figuring them into a proper parabolic curve. Bush's Newtonian reflector, it seems, had been equipped with a pair of 13-inch mirrors, one in speculum metal and the other in glass, with the comment by Renshaw that glass was a much easier medium to polish, especially in its latter stages, when the surface was less likely to 'tear up' than that of speculum. Bush had experimented with astatic, lever-support mounts for his mirror, but found that a heavy cast-iron ring worked best. The whole telescope weighed 15 hundredweight, gave excellent definition of up to $1,400\times$, and could split several optically difficult double stars such as those in the constellations Coronae, Boötes and Hercules [51].

One of the things which brought especial attention to Bush at the Working Mens' Exhibition, though, was the interest which his telescope aroused in the Prime Minister, Gladstone, Queen Victoria, the Prince of Wales, and G.B. Airy, the Astronomer Royal [52]. Very clearly, they all saw Bush as a prodigy, which is no doubt the reason why he received such newspaper publicity. Gladstone and, one presumes, the Royal dignitaries, had been given a demonstration of Bush's telescope by Mr H. Milward, the Nottingham agent. It seems that Airy had visited the Exhibition at Gladstone's request, for the Astronomer Royal's Journal recorded 'I went to the Workmen's International Exhibition to look at a telescope (at Mr. Gladstone's request)', and later sent a report to the Premier [53].

It is interesting to note that during 1870 both Airy and Gladstone had been involved with patronage for working-class astronomers, with Leach and Bush, although Thomas Bush received presents that were considerably better and more expensive than those given to Leach. He was made a gift of a spectroscope, a solar eyepiece, and a gold medal, while the citizens of Nottingham were to present their famous baker–astronomer with a regulator clock [54]. Perhaps Airy felt that as Bush had already displayed great knowledge and industry in building such an excellent telescope – where the only bought-in item was a Steinheil prismatic flat – he deserved a better reward than the Frodsham shoemaker who had merely taught himself French and higher mathematics.

By the Census of 1861, Bush was trading at 102 Canal Street, Nottingham, as a baker and grocer. He was married to a woman named Martha, who rather grandly described herself as a 'Professor of Music', which probably meant a piano teacher. This neighbourhood, whose insalubriousness for astronomical work was to be so pointedly commented upon by the journalists of 1870, also seems to have been the home of some of the poorest occupational groups in the town: their Canal Street neighbours included labourers, washerwomen, and a 'brat man', who was probably a brewery labourer. But the Bushes clearly prospered, for by 1873 Thomas was elected to a Fellowship of the R.A.S., which involved an annual subscription of 2 guineas (£2.10), and was also describing himself as Secretary to the Nottingham General Hospital. Whether this was a part-time job which he combined with the bakery or a full-time post in its own right is not clear, but it signified undoubted social advancement [55].

And then, in 1876, Bush bought a plot of land in the new Thyra Grove property development in the more salubrious Nottingham district of Mapperley, and built an elegant three-bedroomed house, gardens, and an enormous astronomical observatory. On the surviving plans, the observatory took up almost as much ground as the house itself, and the whole site, with house, dome, transit building, and computing rooms, was called 'The Observatory'. One might think that the working-man astronomer had now metamorphosed himself into a Grand Amateur [56].

At Thyra Grove, Thomas Bush observed the Great Comet of 1881, and using the spectroscope given to him ten years before, detected carbon and other vaporous elements in its tail. He also did stellar and planetary work, usually communicated to the *Nottingham Guardian* newspaper, as well as giving papers to the Nottingham Philosophical Society [57]. It is unclear, therefore, why in 1889 he left his Thyra Grove observatory–residence and his Secretaryship of the Hospital, and moved to Willey Park, Shropshire, to become Astronomer in Lord Forester's private observatory. But it is not impossible that Bush had fallen victim to the general economic recession which had set in by the late 1870s, as the great Victorian prosperity began to falter under the impact of German and American economic competition. We know that industrial shares and land values fell alarmingly during this period, and while we have no details of Bush's financial holdings, it is likely that the investments which he had hoped would finance his work at Thyra Grove failed to do so [58].

At Willey Park, he observed with Lord Forester's 8-inch refractor, and even after that scientifically-minded peer died at a great age in 1894, Bush continued at the observatory under the aegis of the Fifth Lord Forester [59]. But in 1909 Bush retired, moved to Lunlendon, Cornwall, and set to work on planning an instrument of truly original design. It had a 24-inch silver-on-glass mirror, and was set upon an ingeniously simple equatorial mount, with a clock drive. Of particular note was its short focal length, for it seemed, from the surviving photographs, to work at an aperture of around f/4. We have no idea how Bush built this remarkable instrument [60].

The telescope came to be set up in a run-off shed at East Grinstead, Sussex, between 1911 and 1915, at the observatory of William Sadler Franks [61]. The exact nature of Thomas Bush's relationship with Franks – tenant, friend, or paid observer – is not clear, but it would appear that Bush was not a poor man when living at East Grinstead, for in January 1924 the 84-year-old astronomer entered the nearby Sackville College, a residential foundation for elderly gentlemen, with a £100 fee, where he died in 1928 in his 89th year [62].

Like so many astronomers who lived within what was patronisingly styled 'humble life' in the nineteenth century, Thomas Bush poses more questions than the records about him answer. It is clear that, in spite of his learning and obviously prosperous baking and grocery business, he still worked long hours, for as one of the journalists describing his telescope in 1870 pointed out, he found his time for astronomy by sleeping only four hours a night [63]. And at the same date, he still regarded himself as a working man rather than as a member of the business-owning middle class, and was happy to put his telescope in an exhibition of working-class ingenuity. But one wonders how much he bristled at the patronising (not to mention insulting) remarks that he and his fellow exhibitors received from lordly journalists who thought that a Workmen's Exhibition by its very nature was a mildly ludicrous thing which only encouraged the lower orders to waste their time.

On the other hand, by the time of his mid-thirties, Bush was aspiring to higher things at Thyra Grove, and one would like to know the circumstances that lay behind his failure to stay in the ranks of the Grand – or at least the well-off – independent amateurs after 1876. For in one single career Thomas Bush seems to have occupied each of the major classes of Victorian amateur astronomers: as artisan, Grand Amateur, professional assistant, and finally, an elderly middle-class enthusiast and F.R.A.S. who observed with a large home-made telescope near the Sussex coast.

11

The day-labourer astronomer

The Victorian working class was anything but an homogeneous social group; rather, it was rather a great cluster of groups, differentiated by occupational skill, income, religion, ethnic perceptions and social attitudes. But what was common to them all was a limitation of opportunities for action within the wider world. And while a small master-craftsman with his slightly better income and more flexible working time enjoyed somewhat more elbow room, the day-labourer, who was an employee or journeyman, had much less. What is truly remarkable, therefore, was the cultivation of astronomy amongst the ranks of those who were obliged to toil for daily wages.

One of the best documentary accounts which we possess of the lives and work of several astronomers within this social group comes from the Scottish–Leeds physician and author, Samuel Smiles. Smiles' best-seller *Self-Help* (1858) had drawn together into a single directed eulogistic study the lives of a variety of rags-to-riches figures as diverse as Benvenuto Cellini and George Stephenson, to illustrate the capacity of the self-motivated genius to change the world. Though Smiles was not interested in astronomy as such, the celebrity which *Self-Help* brought encouraged him to take a serious interest in self-motivated technical people even if they never achieved fame, and merely stayed securely in their humble positions. And by the time that he wrote *Men of Invention and Industry* (1884), Smiles had collected accounts of several astronomers 'in humble life' whom he had interviewed personally. In many respects, therefore, there is a parallel between the verbatim autobiographical accounts which Smiles wrote down from his astronomers and Henry Mayhew's interviews with London street telescope and microscope demonstrators.

But it is always necessary to be on one's guard when using transcribed autobiographical accounts that cross a divide of social class or culture. In some ways, one feels on more firm historiographical ground with a writer like Mayhew, whose primary concern was with the lives of people who earned a living in the streets of London. As his complete roll-call of interviewees includes barrow-boys, thieves, prostitutes, and street-sweepers, we know that he was not particularly hunting out elevating figures in humble stations, which gives a verisimilitude to his accounts of the Microscope Demonstrator and Mr Tregent, especially when the accounts include racy stories or dialect phrases. With Smiles, however, one wonders how much conscious or unconscious tidying up has gone on to make his characters suitable exemplars of working-class enterprise for middle-class readers, even though the spoken narratives are placed within inverted commas. On the other hand, the straightforward factual narratives of Smiles' working-class astronomers, with the enormous sacrifices of time, money, and energy which they relate, have a ring of truth which transcends even the most earnest attempts to present them as model working-class figures.

Smiles is also useful insofar as he often provides the sources whereby he first came to hear of the figures whom he interviewed, along with the circumstances of the meeting. Other useful material is, in addition, generally recorded, such as the astronomer's age, address, and domestic circumstances, whilst precise references to any publications which he might have penned are cited. The narratives which Smiles recorded were obtained after undertaking deliberate and sometimes quite lengthy journeys to meet the individuals in question, and each biography seems to have been the product of several hours of conversation.

I shall be looking at two of Smiles' astronomers of humble beginnings, Samuel Lancaster and Thomas Cooke, in Chapter 12, for by the time of his interview Lancaster had become a college-trained grammar school master, and Cooke one of the most distinguished British astronomical instrument makers of the age, though they had both been involved in the shoemaking trade as boys. But two other interviewees, John Robertson and John Jones, would probably have been lost to historical record had not Samuel Smiles tracked them down, for poor and obscure they remained.

JOHN ROBERTSON: THE RAILWAY PORTER OF COUPAR ANGUS

John Robertson was a porter in the employ of the Caledonian Railway on the station at Coupar Angus, Perthshire, Scotland, where his job included, amongst other things, shouting out the name of the station when the trains stopped. He had been born in the early 1830s, and had lived in the area around Coupar Angus since 1854, when the railway came. He was a figure of considerable celebrity among the 'travelling public' because of his astronomy, and his fame had spread to Perth by 1879. In that year, Sheriff Barclay of Perth had written to Smiles suggesting that Robertson might be worth interviewing, although it was not until 1884 that Smiles eventually arranged it. When Smiles was preparing his journey to Coupar Angus, however, Robertson took the interesting initiative of sending him a photograph to facilitate recognition, and took him home to his wife and family in Causewayend Street, Coupar Angus, for the interview [1].

Robertson told Smiles that it had been his attendance at a lecture delivered near Montrose by the elderly Dr Thomas Dick about the year 1848 which first turned his mind to astronomy. He then came to read Dick's books, specifying his *Treatise on the Solar System* and his *Practical Astronomy*, at the local working men's reading room, as well as receiving instruction in constellation-finding from Mr Cooper, his night school teacher, after spending his working day in a bleach-works [2]. Indeed, the more one looks at working-class astronomers in early Victorian Britain, the more one realises the pervasive influence which Dick had, as a source of both information and inspiration. In addition to Dick, Robertson became familiar with the popular educator series published by Chambers and Cassell, as well as the *Leisure Hour* magazine articles written by Airy's Greenwich Assistant Edwin Dunkin. But after this initial enthusiasm, Robertson's interests languished following his marriage and the bringing up of his family in the 1860s, and it was not until 1875 that his latent passion was again rekindled.

Like all amateur astronomers, John Robertson wanted a telescope, but money was short, and his investment in one of Solomon's '£5 telescopes' only whetted his appetite for something better. He sold it, and augmenting the deficit from his savings, commissioned a beautiful 3-inch aperture refractor, with full fittings, case, and tripod from Thomas Cooke of York, at a cost of £30 [3]. Indeed, this was a handsome instrument, as Smiles testified, and when one remembers that Airy, in his enquiry on behalf of Leach's gift, said that an ordinary 3-inch telescope could be had for £15, and an ordinary 4-inch for £28, one can understand the pride which Robertson took in his instrument. The telescope would have cost the porter a sum that was equivalent to about eight months' wages [4].

But Robertson put it to good use. He observed early in the morning before going to the station to start work at 6 a.m., after he had finished work at 6 p.m., or in the two hours which he had off for meal breaks during the day. Although his hours of leisure were very limited, Robertson was no mere star-spotter, for notwithstanding his own protestation that he was but 'a day labourer in the science' [5], he undertook serious and systematic courses of observations.

Probably because of his hours of work, he had a predilection for solar astronomy, observing sunspots daily by the dangerous technique of shaded glasses, a common one for the period [6]. He made regular drawings of the solar surface after 1878, and corresponded with William Christie, the Greenwich Chief Assistant who was to succeed Airy as Astronomer Royal in 1881. Robertson offered to supply drawings to Christie for such days as the Sun had been obscured at Greenwich, although by that date he was told in 'a very kind letter from Mr Christie' that a photographic record was really necessary. Robertson told Smiles about the 11-year solar cycle, and how some astronomers suspected that this formed part of a longer-term solar cycle, while he believed from his own observations of spots on the solar limb that they were depressed areas in the solar atmosphere. He also spoke authoritatively of the relationship between the Sun, displays of the Aurora Borealis (which he was well placed to observe in Scotland), and the disturbance of the magnetic needle [7].

John Robertson, in addition, observed comets, meteors, and variable stars; and what his conversations with Smiles particularly bring out was his serious standing in the eyes of professional astronomers, and the up-to-the-minute news which they sent to him. Dr Ralph Copeland – who as we saw in Chapter 8, was Director of Lord Crawford's Observatory at Dun Echt, Aberdeenshire – sent copies of the Dun Echt astronomical circulars to Robertson in the same way that he sent them to Grand

Amateur and professional observatories, and Robertson had visited Dun Echt at Copeland's invitation. Robertson was proud of the fact that, when in the early hours of the morning of 4 October 1880, Hartwig of Strassburg discovered a new comet and telegraphed its details to Copeland in Dun Echt, a copy of the quickly-printed Dun Echt circular reached him at Coupar Angus that evening by the 7 p.m. post. Having just finished work at the station, and it being a clear night, Robertson immediately got out his telescope in the garden, and observed the comet 'within fifteen hours of the date of its discovery in Strasburg [sic]' [8]. Similarly, his fortunate observation of the great meteor of April 1878, and his plotting of its coordinates, got him into a correspondence with Alexander Herschel, son of Sir John, who was Professor of Physics at the college which would become Newcastle University.

Like James Veitch, John Robertson did not in any way feel that his working-class job and amateur status should constitute any obstruction to his participation in serious astronomy. Nor did he feel 'out-classed' when dealing with people like Christie, Copeland, or Alexander Herschel. Perhaps, as we saw with Veitch in the previous chapter, it had something to do with Scottish meritocracy and admiration for intellectual excellence.

John Robertson's quiet confidence and healthy self-respect also comes out in two other aspects. Firstly, in spite of his modestly acclaimed status as a mere 'under labourer' in astronomy, he was a published amateur scientist of acknowledged standing. His observations of solar phenomena, meteors, comets, and the like, were reported, under his own name, to *The Observatory* journal, and *The Scotsman*, Dundee *Evening Telegraph* and other newspapers [9]. They are well and clearly written, indicating a man who not only knew his subject, but had no difficulty in expressing himself with a pen. 'Besides', he added, 'I am the local observer of meteorology, and communicate regularly with Mr. Symons', the meteorological 'Recorder and Coordinator'. Smiles also commented upon the meteorological instruments which he saw in Robertson's back garden [10]. Secondly, and very significantly, Robertson had no desire to leave his present occupation and take up positions that could use his talents more effectively than railway portering. His great champion, Sheriff Barclay of Perth, had tried on several occasions to encourage him to improve his position, and had recommended him for jobs in observatories, presumably as a 'professional assistant'. But Robertson would not move, asserting 'I am very comfortable! The [railway] company are very kind to me, and I hope to serve them faithfully' [11]. While on the surface this may sound as though a grateful son of toil was reassuring his betters by posing no threat to the social order, he did go on to qualify his remarks.

Quite simply, by his early fifties Robertson felt settled and at ease with the world, and had no wish to uproot himself to start a new job for a new employer, even in astronomy. While it is true, from what we know of the salaries of 'professional assistant' astronomers of the period, that Robertson could well have doubled his income by letting Sheriff Barclay find him an observatory job (Piazzi Smyth's *second* assistant at Edinburgh, for instance, received £100 per annum), he had no wish to disrupt his life [12]. Railway companies had the reputation for steady employment for life, often with housing, and pensions on retirement, and as he enjoyed his local celebrity in Coupar Angus, and with his 'travelling public', he preferred to remain what Smiles styled 'the contented Coupar Angus astronomer' [13], and in this respect he was at one with Johson Jex and James Veitch. Although Smiles' tone is occasionally patronising, and one wonders how Robertson felt when he read of himself as an astronomer in 'humble life', it is likely that he preferred to enjoy astronomy for the 'labour of love' that it was, and for the independent respect that it brought him, rather than become just another 'professional assistant' who had to do as he was told.

JOHN JONES: THE SLATE COUNTER OF BRYNGWYN BACH

When visiting Yorkshire in the early 1880s, Samuel Smiles received two letters informing him of the activities of the Welshman John Jones. One was from his former Leeds friend, the Revd Charles Wickstead, and the other from Miss Grace Ellis. In consequence, Smiles took the journey from York to Bangor, North Wales, after making the appropriate acquaintance by letter, and called to see Jones at his house in Albert Street, Upper Bangor, on the evening of his arrival [14]. Although in his self-perceived role as a poet–astronomer, John Jones always styled himself in Welsh as 'Ioan, of Bryngwyn Bach' in Anglesey. He had lived and worked on Bangor docks as a slate counter for around thirty-five years at the time of his interview. But one crucial aspect of Jones' life and career which separated him from all the other working-class astronomers in this section was that of language. Being born into a Welsh-speaking agricultural community in Anglesey in 1818, and having received only the most rudimentary schooling even in his mother tongue, he was isolated from the ideas of the wider world in a way that Jex, Leach, and Robertson were not [15].

As a youth, he had 'to work for the farmers', and their arduous demands upon his time and energies 'choked within me' all aptitudes for music and other cultural urges. But part of Jones' job included saddling the horse of the Revd Cadwalladr Williams, the Calvinist Methodist preacher of Pen Ceint, Anglesey, who had a good library of Welsh language books. Working in Williams' library, probably when the minister was away, John Jones began his serious self-education. He acquired a knowledge of English by comparing the scriptural verses in a familiar Welsh New Testament with one in English, and came by this 'Hamiltonian method' to gain not only a good reading knowledge of English, but also of several other languages over the years [16]. Because the affairs of his daily life continued to be conducted in the Welsh language, however, his spoken English was never completely fluent. Indeed, he had acquired the English pronunciation academically, from studying the phonetic alphabet which Thomas Gee of Denbigh published in 1853. Smiles commented on the fact that during the course of their interview, John Jones would sometimes fall back on to his native Welsh, especially when faced with an awkward technical point. But Jones' fondness of philology had led him to acquire twenty-six dictionaries by the date of his interview, mostly second-hand from towns like Liverpool, and from friends who worked the slate ships from Bangor to Liverpool and purchased them for him.

It was the all-pervasive Thomas Dick who first introduced John Jones to astronomy, for his treatise *The Solar System* (1846) had been translated into Welsh by Eleazar Roberts of Liverpool, to become one of the first books to present contemporary astronomical knowledge to a Welsh-speaking audience. Jones implies in his narrative to Smiles, however, that he had found the book in Cadwalladr Williams' library at Pen Ceint, but either Jones or Smiles was getting the chronology confused at this point in the story, for according to the British Library Catalogue, Roberts' translation, *Y Dorspath heulawg (The Solar System)*, was not published until 1850, by which time Jones would have been 32 years of age and working in Bangor. Jones might, on the other hand, have read Dick's *The Christian Philosopher* (*Yr anianydd Christionogol*), which also dealt with astronomy and had been translated into Welsh in 1842, at a time when Jones was still living in Anglesey. But as Jones' knowledge of written English improved he read Sir John Herschel's *Outlines of Astronomy* (1848), and a book which Smiles wrote down as Herschel's *Treatise on the Telescope* – which was probably his *Treatise on Astronomy* [17]. He was also the first British working-class astronomer that I have encountered who mentioned studying the American Denison Olmsted's *Mechanism of the Heavens* (1856) [18].

Jones had come to Bangor from Anglesey when he was about 30 years of age, as a result of illness; and finding light work on the docks counting slates, he stayed. His partial attendance at a Bangor navigation school also added to his knowledge of astronomy. However, as he was earning only twelve shillings (60p) a week in the early 1850s, rising over the years by a few increments, he was, and remained, a very poor man. Rather strangely, Smiles made no reference to any wife or family commitments, although Jones had alluded, somewhat ambiguously to Smiles, that in his younger days, 'to the detriment of my plans; some fair Eve [was] often standing with an apple in hand, tempting me to taste of that.' Yet either this fair Eve, or else another, did succeed in diverting John Jones' attentions away from pure learning, because the 1881 Census return recorded him as living with a 49-year-old wife, enigmatically styled Fa.... [*sic*]; however, it is possible that she might have been absent when he was interviewed in 1884 [19]. It may seem incongruous, though, that a man who saw self-education in astronomy, philology, and other subjects as part of the 'plans' from which he should not allow himself to be 'enticed' should never have envisaged career prospects for himself that went beyond being a dock worker on twelve shillings a week.

Like all other amateur astronomers, John Jones wanted a telescope, and after reading accounts of home-made instruments in the *English Mechanic*, set about making one for himself. Calling upon the services of his friend Captain Owens – of the ship *Talacra* – who had also picked up a Greek lexicon and New Testament for 7s 6d (37½p) for Jones, he acquired from Liverpool the lenses for a simple refracting telescope. The lenses cost 4s 6d (22½p), and Jones mounted them in a cardboard tube to produce a refractor of 36 inches focal length. With this simple instrument, he was delighted to see at first hand those objects about which he had read in books, such as the lunar craters, Jupiter's moons, and some double stars.

'But I was not satisfied with the instrument. I wanted a bigger and more perfect one' [20], Jones told Smiles, in the familiar amateur astronomer's litany for more light that would have been understood equally well by William Lassell and Lord Rosse. The first home-made instrument was sold, and the money used to buy better lenses from Messrs Solomon of London, probably the same firm from which John Robertson had purchased his original '£5 telescope'. (Indeed, the £5 telescopes seem to have been popular instruments with aspiring astronomers, for their merits were discussed at length in the *English Mechanic* in 1876) [21].

It was around 1868, however, at the age of 50, that Jones got around to constructing a reflecting telescope. Speculum, of course, was out of the question for a man of Jones' limited resources, so he opted for the new technology of silver-on-glass mirrors, which was still in its relative infancy in 1868. John Jones, therefore, was one of the first generation of amateur astronomers to fabricate a glass mirror, and without doubt must have been by far the poorest.

Jones obtained a 10-inch diameter disk of rough glass from St. Helens, Lancashire, from what was probably the Pilkington Glassworks, which over the next century would supply countless blanks for amateur and professional telescopes world-wide. It took him nine or ten days to grind and polish it, until his face shone in what had become a working optical diameter of $8^3/_{16}$ inches. Jones then sent his spherical mirror to that pioneer of glass mirror working, George Calver of Chelmsford, to be given a true parabolic curve and a coat of silver. He paid Calver £5, which was a sum equal to more than six weeks' wages. Jones also completed another glass mirror with a diameter of 6 inches [22].

Unfortunately, John Jones said nothing about how he learned to give his mirror a good spherical figure, even if he could not parabolise and finish it himself. It is likely that he had gained his knowledge from the *English Mechanic*, to which he referred in his narrative to Smiles in the context of telescope making, for that publication became a veritable *vade mecum* of technical skills in the late nineteenth century, with an authorship and a readership which embraced labouring men and Oxford dons.

Jones made his own tubes and fittings to mount the mirrors, and the resulting $8^3/_{16}$-inch diameter mirror of 10 feet focal length was found by him to 'have good defining power'. When Jones showed it to Smiles it was mounted in 'his tiny room upstairs' in the Albert Street house, and through the window it was able to reveal the polar caps of Mars when the planet was well-placed. Because of its great size Jones called the telescope his 'Jumbo', in honour, no doubt, of the famous circus elephant which had so captured the late Victorian imagination [23]. Most likely it was the same telescope as that included in the photograph of Jones, its square-sectioned altazimuth tube being of a design still used today by amateur reflecting telescope makers, while the ornate setting wheels used to control its motions look as though they might have been salvaged from a discarded piece of industrial machinery. In their ingenuity in incorporating existing parts that are lying conveniently at hand, amateur telescope makers have not changed in over a century, as I know from personal experience.

In spite of his low income, it is clear that by the age of 65, when Smiles interviewed him, John Jones had acquired quite a collection of instruments. For in addition to the $8^3/_{16}$-inch and 6-inch reflectors to which he makes reference, he mentioned a $4\frac{1}{4}$-inch refractor (which may or may not have been his own property), for he discussed the capacities of the different instruments to bear high magnifications. On the table beside him on the photograph, moreover, is a marine sextant, which he probably acquired on Bangor docks. Though Smiles unfortunately failed to supply any further details, it seems that Jones had also built an equatorial instrument, mounted upon a tripod stand, the right ascension and declination circles of which he had engraved upon slate [24]. But perhaps most remarkable of all was the spectroscope that Jones seems to have been in the process of constructing at the time of his interview. The glass for the prisms came from the old thick sky-lights used on ships, which Jones cut up and ground to the correct angles. It is not clear how he had learned about this relatively new branch of astronomy, and one suspects the *English Mechanic* as a possible source [25].

It does not seem that Jones followed any specific lines of research with his instruments, but probably enjoyed looking at whatever was available. Although the point was not specifically made in the interview, one surmises that John Jones was a man who enjoyed making telescopes as much as he enjoyed using them, which is another trait not infrequently encountered among many modern amateur astronomers. Sadly, Smiles omitted a great deal of Jones' narrative where it pertained to the scientific details of his instruments and their uses, because it was too technical to be included in his account. It would have been fascinating to know the precise details which the slate counter related to his interviewer [26].

While Jones' religious affiliations are not discussed as such by Smiles, it is clear that he was a religious man. For in addition to his familiarity with and use of the Bible as a way of learning foreign languages, and his desire to read the New Testament in the original Greek, Jones made use of Biblical imagery in his autobiographical narrative to Smiles. While he loved all knowledge, he claimed 'of Reuben, Dan, and Issachar; ... I have a favourite, a Benjamin, and that is Astronomy. I would sell all of them into Egypt to preserve my Benjamin.' Here, of course, Jones refers to old father Jacob, whom God named Israel in *Genesis* 42 and 43, who loved all of his twelve sons, including Reuben, Dan, and Issachar, but who was reluctant to let his youngest and most favoured son, Benjamin, go down into Egypt in the time of famine [27].

Attention has already been drawn to John Jones' apparent lack of concern with improving his overall social position, but while contentment seems to have been a declared condition of a number of working-class and artisan astronomers, Jones seems to have been especially poorly paid. His astronomy – or 'Benjamin' – seems to have been pursued entirely for itself, in the evenings after work, and as was Benjamin to Father Israel, so astronomy was the pride and joy of Jones' old age. In the great Welsh tradition of self-education – be it singing, Biblical scholarship, poetry, or astronomy – these studies seem to have been pursued for a mixture of personal and religious motives, but rarely for career advancement. Mention has also been made of that great obstacle which working-class Welsh people encountered when dealing with astronomy, especially before the 1850s: namely, a lack of printed material in their mother tongue. Until the translation of Dick's books in 1842 and 1850, there was virtually no basic astronomically-related material in Welsh, with the exception of Robert Roberts' astrological almanacs [28]. The lack of a tradition of astronomical writing in the Welsh language was probably another reason why Jones did not communicate his observations to local English language newspapers in the way that John Robertson did in Scotland.

Indeed, one cannot help but be struck by the difference in attitudes to popular education in Wales, England, and Scotland. John Robertson and James Veitch wrote with confidence to newspapers, passed on their observations to scientists of eminence, and did not see it as strange within the society in which they lived that men who worked with their hands could also have first-class minds which deserved to be taken seriously. In England, by contrast, it was assumed that the manual working classes would keep their places, although great sacrifice and ingenuity might sometimes be deemed worthy of a modest present. In Wales, the English rules applied, while the prevalence of the Welsh language among the native population imposed yet another barrier, and generally made Welsh culture more inward-looking. Perhaps it was this latter reason which gave John Jones the remarkable self-containment which so shines through his narrative, and which also expressed itself in privately-composed Welsh poetry, which Smiles published in English translation [29].

John Jones lived another fourteen years or so after being interviewed by Smiles, dying at the age of 80 in 1898. In the latter part of his life, moreover, he became something of a celebrity, as the working-man astronomer whose diligence and industry had enabled him to do so much with so little. He was written about by the Cardiff journalist and amateur astronomer Arthur Mee, and by Silas Evans [30]. And it was no doubt through the activities of Mee that he became a member of the essentially middle-class Astronomical Society of Wales. For the ladies and gentlemen of that Society could not have denied that to Welsh amateur astronomy John Jones – whom they graced with the title 'Seryddwr', or Astronomer – was truly old father Israel [31].

ROGER LANGDON: A VILLAGE STATION MASTER ASTRONOMER

All of the artisan or labouring astronomers that have been examined so far are known, for the greater part of the available narrative of their lives and works, from what persons occupying superior social positions have recorded. And while James Veitch is remarkable insofar as he corresponded on equal terms with people of eminence, it is still through those people's records, especially Sir David Brewster's, that we have obtained the principal facts of his achievement. With Roger Langdon, however, one encounters a working-class astronomer who wrote autobiography under his own hand, the latter part of which was completed by his daughter, and then published. He also published various astronomical pieces in his own right, so that unlike everyone else in his social position that we have seen, his own words greatly outnumber those of his middle-class admirers. Because Langdon's autobiography is unique amongst working-class astronomers, therefore, it is worth looking at his circumstances before he came to be seriously involved in the science, for the obstacles which he had to surmount would have had something in common with those faced by other astronomers in his position.

Roger Langdon was born in poverty in 1825, and at the age of 8 went to work pulling up turnips for a farmer at the wage of one shilling (5p) a week. He was sufficiently skilled as a ploughboy by the age of 12 to win third prize in a ploughing competition, and two years later, in 1839, walked the 33 miles from his home, near Crewkerne, Somerset, to the port of Weymouth, with the intention of going to sea. He chanced upon a ship sailing for Jersey, where he was to spend the next eight years of his life doing a variety of jobs [32].

While still only a small boy, however, he had been drawn to intellectual matters. He tried, and failed, to save up the sum of ninepence to purchase a copy of Pinnock's *Catechism of Astronomy* [33] which he saw in a Crewkerne printseller's window, and so impressed the local curate with the scraps of the New Testament that he carried in his pockets that the Revd Peter Cornwall presented him with a new Bible [34].

Like many working-class figures of the period, such as John Jones, and those alluded to by Thomas Dell when lecturing to the young men of Aylesbury, the narrative of Roger Langdon drew attention to the appalling poverty that prevailed, especially during 1846 and 1847. Bad harvests and industrial recession hit the poor severely during those years, in spite of Peel's abolition of the tax on bread, so that in preparation for his marriage, which took place in Bristol in 1850, Langdon made every effort to get a steady-paying job, either in the Post Office or on the railways. While these occupations did not pay the highest wages, they were, as we saw with John Robertson, regular, reliable, and cushioned against the vagaries of economic swings. He was taken on as a porter at the Bristol station by the Bristol and Exeter Railway (a company later to be amalgamated with the Great Western Railway, or G.W.R.), at the rate of 12 shillings (60p) a week, which was exactly the same wage that John Jones was then earning counting slates on Bangor docks [35].

Over the next few years, Roger Langdon crept up the company's promotion ladder, working twelve hours a day as a line-side signalman at Durston, Devon, and then as Station Master at the village of Silverton, near Exeter. Here he came to receive the princely wage of 30 shillings (£1.50) a week, with the added advantage of no all-night shifts [36].

Langdon's ingenuity and manual dexterity were expressed in the building of harmoniums, magic lanterns and painted slides, and a model of the nearby Stoke Canon church wherein a penny in the slot rang an authentic peel of miniature bells [37]. He also attended evening classes in Greek, offered by the local curate, the Revd Mr Jackson, at a halfpenny a week, as he wished to be able to read the Scriptures in their original languages, as well as have access to modern scientific publications [38]. He had an excellent partner in his wife Anne – whom he admitted had made his achievements possible – for she was an intellectual woman who augmented his modest earnings by running, in their house, a small school for local children. They would have had quite a walk, however, for Silverton station was about two miles by road from the village [39].

We do not know exactly how Roger Langdon came to develop his interests in astronomy, but in 1865 he built his first telescope, a simple refractor constructed from second-hand lenses. He sold it for 7s 6d (37½p), bought some better optics, and produced an instrument which would show the rings of Saturn and other features. Unfortunately, Langdon did not record the details of his telescope under his own hand, for this part of his narrative was completed by his daughter Ellen, and she speaks of this second telescope as possessing 'reflectors' 4 inches in diameter [40]. But as he did not figure mirrors until his third telescope, one presumes that Ellen Langdon was really talking about a refractor.

Langdon's mirror-working activities began when he left his £1-a-week signalman's job to become Station Master and sole member of the station staff at Silverton in 1867 [41]. His first essay in mirror-working was a 6-inch reflector of 5 feet focal length. And like his Welsh contemporary, John Jones, it was to the *English Mechanic* that he turned for advice. And through that magazine, he entered into correspondence with a Dr Blacklock, an already experienced glass mirror figurer who was extremely forthcoming with his help. It was Blacklock's advice, conveyed through this correspondence, which enabled Langdon to silver and test his finished mirror [42]. But another of Langdon's correspondents was James Nasmyth. Whether or not the connection was made via the *English Mechanic* is not recorded, but it obviously extended over several years, although Nasmyth's advice once led Langdon inadvertently astray, until he realised that a particular tool which Nasmyth was recommending was only suitable for speculum metal, not glass mirrors. Even so, he was proud to enlist Nasmyth among his correspondents, and knowing as we do how the Brethren of the Big Reflecting Telescope exchanged ideas between themselves, it is gratifying to see one of them being so free with his advice to a poor man whom he took seriously [43].

Langdon's fourth and finest instrument was an 8¼-inch silvered glass reflector, part of which he financed by selling his 6-inch instrument for £10 in 1875. Set up on an equatorial mount, with accurate setting circles, and fitted with a Ramsden eyepiece, it was an instrument possessing excellent defining power [44]. Its permanent mount inside a sheet-iron observatory, with a rotating conical 'dome' with opening shutters, enabled Roger Langdon to do serious astronomy, including wet-plate photography of the Sun, Moon and planets. And on one occasion when he was visiting the Royal Observatory some years later, he was delighted to find photographic dark-slide arrangements on some of the Greenwich telescopes which were exactly the same in principle as that which he had at home in Silverton [45].

In many respects, however, Roger Langdon had a job which made it possible to fit his astronomy into his normal working day. As the G.W.R.'s jack-of-all-trades at Silverton until he acquired an assistant in 1875, the limits of his working day could extend from 6 a.m. until the arrival of the last coal train about twenty hours later. Meals, relaxation and sleep were fitted around the comings and goings of the trains [46]. And as he and his family lived on site, he probably had little difficulty in incorporating astronomy into his broken-up days and nights. As a rural Station Master, moreover, he would also have had much more physical space at his disposal than most other working men. Unlike the cramped back yards available to Robertson and Jones, he had the space upon which to erect an observatory with a floor area that was as large as the sitting-rooms of most industrial workers' houses. A photograph of Silverton station, preserved in the archives of the National Railway Museum at York, shows that Langdon and his family lived in an elegant detached house, with the observatory, with its conical 'dome', situated some yards away in the garden.

Like all working-class astronomers, Roger Langdon no doubt enjoyed the local kudos and reputation within a small community which his knowledge inevitably brought with it. Second to the vicar, he was the most learned man in the village, with his knowledge of dead languages and collection of books, while the local folk were convinced that their Station Master could 'rule the planets' with his strange tubes. Had he not been renowned for his staunch Churchmanship, and his opinion sought by travellers as to the suitability of the radical new Bishop of Exeter – Frederick Temple – for his office in 1869, he could have been feared as a wizard! [47]

Silverton station became something of a place of pilgrimage from several miles about, as people came at all hours to be shown the sky or see the Station Master's 'planet-ruling' contraptions. And Langdon seems to have been remarkably patient in talking to people and showing them around, so that one cannot avoid drawing parallels between him and John Robertson, in their dispensation of astronomical knowledge to the travelling public. On some occasions, it was not only the poor and simple who enquired about his *astrological* skills; though when the better-educated did so 'he would say that he was ashamed to hear them ask such a question' [48].

One educated man who did visit Silverton station was Clifton Lambert, the son of the General Manager of the G.W.R., who came in 1889. By this time, the 64-year-old Langdon was white-bearded and patriarchal (the death of one of his sons in a railway accident had caused his hair to turn white), and while some of Lambert's remarks might seem patronising, when one penetrates the linguistics of Victorian cross-class writing one sees how truly impressed he had been. For 'the career of Roger Langdon provides for all of us a striking illustration of what force of character will accomplish even in the humblest surroundings and in the face of the most serious obstacles' [49]. Clifton Lambert went on to write articles about the Station Master of Silverton in the *Great Western Railway Magazine* and other periodicals. Langdon had similarly impressed Sir Thomas Acland and the Revd F.H. Fox Strangeways among the local Devon gentry, and Sir Thomas permitted the body of Roger, and other Langdons, to be interred in the Acland family private burial ground [50].

Roger Langdon was a fully literate man with published works to his credit, in addition to the autobiography. In 1871 he read a paper on his observations of Venus to the R.A.S., though prior to the event he was so nervous, as he later confided to his daughter, that 'he wished the earth would open and swallow him up' [51]. His work on Venus was published in the *Monthly Notices* of the R.A.S. in 1872 and 1873.

These papers indicate that even by 1872 Langdon had 'observed Venus a great many times besides this mentioned above, having made it my special work' [52]. When observing the planet in 1870 and 1871 with his first 6-inch reflector, for instance, he obtained heightened definition by inserting a pinhole diaphragm into the eyepiece, which he claimed enabled him to detect a pear-shaped marking near the planet's terminator, over the period May to November 1871. Then in April 1873, using 'one of Mr. Browning's excellent achromatic eyepieces', he reported the raggedness of the Venusian terminator and the exceptional brilliance of the horns to the R.A.S [53].

And aiming at more popular audiences, Langdon published astronomical pieces in the *Exe Valley Magazine* and *Home Words*. These articles, the first of which is a description of the structure of the lunar surface, as told by 'The Man in the Moon' himself, and the second an account of Coggia's Comet of 1874, display real didactic power, combined with humour and an easy style of explanation [54]. It is a pity that, unlike his railway colleague of Coupar Angus, Roger Langdon had no-one urging him to become something more than a village Station Master, for one senses that he could have been an inspired school teacher or journalist.

ASTRONOMERS IN ISOLATION

When one looks at the poor, obscure, and self-taught individuals discussed in the present and the two preceding chapters, a number of important points emerge. All of them, with the possible exceptions of John Leach and Thomas Bush, seem to have been content with their places, and did not attempt to rise socially even when encouraged to do so. One is also impressed by the high level of religious motivation present in their lives – not just in their astronomy – as shown in the desire of men like Veitch, Jones, Bush and Langdon to master the original Biblical languages and suffuse their very speech with Christian metaphor. Only Johnson Jex appears to have been overtly indifferent to religion. And it is possible that it was their belief that God had put them into a particular social place to fulfil His design which encouraged them to accept their lot cheerfully, and find contentment in private study and in the contemplation of the heavens.

On the other hand, it would be incorrect to assume that contentment was the same thing as supine acceptance, for many of these working-class astronomers made remarks which indicated their acute perception of what was socially fair and just. Three of the astronomers who were active around 1870, for instance, had some kind of dealings with the radical Prime Minister, William Ewart Gladstone, either through receipt of his government's patronage or, as in the case of Roger Langdon, by avowed support for the Liberal Party. For Gladstone, we must remember, was not only a scholar–statesman committed to social reform, but one who saw such reform as part of the duty of a Christian gentleman. And while Roger Langdon might have been content in his place as a Station Master astronomer, he made no secret of his warm approval of Gladstone's controversial appointment of Temple to the Bishopric of Exeter, as Temple would 'make the clergy work' [55] and compel idle incumbents to serve their parishioners.

While it is true that the poor and obscure astronomers had some aims in common with the Grand Amateurs, such as a desire to obtain as much light as either their pockets or the available technology could afford, one of the most striking differences between the two sets of scientists, as mentioned above, lay in their peer relationships. For while Lassell, Dawes, Rosse, Smyth and so on were fully informed about each other's scientific activities, met, corresponded, visited, and dined together, there is no indication that any one of the men discussed in the last three chapters – except William Knock, Mayhew's Microscope Demonstrator, and Robert Children, who were acquainted through the context of employment – even knew of any other's existence. Leach and Bush might correspond with Airy, Langdon with Blacklock, Nasmyth and the R.A.S., and Robertson with Copeland and Alexander Herschel, yet I have found no evidence of two working-class astronomers corresponding with each other. All the contacts seem to have been vertical, and not horizontal. It is also worth noting that in the year 1870, when Bush, Langdon and Jones were all busy figuring silver-on-glass mirrors, and Airy was recommending useful official presents for Leach and Bush, the penny post and the fourth-class railway ticket could have brought together all of these men, and no doubt many more. Yet until the founding of the first local and national amateur astronomical societies after 1881, the necessary infrastructure of communication for working-class scientists did not exist.

And finally, it is hard to over-emphasise the crippling financial burden that a passion for astronomy could inflict upon a working-class family. The purchase of a £5 telescope, or one of Horne and Thornthwaite's 'Star Finders', would have consumed over two months' wages for John Jones or the newly-married Roger Langdon; and when the Revd Thomas William Webb was speaking of the 'trifling outlay' that was necessary to pick up a 'common [achromatic] telescope' in 1859, he was clearly not thinking in terms of the working man's hoarded shillings [56]. It is true that by 1908, fifty-odd years later, when F. Burnerd of Putney, London, was advertising 'Cheap 2-inch Astronomical Telescopes' with achromatic lenses and a stand for 43 shillings (£2.15), the real cost of astronomy had come down considerably, yet instruments of decent quality were still exorbitantly expensive for working people to purchase [57].

When one considers, therefore, the cultural isolation, limited leisure time, and great expense against which a working-class scientist had to strive, one cannot but marvel at what the poor, obscure and self-taught of the Victorian amateur astronomical community actually achieved. They were, indeed, the true 'lovers' of astronomy, to an extent which made the classic definitions of Herschel and Pogson almost pale in the degree of sacrifice which they implied.

Part 3

The rise of the leisured enthusiast

Prologue

Over the last twenty years of the nineteenth century, the social complexion of British astronomy changed, as research initiatives at the cutting edge of the science moved into the hands of the paid, and frequently academically-trained, professionals, who generally worked with equipment that was not their own property. As we saw in Chapter 7, the emergence of figures like Lockyer, Gill, Copeland and Ball represented a new social type in British astronomy: men who were neither astronomical servants nor Grand Amateurs, yet whose primary occupation was original research. And at the same time, one sees the Grand Amateur gradually fading from the scene in terms of fundamental research importance, and his metamorphosis into a distinct and new type of amateur astronomer. These persons did not necessarily lack financial means or education, but in the economic recession of the 1880s and 1890s, when *laissez-faire* individualism was being called increasingly into question across society in general, the rising generation of amateur astronomers tended to be people of more modest fortune and ambition. They were what might be called the 'leisured enthusiasts'.

This was also a period when 'professionalism' was spreading across many aspects of national life where it had not been in 1850, and the social definition of expertise in a wide variety of subjects was shifting from the great amateur to the academically trained and institutionally related professional. The antiquary was being superseded by the archaeologist, the connoisseur by the museum curator, the bibliophile by the librarian, and the naturalist by the biologist. Even politics was moving in the same direction, as middle-class career politicians like Benjamin Disraeli and William Ewart Gladstone stole the reigns of government from the hereditary élite of ancestral peers, and forged tightly-controlled modern-style political parties.

One factor, of course, which brought about this growth of specialisation was the great increase in all branches of knowledge which took place during the nineteenth century, combined with a broad and on-going democratising tendency within British society. Social and political reform had come to predicate not only the widening of the electoral franchise, but also the provision of facilities for the general public which had once been the preserve of select individuals or restricted corporations. The Public Libraries Act of 1850 and subsequent enabling statutes had created a framework for the establishment of civic museums, libraries and art galleries, while a variety of acts after 1870 had sought to provide improved school and local college facilities out of public money. Similarly, the founding of the 'Victoria' Universities of

Manchester, Liverpool, Leeds and Belfast in the 1880s, with further nascent universities elsewhere, heralded a new type of university, linked not to Oxbridge-style colleges but to the needs and patronage of great industrial cities. And none of this could be conducted by a select coterie of private gentlemen, but demanded new types of expertise and management which a growing army of professionals was coming to provide by 1900.

Professional astronomy in Britain, therefore, grew up as a part of this wider social process, while the people who made up the new amateur astronomical community were of a different type, by and large, from the amateur Fellows of the Royal Astronomical Society of the 1840s. For one thing, these new amateurs did not control astronomy politically, sit on national scientific committees, become Fellows of the Royal Society, or frame the policies or raise the finances which would direct research. Most of them, furthermore, were not gentlemen of leisure, but people who combined their astronomy with a continuing professional occupation – being solicitors, doctors, schoolmasters, clergymen, managers, and journalists. And as they lacked the time and the surplus wealth of their scientific ancestors, they could not immerse themselves entirely in their hobby or engage full-time assistants, for even the traditional professions were tightening up on standards by 1880. The Law Society, for instance, was shaking the dilettantism out of provincial legal practice, while the more socially-aware bishops and Church Commissioners would no longer tolerate pluralism and pastoral idleness on the part of the clergy, so that many country rectors of 1890 lacked the time and the money to be Grand Amateurs in the old sense.

The new astronomical enthusiasts, moreover, created fresh demands in their own right, as well as providing facilities that would turn out to be especially enriching to the science. One important new development came in the organisations which they created to cater for their particular needs. For these people were what Captain Maurice Anderson Ainslie described as 'Amateur[s] who take up practical observing, as distinct from the study of theory' [1], to constitute a body of people with different objectives from either the old Grand Amateurs or the contemporary professionals. Amateur astronomical societies, operating both locally and nationally, were their particular institutional creation, where observational information could be compared and published in a context that was different from the higher mathematics found in R.A.S. meetings and publications. Furthermore, these societies were of enormous importance insofar as they admitted women to full and active membership, to break down the male exclusiveness of the old Royal Chartered Societies, and give women their first institutional footing in British science. They also provided an environment in which working-class astronomers could escape from their old isolation, to meet and exchange ideas with people of similar interests but occupying different social brackets.

The new amateurs also created demands for information. Along with the *English Mechanic* and *Astronomical Register*, authors like Richard Anthony Proctor and Sir Robert Stawell Ball began to produce books which explained the latest astronomical knowledge, without mathematics, to people who could afford to spend 12 shillings (60p) or a pound on a handsome volume.

They wanted, in addition, good-quality, affordable telescopes with which to do 'some really useful work ... advancing knowledge, even in a small degree' [2], as they meticulously drew the surfaces of the planets, studied comets, and became intimately acquainted with the features of the lunar surface. Firms like Cooke of York, Calver of Chelmsford, and to a lesser degree Grubb of Dublin, began to meet this lucrative market for good amateur telescopes, while part-time manufacturers like George Henry With of Hereford were willing to supply excellent glass mirrors for amateurs to make up into telescopes themselves.

By the last decade of the nineteenth century, therefore, an amateur astronomical community had come into being which was quite distinct from anything that had existed fifty years before, and which in many of its essential aspects was the same as the amateur astronomical world which has come down to us today.

12

A goodly pursuit for a Godly mind: Thomas Webb and his influence

THOMAS WILLIAM WEBB

Modern amateur astronomy, as a pursuit for serious observers whose principal motivation was pleasure, fascination, or the glory of God, as opposed to fundamental research, began in a Herefordshire vicarage in the 1850s. For the Revd Thomas William Webb, vicar of Hardwicke and later prebendary of Hereford Cathedral, was reckoned by 1893 to be the 'father of all amateur astronomers', and a man who had started a movement [1]. Though an Oxford mathematics graduate, with a firm grasp of both theoretical and practical astronomy, Thomas Webb spoke not to the Grand Amateurs, but to those men and women in various walks of life who were enthusiastic to take up the science as a serious and instructive hobby.

His influential book, *Celestial Objects for Common Telescopes* (1859), acknowledged Admiral W.H. Smyth as its inspiration and dedicatee, but as Webb pointed out, Smyth's *Cycle of Celestial Objects* (1844) by 'its very superiority, to say nothing of its bulk and cost, renders it more suitable for his [the advanced amateur's] purpose, than for the humble beginnings which are now in view' [2]. The growing fashion for amateur astronomy, pointed out by John Weale and others, had meant that a new market had come into being by 1859 which had not been significant fifteen years earlier: a market aiming at more financially modest, and perhaps less highly educated readers. For as opposed to the two hefty octavo volumes of Smyth's *Cycle* (which had been long out of print by 1859), Webb's *Celestial Objects* in its first edition was a little book that packed a great deal of information into its 247 pages of small print [3]. In many respects, however, it went through a structure that was similar to that of the *Cycle*, beginning with telescopes, observing procedures, and useful tips, before going on to a detailed descriptive catalogue of planetary and stellar objects which the amateur could observe.

Webb's *Celestial Objects* brought twenty-five years' worth of personal experience to the beginner's aid, for he had made his first recorded observation of Jupiter in 1834. He had acquired a 3.7-inch fluid lens achromatic refractor by 'the younger' (either William or Thomas) Tulley [4], and between 1847 and 1856 began to work his way systematically through the objects in Smyth's *Cycle* to acquire a thorough and first-hand knowledge of the sky. For as Webb urged his readers, one had to look at the heavens at first hand to truly appreciate their wonder, otherwise it was like relying on the 'narrative of the traveller instead of the direct impression of the scene' [5]. And by a 'common telescope', Webb stipulates an instrument of similar dimensions to his Tulley, being of around 3 inches aperture and of 5 feet focal length. Such instruments, he went on to remind his readers, were 'far less expensive than formerly; a trifling outlay will often procure them of excellent quality, at second hand' [6]. As we saw with astronomers like John Jones and John Robertson, however, what one deemed as 'trifling' was very much dependent upon one's circumstances in life, which says a lot about Webb's expected readership.

Thomas Webb was the son of a clergyman, the Revd John Webb, and was born in Ross-on-Wye, Herefordshire, a member of a family with London and West Country affiliations. His marriage to Miss Henrietta Montagu Wyatt, of Mitchel Troy, Monmouth, brought him connections in the Welsh border gentry to add to his own, and in 1874, following the death of the seventh baronet, Sir Henry Webb of Odstock, Wiltshire, he became head of the powerful Webb family. On his own death, on 19 May 1885, Thomas Webb left farms and landed properties in Herefordshire and numerous personal and charitable legacies from an estate valued at £16,986. In short, he was a wealthy and well-connected clergyman, who could well have entered the Grand Amateur league had he chosen to do so, and one must ask why Webb took the much more modest approach to astronomy that he did [7].

The reasons lie within the character of the man, and in his sense of vocation. Webb's astronomy was not the sacrificial activity of a man driven to push back intellectual frontiers, like Herschel, South, Lassell, and Rosse, or of one who was willing to go to great lengths of financial and social inconvenience to see where no one had seen before. Instead, he was an *observer*, who took the deepest pleasure in the contemplation of the sky through instruments which by fundamental research standards were very modest. For as his *Celestial Objects* and other publications make clear, his main motivation lay in the sheer divine majesty of the heavens, and in his wish to urge others to 'taste and see' that majesty for themselves. This fitted in perfectly with his calling as a country clergyman, for despite his good social connections and participation in the social round of a country gentleman (which comes over in the appearances of Webb in Francis Kilvert's *Diary*), he remained a conscientious parish priest to the end of his days [8]. As his friend, protégé and executor Thomas Henry Espinall Compton Espin recorded, he liked to go off in an afternoon 'with a knapsack laden with all kinds of little comforts for the sick', and 'he would walk with vigorous stride up the hills to see some distant parishioner ... And then, when the cottage was reached ... impart the sunshine of his life to theirs.' Webb sought 'neither offices in societies nor Church preferment' but found fulfilment in his Christian ministry and in his astronomy [9].

Although his *Celestial Objects* was intended to be a more accessible work than Smyth's *Cycle*, it was by no means a simple book, for once one had passed through the introductory chapters on the beauty of the sky and basic telescope management, one encountered a great deal of tightly organised erudition, including Latin quotations, which are cited without translations. Yet one is immediately impressed by the countless hours of observing experience that must have lain behind it all, for no matter whether he is describing the Moon, the rings of Saturn, or a nebula, Thomas Webb's practical familiarity with the objects is evident. And what is more, he provides his readers with a full historical account of the previous astronomers who had also viewed the object under discussion. For instance, he relates that Galileo, in spite of the careful attention which he paid to the constellation Orion, failed to observe the Great Nebula [10]; and that in his derivation of the solar periodicity law, Heinrich Schwabe had observed the solar surface on an average of 300 days in the year for over thirty years and had noted 4,700 separate sunspot groups [11]. One of his most interesting historical asides concerned his argument that the Cassini Division in Saturn's rings should really be known as *Ball*'s Division, in honour of the two Minehead, Devonshire, brothers who first saw it with a 38-foot refractor in October 1665, a full decade before Cassini [12].

Thomas Webb was a leading expert on the amateur astronomer's telescope, and while his *magnum opus* was based upon observations made with a 'common' 3.7-inch instrument, he sold it in 1858, and ordered a 5½-inch Alvan Clark refractor on the advice of his friend W.R. Dawes [13]. Then a few years later, in 1864, Webb purchased a new silver-on-glass reflector of 8 inches aperture from the Hereford schoolmaster and mirror-maker, George Henry With, who also became a friend, and in 1885 his co-executor along with Espin [14]. He tested the Clark refractor alongside the With reflector over the following years in an attempt to quantify the respective light-grasps of the two classes of instrument, for Webb possessed that essential attribute for an observing astronomer: keen eyesight, which continued down to the time of his death, though Espin recorded that in the latter part of his life Webb had some astigmatism [15].

Then over the last nineteen years of his life, Webb observed with another glass reflector with optics by his friend With, which had an aperture of around $9\frac{1}{3}$ inches and revealed superb planetary detail [16]. This instrument was mounted on a 'Berthon' equatorial stand after the design of another clerical amateur, the Revd Edward Lyon Berthon of Romsey, Hampshire, which avoided some of the awkwardness of the German mount, and was both compact and elegant. This telescope, and its mount, were set up in the front garden of the Hardwicke vicarage, in an observatory which was also of Berthon's design, and which he described in the *English Mechanic* in 1871 [17].

When one looks through the manuscript observing books which Webb continued to lay out and rule off in the same firm and calligraphically exquisite hand until 19 March 1885, shortly before his death, one gets a sense of what astronomy meant to him: it was an exact and meticulous act of contemplation [18]. In many ways, Webb's use of language in his observing books is reminiscent of that of the naturalist when beholding a beautiful or rare flower; and within the culture of amateur science in the nineteenth century, one can understand how Robert Ward (see p.182) and others spoke of 'the flowers of the heavens' as objects of sensory and spiritual delight. On 21 August 1857, for instance, Webb recorded that 'The field is beautiful with 64 [magnification]' when describing M15, the nebula (globular cluster) in Pegasus, while on 1 May 1869, he characterised M13 in Hercules as 'A most superb object, to which one always returns with wonder & pleasure.' The yellows, reds, and blues of the stars, the corruscating glory of comets, and the craggy diversity of the lunar landscape – 'the view was charmingly distinct – reality itself, not a mere image' – always inspired Webb with wonder [19]. His last recorded observation, made on 19 March 1885, notes star positions and colours, on a night of indifferent weather, when the wind blew out his candle as the 78-year-old gentleman was trying to read the setting circles of the $9^1/_3$-inch reflector [20].

Indeed, the contemplative dimension is crucial when looking at all those astronomers who lacked the time or resources, and perhaps the inclination, to do cutting-edge, fundamental research, and helps us to answer the question why so many people from all walks of life spent so much time and energy in examining the sky if they could not hope to seriously influence the course of original scientific thinking. And because observational astronomy fitted perfectly into the religious culture of the age, it is easy to see why clergymen and pious lay men and women were so prominent among its devotees, with figures like Espin, Berthon, With and Henry Cooper Key in Webb's own time, down to the Revd T.E.R. Phillips and others in the twentieth century.

According to the survey conducted by the *Astronomical Register*, by 1866 there were 48 private observatories operating in Britain, 32 of which were equipped with equatorial refractors, and 7 of which had reflectors. Most of them would have been owned and operated by men like Webb and his friends [21].

THE AMATEUR ASTRONOMER'S TELESCOPE

In many respects, the astronomical enthusiast had different optical requirements from those of the Grand Amateurs. A capacity for making critical positional measurements was not the dominant concern, but rather the possession of a large, clear visual field. An authentic representation of stellar and planetary colour, combined with an ability to detect subtle surface details, was sought after, which goes towards explaining the popularity which the reflecting telescope was to have among the new generation of practical observers. And as these individuals were neither very wealthy nor full-time amateur astronomers, as many of the Grand Amateurs had been, certain features of economy and convenience were important. Observational astronomy also appealed to amateur inventors, and it is not for nothing that the pioneering organ of the Victorian do-it-yourself enthusiast, the *English Mechanic*, regularly carried astronomical articles and designs, a remarkable number of which were written by clergymen.

A particularly noteworthy astronomer–inventor–parson was the Revd Edward Lyon Berthon, for in addition to his purely astronomical innovations, including the 'Berthon dynamometer' – a device for measuring the diameter of the exit pupil of an eyepiece – he invented several items of nautical and naval equipment, among them the 'Berthon collapsible boat'. Indeed, this latter invention still constituted one of the major manufacturing industries of his old parish of Romsey in Hampshire over forty years later, in 1911 [22]. The Berthon mount, mentioned above, was a compact and elegant arrangement for a silver-on-glass reflector of relatively short focal length. Very clearly, it was aimed at the astronomer who was likely to be using a lightweight glass rather than heavy speculum metal mirror, working at f/10 or so. It was ideal for a 6- to 12-inch mirror, and unlike a German or Lassell-type equatorial mount, it could be made to work in a confined space. The tube, usually of thin sheet iron, was counterpoised by a pair of stirrup-like counterweights that were hung equidistantly across the support pillar, around which they moved in right ascension, while declination was controlled by means of a semicircular plate set upright upon the right ascension circle. The Berthon mount allowed a great deal of telescope to be packed into a relatively small space, while leaving ample room for access to a Newtonian eyepiece [23].

Although the wooden observatory which Berthon invented around 1863 formed an ideal 'telescope house' for his own mounted 9¼-inch silver-on-glass reflector, it could take other types of instruments and stands. And by the time that he came to describe his 'Romsey' observatory in the *English Mechanic* in 1871, it had already been replicated and tried over several years by other amateurs including Thomas Webb, who could not 'speak more highly ... of the comfort that this telescope house has given me' [24]. Indeed, this was probably the first observatory prototype ever to have been specifically designed and published for modest amateur usage. Like the Grand Amateur observatories of South, Smyth, Lee, and others, the 'Romsey' observatory had a round dome-chamber for the main instrument, to which was adjoined a rectangular transit and computing room – an essential requirement if a rural astronomer was going to regulate his sidereal clock. But unlike the Grand Amateur structures, the 'Romsey' observatory aimed at economy, being based upon thirteen strong vertical posts driven into the ground, to which the 'thinnest weather boards' were securely fixed to enable it to resist severe gales. It did not possess a dome as such, but an asymmetrical conical structure, made of radiating wooden rafters, upon which 'nothing but painted canvas' was stretched to keep out the weather. Shutters in the asymmetrical cone gave access to the sky from the horizon to beyond the zenith. This 'dome' rotated upon twenty window sash rollers which the aspirant builder could 'procure from any iron monger'. Depending on the amount of decoration which one wished to lavish upon it as a garden ornament, the basic 'Romsey' observatory, complete with ornamental finials, could be built in a fortnight 'by any village carpenter' for £10 or £12, or £8 10s (£8.50) minus the transit room extension. Its dome-chamber was 10 feet in diameter with a 10-foot head-clearance. Whether the Revd Mr Berthon built the original observatory with his own hands is not clear, but from the detailed carpentry instructions which his article contains, one suspects that he could have done, for he displayed the accomplished do-it-yourself enthusiast's instinct for how to incorporate commercially-available parts, such as window sash rollers, and make them serve an original function in a home-made artefact [25].

Considering the size, elegance and cost of this little wood and canvas observatory, it would be interesting to know what resemblance it had to the building constructed of identical materials from which Thomas Dell observed in Aylesbury in 1851 (mentioned in Chapter 5). And if so much could be done for £12 in 1871, one is left wondering what advantages the 'economical' observatory which Thomas Maclear had built in Biggleswade for £50 in 1834 had over it [26].

The burgeoning world of enthusiastic observing astronomers created a new and diverse market for instruments and accessories, for amateur reflectors could be assembled from a variety of sources by 1875. One might figure a glass mirror from scratch, like Roger Langdon; part-finish and send it off to a specialist for optical completion, like John Jones; or buy a finished mirror from a supplier like George H. With or George Calver. The telescope could then be built by oneself, or else a third party could be paid to mount the mirror. Even so, good eyepieces had to be purchased, like the Browning eyepiece mentioned by Langdon for use with a home-figured mirror [27], while the manufacture of optical flats became something of a professional specialisation of firms like John Browning. The *Astronomical Register* was coming by the 1870s to carry regular advertisements for the supply of parts and accessories for the assembly of reflecting telescopes [28].

Almost all of the enthusiastic observing amateurs from the 1860s onwards used reflecting telescopes for at least some of their work, and it is hard to overestimate the impact of the silver-on-glass reflector upon the increasing popularity of practical astronomy. For the glass mirror cut out a whole series of obstacles which had been implicit in the use of speculum metal, and transformed a costly, time-consuming ordeal into a relatively straightforward craft exercise. No longer was it necessary to go through all the metallurgical preliminaries of obtaining pure copper, tin, and arsenic, building a furnace, acquiring crucibles, making moulds, and in the process risking serious accident. Instead, one simply sent off to a glass manufacturer for a finished piece of thick plate glass – as did John Jones in 1868 when he obtained his blank from St. Helens, Lancashire [29] – confident that it would not be semi-porous, excessively brittle, or dull, as was often the case with a metal mirror casting. The finished mirror could then be silvered by the astronomer himself, using techniques that were already part of the accessible amateur literature by 1870, or else sent off to Calver, G.F. Lemmon of Hastings, or some other professional supplier, to be given its reflective coat [30]. As John Browning was to argue in his *A Plea for Reflectors* in 1867, good silver-on-glass reflectors had tubes that were only half as long as those of refractors of similar aperture, they had a superior resolving power when used on dim double stars or planetary surfaces and often gave crisper star images, while unlike large-aperture refractors, they were not 'beyond the reach of all but wealthy persons' [31].

In 1859, the Revd Henry Cooper Key, incumbent of the parish of Stretton Sugwas, some four miles from Hereford and only a few miles away from Webb's parish of Hardwicke, built a mirror-figuring machine that was based on the one described by Lord Rosse in *Philosophical Transactions* [32]. It produced excellent glass 'specula', some with apertures as large as 12 and 18¼ inches, the former of which later came to be used in Sir David Gill's observatory near Aberdeen (prior to his work at Dun Echt and the Cape), and the latter of which was mounted on a 2-ton Berthon stand [33]. Indeed, one is forced to ask what it was about clerical, rural Herefordshire in the late 1850s which provided such fertile soil for amateur astronomy, for contemporary with Cooper Key and Webb, George H. With, schoolmaster and Anglican lay reader of Hereford city, was also turning his hand to mirror-making.

George Henry With was born in London, and came to Hereford in 1851 at the age of 24 to take up the Mastership of the Blue Coat Boys' School in the city. With was not a university graduate, but belonged to the new profession of Certified School Master, having been trained at the St. John's Teachers' Training College, Battersea. Certified Masters, however, came several rungs below the ordained, Oxbridge-educated Masters of Arts who formed the teaching establishments of the major public schools and great grammar schools, and generally worked with pupils from more humble backgrounds. With's salary of £85 a year plus a house was well below that of a public school headmaster [34], under-master, or even one of Airy's junior Warrant Assistants at Greenwich, and one can understand how the manufacture and sale of excellent glass mirrors – a 12-inch aperture cost £55 – could substantially augment his income.

As Robert A. Marriott has shown in his splendid and ground-breaking study of the life and work of With, which I have found invaluable to my own understanding of this mirror-maker, he was brought to Hereford by the Very Revd Richard Dawes, who became Dean of Hereford Cathedral in 1850, and who had recognised With's abilities when he was working at Alverstoke National School, Gosport. Dr Dawes (who was no relation of W.R. Dawes the astronomer) was an educationalist of advanced views, who in the days before the creation of a national elementary education system after 1870, was aware of the obstacles which stood in the paths of ordinary people when it came to the acquisition of a good basic education. Dawes was also interested in science, and probably exerted a decisive influence upon the shape of With's scientific interests [35].

In addition to his Mastership of the Blue Coat School which he held up to 1876, however, George With came to teach at Hereford Cathedral School and Ludlow Grammar School, and held the title of Professor of Science at Hereford Ladies' College. He also conducted correspondences with such internationally distinguished scientists as Thomas H. Huxley and John Tyndall. Though With was never an ordained priest in the Church of England, his lay readership involved him in liturgical duties in Hereford Cathedral, especially as he read the Biblical languages of Latin, Greek and Hebrew [36].

With started to make mirrors around 1860, and by 1863 was being congratulated in print by Cooper Key for already having 'completed three or four glass specula of the highest class', while the following year, as we have already seen, he supplied an 8-inch mirror to Webb [37]. And all the evidence suggests that Cooper Key knew what he was talking about, for he was probably the man who first directed With's attention to silver-on-glass mirror-making. For as With later told George Calver of Chelmsford, when it came to figuring mirrors he had been in the dark, until ' ... Cooper Key and I worked together. *He* discovered the graduated tool value, I only worked out the matter' [38]. In many respects, With fitted perfectly into the new reflecting telescope, amateur astronomy culture, for not only was he in at the beginning of the new glass mirror technology, but his mode of manufacture and supply remained linked to the observing amateur market, where the new type of telescope demand lay. As R.A. Marriott has shown, George With was an optical specialist who did not make complete telescopes, but concentrated upon primary mirrors. His mirrors were set upon Berthon and other mounts by various makers, and were often matched with optical flats by the professional optician John Browning, who also supplied tubes and mounts [39]. In the time-honoured tradition of scientific instrument-making, the Victorian amateur astronomer's telescope, just like the marine chronometers of John Arnold or the sextants of Jesse Ramsden, generally brought separate specialist-made component parts together into one finished structure [40].

With's mirrors soon came to be renowned for their excellence. As early as 1863, Henry Cooper Key was praising their image clarity and defining power, as was Thomas Webb soon afterwards; and when tested on difficult compound stars, like γ^2 Andromedae, they often showed themselves to be equal to those refractors, aperture for aperture, upon which W.R. Dawes had published his test results in 1867 [41].

In 1893, With sent a detailed account of his mirror-making techniques to George Calver. Although Calver was in some respects With's rival in the manufacture of fine mirrors for amateur reflectors, the two men were on very amicable terms, and shared ideas. With told Calver that for the grinding of his glass mirrors he used a machine which 'was a simple form of Lord Rosse's machine, with stroke and slide motion', which was probably influenced by Cooper Key's machine of 1859. Using pitch-tempered resin cut into graduated squares to form the tool, it 'generally took about from thirty to forty hours' to polish a mirror on the machine. The spherical mirror was then figured to its desired parabolic form by means of three separate tools, each of increasing delicacy, using a measured sequence of circular and cross strokes [42]. With seems never to have used the Foucault 'knife edge' test to determine the eventual perfection of the figure, but like William Lassell in the 1840s used a watch dial, set up about 100 feet away. When he could see all of the details on the dial, in all parts of the reflecting surface, the mirror was ready for its warranty [43].

One unused With mirror came to light in 1922, still wrapped in a copy of the *Herefordshire Times*, and warranted 'absolute perfection'. Another which was obtained by J.L. Cummergen in 1942 still carried its certificate: 'Focus 5 feet 7 inches. This 6½ inch easily divides γ^2 Andr.[omedae] and μ^2 Boötis. It carries a power of 1000 easily on stars. Wonderful perfection. August 1872' [44].

In 1887, the 60-year-old With decided to sell off his entire stock of around 100 of his 'reserves' and 'choicest reserves'. They accounted for about half of his total output over his thirty years as a mirror-maker. It is not clear why he did this, but his published price list – very obviously headed 'A Remarkable opportunity for Amateur Astronomers' – offered the mirrors at less than one third of their originally advertised prices: a 10¼-inch, for instance, was reduced from £38 10s (£38.50) to £10 10s (£10.50) [45]. No doubt this action accelerated the circulation of his products, although his mirrors had long been in distinguished amateur and professional hands. James Francis Tennant, for example, had used a Browning-mounted With 9-inch to observe the Indian eclipse of 1868, while in 1872 Joseph Norman Lockyer had one which produced 'exquisite definition'. The With instrument in the Temple Observatory at Rugby School and one in a privately-owned observatory in Sydney, Australia, were found superior to Clark and Merz refractors of similar aperture. By 1890, With's mirrors were in use in Europe, Canada, Australia, Asia and elsewhere [46].

In Britain, they became status symbols within the amateur world. William Frederick Denning had a 10-inch (also mounted by Browning), while the wealthy Leeds amateur, Scriven Bolton, and the Revd Theodore Evelyn Reece Phillips were proud possessors down into the twentieth century [47]. In 1877 With made his largest mirror, an 18-inch, which was acquired by the vicar of St. Paul's Church, Alnwick, Northumberland. The Revd Jevon J. Muschamp Perry was not only an observer, but also seems to have collected telescopes, along with being a keen reader of the *English Mechanic*, for he bought Cooper Key's 18¼-inch after his reverend colleague's death in 1880 [48]. He already had a 6½-inch Calver, and only regretted that he could not afford to order Calver to make him a duplicate of the 37-inch which he was finishing in 1879 [49]. Even so, one suspects that Muschamp Perry was a long way from being an impecunious clergyman. But he was by no means the only collector of fine reflecting telescopes by the 1880s, for when Nathaniel Everett Green bought the With 18-inch from Muschamp Perry, he added it to his existing arsenal of three With mirrors. The four With telescopes were used by Green in his detailed studies of Jupiter, and he also took the 13-inch to Madeira to observe Mars during the favourable opposition of 1877 [50].

Several With mirrors came into the possession of the British Astronomical Association, so that they could be loaned out for use by serious observers, and some of them, such as the 18-inch (presented to the B.A.A. in 1897), and the 9¼-inch Berthon-mounted mirror, are still in regular use at the end of the twentieth century. Both of these instruments, for example, have been on loan for use in Denis Buczynski's Conder Brow Observatory, near Lancaster, and Mr Buczynski's own researches have led him to believe that the latter instrument could well be Webb's original With–Berthon of 1866 [51].

Although With made silver-on-glass mirrors for sale, he remained in many respects an optical amateur, to whom schoolmastering was his real profession. But this was not the situation with George Calver, who practised as a professional mirror-maker at Widford, near Chelmsford, Essex, between 1870 and 1904, and carried on working into old age after his return to Walpole, Suffolk, making several thousand excellent silver-on-glass specula in the course of his working life. But as was indicated above, his relationship with With was always friendly, and this might have had something to do with the fact that it had been the challenge of a With mirror which had set him upon his career as a manufacturing optician. It seems that when living at Yarmouth, Norfolk, in the 1860s, another clerical enthusiast, the Revd Mr Matthews, showed George Calver his newly-acquired With reflector, and challenged him to make a mirror as good as With's [52]. Calver's business expanded considerably, and by the end of the century he was employing assistants and sub-contracting metal and other components of his instruments to specialist firms, so that Calver optics were being sold as the essential ingredients of telescopes of polyglot manufacture. He also finished, tested, and silvered mirrors for amateur telescope-makers who could not get it right for themselves, as when he charged John Jones £5 to parabolise and coat his $8^{3}/_{16}$-inch mirror [53].

Between With, Calver, and the do-it-yourself instructions to be found in the *English Mechanic*, one appreciates the enormous impact which the silver-on-glass mirror had on the development of amateur astronomy in the latter half of the nineteenth century. Indeed, it liberated the science of astronomy from its old limitations of cost and essential mathematics, to render it accessible to busy people of modest means and dexterous touch. The relative cheapness, compactness, and ease of maintenance of the glass-mirrored reflector, with its smaller focal ratio, made it possible to pack all the power of John Herschel's 20-foot telescope of 1820 into a modest back garden shed observatory by 1870, and without the endless fatigue of having to keep three heavy metal mirrors polished and ready. And as freshly-polished speculum metal rarely exceeds 70 per cent light reflectivity, whereas silver on glass can exceed 90 per cent, one immediately sees the source of the instrument's spectacular success.

Amateur astronomy would not experience a similar quantum leap forward in its technology for another 125 years, when in the late 1980s new mass-produced catadioptric telescopes, charge coupled devices (CCDs), and powerful personal computers enabled anyone with a couple of thousand pounds to spare to see from their back gardens things which only professional observatories could have seen forty years before.

It would be incorrect, however, to place the rise and success of observational amateur astronomy entirely at the door of the silver-on-glass reflecting telescope, for the refractor also tumbled in relative price and ascended in quality over the same period. Likewise, spectroscopes, photographic equipment, and other accessories aimed at the amateur market became increasingly accessible by the 1880s, not to mention a widening range of interesting and informative popular books.

Without doubt, the man who developed the high-quality amateur refracting telescope was Thomas Cooke of York. Up to his appearance on the scene around 1840, however, the manufacture of achromatic object glasses was an arcane trade that was largely controlled by a handful of firms worldwide. There was Merz and a couple of other leading firms in Germany – mainly around Munich – some lesser ones in France, and the Dollond–Tulley–Simms clique in London. It was Thomas Cooke in England and Alvan Clark in America who barged as bulls into the optical china shop and brought in their wake those refreshing winds of change which shook up old monopolies, old technologies, and old pricing policies to bring the refracting astronomical telescope out of the closet of exclusiveness [54]. For both Cooke and Clark were outsiders in the optical instrument-making trade, being gifted amateurs who found that they could beat the professionals at their own game. In the same way that, in 1781, the Astronomer Royal was obliged to admit, when William Herschel discovered Uranus, that a Bath organist could figure metal mirrors that were superior to anything that came out of a West End commercial workshop, so a Boston U.S.A. portrait painter and a Yorkshire shoemaker-turned-schoolmaster revolutionised and popularised the achromatic object glass.

Thomas Cooke was born in Allerthorpe, in the East Riding of Yorkshire, and after working at his father's trade of shoemaking, and being dissuaded by his mother from running away to sea, he used his self-acquired knowledge to earn his living as a teacher. After teaching in Stamford Bridge – from whence he was to marry one of his pupils – and elsewhere, Cooke came to teach mathematics at the Revd Mr Shapkley's school in Micklegate, York [55]. During this time – in the late 1820s and 1830s – Cooke's broader interests in mathematics and science had inclined him towards astronomy and telescopic optics, so that he cast and figured a 6-inch speculum-metal mirror. Unfortunately it broke, although Cooke was able to salvage a piece that was large enough to enable him to turn it into a 3½-inch optical surface which gave a clear image. But one suspects that this experience with brittle speculum metal caused him to turn to glass, which in the 1830s meant the achromatic lens [56].

One of the main historical sources for Cooke's early life comes from the narrative reported to Samuel Smiles by James Nasmyth, who had known Thomas Cooke well, while the more personal history comes from Mrs Hannah Cooke, who was interviewed by Smiles at Saltburn, Yorkshire, at some time after her husband's death in 1868 at the age of 61. Yet neither Nasmyth nor Mrs Cooke related the avenue by which this self-taught outsider came to master the complex practical business of constructing a corrected astronomical achromat once he had found from the published literature how to 'calculate his curves from data depending upon the nature of his glass' [57].

Even so, Thomas Cooke made himself a 4-inch refractor possessing an 'admirable defining power' which seems to have excited considerable interest in York, which in the 1830s was a leading centre of culture outside London. The first telescope which Cooke seems to have sold, however, had a lens made from the thick bottom of a glass drinking tumbler, and it went to Professor John Phillips, the geologist, astronomer, and later Oxford scientist, who then lived in York where he was Curator of the Museum. Phillips was to purchase further, finer Cooke telescopes, and he became one of his champions in the wider world of science [58].

Cooke opened a shop in Stonegate, York, in 1836, to concentrate on the manufacture of refracting telescopes, and soon received orders from various local dignitaries with scientific tastes, including the Revd William Vernon Harcourt, Canon of York Minster, who was also an F.R.S. and in 1831 had been an inspiration behind the founding of the British Association for the Advancement of Science. Cooke's reputation was established upon the early 4- and 4½-inch achromats which he produced for Harcourt, William Gray, and John Phillips, and by 1850 he was master of a flourishing business, while his Buckingham Street Works was fast becoming a world leader. By the time of his death, Cooke's firm was making a wide range of high-quality refracting telescopes, from portable tripod instruments, such as 3-inch refractors of the type which John Robertson purchased for £30, up to the 25-inch equatorial, built for Robert Stirling Newall, the Gateshead steel-wire manufacturer. His clientele covered the complete astronomical spectrum, including friends like Nasmyth and Phillips, amateurs turned professional like Professor Charles Pritchard, Grand Amateurs of the standing of W.R. Dawes, and dedicated observing amateurs of modest means, like John Robertson [59].

Not least among Thomas Cooke's contributions to telescope design and manufacture were the major improvements which he made to the German mount, for having 'found the equatorial comparatively clumsy; he left it nearly perfect' [60]. By the 1850s, he was manufacturing all-metal telescope mounts, abandoning wood entirely, and using cast-iron pillar stands, cast-iron declination axes, and tubes which combined relative lightness with rigidity. His equatorials were not only symmetrical from an aesthetic point of view, but concentrated the weight of the instrument on a series of balanced points, to obtain rock-solid rigidity that minimised vibration and ensured even tracking in right ascension. These very distinctive Cooke features went into all of his equatorials, whether they carried telescopes of 3, 8, 10 or 25 inches aperture [61].

But Cooke did not dominate the entire refracting telescope trade, even within Britain. Alvan Clark's instruments were also renowned both for the excellence of their optics and the firmness of their equatorial mounts, and Clark indeed virtually tied up the American market by 1870, especially in the production of telescopes for the Liberal Arts College teaching observatories. And in Britain and the Empire, Thomas Grubb of Dublin and then his son, Sir Howard, were often neck-and-neck with Cooke, both optically and mechanically, though Grubb's market tended to concentrate more upon the manufacture of big institutional telescopes, such as those supplied to the Greenwich, Dunsink, Melbourne, Cape and Vienna Observatories. Though both Thomas and Sir Howard Grubb did supply instruments to Grand Amateurs, such as E.J. Cooper, William Huggins, and Isaac Roberts, all of these commissions, like those from the professional observatories, were for significant research instruments rather than for the more modest refractors which Cooke manufactured in such abundance.

While Thomas Cooke, and the expanding company which his sons continued after his death, built many capital instruments for professional observatories, as Dr Anita McConnell has shown, it was nonetheless from the well-off amateur market that he drew much of his business [62]. And it was word of mouth within the community of scientists that generally sold them, especially in the early days. One of his first instruments – a 4½-inch, sold to William Gray, a wealthy York solicitor, in the 1830s – enlisted Gray as his friend, adviser and champion. A 7¼-inch refractor sold to Hugh Pattinson, a retired Newcastle chemist, in 1851 caused an order for a 9¼-inch to be placed by his friend Isaac Fletcher, who came and saw and admired [63]. Cooke was also subject to royal patronage, as in 1859 Prince Albert purchased a 5½-inch refractor, which was set up at Osborne House, Isle of Wight. (In 1935 this instrument was presented to the British Astronomical Association by King George V.) It was likewise this spreading fame, however, which led Cooke to make one of his biggest professional mistakes, when he accepted the order in 1862 to work up the two 25-inch glass blanks into the world's largest achromatic lens for R.S. Newall, (mentioned above). Cooke's low quote and short estimate of completion time almost ruined him [64].

But more significant as far as the observing amateur is concerned was Cooke's routine manufacture, at his Buckingham Street Works, of excellent 3- to 6-inch instruments of excellent quality For it was these instruments which provided the alternative to the reflecting telescope, and when combined with it, made serious observational astronomy a viable pursuit for those members of the middle classes who could afford to spend several tens of pounds, but not hundreds or thousands of pounds, on their passion.

Although good-quality finished refracting telescopes were becoming more accessible by the 1880s, it would be interesting to know how many astronomical enthusiasts attempted, and succeeded, in making their own achromatic lenses, in the same way that Cooper Key, Roger Langdon, John Jones, and others found that they could make good mirrors. The making of a lens is fraught with a range of difficulties which do not face the mirror-maker. For one thing, an achromatic lens needs two perfectly transparent blocks of glass, each of which has a different refractive index. The dense glass blocks of flint glass were usually manufactured on the Continent, obtained through Paris, and in the earlier part of the nineteenth century were heavily taxed upon their import into England. The less dense blocks of crown glass were sometimes manufactured in Britain, but until firms like Chance Brothers of Birmingham and Pilkingtons of St. Helens started up in business around the mid-century the best still came from continental Europe.

For a reflecting telescope, one only needed to figure a single optical surface – that which would be silver-coated on completion – so that bubbles or striations in the body of the glass did not matter. When making an achromatic object glass, however, the surfaces of the two blocks had to be figured with a set of four exact curves, both convex and concave, to make the two refractive indices cancel out each other's aberrations. Only then, when perfectly matched, would they produce an even and distortion-free field of view. The real secret lay (as the young Thomas Cooke came to appreciate) in measuring the precise refractive index of each glass block, and then calculating the individual geometrical curve that had to be imparted to each of the four optical surfaces. Achromatic lens-making was much less empirical than that of mirrors, and demanded a thorough scientific mastery at every step. On the other hand, as we have seen, a Yorkshire schoolmaster and a Boston portrait painter did master this art with spectacular success and advantage to astronomy by 1860. And so too did Samuel Lancaster in the early 1880s, though unlike Cooke and Clark he never went on to win renown.

Samuel Lancaster, of Bainbridge, in Wensleydale, Yorkshire, was born in 1853, the son of a village schoolmaster, and as with the careers of John Jones, John Robertson, and the early life of Thomas Cooke, we have Samuel Smiles to thank for seeking him out and twice interviewing him at some length, though he was already familiar to readers of the *English Mechanic* [65]. According to the autobiography which Lancaster supplied to Smiles, he seems to have been a sickly child, although it was not from his father, but from one William Farrel, the local shoemaker – that ubiquitously learned tradesman of the early Victorian age – that he first learned his alphabet, and by whom he was fired with intellectual curiosity. Fortunately, Lancaster's father had a collection of books on science and was an amateur chemist and wet-plate photographer, so that Samuel soon found encouragement for both his scientific and artistic tendencies. And like virtually all amateur astronomers, both Grand and modest, he had a fascination with making things, including a simple telephone and a petroleum-fired metallurgical furnace which could melt a piece of steel in 7½ minutes [66].

Samuel Lancaster followed a similar professional path to that of George With, and became a Certified School Master after attending colleges in Darlington, Newcastle-upon-Tyne, and Durham. Although Lancaster made no mention of any religious motivation behind his work, he seems to have been of good credit with the Church of England, as he had been granted his qualifying Certificate from the Durham Diocesan College for schoolteachers in 1877. By early 1881 – when Smiles first encountered him – he was working as Assistant Master in the Yorebridge Grammar School, at Askrigg, Yorkshire [67].

Although Lancaster wanted a telescope with which to observe the heavenly bodies, it is clear that his inspiration to astronomy came from the directions of mathematics and instrumentation rather than straightforward contemplation. He was familiar with the writings of the Cambridge mathematician Isaac Todhunter (like the self-educated John Leach in Chapter 11), Augustus De Morgan, G.B. Airy and others, and had become fascinated by algebra and geometry after his initial examination failures had caused him to take these subjects 'by the horns' [68]. While Lancaster was self-taught in many respects, he had nonetheless received an academic training, even if it was not as comprehensive as a full university degree, and had obtained tickets to attend lecture courses in mathematics and physics, at an unspecified institution in Newcastle, which was probably the collegiate foundation which would later become the university, and at which Alexander Herschel taught.

It was this grounding which enabled Samuel Lancaster to take a serious interest in optics. Having read Airy's *Geometrical Optics* for the theory, and searched through all his back numbers of the *English Mechanic* for the practice, he acquired a pair of 6½-inch glass disks, and after much struggle, eventually figured a good mirror with a 12-foot focal length. He followed it with an 8-inch reflector, on an iron equatorial mount. It was the success of these instruments and the accuracy of his equatorial mounts which enabled him to pinpoint celestial objects using the setting circles that he built for them, and this spurred him on to design a polishing machine on which he could figure an achromatic object glass [69].

Lancaster regarded the achromatic refractor to be 'the King of instruments', and by September 1884 – about a year after Smiles' first interview – he had completed a lens-polishing machine. But the biggest drawback faced by 'amateurs of my class' when attempting to work an achromatic lens, lamented Lancaster, was the very high cost of the flint-glass blanks [70]. Even so, it is clear that the young schoolmaster-astronomer was particularly fascinated by the problems inherent in the construction of such lenses, for he described in detail the obstacles that had to be overcome. By the time of his interview with Smiles, however, it is clear that he had already gone a long way towards mastering the practical art of object-glass construction, for even by February 1880 his name was being cited in the pages of the *English Mechanic* with relation to achromatic lens making [71].

One of the major obstacles was that of ensuring an even pressure-action of the figuring tool, which involved a 'three bodies' problem in dynamics, as one tried to replicate exact conic sections as it moved across the glass, so Lancaster told Smiles, while this was further complicated by a varying 'coefficient of attrition', as the abrasive agent between the tool and the lens surface could vary in its bite action. If this happened, it resulted in 'rings of unequal wear' on the optical surface, which produced an aberrated image. Lancaster believed that a completely rigid machine, and uniformity of those moving parts which rotated the lens and moved the figuring tool above it, would provide the solution [72].

Apart from some references to his observations of Jupiter, Smiles records nothing about what Samuel Lancaster actually looked at through his telescope. On the other hand, Smiles' own interests, one must not forget, were not primarily astronomical, but concerned with individuals whose ingenuity and determination enabled them to transcend the limitations of their circumstances. It had been Smiles' hope that Lancaster would go on to publish the results of his 'interesting investigations, from his own mind and pen' [73], though I am not aware that he ever did so, to any significant degree.

It is clear, however, that by 1882 amateur astronomical observers and the passion which they displayed for their telescopes had become sufficiently well-known in society at large that Thomas Hardy could use both 'to make science, not the mere padding of a romance' but its 'actual vehicle' in his novel *Two on a Tower* [74]. Indeed, this is the only example, as far as I am aware, in which an astronomer and his telescope supplied the grounds for both the characterisation and the plot action of a major piece of English fiction.

But *Two on a Tower* is far from being a simple tale of astronomical enthusiasts. Instead, it is a complex and cross-currented romance about an oddly matched couple who are brought together by the tower upon which the young astronomer has set up a small telescope without permission from the tower's owner. The owner himself has been absent for several years in Africa, indulging his passion for big-game hunting, and has left his beautiful 29-year-old wife, Lady Viviette Constantine, back at home with instruction that she must live in isolation until his return.

Chancing one day to visit the supposedly deserted tower on her estate, Lady Constantine encounters Swithin St. Cleeve on its upper platform engrossed in observing 'a cyclone in the sun' with his modest telescope [75]. The lonely woman soon finds herself doubly attracted: by the appearance of the classically handsome, golden-haired youth who has only recently left school, and by his philosopher's sense of human insignificance in the great vastness of the stellar Universe. They become friends, and start to observe together; and soon Lady Constantine falls in love with 'her astronomer' and with the answers he has to offer to the profoundest questions about the beginnings and endings of things.

Swithin St. Cleeve is a gentleman of modest means, the son of a deceased curate, who enjoys the benefit of a small private income. Although apparently not destined for a university education, his ambition is to become Astronomer Royal, as he naïvely believes that the holders of that Office enjoy an abundance of leisure for private observing. He has built telescopes and stands, and by the time that he is 19, has constructed a large instrument with a milled board tube, which only wants an object glass. Raising the final £12 necessary for its purchase against his forthcoming annuity payment, Swithin St. Cleeve brings his 'magnificent eight-inch first quality object lens' triumphantly home from London, and takes it round to Viviette. Then he accidentally smashes it [76].

Lady Constantine not only replaces the lens but orders for 'her astronomer' a large equatorial refractor from the famous London firm of 'Hilton & Pimms' (Troughton and Simms?), and in spite of her wealth is taken aback to learn how much it all costs [77]. Even so, everything is mounted on top of the tower – Rings Hill Spear – and a large dome rotating on rollers is used to cover it.

Thomas Hardy was as good as his word when he said that the astronomy would be much more than the 'padding' of his story. In addition to his more general interests in astronomy, he had read R.A. Proctor's *Essays in Astronomy* (1872), along with other works, and may have made an official visit to the Royal Observatory in 1881 with the purpose of seeing how large telescopes and their domes operated. He certainly corresponded with Edwin Dunkin, who had been promoted to the rank of Chief Assistant at Greenwich in 1881 when Airy retired. While an already famous writer by 1881, we must not forget that Hardy was an architect by original training, and would have had a familiarity with technical subjects [78].

One of Hardy's spurs towards using amateur astronomy as a vehicle for a novel could well have been his sightings of Tebbutt's Comet (the Great Comet) from Wimborne, Dorset, in June 1881, while the 1874 and forthcoming 1882 transits of Venus had excited great public as well as scientific interest in astronomy [79]. Both comets and transits are referred to in the story, along with planetary observation. There is also Swithin St. Cleeve's passion for double and variable star work, and several aspects of contemporary cosmology, while Viviette is astonished (in Chapter 4) when she is told that the 3,000 stars which she can see with the naked eye will be increased to 20 million in a telescope. Indeed, it is through the cosmology that Hardy finds the avenue which he wishes to explore in the novel: the ephemeral character of human life and passion when set against the vastness of space.

Swithin St. Cleeve, however, is a very different type of astronomer from the clerical enthusiasts whom we have seen in the earlier part of this chapter. It is not the spectacle of God's celestial majesty which moves him, but the nihilism of human brevity seen against a Universe which is infinite and purposeless. If all of these stars were made by God for human delight, he argues, why had he kept them concealed for so long? 'Nothing is made for man', Swithin warns Viviette – for she is a woman who is temperamentally attracted to religion – 'I think astronomy is a bad study for you. It makes you feel human insignificance too plainly' [80]. Indeed, this gloom of what Swithin called a 'tragic' science touched upon a sensitive nerve within late Victorian intellectual culture. It was the nerve of nagging agnostic doubt that had been exposed over the previous forty years, by Comte's Positivism, the decline of 'the argument from design', and by Darwinism, to cast a grim shadow across *fin de siècle* romanticism.

Yet Swithin St. Cleeve is a literary creation. Real-life observing amateur astronomers, as we have seen, approached the science not from the direction of romantic *angst*, but from the rather healthier ones of contemplation, curiosity, and technical delight. And at the same time as Swithin and Viviette were on the top of their tower, other amateurs around Britain were seeking each other's company in the first amateur astronomical societies, and comparing observations and the performance of their home-made or purchased telescopes. And while they read the *English Mechanic*, *Nature*, or *Knowledge*, and bought the books of J.N. Lockyer and R.A. Proctor, that greatest of all the late Victorian and Edwardian popularisers, Sir Robert Ball, hit the nail upon the head in 1905 when he said that the observing astronomer 'will greatly increase the interest of his work if he uses the charming handbook of the heavens known as Webb's *Celestial Objects for Common Telescopes*.' For forty-six years after its first edition, Webb's book had become a classic of serious amateur astronomical literature [81].

13

That clubbable passion: the amateur astronomical society

In the same way that fundamental research in British astronomy and the other sciences had led to the establishment of bodies like the Royal Society and the Royal Astronomical Society, so amateur observational astronomy produced its own corporate and social institutions. The real ancestors of the Victorian astronomical societies, however, were the broader-based and non-specialist Literary and Philosophical Societies and Mechanics' Institutions.

As part of the growth of cultural life in late eighteenth-century Britain, many towns such as Edinburgh, Manchester, Liverpool and York began, especially from the 1780s onwards, to establish Literary and Philosophical Societies. They were genteel affairs, aimed at the comfortably-off classes, who could afford the necessary subscription to acquire elegant town-centre properties which would contain good libraries, meeting rooms, and general social facilities. Literature, art, music, politics, science, and a wide range of subjects would be lectured upon and discussed, and many Societies provided bases from which scientific work of fundamental importance could be conducted. William Herschel's first serious astronomical observations were presented in the 1770s to the Bath Philosophical Society. The Manchester 'Lit. and Phil.' formed the intellectual home of John Dalton, the schoolmaster who laid the foundations of the atomic theory of chemistry; the young William Lassell attended lectures at the Liverpool Society in the early 1820s; while it was the Yorkshire Philosophical Society, with its grand premises and Museum, which both employed John Phillips as Keeper and gave birth to the British Association for the Advancement of Science in 1831 [1].

In many ways, the Mechanics' Institutions, which were starting to be formed by the 1820s, were very similar in character, though in that world of clearly differentiated social ranks, they catered for members who wore fustian jackets rather than broadcloth coats. Yet both organisations aimed to provide cultural facilities to a paying membership, and astronomy could be provided if the members wanted it [2].

As in many other aspects of nineteenth-century culture, it was the Scots who took the initiative when it came to founding astronomical societies. As Dr David Gavine has shown, cities like Dumfries, Edinburgh, and Glasgow were offering specifically astronomical facilities within the wider context of Literary and Philosophical Societies, while Thomas Dick's Directorship of the Dundee Watt Institution ensured a solid astronomical emphasis within that Society [3]. The Edinburgh Astronomical Institution in 1820 was really a club for scientifically-minded gentlemen, and included within its ranks some of the leading dignitaries of both city and university. But when they presented a Loyal Address to King George IV, at the royal visitation to the city in 1822, they got their empty buildings on Calton Hill dignified with the title 'The Royal Observatory'. The Institution's subsequent collaboration with the University, and the Regius Professorship of Astronomy, led to the initiation of regular academic astronomical research in Scotland [4].

Yet this type of astronomy was closer in its social milieu to that of the Grand Amateur figures who created the Royal Astronomical Society than to those more modest individuals who comprised the ranks of observing astronomers after 1860. Attempts had been made, however, to found more popularly-based societies from the 1830s onwards, such as the ill-fated 'Uranian Society', which one Mr W. White tried to establish in June 1839 'for the practical cultivation of astronomy' [5]. But none of these societies succeeded.

In the 1830s, Sir David Brewster had suggested that a public observatory should be set up, while around the same time his expatriate fellow-Scotsman, Ebenezer Henderson, further proposed that a public observatory should be founded in Liverpool, but neither scheme came to anything [6]. Then by the 1850s efforts were again being made in some towns to set up civic observatories where members of the public could come and observe the heavens with a capital instrument, and perhaps consolidate their interests by forming an astronomical society. And as we saw in Chapter 10, in December 1851 the city of Nottingham had been offered instruments worth £10,000 by Henry Lawson, in the hope of establishing a Midlands Observatory, although the donation lapsed because the town council could not provide a site for the observatory, nor come up with a maintenance endowment and £200 per annum for a keeper's salary [7]. Then Liverpool, that most astronomically vibrant of all provincial British cities during the Victorian age, started the ball rolling again, when in October 1859 the *Liverpool Mercury* launched a 'Proposal to establish a cheap Public Observatory', providing equipment 'unattainable by private individuals'; however, nothing concrete came from it [8].

The most successful public or civic observatories to get as far as foundation and operation, however, were the somewhat later ones of Airdrie, Paisley, and Dundee in Scotland, along with the Jeremiah Horrocks Observatory, Preston, Lancashire, all of which still remain in operation down to the present day as educational facilities [9]. On the other hand, civic or publicly-accessible observatories quickly became a part of American scientific culture, starting with the Cincinnati Observatory in 1842, though they never caught on with the same vigour in England.

The first distinct and specifically amateur astronomical society in England was that which was founded in Leeds in 1859. It was conceived in the last days of September 1858, when the British Association for the Advancement of Science was holding its annual jamboree in Leeds, and Donati's Comet was blazing across the heavens. George Biddell Airy, Sir John Herschel, Dr John Lee and all the leading figures of British astronomy were in Leeds, and on the last night, 30 September, Sir John Herschel delivered a public lecture to a packed meeting in the rooms of the Leeds Philosophical and Literary Society [10]. Airy and Herschel were afterwards approached for their support in the founding of an astronomical society, which later dated its foundation to 1859.

Unfortunately, this first founding of the Leeds Astronomical Society seems to have left few recorded traces, although around April 1860 the Society had come to possess 'an iron round building', in Love Lane, off North Street, Leeds; but the Society's shortage of cash might be inferred from the fact that this £5 observatory was not paid for until 1863, and was sold in 1867 [11]. On 7 March 1861, however, the commencement of a weekly class in 'Practical Astronomy and Meteorology' was being advertised from this observatory, that was to be 'conducted by efficient teachers'. The Society then boasted among its officers an 'Observer', one Mr T. Hick, and a 'Mathematical Tutor', Mr T. Gregg, along with Herschel and other dignitaries among its acknowledged patrons [12]. The above, and the Society's subsequent history, is preserved in a large folio 'Album' which formed the principal archive of the Society from 1863 to the 1930s, and which is still in the hands of the Society. It is on a printed prospectus sheet, with a handwritten date of 31 March 1863, that it is plainly stated that 'This Society was established in the year 1859.' The same document makes clear the goals and purposes of the Society. It was the intention that it would diffuse astronomical and meteorological knowledge in Leeds, and 'foster a taste, and afford opportunities for study and research in these departments of Science' [13].

The Society clearly aimed at a large membership, cutting across a wide range of economic groups, for ordinary membership cost only four shillings (20p) per annum, which was lower than that of a Mechanics' Institution. Only half that sum was charged for people who were already members of the Mechanics' Institution, Literary and Philosophical Society, or other organisations sympathetic to the Society. It was hoped that the 'very low sum' charged for annual membership would be counterbalanced by 'a large increase in the number of Members'. A flood of members was clearly expected, for a number of elegant tickets were printed, with blank spaces into which the subscriber's details could be written [14].

In 1863, the Honorary President was G.B. Airy, the Astronomer Royal, who was joined by such luminaries on the Society's list of patrons as Sir John Herschel, Lord Wrottesley, the Earl of Carlisle, Canon Professor Temple Chevallier, and the industrial philanthropists, Titus Salt and Benjamin Gott. The Leeds surgeon William Clayton F.R.C.S. served as the active President on a daily basis [15]. While Airy was willing to act as Honorary President, he was nonetheless quick to recite the familiar litany which he repeated to all private persons who presumed upon his time: 'The constant pressure of my own official employments makes it almost impossible for me to offer any real assistance to the Society' [16].

The Society's great *coup*, however, was the remarkably active support which it received from the 71-year-old Sir John Herschel. At some stage prior to March 1863, moreover, Sir John had even gone as far as to advise the Leeds Astronomical Society about the purchase of a telescope, and had been instrumental in their acquisition of a handsome 3-inch refractor, with a brass tube, equatorial mount, and fittings, from Ross and Company in London, which one assumes was mounted in the Love Lane Observatory. The instrument is still in the possession of the Society [17].

Herschel was then was prevailed upon to present a lecture on 'The Yard, the Pendulum, and the Metre', which was delivered on 27 October 1863 [18] in the Theatre of the Leeds Philosophical Society – one of the venues which the Astronomical Society used for its large meetings – and received a great deal of newspaper coverage. Yet when one looks at his private Diary, one finds that Sir John and Lady Herschel were entertaining a large group of friends to dinner at Collingwood, their Kent mansion, on 27 October 1863, instead of being in Leeds. Indeed, Sir John's health, which at best was not robust, was particularly troublesome in 1863. Violent bronchitic attacks, at which he coughed up 'brown creatures', combined with painful joints made the elderly and rather frail patriarch of British science reluctant to undertake long journeys. Instead, he wrote the lecture at Collingwood between 15 and 30 September 1863, and on 1 October 'Dispatched the MS. of the Leeds Lecture to Mr. Trant, the Hon. Sec. Leeds Ast. Soc.' On 27 October, the lecture was read to the assembled audience by Mr. P. O'Callaghan, B.A., the Vice-President of the Society [19].

One would like to know what it was about Leeds and its Astronomical Society which made it possible to arouse such interest and patronage from figures like Airy, Herschel, and Chevallier. Of course, as we saw in Chapter 9, these men were concerned with the furthering of popular astronomical knowledge, and Airy in particular was interested in encouraging working-class scientific endeavours. However, one feels that the Leeds *annus mirabilis* of 1863 would not have taken place but for the driving energies of one man. This was William Trant junior, who wrote to Airy and the other patrons, and had entered into a substantial correspondence with Sir John Herschel. It was all the more astonishing, therefore, in August 1864, when in receipt of wider advice about how he might become a professional astronomer, that Trant revealed to Herschel '... I am twenty years of age' [20]. In the following month, Trant went on to tell Herschel that he regretted the little leisure that his employment in the textile trade made available to him, but he was resolved 'to know astronomy not only as a 'pleasurable companion' but also as a *bosom friend*' [21].

Indeed, this was an all too familiar story, seen in real-life characters like John Leach, Charles Todd, and Thomas Bush, and fictional ones like Swithin St. Cleeve, as able young men aspired to rise to scientific eminence from obscure backgrounds. While William Trant never became a professional astronomer, his ambitions later drove him from Leeds to London, India, and Canada, where he worked as a journalist and magistrate, which at least took him many steps up from working in a factory [22].

It is unfortunate that no membership lists survive from these early days of the Leeds Astronomical Society, for it would be interesting to know how many people took advantage of the four-shilling subscription charge. After all, Leeds was one of the great industrial cities of Britain by 1863, with a large population, while it is also clear that the Astronomical Society was on friendly terms with the Literary and Philosophical Society, Mechanics' Institution, and the Young Men's Christian Association, whose several premises they sometimes used, and whose members could join at a concessionary rate.

As far as one can tell from the surviving 'Album' volume, and from what can be reconstructed from the post-1892 *Journal* which the Society issued, things seem to have gone rather flat by the later 1860s. On the other hand, one can to some extent trace the changing impact of the Astronomical Society upon the city of Leeds from the frequency of astronomical articles which appeared in the local newspapers, and which an unnamed person dutifully pasted into the 'Album'. But it is difficult to know whether any of this quite substantial astronomical coverage in the local press had anything to do with an organised astronomical society, or whether it was scientific reportage written by journalists or by local people who had an independent interest in astronomy. Similarly, it is hard to know exactly when the present 'Album' was started, and how many of its cuttings from the 1860s and 1870s were pasted in after 1892.

What is certain, however, is that both the 1859 and 1863 incarnations of the Leeds Astronomical Society were relatively short-lived as far as their influence was concerned, for when the Society did take on its final form in the 1890s, it proclaimed itself as having been 'Founded 1859, Re-Established, 1863. Re-Organised 1892' [23]. But at least one man had continued with the Society from its first flourish in 1859 down to its reorganisation in 1892 and successful growth thereafter. This was William Donald Barbour, an expatriate Scotsman who had been 26 in 1859, and whose 1903 obituarist praised his unwavering support for Leeds astronomy and his Honorary Secretaryship of the Society [24]. It is only unfortunate that Barbour, whose wife and daughter were also active in the Society by 1903, never published his own account of its early history in the Society's *Journal*.

No matter how active or quiescent the Leeds Astronomical Society might have been between 1863 and 1892, however, it cannot be denied that the Society which came to be founded in Liverpool in 1881 had a decisive impact upon the organisation of amateur astronomy in Great Britain and the wider world. For it was that tide of enthusiasm engendered in Liverpool in the 1880s which led to the establishment of the British Astronomical Association in 1890, the 're-organisation' of the Leeds Society in 1892, and the creation of new civic or local branch societies in Manchester, Wales, Australia, and elsewhere in the years that followed.

As we have seen already, Liverpool had long since enjoyed a prominence in astronomical research by 1881, with figures like William Lassell, John Hartnup, and the Dock Board and Bidston Hill Observatories. And co-terminously with Leeds, the year 1859 had seen the emergence of an astronomical section within the Liverpool Literary and Philosophical Society, along with an attempt to found a public astronomical observatory in the city [25]. But it was that growing momentum of observing amateur astronomers which we saw in the last chapter which brought the Liverpool Astronomical Society into being.

It was the Revd Thomas Webb's young friend, protégé and eventual executor, the Revd Thomas Henry Espinall Compton Espin, whose father was Rector of

Wallasey, Birkenhead, just across the river Mersey from Liverpool, who, in 1880, wrote to the *English Mechanic* to advocate the formation of an association of amateur astronomers [26]. Its purpose would be to give coordination and direction to amateur work and strengthen its scientific value through the publication of results. Then following an astronomical lecture given at some venue in Liverpool in 1881, three or four people proposed the setting up of a 'popular society' to watch the heavens and report their findings. According to William Henry Davies, who later reported the incident to *The Observatory*, one of the greatest obstacles facing these enthusiasts was their lack of access to telescopes, beyond 'such as might be obtained any fine night (particularly at full moon) for a penny, at the street corner' [27] (in a service similar to that made available by Mr Tregent, as seen in Chapter 9).

To help solve the telescope problem, however, the Liverpool enthusiasts acquired a stock of 2-inch aperture lenses and some lengths of brass tubing, and made up their own instruments in the way that Roger Langdon and John Jones had done before them, and countless others would do thereafter [28]. Within a few months, the Liverpool amateurs had recruited seventeen people into their group, less than a half of whom were committed observers, the others apparently being 'armchair astronomers' who enjoyed the talks and discussion, but who were less fond of long night watches at the eyepiece. Indeed, even by the time of Davies' letter in 1883, the basic split between active observers and armchair enthusiasts still found in present-day amateur societies across the world was in existence, as Davies recognised the need for that 'kind of halfway resting-place between the amateur public and the Royal Astronomical Society' which an organised amateur society provided [29]. Like the old Leeds Society, it fixed a low subscription charge, which in 1886 stood at five shillings (25p) per annum. And as its original constitution of 1881 proclaimed, its aim was 'To advance the science of Astronomy and promote research ... to circulate information ... by lectures, the publication of articles, books and transactions ...', and this broad range of intentions brought in a flood of applications for membership [30]. Yet only a part of them came from the Liverpool area, as amateurs in various parts of Britain wanted to enjoy the benefits of the Society's circulars and *Journal*, even if they could not attend the meetings.

The exact numbers of people who were in this rapidly growing Society at any one time are not easy to ascertain, although in 1886 it was announced in *The Observatory* that the Liverpool Astronomical Society had authorised its first affiliated foreign branch, in Pernambuco (now Recife), South America [31]. To qualify as an affiliated branch, moreover, it was necessary to have over fifty registered members, and Mr G.W. Nicholls, the Secretary of Pernambuco Astronomical Society, had just signed up the fifty-first, and 'anticipates a membership of a hundred before long' [32]. And soon, there were branches in the Isle of Man and Australia [33].

Indeed, what Espin, Davies and their friends had done was create something that was very much more than a local club, for the virtually instant national and international membership of this Society showed how many people there were who wanted a 'halfway house' between private or dilettante celestial viewing and the Royal Astronomical Society. Everyone, it seems, wanted to join, including professional scientists of world repute, Grand Amateurs, and ordinary enthusiasts.

Ordinary and Associate Members included Sir Robert Ball, Charles Piazzi Smyth, Giovanni Schiaparelli (the Milanese discoverer of the martian 'canals'), the telescope-makers George Calver and Sir Howard Grubb, Otto Struve (Director of Russia's celebrated Pulkowa Observatory) and R.S. Newall, whose 25-inch Cooke was still the biggest refractor in Britain. Then in February 1887 His Imperial Majesty Dom Pedro II of Brazil was made a member, when visiting England, though he was assassinated upon his return to South America [34].

But perhaps most telling of all, in this world of male-dominated science, the Liverpool Astronomical Society in the 1880s contained at least eight women members [35]. One of them, Miss Elizabeth Brown, became the first Director of the Society's Solar Section, while the eminent historian of astronomy, Miss Agnes Clerke, also joined. In this respect, the Liverpool Society became the first scientific society, at least in Britain, to grant full, and in the case of Elizabeth Brown executive, membership to women, nearly forty years before they would be admitted to the Royal Society and Royal Astronomical Society. The role of women in British amateur astronomy, however, will be looked at in more detail in the next chapter.

While the Liverpool Astronomical Society held regular meetings in Liverpool, which would have provided a social as well as a purely scientific dimension for the local astronomers, it is clear that by 1885 the greater part of the membership derived its benefit from the circulation of its publications, such as an excellent monthly magazine, classes, and ancillary services, rather than from the actual meetings. Since 1884, for instance, Mr J. Gill had been running astronomy classes in Liverpool and awarding Certificates of Competence, while at least one other set of classes was offered, at the Douglas, Isle of Man branch [36].

And very quickly after 1881, the Society developed specialist sections to cultivate specific branches of observational astronomy, and keep members informed of discoveries, or notable events taking place within their interest area. The fast Penny Post and railways made it easy to have Section Directors who lived miles away from Liverpool, yet who still remained in close touch with the 'home' city. Elizabeth Brown, for instance, directed the Solar Section from her private observatory in Cirencester, Gloucestershire, over 150 miles away, while Thomas Gwyn Empy Elger, Director of the Lunar Section, lived in Bedford, where he was a leading citizen and Mayor [37]. Double-star, variable-star, comet and individual planetary sections soon came into existence, under the direction of the leading amateurs of the age. James Gill, Captain William Noble, and William Frederick Denning all took leading roles within the organisation of the Society, while the Grand Old Man himself, Thomas W. Webb, became a member, and wrote for its *Journal* up to his death in 1885. Rather sadly, William Lassell died a year before the Society's formation.

It is possible to reconstruct a great deal of what was happening within the Liverpool Astronomical Society during the 1880s from the attention which it received in the periodical literature, while the published abbreviation of its name to L.A.S. indicated its already familiar standing as early as 1883. It would also appear that by this date no-one felt that the acronym might have confused the Liverpool with the quiescent Leeds Society. Journals like *The Observatory*, which in some respects was an unofficial organ of the Royal Astronomical Society, carried the abovementioned articles of Davies and others, although the *English Mechanic* is one of our best guides – beyond the Society's own publications – as to what was going on. Between 1881 and 1890, the *English Mechanic* published over 36 letters and articles which reported upon or otherwise discussed the research and teaching functions of the L.A.S [38]. These articles reported aspects of the Society that were as diverse as the discussion of techniques for stellar position reduction and planetary observations, 'A Lady's Observations of the Eclipse of 1884 ...', and the announcement of the membership enrolment of the Emperor of Brazil [39]. The activities of the Society produced responses across the astronomical community, and clearly placed it in the limelight.

By 1886, however, many people were aware of the paradoxical situation of an ostensibly local astronomical society being in reality an organisation of international standing, with a body of officers who were not Liverpudlians. In particular, it worried the Grand Amateur astronomical photographer Isaac Roberts, who lived and observed in Liverpool, and who felt that it should primarily be a society for Liverpool astronomers [40]. W.H. Davies, now signing himself 'A Vice-Pres. of the L.A.S.', wrote a stiff defence of the international policy of the Society in *The Observatory*. Not only did he round upon points which Roberts had made about the Society's finances and internal organisation, but proclaimed that 'It has never been found necessary that active workers should reside in Liverpool.' Indeed, 'The necessity for paid assistance is negatived by the fact that such men as Denning, Franks, Gore, Espin, Elger and others have cheerfully volunteered regular literary and scientific work', which would minimise the Society's overheads and add to its breadth of base and expertise [41].

The year 1886, indeed, seems to have been one of identity crisis for the Liverpool Astronomical Society, for its Annual General Meeting of the 1886–1887 Session was held not in Liverpool but in London, at the rooms of the R.A.S. in Burlington House. By this date, the Society had over 400 active members, and its President, the Bedfordian T.G.E. Elger, emphasised its national character, and the need to form local branches around Britain [42].

One cannot help being amazed at the speed with which the informal get-together of a handful of ill-equipped enthusiasts after an astronomy lecture in Liverpool in 1881 had led to the burgeoning organisation of 1886. Very clearly, amateur observational astronomy was all set for expansion and unification in a similar way that Grand Amateur astronomy had been in 1820. And as in 1820, what was needed was not a central directing agency in the bureaucratic sense, but a clubbable body – even if for most of the members it was a corresponding club – where letters, ideas, and observations could be recorded, exchanged, and re-directed. Following the demise of the *Astronomical Register* in 1883, moreover, there was a need for a new astronomical periodical. There was, of course, that *vade mecum* for all practical technological enthusiasts, the *English Mechanic*, which carried so much of the correspondence surrounding the early Liverpool Astronomical Society, and that new journal after 1877, *The Observatory*; but even so, the *Journal* and other publications of the Liverpool Astronomical Society fulfilled a major need in serious amateur astronomical literature.

It is very clear, however, that something happened in British society between 1840 and 1881 as far as its awareness of popular astronomy was concerned, and especially over the two decades after 1863. For as we saw above, the proposed 'Uranian Society' of 1839 had sunk without trace, although by the 1850s public observatories were being proposed. The slow fuse of the Leeds Astronomical Society was lit in 1859, culminating in its *annus mirabilis* of 1863, though fading thereafter. But over the next eighteen years, a growing body of fuel was being piled onto the amateur astronomical fire in the wake of Webb's *Celestial Objects*, in the form of cheaper and more accessible telescopes and books, along, of course, with journals like the *Astronomical Register* and the *English Mechanic*.

In short, a community of interests was being defined and provided for, though by the 1880s the astronomical community had escalated to a point which required formal organisation. This was, indeed, a formalisation through which a number of other leisure activities were passing at the time, as clubs devoted to tennis, cricket, football, numismatics, philately and other pursuits were feeling the need for rules and organisational structures whereby their members could relate to each other. For in much the same way as the emerging middle classes were defining their professional power through the new qualification-granting occupational associations, so they defined their leisure power through clubs devoted to their various interests.

By 1889, however, the Liverpool Astronomical Society was in serious difficulties. Its finances were in the red, and money was owed to its printers, as the Society found that its small membership charge could not sustain an organisation of the magnitude that had grown up. The resignation of the Treasurer, the illness of the Secretary, and the death of the President, all around the same time, seemed to give weight to the warnings about over-expansion raised by Isaac Roberts three years earlier.

The difficulties which the Liverpool Astronomical Society was experiencing by 1890 only emphasised the need for a more enduring amateur astronomical society, which many people who did not live in the north of England felt should be based in London. This mood was expressed publicly in July 1890, when William Henry Stanley Monck, the pioneer photometrist, who was a barrister by profession, wrote a letter to the *English Mechanic* advocating the formation of a national amateur astronomical society [43]. Apart from the proposal that this new society should be based in London, it was to possess most of the original characteristics of the Liverpool Astronomical Society in its heyday. It was to be relatively cheap in terms of its subscription, open to both men and women, and catering for people who found the R.A.S. both too expensive and too technical. In short, it was to be virtually identical to the 'halfway house' which W.H. Davies had spoken of with relation to the Liverpool Astronomical Society in 1883.

A number of other people – many of whom were old Liverpool members – expressed their warm approval of Monck's idea. One of the quickest off the mark was Miss Elizabeth Brown, Solar Section Director of the Liverpool Society, who got her letter into the next week's number of the *English Mechanic* after Monck [44]. Elizabeth Brown, furthermore, began to encourage others along a similar way of thinking, and Edward Walter Maunder, who was on the staff of the Royal Observatory, Greenwich, wrote a letter of his own in early August 1890, informing the readers of the *English Mechanic* that a Society along the lines proposed by Monck was already in the process of formation. Furthermore, Maunder approached several professional and Grand Amateur astronomers for their support, and was pleased to receive that of William Huggins [45].

The Provisional Committee of what would soon become the British Astronomical Association (B.A.A.) looked in many respects like a roll-call of officers from the Liverpool Astronomical Society, with figures of such stature within the amateur astronomical world as George Calver, T.G.E. Elger, T.H.E.C. Espin, W.S. Franks, Captain William Noble, and Sir Howard Grubb [46]. And right from the start, there were women in prominent positions: Elizabeth Brown, for instance, carried over her old Liverpool job, to become Solar Section Director; while the astronomical writers Agnes Giberne, Agnes Clerke and Mrs (later Lady) Margaret Huggins occupied influential positions. The future Mrs Annie Scott Dill Russell Maunder, moreover, would soon come to play a decisive role in the Society which would extend down to 1946 [47].

By October 1890, the new Association already had 283 members, and when the formative meeting was held at the Hall of the Society of Arts, Adelphi, London, on 24 October, the Treasurer, William Henry Maw, announced that there was already more than £200 in hand. The first 'Ordinary' meeting of the British Astronomical Association was held a month later, on 25 November 1890 [48].

While the B.A.A. had the laudable intention of being available to all amateur astronomers, and its *Journal* and other published notices enabled it to reach members anywhere in Britain or the Empire, its guiding council and metropolitan organisation imparted a distinctly 'establishment' tone which might well have scared off humbler members who could still have afforded the remarkably reasonable half-guinea (52½p) annual subscription. Its founding President, Capt William Noble – immortalised in a snapshot wearing boots and carrying a shotgun – was a retired army officer, of legendary eccentricity, who had been writing a fortnightly astronomical column in the *English Mechanic* since the mid-1860s [49]. One of its Vice-Presidents was an Irish Earl, while clergymen, knights, professional people, and ladies and gentlemen of independent means dominated its affairs. The fact that it met on Wednesday afternoons at 5 p.m. at Barnard's Inn, amidst the Inns of Court of London's barrister community, says something about the type of people who were expected to be able to attend the meetings [50]. Even so, the B.A.A. gave a structure to the social organisation of British amateur astronomy at the national level that was to remain virtually intact until the formation of the Federation of Astronomical Societies in 1974.

In addition to the monthly meetings and publications, the B.A.A. continued to develop those three aspects of organisation which the Liverpool Astronomical Society had pioneered so successfully and which would prove to be invaluable to the wider development of amateur astronomy: provincial and overseas branches, specialist research sections, and foreign expeditions. And where the early Liverpool Society had laid up a stock of lenses and brass tubes for assembling into telescopes for members, so the early B.A.A. began to acquire a loan collection of instruments, many of which, indeed, were telescopes possessing serious scientific potential.

By February 1891 the B.A.A. had established its first branch, in the Midlands at Birmingham, followed in January 1892 with one in Manchester. Two Scottish branches were founded in 1894 and 1896 respectively, while over the same period two were founded in Australia, the New South Wales branch of which was presided over by that patriarch of Australian amateur astronomy, John Tebbutt. With its regional and overseas structure, the B.A.A. was well placed to stimulate amateur astronomy in distant places, and always had the resources of the London base to provide support [51].

By 1891, within a year of its foundation, the B.A.A. had established ten specialist observing sections, devoted largely to aspects of solar, planetary, variable and coloured star observation, and over the next few years several more sections would be added. They were intended to cater for and coordinate the needs of serious observers who were committed to the systematic monitoring of such things as solar flares or marks on planetary surfaces, and right from the start there were astronomers of distinction directing them. T.G.E. Elger and Walter Goodacre, for instance, who were amongst the finest lunar cartographers of their day, directed the Lunar Section; while the distinguished priest–schoolmaster–astronomers, Father Aloysius Laurence Cortie and Father Walter Sidgreaves of the Stonyhurst College Observatory, took over the Solar Section after the death of Miss Brown. The greatest Mars observer of the twentieth century, Eugène Michel Antoniadi, was Director of the Mars Section between 1896 and 1917 [52].

In 1896, 1898, 1900 and 1905, the B.A.A. sent expeditions to observe and photograph eclipses in a variety of countries, including Norway, India, Majorca, and America. It was as an eclipse photographer that E.W. Maunder's wife, Annie, was to continue to develop her work as a solar astronomer; while the Revd John Mackenzie Bacon, accompanied by his daughter Gertrude, succeeded in photographing the 1900 eclipse with an early cine-camera attached to his telescope [53]. Of course, such foreign expeditions were becoming increasingly feasible to amateurs by the 1890s, as timetabled steamships, trains, and telegraphs shrank the world; even so, their great expense, in terms of time and money, gives an indication of the personal circumstances of many of these early B.A.A. members [54].

It is also important to remember that by the 1890s there were many servants of the British Empire scattered across the globe who were keen to join the B.A.A. to help them in their private observing. One such figure was John Willoughby Meares, whose 1898 Observing Book gives a detailed account of his observation of the Indian eclipse of that year. Meares had been a founder member of the B.A.A., and when he went to live and work in India, in September 1896, he took with him a 3¼-inch refractor and a 9¼-inch With–Browning–Calver reflector (the latter being a gift from his father). On 22 January 1898 he joined the B.A.A. party at Buxar, and obtained a photograph, but more importantly, made a beautiful drawing of the coronal streamers that were visible during totality [55].

It would be incorrect, however, to assume that the early B.A.A. swept all before it as far as organised amateur astronomy was concerned. For one thing, developments were taking place in other countries, especially the United States. As early as 1882, for instance, the Scientific Society of Boston, U.S.A., wrote to John Tebbutt at his home in Windsor, New South Wales, Australia, suggesting that he might form a group of antipodean astronomers who would search the southern skies for comets, to match the efforts of those American amateurs who were active in the northern hemisphere [56]. And in 1889 the Astronomical Society of the Pacific was formed in California, which was to give such an impetus to American and Pacific basin astronomers [57].

Yet in addition to what was happening in America, Australia – and, perhaps, Pernambuco – there were other locations in the 1890s where amateur astronomy was taking place in Britain outside the direct orbit of the B.A.A. The most notable was Liverpool, where in the wake of its collapse and apparent abduction to London, the L.A.S. was continuing as a purely local society. It was greatly assisted in its continuing independent existence, however, by the generosity of Thomas Glazebrook Rylands, who in the faltering days of 1889 had given his 5-inch Cooke equatorial refractor and a 2-inch transit instrument to the Society [58]. With the help of the City Council, an observatory to house the Rylands refractor that was also accessible to students and members of the public was built at the Liverpool Nautical College, and during the Edwardian period there was sufficient activity within the Society to send an expedition to Spain to photograph the solar eclipse of 1905, while in 1910 it was successful in informing the public of and showing Halley's Comet at its observatory [59]. Yet with the founding of the North-West branch of the B.A.A. only thirty miles away in Manchester in 1892, one senses that the late Victorian L.A.S. must have felt under a cloud as an institution with its own identity.

On the other hand, the year 1892 saw the beginning of a vigorous 're-organisation' of the Leeds Astronomical Society, due largely to the efforts of the previously mentioned William Donald Barbour. The revival was heralded by a leader in the *Leeds Mercury* Supplement on 3 December 1892, which took up the point that, as Sir Robert Ball had recently commented, an Astronomical Society existed (or had existed) in Leeds [60]. The 'few survivors' of the old 'Leeds Astronomical Society' (one of whom was Barbour) were then moved to hold a public meeting, presided over by the renowned Leeds amateur astronomer Washington Teasdale, and with Barbour as Treasurer [61]. It is not clear whether Teasdale, H. J. Townsend and others behind the revival of 1892 were members of any continuing society that had quietly existed after the glories of 1863, or whether their 're-organisation' of the old Leeds Astronomical Society was in all but name a new foundation.

Even so, it is clear that Barbour had an active and continuing interest in amateur astronomy that connected his 1859–63 membership of the L.A.S. with its resuscitation in 1892. In 1876, for instance, he was collecting cuttings from the *English Mechanic* that discussed the advantages of the new £5 telescopes, pointing out that for a total of £8 13s (£8.67) one could purchase such a telescope, a tripod, some eyepieces, and two of R.A. Proctor's beginner's books prior to graduating on to Webb's *Celestial Objects* [62]. And at some time before April 1884, he had travelled up to Gateshead to visit R.S. Newall and look through his famous and sadly under-used 25-inch Cooke refractor; while in 1887, when the Liverpool Astronomical Society was still going strong, he had met with Herbert Sadler at W.H.

Davies' house in Liverpool to discuss astronomy [63]. It is quite likely, in fact, to have been his acquaintance with Davies, and the example of the Liverpool Astronomical Society, which had led Barbour to re-form the Leeds Society in December 1892, especially as the success of the B.A.A. was showing how popular amateur astronomy had become.

Fortunately, the resurrected Leeds Astronomical Society thrived, and by October 1896 was advertising its winter syllabus of lectures and meetings. They included, amongst other things, a lecture on the life and work of Caroline Herschel by the first lady member of the Leeds Astronomical Society, Miss Florence Taylor, and in the following year she lectured on Mary Somerville [64]. As an organisation, the Leeds Astronomical Society was clearly strong, confident, and willing to enter into correspondence with the eminent scientists of the world, just as it had been in 1863. In 1897, for instance, Barbour was corresponding with Percival Lowell in Boston about the properties of reflecting telescopes – and receiving detailed replies [65] – while in 1898 he was writing to the Astronomical and Physical Society of Toronto [66]. It is also interesting to note that in 1893 William Trant wrote from his new home in Canada to inform Barbour of his role in the original foundation of the Society, over thirty years before, and of the great difficulties which he had experienced in raising the money necessary to purchase the Ross telescope, on which Sir John Herschel had advised [67].

After two false starts in 1859 and 1863, therefore, the Leeds Astronomical Society had developed sufficient social momentum by 1892 to enjoy a stable, continuing existence. Within its ranks it included such local figures as the industrialist–amateur, Scriven Bolton, with his 18-inch reflector, and Mr Greevz Fysher, astronomer, polymath, and all-round Leeds eccentric [68]; while in the wider world it enlisted the famous professional astronomer, Herbert Hall Turner. Turner had become Professor of Astronomy at Oxford and President of the R.A.S., but never forgot his Leeds roots, and seemed proud to play a prominent role in the Society's seventieth birthday celebrations in 1929 [69].

In 1898, W.D. Barbour acknowledged the receipt by the Leeds Society of the first volume of the *Journal* of the Astronomical Society of Wales from the Cardiff journalist, antiquarian, and amateur astronomer, Arthur Mee [70]. According to the Revd J. Silas Evans, writing in 1923, Arthur Mee, who was a Scotsman by birth, had begun to speak to local amateur astronomers about the possibility of forming an astronomical society almost as soon as he arrived in Cardiff to work as a journalist on the *Western Mail* in 1892. He was told by one of the people whom he approached, Norman Lattey, that the chances of forming an astronomical society in Cardiff were small, though Lattey later changed his mind, and in 1894 became one of Mee's staunchest supporters in getting the Astronomical Society of Wales off the ground [71]. Within a short time, it had enrolled 100 members, with a subscription rate set at five shillings (25p) for men living within 15 miles of Cardiff, and 2s 6d (12½p) for ladies and people beyond Cardiff [72]. It would be interesting to know who were the 'eminent astronomers' around Cardiff to whom Mee was speaking between 1892 and 1894, for the remark implies the presence of a significant yet unorganised body of observers. In addition to Norman Lattey it probably included that dozen or so people who formed the early Council of the Society, such as James Waugh M.A., Professor Lloyd Tanner, and Miss E. Graham Hagerty: people who all seem to have had some connection with education in Wales.

One invaluable guide to the character and composition of the Astronomical Society of Wales (A.S.W.) was its quarterly journal, the *Cambrian Natural Observer*, which was edited by Mee. These little volumes of about sixty pages apiece, running from 1898 to 1911, followed a similar format to the *Journal* of the Leeds Astronomical Society. They reported members' observations of the Sun, Moon, planets and other phenomena, discussed telescopes, described members' observatories – such as that belonging to Colonel Ernest Elliot Markwick – and informed readers of forthcoming celestial events [73]. The Society does not seem to have had specialist sections like the old L.A.S. or new B.A.A., and was perhaps less oriented to original observation. Although it aimed to promote 'the Study of Astronomy and the Allied Sciences' in Wales, it had several members – usually with Welsh names – living elsewhere in Britain and overseas, and probably had a substantial 'armchair' membership. Yet one also spots familiar names that appear on the lists of the old L.A.S. and the new B.A.A., such as Elizabeth Brown, Captain William Noble, T.G.E. Elger, Scriven Bolton, and other members of the amateur astronomical 'establishment'. J. Willoughby Meares – who was mentioned above with relation to the Indian eclipse of 1898, and who was not a Welshman – also joined the A.S.W. along with the B.A.A. and the R.A.S [74]. Very clearly, if you took your amateur astronomy seriously in 1905, you joined the Astronomical Society of Wales, especially if you felt that you had any Celtic blood.

As with the Leeds and other societies, the detailed lists of members that the Astronomical Society of Wales published during the late Victorian and Edwardian era tell us a great deal about its social composition. In January 1901, for instance, there were 114 full members of the Society. No less than 35 of them possessed some sort of academic or professional designation, such as the title Reverend, a university degree, or an F.R.A.S.. Very significantly, of these 114 members 22 were women. There were also 23 associate members, who were part of the wider amateur or professional scientific world and who generally lived elsewhere than in Wales. What is more, the Society still drew attention to its glorious dead, mentioning its working-man member John Jones – 'Seryddwr' (Astronomer) – who had died in 1898, and the associate member Elizabeth Brown, who had died suddenly in 1899 [75].

The Society, indeed, was a thriving concern with a rising membership. In 1899 there were 88 members, then 1900 saw an increase to 103. And as we have seen above, by 1901 there were 114, while in 1908 it reached 242 members, some 41 of whom were women. E. Walter Maunder, W.F. Denning, and Father Cortie of Stonyhurst were among the 10 associates. Even though numbers had fallen slightly by 1910, when there were 226 full members, including 27 women, plus nine associates, it is clear that the Society was a going concern almost to its sudden demise at the time of the outbreak of war in 1914. As with the L.A.S., the early B.A.A., and the reorganised Leeds Astronomical Society, one sees from their degrees and letters that the prominent figures in the Astronomical Society of Wales were middle-class professional people. Indeed, one early Council Member of the Astronomical Society

Plate 51. Messier 31 and 32 in Andromeda, photographed by Isaac Roberts on 29 December 1888 after a 4h 7m exposure. (Roberts' working photographer was W.S. Franks.) Roberts, *A Selection of Photographs* (pl.44), 31. Roberts believed that this photograph confirmed G.P. Bond's visually-drawn conclusions of 1848 that M31 was 'condensing' out into a series of annular dust bands, around a central nucleus. From a superior photograph, taken on 17 October 1895, however, he changed his mind, recognising 'that the Nebula is a left-hand spiral, and not annular as I at first suspected': Roberts, *Photographs of Stars, Star-Clusters and Nebulae* II (London, 1899) 63.

Plate 52. Amateur astronomers. D. Olmsted, *The Mechanism of the Heavens* (London and New York, 1856). These young men, with their globe, books, navigator's octant, bust of Galileo, and small refracting telescope pointing through an open window, were probably American amateurs.

Plate 53. An astronomy lesson, perhaps by an early Victorian father to his children. J. Joyce, *Scientific Dialogues* (Manchester, 1844), frontispiece.

Plate 55. Mr Tregent and his street telescope demonstration. H. Mayhew, *London Labour and the London Poor* III (London, 1861) 82.

Plate 54. 6-inch aperture reflecting telescope by James Veitch, mounted in a hexagonal mahogany tube with brass bands and fittings. Optics missing, originally working at the Gregorian or Cassegrain focus. The brass band around the tube is stamped 'James Veitch, Inchbonny'. Examined personally by the author, 26 October 1993, for its private owner and restorer. Present location unknown. Scale drawing by A. Chapman.

Plate 56. Roger Langdon (1825–1894), the Station Master of Silverton. Courtesy of Tiverton Museum, Devon.

Plate 57. Silverton Station, Devon, *c*.1890, showing Langdon's conical observatory in the front garden. Exact tracing from a copy of an original photograph in the National Railway Museum, York Tracing by A. Chapman.

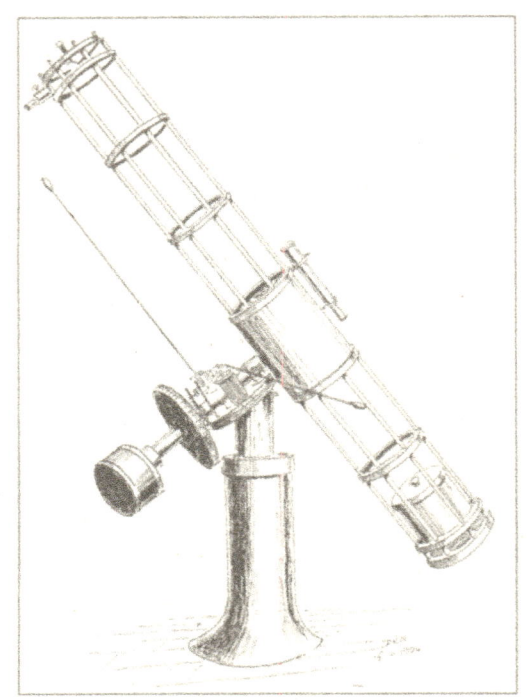

Plate 58. The Newtonian reflecting telescope of 13 inches aperture built by Thomas Bush, as it appeared at the Working Men's Exhibition, in London, 1870. Drawing by David K. Northrop of the original telescope depicted in the *Nottingham Daily Guardian*, 13 July 1870.

Plate 59. Newtonian telescope of 24 inches aperture by Thomas Bush, c.1920. Bush is probably the individual sitting with the instrument. Drawing by David K. Northrop from a photograph held by Nottingham University.

Plate 60. John Jones and his 'Jumbo' Newtonian reflector of $8^3/_{16}$ inches aperture. Originally in *Young Wales* (1898) and reprinted in S. Evans, *Seryddiaeth* (pl.43)273. This is the quintessential depiction of a working-class scientist of the Victorian age, including the contemplative face and skull-cap of the philosopher, and the heavy boots of the day labourer.

Plate 61. Charles Paton (1849–1938), from a photograph in the possession of his descendant, Don Simpson, of Sunderland.

Plate 62. Caroline Lucretia Herschel (1750–1848). Ball, *Great Astronomers* (pl.9) 207.

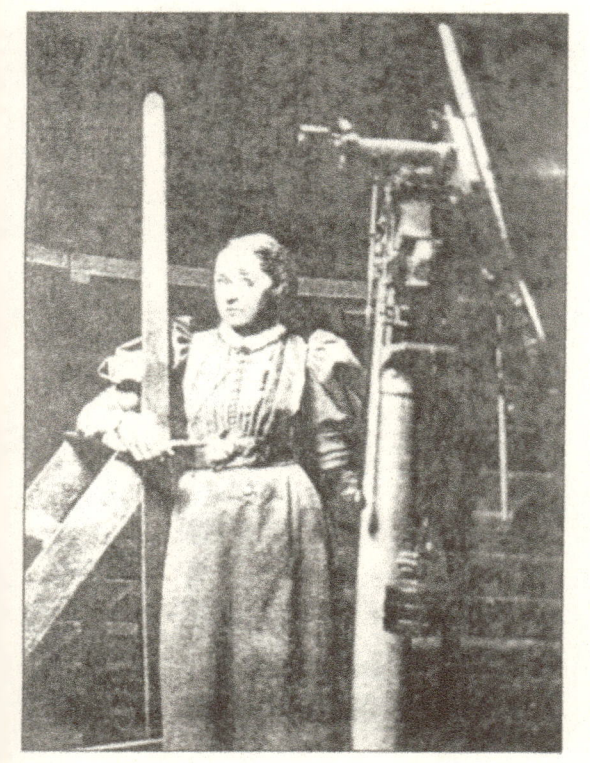

Plate 63. Elizabeth Brown (1830–1899), *J.B.A.A.* **9** (1898–9) 214. This is the earliest photograph, with which the author is acquainted, of an English woman within an astronomical observatory. (Earlier photographs of Maria Mitchell and her students survive in the Vassar College Observatory, U.S.A.).

Plate 64. Florence Taylor of Leeds, later Mrs. Hildred of Nobles County, Minnesota, U.S.A. Original photograph c.1898, Leeds Astronomical Society 'Album', still in the possession of the Society.

Plate 65. The Revd Thomas William Webb (1806–1885). Hutchinson's *Splendour of the Heavens* II (pl.28) 713.

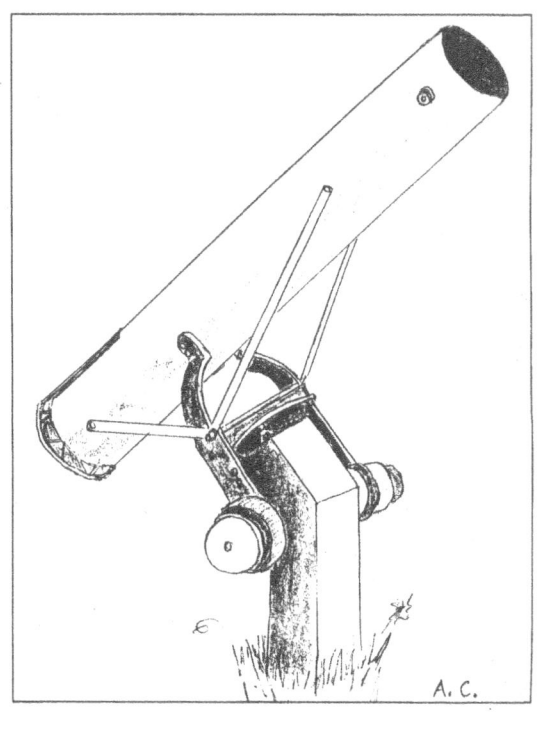

Plate 66. Silver-on-glass Newtonian reflecting telescope of 9¼ inches aperture, mounted upon a Berthon stand. This iron instrument – now in the B.A.A. instrument collection (no.83) – has been used, and its provenance researched, by Denis G. Buczynski, M.Sc., of Lancaster. This is almost certainly the instrument, with a mirror by G.H. With, which was built for the Revd. T.W. Webb in the 1860s. The tube sits on a circular plate set in the equatorial plane, and is counterpoised by a pair of weights, mounted stirrup-fashion, around the support pillars, hence Berthon's designation – the 'equestrian' mount. Drawing by A. Chapman.

263

Plate 67. The Revd Edward Lyon Berthon's 'Romsey' observatory, of the same design as that used by T.W. Webb, and which could be constructed for £8 10s. or for £12 with transit room. *English Mechanic*, 13 October 1871, 83.

Plate 68. The Temple Observatory, Rugby School. Crossley *et al.*, *Handbook of Double Stars* (pl.30) 1.

Plate 69. The University of Cork Observatory. Note its similarity to an early Gaelic church, the lancet windows forming part of the transit shutters, and the tower carrying a dome. H. Grubb, 'On the Equatorial Telescope... of The Queen's College, Cork', *Scientific Proceedings of the Royal Dublin Society* **2** (1880) 347-69: pl.24.

Plate 70. Advertisement for astronomical lectures in Leeds, 1861. Leeds Astronomical Society 'Album'.

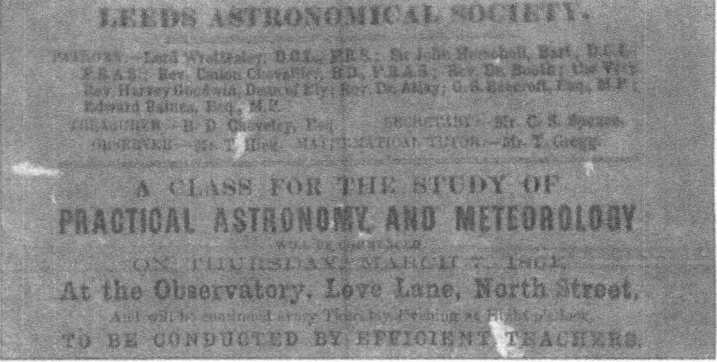

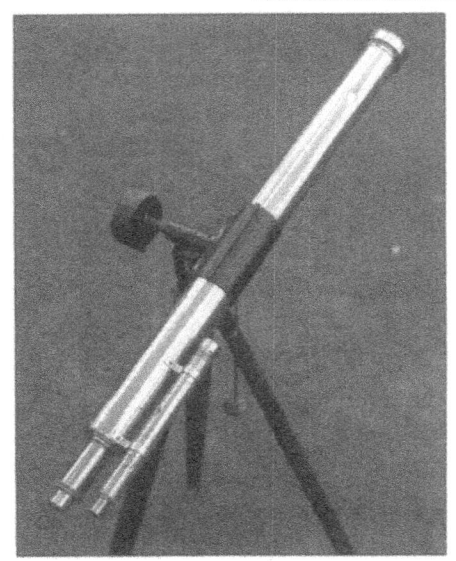

Plate 71. Refracting telescope by Ross of London, ordered on the suggestion of Sir John Herschel for the Leeds Astronomical Society, 1863. This immaculately preserved instrument is still owned by the Leeds Astronomical Society.

of Wales, Miss E. Graham Hagerty of the Higher Grade School, Cardiff, sported the letters A.R.C.Sc. (Associate of the Royal College of Science, now Imperial College, London) – one of the first women to indicate the possession of a professional qualification [76].

Yet it would be interesting to know how many of these members of either sex were, like John Jones, working people. Sadly, nothing specific is recorded about working-class members, although the Society's very cheap annual subscriptions, of five shillings and 2s 6d, would suggest that they were welcome, at least in theory. One other man besides Jones who had done much to educate himself was Thomas Harries of Llanelli. He had taught himself astronomy from the books of Thomas Chalmers, Thomas Dick, Richard A. Proctor and Sir Robert Ball, had acquired sufficient French to read Camille Flammarion, and then set himself the goal of 'spreading knowledge among the working classes' [77].

It is possible to put some flesh onto the bare bones of the Society's membership lists, however, from the Revd Silas Evans' *Seryddiaeth a Seryddwr (Astronomy and Astronomers)* of 1923. Silas Evans, who himself had been a member of the Astronomical Society of Wales, compiled a patchy and distinctly uncritical history of Welsh astronomy from the days of yore down to 1922, which included biographical notes upon some fifty men who had been active in Wales over the preceding thirty-odd years. With the exception of the Smilesian hero, John Jones, most of the individuals listed by Evans were men of some social standing, including 22 Fellows of the R.A.S., eight members of the B.A.A., 18 holders of M.A. or B.Sc. degrees, and six men with doctorates. Some are specified as teachers, medical men, or retired army officers, while there are nine clergymen. Some of Silas Evans' figures spanned several categories, such as Tegfan Davies, who was a clergyman with a doctorate and an F.R.A.S. Others had their occupations defined, such as J.F. Jones Tybrith, who was a farmer, and G. Parry Jenkins, who had known G. H. With when he lived in Hereford and had been a Colwyn Bay bank manager prior to his emigration to Toronto. A few, such as the scientific antiquarian and collector Lewis Evans, did not live in Wales, but were claimed as Welshmen. It is clear, moreover, that those men of whom Silas Evans included photographs were very proud to pose with their telescopes. Some of these instruments, such as those of John Jones and J. Alun Lloyd, were of extremely ingenious home-made construction [78]. The driving force behind the Astronomical Society of Wales, however, was Arthur Mee, whose popular publications, such as his annual *Heavens at a Glance* wall-chart, and contributions to the Society's and the B.A.A.'s journals did so much to advance the subject.

Yet not a single woman is mentioned in Evans' book. Indeed, this was all the more surprising when one considers that women consistently formed between one-fifth and one-sixth of the membership of the Astronomical Society of Wales during the first decade of the twentieth century. There is no explanation for his omission, and one suspects that Silas Evans did not consider a woman's place to be in the observatory.

After the cessation of hostilities in 1918, the Astronomical Society of Wales was not resurrected in its old form, but in 1922 a Cardiff Astronomical Society was founded. It soon acquired forty members, who either built or inherited via the old A.S.W. an observatory mounting a large equatorial reflector by George Calver, which had been given to the City of Cardiff by the late Franklen G. Evans in 1906 [79].

Another society was also in existence at the nearby town of Barry by 1923, though one of the most noteworthy of all local societies was the one founded in the Llyfni Valley in 1921, and this was because its fifty-strong membership comprised 'mainly miners' who nonetheless enjoyed the loan of a 3½-inch Cooke refractor. Although its President was the Revd William Edwards, M.A., F.R.A.S., it is the only society that I have encountered, in a *milieu* still overwhelmingly dominated by middle-class figures, which was acknowledged to comprise a predominantly working-class membership [80].

There was also the Ulster Astronomical Society, which, like the others that we have seen, was founded in the 1890s. At its peak it had 119 members, mostly from around Belfast, ten of whom were women, but I have not been able to find out much more about the Society's activities [81].

Each of the above societies, in Liverpool, Leeds, Wales and Ulster, enjoyed an independent existence that was distinct from B.A.A. branch status, although some of their members would have been B.A.A. members in their own right. On the other hand, the North-West branch of the B.A.A. soon began to experience some tension with the parent society in London on the ground of subscriptions.

The North-West branch of the B.A.A. had got off to a flying start in January 1892, with a visit from E.W. Maunder [82]. Its activities were conscientiously reported in the *City News* and other Manchester newspapers, as the Jesuit astronomer, Father Walter Sidgreaves, took up the Presidency and delivered the inaugural lecture on 26 February 1892 [83]. It is clear that many of the early branch members were already serious astronomers, for in addition to Sidgreaves, Samuel Okell, the Vice-President, owned a 20-inch reflector, Dr Steele Sheldon of Macclesfield was an active astrophotographer, while Mr R. Wilding of Preston, some 25 miles north of Manchester, owned a 19-inch and had access to an 18-inch reflector [84]. Astrophysics, spectroscopy and astrophotography seem to have predominated among the interests of the leading (and no doubt well-to-do) members of the branch, and spectral plates were sometimes sent to Huggins in London [85].

By early 1893 there were 65 members active in the North-West branch [86]. Among this number were five clergymen, including Archdeacon Wilson, Vicar of Rochdale, Professors from Owen's and other colleges in the city, doctors and various professionals. One presumes that at least one influential Manchester accountant was a member, as the branch met at the Chartered Accountants' Hall in the city. And right from the start – as was only fitting in the city which was at this same time laying the foundations of the British Suffragette movement – there was one woman member on the branch council, Miss Mary Holehouse [87].

By November 1895 the membership had risen to 76. Quite apart from the astrophysical observers, however, it is clear that by this date the social, clubbable dimension of the Society was strong, for in his Report to the 1895 Annual Meeting, Thomas Weir, the Secretary, complained that while only 25 people on average attended the Society's meetings, between 55 and 60 came to the social *conversaziones*

[88]. The social dimension of the branch's activities was also apparent in the 1897 remark that members 'find other friends at Schofield's Restaurant, No 86^A King Street... almost opposite our place of Meeting', as The Accountants' Hall was at 65 King Street, Manchester [89]. Dress codes were observed at the *conversaziones*, for when Thomas Weir issued a circular to announce one at which the newly-discovered 'Röntgen Rays' (X-rays) would be explained on 27 April 1896, morning dress was specified [90].

The overall social tenor of the North-West branch of the B.A.A. which comes through the records for the 1890s is soundly middle-class. And even where a member's social position cannot be ascertained from the appendage of professional letters of qualification, it can often be inferred from their address. Some years ago, Mr Kevin Kilburn, of the Manchester Astronomical Society, transcribed the addresses of all the early North-West branch members between 1891 and 1901 from B.A.A. and other archives, and this list only confirms the proposition outlined above regarding the social status of late Victorian Manchester's amateur astronomers. For of the 88 men and one woman who joined the North-West branch of the B.A.A. during 1891–1901, the overwhelming majority lived in the prosperous districts of Manchester, Lancashire and Cheshire [91]. Some lived in central Manchester itself, in the commercial districts like Deansgate, or in its then fashionable southern suburbs of Whalley Range, Fallowfield, or Alexandra Park. Few names are found of people living in the coal and cotton districts of north and east Manchester, and those that do occur are usually of clergymen, justices of the peace, physicians, or gentlemen living in named residences that were sufficiently prominent as not to need a street number. Very few of the members' names are associated with the working-class residential districts of the city or the adjacent Lancashire towns [92]. Nonetheless, it would be interesting to know if any working people did join the branch.

Considering the generally well-off social composition of the North-West branch, therefore, it might seem strange that the leading bone of contention between the Manchester astronomers and the B.A.A. central committee in London concerned the subscription charge. The B.A.A. required each North-West branch member to pay the standard annual subscription of half a guinea (52½p) to the Association, while the local committee in Manchester levied a further half-crown (12½p) to cover local expenses, thereby charging each of its 70-odd members 13s 6d (67½p) a year [93]. Perhaps this hard financial fact has something to do with the absence of members with addresses in the poorer districts of Manchester, for 13s 6d was equivalent to more than half a week's wages for a skilled worker in 1894. The Liverpool, Leeds, and Welsh Societies, by contrast, charged only 5 shillings (25p) a year.

In 1894, the subscription issue became a subject of some plain speaking between Thomas Weir, Secretary of the North-West branch, and Philip F. Duke, Secretary of the B.A.A., which provides invaluable information about the financing of amateur astronomical societies in the 1890s. Thomas Weir argued that, as the meeting hall in Manchester consumed £12 a year from the North-West branch's revenues, and with 72 members the local half-crown subscription only brought in £9, the B.A.A. should return 3 shillings (15p) from each member's subscription back to Manchester, to balance the books. This was especially pertinent as the Manchester people still had to contribute towards the £16 9s (£16.45) fee for the annual hire of the London meeting room, while having no help with payment for their own premises [94]. Philip F. Duke recommended that Weir and his colleagues might consider local fundraising in Manchester! One had to take into account, as he told Weir, the fact that 9s 6d (47½p) out of every half-guinea (52½p) paid by each of the 817 members of the B.A.A. nationwide in the 1893-4 session had been spent on journals and circulars sent to them individually, and the remaining ninepence (4p) had gone on premises and running costs [95]. Indeed, these figures give us a glimpse of why the original Liverpool Astronomical Society of 1881 was in financial difficulties by 1886, when one sees that it was trying to run meetings, and publish and circulate an international journal, on a five-shilling annual subscription.

Even so, the Manchester members, in spite of the excellent B.A.A. *Journal* and *Circulars*, felt that they were subsidising London, and with the shrewd men and women of 'Cottonopolis' this rankled. By 1902 very few new members were coming forward in the north-west [96]. Then, in September 1903, a meeting was held in 'the Godlee Observatory, Municipal School of Technology', Manchester – later the University of Manchester Institute of Science and Technology (U.M.I.S.T.) – under the Presidency of Professor Thomas Core, with the intention of forming an independent Manchester Astronomical Society. Now separate from B.A.A. branch status, though many individual members stayed in the Association, it could run its own affairs and finances, and it promptly declared a five-shilling annual subscription, and half a crown for students [97].

The new Manchester Astronomical Society became a success, and the surviving 'Minute Book', covering the years 1903 to 1909, gives a sense of how meetings were organised, speakers found, and business transacted. That essential artefact for all astronomical societies, however, the 'lantern' or slide projector, became the subject of considerable discussion and financial outlay in 1906, when two guineas (£2.10) was spent in electrifying the existing 'limelight' instrument. Even then they were having to pay a Mr Hoyle from Bury, Lancashire, 3 shillings (15p) per night for operating it [98]. One enormous asset to the Society, however, was the donation of an observatory, worth £10,000, to the Science College where the Society met in 1903. This came through the munificence of Francis Godlee, a Manchester cotton spinning magnate and philanthropist, who was also a Vice-President of the Society. The Godlee Observatory, with its 8-inch refractor and 12-inch reflector, mounted by Grubb of Dublin to counterweigh each other on a large steel equatorial (in the arrangement which Grubb had pioneered for William Huggins and Isaac Roberts), still forms the home of the Manchester Astronomical Society, almost a century after its creation [99].

By 1903 the North-West branch of the B.A.A. had grown to a size where it demanded its independence. It is surprising, therefore, that at the same time the Association's first branch in the Midlands was lamenting in the B.A.A. *Journal* that 'The Committee regret that they have to report a general lack of astronomical interest in the Midlands, which makes it difficult to carry on the work of the Branch successfully' [100]. Indeed, this was in spite of the efforts of William Henry

Robinson, F.R.A.S., of Walsall, who had served as Vice-President of the Branch, and who in his capacity as author and publisher would continue to promote astronomy in the Midlands for many years to come. In this respect, he had carried on the work of the Revd Alfred Adolphus Cole, also of Walsall [101].

The year 1904 also saw the formation of the Newcastle-upon-Tyne Astronomical Society, with the Revd T.H.E.C. Espin – whose Tow Law vicarage was some miles south-west of the city – as its President. Espin had moved to the north-east from Liverpool to act as his father's curate upon the occasion of the elder Espin 'retiring' to the easier parish of Wolsingham [102]. But in 1888 T.H.E.C. Espin was given his own living at nearby Tow Law, where he was to set up the major private observatory which was to form his scientific base down to his death in 1934. In addition to his presidency of the Newcastle society, however, Espin was to remain a dominant figure in British amateur astronomy down to his death, with memberships of the Liverpool Society, the B.A.A., and the R.A.S. More will be said about him in Chapter 15. Under Espin's presidency, however, the Newcastle-upon-Tyne Astronomical Society displayed a similar social tone to that of the other big amateur societies of the late Victorian age, with its dress codes for social events, meals in fashionable Newcastle restaurants, and lectures from distinguished scientists such as the Astronomer Royal [103].

On the other hand, it is very difficult to know why a science like astronomy took off in Liverpool, Cardiff, Manchester and Newcastle, yet failed to produce a similar sustainable growth in the Midlands. Even more surprising – when one looks at the numerous classes that were on offer at the Walsall Science and Art Institute over the winter of 1889–90 – was the superabundance of instruction in chemistry, magnetism, and electricity, but a lack of anything on astronomy [104]. From an historical point of view, it would be interesting to know what drove some sciences more than others when it came to their popular exposition.

Although the regional dissemination of astronomy across Britain might have been patchy during the 1890s, the astronomical lecture circuit emerged over the same period as a fact of amateur scientific life. While there were still popular lecturers who earned a living delivering 'Grand Popular Astronomical Entertainments' with dioramas and planetaria, such as the lecture given by B.J. Malden in Walsall in 1885 [105], the new lecturers were of a very different breed from the old astronomical educators of the people whom we saw in Chapter 9. The men and women whose names occur in the B.A.A. *Journal* and in the publications of the Welsh, Leeds, Liverpool, and Manchester Societies of the 1890s are active scientific participants, and not mere popular entertainers. Many of them – such as William Robinson, Samuel Okell and Dr Steele Sheldon – were themselves dedicated amateur astronomers. Others were professional scientists who worked in academic observatories, such as Oxford or Greenwich; while the priest–schoolmaster–professionals of Stonyhurst College, Fathers Sidgreaves and Cortie, spanned both categories. And perhaps most remarkable of all within their time were the new lady lecturers, Mrs Annie Scott Dill Russell Maunder, Gertrude Bacon, Florence Taylor and R.A. Proctor's widow Sallie, who both observed and perhaps made some money from their lecturing. More will be said about these women lecturers in Chapter 14.

What all of these people shared in common, however, was a high level of astronomical understanding in their own right, and the knowledge that, unlike the old popular lecturers, they would be addressing informed audiences.

Sir Robert Ball, now the successor of J.C. Adams as Director of the Cambridge Observatory, was clearly the lecturer to get to your society if you could persuade him. He lectured to the Astronomical Society of Wales in 1901, and his well-publicised visit was marked by the appearance of his portrait in their *Cambrian Natural Observer* [106]. It had been Ball who had summoned the Leeds Society out of the doldrums in 1892, though his invitation to Manchester in 1908 lapsed, 'Sir R. Ball's fee being too high for the Socy. to meet' [107]. Second to him came figures like Frank Arthur Bellamy of the Oxford University Observatory, who lectured on Jupiter in Manchester in October 1906, and Herbert Hall Turner, the eminent ex-local lad who was the darling of Leeds [108].

The B.A.A. was often successful in obtaining the services of top-class scientists as lecturers, including at the provincial branches, and in 1903 Eugène Michael Antoniadi not only lectured on his researches to the Midlands Branch in Birmingham, but confided to his audience that 'it was in 1888 that he started observing Mars with a 3-inch O.G. by Steinheil' [109]. In 1901 they had enjoyed the services of the illustrious Sir Oliver Lodge; while along with Manchester the Midlands astronomers in the 1890s had been graced by visits from E.W. Maunder [110].

It is also interesting to note that, even at this early date, there was a demand for information about the history of astronomy, and a small but significant number of historical lectures were offered to the new societies. There was, of course, Agnes Clerke's monumental *A Popular History of Astronomy in the Nineteenth Century* to provide a coherent published foundation to the subject, and while lectures on such relatively distant figures as Jeremiah Horrocks were on offer — especially in the north [111] — it was the profound changes through which the science had passed in their own century which seemed to fascinate most people. Caroline Herschel, as we have seen, was lectured on by Florence Taylor [112], while in October 1897 Professor Thomas H. Core treated his own Manchester Society to his personal perspective on recent astronomical change. Core drew a distinction between the tradition of Newtonian celestial mechanics, such as lay behind the discovery of Neptune in 1846, and the revolution in astrophysics which had begun with Kirchhoff's spectroscopic analyses and the applications of photography [113].

It is clear that by 1900 British astronomy had undergone a social and technical change which was fundamentally different from the practice of the science by the Grand Amateurs. And what had been so significant in the organisation and articulation of that change was the emergence of societies of amateur astronomers, who came together not only in scientific, but also companionable, relationship. For by 1900, the basic features of amateur astronomy as it would continue for at least the next century were firmly in place: regular meetings in a social environment, a mixture of home-grown and guest lecturers, published magazines, and a two-tier system of national and local organisation. While historians of the twenty-first century will not, unfortunately, be able to reconstruct the contents of lectures and the goings-on at present-day astronomical society gatherings from extensive local press coverage, as a modern scholar can from the newspapers of Victorian Leeds, Manchester, and Cardiff, one cannot help but be struck by the way in which the expressive format of amateur astronomy has changed so little since the 1890s.

But the most extraordinary thing which the clubbable passion of the Victorian astronomical society facilitated in Britain was the entrance of women into active and extensive scientific life as observers, writers, and lecturers. For whereas the public profile of women astronomers was virtually non-existent in 1875, they had become an established, and even prolific, part of the astronomical scene by 1905.

14

Now ladies as well as gentlemen

The appearance of women upon the British astronomical scene came with something of a rush after 1880. In 1875 or so, there seemed on the surface to be little more than a handful of female figures who were active in the science, comprising people like the solar observer Elizabeth Brown of Cirencester, and the astronomical writer Agnes Giberne. In some respects, this scarcity may seem all the more surprising, for the nineteenth century had seen women becoming involved in a widening range of intellectual pursuits that went beyond the more traditional ones of literature, music and art. Frances Trollope (the mother of Anthony) made her mark as a topographical writer across Europe and America, Mary Ward was a microscopist of note, while the explorers Mary Kingsley and Gertrude Bell penetrated respectively into Africa and Arabia [1].

Almost all of these women, however, shared one or two significant characteristics. They were either independent women of means, like the astronomer Elizabeth Brown; or else they used their knowledge to earn their own livings, and in the case of Frances Trollope, support a family [2]. Virtually all of them, moreover, came from the better-off classes of society, and had received a decent education as girls, combined with access to books, though one of the few exceptions to that rule was Mary Anning of Lyme Regis. She had started to dig fossils out of the local rocks to sell to visitors in her small shop, but later came to be recognised as a field palaeontologist of great significance [3].

In their respective independent or money-earning careers in science, therefore, one might consider that these women formed parallels with their male counterparts, although there were two very fundamental differences between the sexes. For one thing, working women could not enter into the forms of regular scientific employment which were open to men. They could not become museum curators or paid assistants to Grand Amateurs, and had only three avenues of paid work open to them: writing, school teaching, and after 1880, lecturing. And secondly, the main scientific societies did not admit them to membership, so that women lacked those venues of meeting, exchange, and dining which bodies like the Royal Society and the R.A.S. made available to men – even to certain prominent assistant astronomers, such as J.R. Hind and Edwin Dunkin.

Female scientists, rather like working-class male scientists, suffered from isolation. It was all the more momentous, therefore, that when the big amateur astronomical societies came into being, starting with Liverpool in 1881, women were admitted to full membership. From a mere handful of women astronomers whom one could name and locate in 1875, one suddenly finds the existence of well over a hundred by 1905, largely through the lists of the new amateur societies [4]. This naturally begs the question as to how many women astronomers owned telescopes or read scientific publications before the amateur societies gave them a recorded presence, though one suspects that they were not rare. Indeed, this informed presence can be gauged somewhat by the numbers of women who began to take active roles in the organisation of the new societies, and who wrote articles for their magazines, all of which suggests the existence of a force only waiting to be tapped.

In 1896 and 1897, Miss Florence Taylor, who had been the first lady member of the Leeds Astronomical Society, gave a pair of papers to the Society on the lives and works of those two scientists who clearly inspired so many later generations of women astronomers: Caroline Lucretia Herschel and Mary Somerville [5]. Indeed, these two women, both of whom possessed established reputations and unique honorary memberships of the R.A.S. even before the Victorian age began, became icons of scientific achievement for women, and Mrs Somerville's name was given to Oxford's new women's college in 1879. And very significantly, both of these ladies enjoyed the advantages of great longevity, as ideas encountered and even discoveries made in the eighteenth century were carried deep into the Victorian age by two minds each of which remained sharp for the best part of a century apiece.

Caroline Herschel was born in Hanover, and in 1772 at the age of 22, came over to England to act as her brother's housekeeper in Bath. After starting on a career as a professional musician in her own right in Bath, Caroline began to assist her brother in his 'Surveys' of the heavens, and even learned how to polish speculum-metal mirrors with her own hands. It says much about her gifts, her versatility and her self-perceived role as her brother's assistant in all things, that she too soon became an accomplished practical astronomer, and a person who also enjoyed the painstaking work of astronomical calculation [6].

Between 1786 and 1797, Caroline Herschel discovered no less than eight comets in her own right. She owned, or at least had access to, a variety of telescopes including a small refractor and several reflectors, one of which is preserved in the Science Museum store. Her 'comet sweeper', however, was a reflector of about 2 feet focal length that pivoted upon a vertical board which in turn rotated around a heavy tressle support. The eyepiece, at the Newtonian focus, very conveniently stayed in one position, while a simple pulley and handle arrangement made it possible to raise and lower the tube around the eyepiece [7]. In this way, Caroline could work with a wide-field instrument to 'sweep' the sky and pick up comets. It goes without saying, however, that one could not search for comets which, as they approached the solar system, appeared as mere smudges of light, unless one already had an intimate knowledge of the sky. But possessing a knowledge of nebula distribution second in the world only, perhaps, to that of her brother William, she was ideally qualified to become a highly successful cometographer. She also herself discovered several nebulae.

In the 1790s Caroline undertook the massive task of reducing and indexing the 1725 *Historia Coelestis Britannica* of John Flamsteed, which in spite of its age was still the most comprehensive catalogue of the stars of the northern heavens. She compared the published figures which Flamsteed's assistants had prepared from his original observations (Flamsteed himself had died in 1719) with the surviving observations themselves, then reduced and indexed them. When her results were published in a folio volume by the Royal Society in 1797, it soon became apparent that the *Historia* catalogue was full of mistakes, while Flamsteed had accurately observed some 500 stars that had never appeared in the *Historia* [8]. Flamsteed had even observed the Herschel-discovered planet Uranus in 1690, although he had logged it as a star [9].

But in addition to being an observing astronomer, Caroline Herschel was the chronicler of the work of both herself and her brother. Her diary is a remarkable document, and is the source for much of the anecdotal material that comes down to us concerning the work which brother and sister performed at Bath and Slough. These include the story of Caroline feeding William by hand while he performed a 16-hour polishing marathon on a mirror [10], and the near tragedy that occurred on 31 December 1783. Assisting her brother, who was aloft at work on the new 18-inch aperture, 20-foot focus reflector, Caroline slipped in melting snow, and impaled her right thigh upon one of the large iron hooks – 'such as butchers use for hanging their joints upon' [11] – used for moving the instrument. The hook tore out two ounces of her flesh, and, considering the high risk of gangrenous infection in that unsterile age, she was lucky not to require an amputation. No doubt the cold weather worked in her favour.

Caroline Herschel's diary and private writings clearly display the rather paradoxical nature of her position. On the one hand, she sees herself as her brother's loyal assistant, who sits with pen poised above a semi-frozen inkwell in the middle of the night, as William shouts down the coordinates of new nebulae from the top of the telescope tube. On the other hand, she is a scientist who makes significant original discoveries of her own, and has the intellectual assurance to correct the errors of a former Astronomer Royal, and have her findings laid before the Royal Society. She was a Grand Amateur's assistant, and then co-worker, the daily recorder of the first deep-sky survey ever undertaken in the history of astronomy, and then a cometographer, nebula discoverer, and celestial cataloguer in her own right. And when this achievement was crowned in old age, by the award of the R.A.S. Gold Medal in 1828, and honorary membership of the R.A.S. in 1835, it is hardly surprising that she became an inspiration to later generations of scientific women, to be written about by Agnes Clerke, and lectured upon by Florence Taylor [12].

The other woman who was also honoured by the R.A.S. in 1835 was Mary Somerville. Like Caroline Herschel, she also displayed that curious mixture of intellectual certainty and social caution which derived from being a woman in a man's world. For while one could argue that a man from humble origins might feel a similar social ambivalence in the world of science – men like John Dalton, Michael Faraday, or William Whewell – the openness of the scientific community ensured their complete acceptance once they had showed their intellectual mettle, and each of the above men, along with many others, rose to fame, an F.R.S., and in the case of Whewell, the prestigious Mastership of Trinity College, Cambridge. Yet women were disadvantaged simply because they were women, irrespective of gifts or social background, as the universities and the Royal Societies remained closed to them. Honorary memberships might be granted in conspicuously exceptional cases, but not the full, routine memberships which made someone an accepted *savant* as opposed to a venerated prodigy.

Mary Somerville was born in 1780, the daughter of a Scottish admiral in the Royal Navy, Sir William Fairfax, whose century-long life (1713–1813), and fathering of Mary at the age of 67, goes some way towards explaining the source of her own prodigious energy and longevity. And yet when Ellen Mary Clerke – sister of Agnes Clerke the astronomer – wrote Mary Somerville's 'Life' for the *Dictionary of National Biography* in the 1890s, we obtain an insight into the way in which women intellectuals were perceived in that century, and how one woman perceived the role of another. For when Mary Somerville (or Miss Fairfax) had been a young woman, a century before her 'Life' was written, her beauty had won her 'the soubriquet of the Rose of Jedburgh [which] formed a piquant contrast to her masculine breadth of intellect' [13]. One wonders how far her beauty, personal charm, and good social connections enabled her to gain a hearing in those learned circles which would perhaps have been closed to a woman who lacked them. By contrast, no-one commented upon the role played by Michael Faraday's good looks in rising from a bookbinder's workbench to a Royal Institution Professorship.

Mary was lucky, however, in having an uncle, Dr Thomas Somerville, who encouraged her intellectual pursuits, and then her first cousin and second husband, Dr William Somerville. When her army doctor husband was appointed Physician to the Chelsea Hospital in 1817, and elected F.R.S. soon after, Mary Somerville found herself in a highly congenial environment. With access to the Royal Institution, and via her husband the social world of the Royal Society, she came to know the leading scientists of the age: Thomas Young, William Wollaston, Charles Babbage, the young John Herschel, and his aged father Sir William. She also became acquainted with Caroline Herschel, and Maria Edgeworth, who was herself a popular scientific writer of note; while the elderly Dr Somerville, in acknowledgement of his wife's talents, became her amanuensis and general assistant [14].

But Mary Somerville was never a regular observing astronomer in the way that Caroline Herschel was, in spite of owning a reflecting telescope made by her fellow-Jedburghian, James Veitch. Mathematics and their application to celestial mechanics and physics were her real love, and in 1827 it was suggested to her by England's scientific Lord Chancellor, Henry Brougham, that she might write an English treatise upon Pierre Laplace's *Mécanique Céleste* (1799–1827) [15]. She was warmly supported in this, and in her subsequent endeavours, by Sir John Herschel, and when her *Mechanism of the Heavens* came out in 1831, it was acclaimed not merely as a translation but as an original critical exposition of Laplace's work. The reviewers were amazed by the female mathematician's insight and powers of exposition, and William Whewell gave it the ultimate accolade by recommending it as a textbook on the complex subject of celestial mechanics to Cambridge science undergraduates [16]. It was followed by her *Connexion of the Physical Sciences* (1834), and *Physical Geography* (1848), in which she examined the mathematical connectedness of science. It was the *Connexion* which led to her honorary elections to the Royal Irish Academy (1834) and the R.A.S. (1835), and a Civil List Pension, in recognition of her achievements. The Royal Society could not grant her a Fellowship, but paid her the compliment of commissioning a life-size portrait bust by Sir Francis Leggatt Chantrey, to stand amongst their worthies [17]. At the time of writing (in 1997) it stands by the door of the Royal Society Library.

The illustrious standing which Mary Somerville had come to enjoy by her fifties meant that she could correspond with scientists across Europe and America on equal terms, although her retirement to the milder climate of Italy with her ageing and long-lived husband brought its drawbacks. For as she lamented in a letter to Lord Rosse, sent from Rome in 1843, one felt so cut off from books, conversation, and news of those rapid advances in science going on in London and Paris when living in Italy [18]. Even so, she published her last book, dealing with the age's latest work in chemistry and biology, at the age of 89, after which she resolved to leave the life sciences alone and return to her beloved mathematics. She was still active, and keeping herself abreast of the new sciences of spectroscopy and solar physics, right up to her death at the age of 92.

Yet in many ways Mary Somerville was what Dr Mary Brück has so aptly styled a 'mathematician and astronomer of underused talents' [19]. Her published work lay not in original scientific research, but in a lucid exposition and brilliant interpretation of the work of others; and many people have wondered, along with R.A. Proctor, what these formidable talents might have achieved had they been devoted to pure research in mathematical astronomy [20]. But was she 'underused' simply because some university could not offer her a chair as might have been the case had she been a man? For Mary Somerville fitted perfectly into the cutting-edge culture of British science in the early nineteenth century, as a Grand Amateur moving in a world of Grand Amateurs. Being a lady of independent means (especially after receiving her £300 a year Civil List Pension), she moved with ease amongst those to whom science was a way of life and an intellectual passion, rather than a job – among people like Sir James South (who also had a £300 per annum Civil List Pension), Charles Babbage (who would have liked one), Sir John Herschel, Lord Rosse, and Ada, the mathematical Countess of Lovelace [21]. Indeed, far from being a curious female appendage to the astronomical establishment, Mary Somerville was able to become part of its cultural mainstream in a way that would have been much more difficult in a country like Germany or Russia, where a Ph.D. and a university post were essential prerequisites to being taken seriously in the world of science. It could also be argued that her work as a high-level scientific expositor made her influence wider than it would have been had she confined herself to pure science. Her books were widely read, especially the *Connexion of the Physical Sciences*, of which she had just completed the revision of the tenth edition when she died in 1872. And it is hard to believe that those later nineteenth-century women amateur astronomers, such as Florence Taylor, would have read and been so influenced by her writings had they been confined to pure research.

As for women in professional astronomy, however, it was Maria Mitchell of Nantucket, U.S.A., who took the lead. Being the daughter of an amateur astronomer, the sister of a leading hydrographer, and the product of a region of America where navigation was in the blood, she became skilled in both observational and computational astronomy. Her discovery of a comet in 1847 brought her fame, while her medal from the King of Denmark in 1849 when she was still only 31 years old constituted the first astronomical honour ever granted by a foreign state to an American [22]. She was later employed as a Calculator in the American Nautical Almanac Office, and in 1865 became Professor of Astronomy at Vassar College, a newly-founded women's college in New York State. At Vassar she had charge of a 12½-inch Fitz refractor, which the Alvan Clark firm re-worked and mounted, while she gave her own 5-inch Clark to be used by the students. The Vassar College Observatory began to produce a succession of women academic astronomers, most notably Maria Mitchell's successor, Mary Watson Whitney, and Antonia Maury, whose careers, like those of most late nineteenth-century American and European astronomers, were primarily in the new research field of astrophysics [23].

But as was shown in Chapter 3, in spite of its strong amateur tradition American astronomy differed in many key respects from that of Britain: most notably in the development of Liberal Arts Colleges like Vassar, and in its abundance of wealthy benefactors. British city colleges, like Owen's, Manchester, often tended to be more inclined towards vocational or industrially-related pursuits, rather than combining science with classics to produce a broadly well-educated lady or gentleman, as was the case in an American Liberal Arts College. It is true that Oxford and Cambridge Universities, with their new women's colleges after 1868, catered for the more traditional classical education, though this was at the expense of an almost total exclusion of science from the undergraduate curriculum, combined with a very limited social access [24]. Things were more open in America, where it was possible by 1870 for a young man or woman to obtain an excellent rounded education, including classics and science in the same four-year course, which could lead on to postgraduate studies, and perhaps a job in astronomy.

English industrial money tended generally but not exclusively to be spent directly upon its possessor's own passion, rather than on the endowment of a college or public institution. The New York brewer, Matthew Vassar, spent his money on founding a women's college and observatory, in a similar way to which William Horatio Crawford spent his Irish brewery profits on founding an institution and observatory at Cork [25]; whereas the alcohol profits of William Lassell, Richard Carrington, George Bishop and others went to pay for their own private researches. Some of this fermented money, of course, could be of direct benefit to astronomical women, as was the case with Elizabeth Brown, whose astronomical career derived from the wealth of her wine-merchant father [26]; but the results were much more fortuitous than when one spent the money to found a college.

These were among the reasons why Britain still failed to produce those institutions of higher education where a woman could obtain a first-class scientific training in a way that had become possible in America by 1875. With the waning of the Grand Amateur tradition by this time, moreover, and the gradual disappearance of that world in which Caroline Herschel and Mary Somerville had moved so easily, one could argue that there were fewer avenues through which British women could find serious astronomical expression. It was all the more important, therefore, that when that wave of amateur astronomical society foundings took place after 1880, the new societies went out of their way to enrol women members, to give British scientific women that opportunity which they otherwise lacked.

These societies soon had within their ranks women displaying a diverse range of astronomical interests, including the observers, writers and lecturers mentioned above. The early Liverpool Astronomical Society after 1881 enrolled at least eight women within its ranks, some of whom were, and still remain, little more than names, whereas others would go on to become leaders of the amateur astronomical movement, and make serious contributions to the science [27]. Yet we must not forget that those women who did remain 'little more than names' in the half-dozen major amateur astronomical societies and their regional branches that came into being between 1881 and 1904 were not necessarily without significance in the wider educational context. Some of them, such as Miss E. Graham Hagerty of Cardiff – an Associate of the Royal College of Science (now Imperial College, London) – were schoolmistresses, and must have been active in passing on their astronomical knowledge and interest to their pupils [28]. Others were wives or daughters of members, such as Mrs Thorp and Mrs Sheldon of Manchester, or Miss C.A. Barbour of Leeds, whose names appear in the respective society journals. It tells us a great deal about attitudes towards the wider role of women in education and in society at large when wives and daughters not only joined societies but sat on their councils and delivered lectures, as did the Thorp, Sheldon and Barbour women, along with many others elsewhere.

Virtually complete membership lists survive for the six big late Victorian and Edwardian astronomical societies, and one is struck by the size of the female membership by 1905. The Leeds Astronomical Society, for instance, averaged a 10–20 per cent women membership in the first decade of the twentieth century, and one notes that when overall membership of the Society began to decline, the female membership still increased as a percentage. In 1900, for instance, the Leeds Society had a total of 98 members, eight of whom were women, although of that eight one lived in Wales, and another in Minnesota, U.S.A. By 1910, however, there were only 74 members in the Society, but 15 (20.27 per cent) of them were now women, five of whom sat on the 17-member Council. When the Leeds astronomers got going again after the horrors of World War I, there were still 15 (but now 27.27 per cent) women, including their Minnesota stalwart, Mrs Florence Hildred (the former Miss Florence Taylor), who had retained her membership since her marriage and her emigration to Wilmont, Nobles County, Minnesota, U.S.A., in 1898 [29].

One might almost suspect, moreover, that the Cardiff-based Astronomical Society of Wales, founded in 1894, predicated a female membership by the incorporation of a beautiful woman looking through a telescope – a modern personification of the goddess Urania, perhaps – into the emblem of the Society. Though I have not been able to locate a membership list for the Welsh Society earlier than that printed in their journal the *Cambrian Natural Observer* for January 1901, it is clear that women were prominent in it. Within a total subscribing and honorary membership of 137 persons there were 23 (16.78 per cent) women in the Astronomical Society of Wales in 1901. By 1908, there were 252 members, 41 (16.26 per cent) of whom were women, though by 1910 the total membership had fallen to 235, some 27 (11.48 per cent) of whom were women [30]. And in May 1893 (as we saw in Chapter 13) the newly-formed Ulster Astronomical Society included ten women within its total membership of 119 [31].

Upon its foundation in 1890, the British Astronomical Association aimed specifically to accept women on equal terms with men, which is no less than what one would expect when one of its principal instigators was Elizabeth Brown. As we saw in the previous chapter, women were active in the B.A.A. right from the start, not only in matters of organisation, but in observational work, and as travellers on the Association's eclipse expeditions. And that tiny handful of British women who earned a meagre 'short term contract' living as professional astronomers in the early 1890s – the 'Lady Computers' on the staff of the Royal Observatory, Greenwich – also joined. Alice Everett and Annie Scott Dill Russell (who would later become Mrs Maunder), armed with their Cambridge science degrees, both became members of the Association because the R.A.S. would still not accept women Fellows in 1892 [32]. The two Cambridge graduates were soon elected onto the Council of the B.A.A., so that including its Original Council Member, Elizabeth Brown, the Association quickly acquired three women, and then more, on its governing body.

It has recently been pointed out by Peggy Kidwell, however, that in spite of their admission on equal terms with men into the B.A.A., women still constituted only 8.5 per cent of the membership over the period 1901–1930 [33]. Yet instead of seeing this figure as indicating an implicit resistance to women within male-dominated British astronomy, it can also be interpreted as an astonishing progress within a subject where their presence had been negligible only twenty years earlier. And as we saw above, the female membership of local societies such as Leeds, or regional ones such as Wales, could often be almost double the B.A.A. figure.

Both Caroline Herschel and Mary Somerville were integral to the Grand Amateur world, and while Elizabeth Brown was a lady of ample means and independence, her entry upon the serious astronomical scene came at a time when that tradition was waning in its overall significance within British astronomy. In many ways, therefore, she stands on the cusp between the tradition of the Grand Amateurs and that of the leisured enthusiasts. This status is perhaps confirmed by the fact that by the early 1880s – when she came to prominence as a visual solar observer – the Royal Observatory and other professional institutions world-wide were already making daily photographic records of the Sun. This does not in any way minimise her importance within the amateur tradition, but it does, perhaps, bring out more clearly the distinction that was coming to emerge between those who observed the heavens for inspiration and delight, and those who were engaged in the pursuit of fundamental discovery.

It has already been mentioned that Elizabeth Brown was the daughter of a Cirencester wine merchant. Her father, Thomas Crowther Brown, was a Quaker and a local philanthropist with strong scientific interests. The meteorological recorder for Cirencester, in that national web of amateur meteorologists, he encouraged his daughters Elizabeth and Jemima to share his interests [34]. We do not know how far back Elizabeth's scientific interest extended, but in 1871, at the age of 41, she took over her aged father's work, continuing it over the remaining twelve years of his life, and then on to the end of her own life in 1899. It is not clear when she first became seriously interested in astronomy, but in 1883, the year in which her father died, she read a paper on sunspots to the new Liverpool Astronomical Society, and was invited to direct its Solar Section, which presupposed an established reputation in visual astronomy [35].

As Mary Creese has indicated in her recent study, Elizabeth Brown was 'probably the only woman astronomer of her time to have her own observatory' [36]. It contained a clock-driven equatorial refractor of 3½ inches aperture, a grating spectroscope, and an astronomical clock. Though the equipment might seem modest, it was very appropriate for her chosen field of solar study. Her observations were made with various research criteria in mind. For one thing, she was interested in the relationship between sunspots and faculae, and made numerous painstaking drawings which aimed to correlate the two. She was also concerned with the appearance and movement of the spots upon the solar surface with reference to the Sun's polar and equatorial coordinates, their proper motions, and the exploration of other phenomena first noted by R.C. Carrington and others. This kind of planetary-surface work, indeed, was the branch of astronomy to which amateurs were, and still are, ideally suited: people with free time to accumulate extensive runs of visual data from which a wide variety of conclusions can be drawn. For while professional astronomers using sophisticated cameras might have routinely gathered data of this sort by 1880, they rarely had the time to ponder over it in the way that an enthusiastic amateur had [37].

Elizabeth Brown was keen to encourage other women, 'many of whom have ample time at their disposal, and who are often skillful in the use of the pencil' to take up astronomy [38]. And in particular she recommended solar studies to them, because 'No exposure to night air is involved' [39], in spite of T.W. Webb's firm assurance that night air was no more deleterious than day-time air [40]. Yet Elizabeth Brown was clearly not afraid of night air, for she also did work on coloured and other stars, though one wonders how far the damp cold of Gloucestershire winter nights contributed to the bronchitic complications which killed her on 5 March 1899 [41].

Elizabeth Brown was the first woman to follow in the footsteps of those male astronomers from the seventeenth century onwards who had travelled to distant places to make astronomical observations. In the summer of 1887, accompanied by an unnamed lady companion, she travelled to Russia with Ralph Copeland of Dun Echt and Father Stephen Perry of the Stonyhurst Observatory. They travelled by ship and by train to the elegant summer residence of the Russian astronomer Professor Theodor Alexandrovich Bredichin, to observe a total eclipse of the Sun. Though cloud made the observations of indeterminate value, the expedition clearly indicates the seriousness with which Elizabeth Brown – not to mention her unnamed companion, who was drawn in to assist Father Perry with his spectroscopes – was taken by solar observers of international standing [42].

Two years later, she went on another expedition to observe an eclipse at Trinidad, and obtained excellent views of the corona and prominences. Unfortunately, however, Father Perry, who was again one of the party, died on the voyage home [43]. After the B.A.A. had come into being, she travelled in 1896 to Vadsö in northern Norway as part of a group of 58 Association members to observe the total eclipse of the Sun, and was already making preparations to travel to Portugal to observe the 1900 eclipse when she died suddenly in March 1899 [44]. Being a lady of ample independent means – at a time when not only was world travel becoming faster and more comfortable, but women were finding their scientific voice across Europe and America – she visited scientific institutions in Canada and America in 1884, and was received at the Spanish Royal Observatory in Madrid in 1893. In 1893 she was also honoured for her meteorological work, when she became one of the first women to be elected to the Royal Meteorological Society [45].

Upon her death, Elizabeth Brown became an important benefactress of the B.A.A., leaving her observatory, her instruments and £1,000 to the Association. The magnitude of her independence, moreover, can be gauged from the fact that when her unmarried sister and principal beneficiary, Jemima, died in 1907, the Brown estate was valued at £63,182 [46].

Solar astronomy was also the particular speciality of Annie Scott Dill Russell, who after graduating Senior Optime (the highest mathematical honour then available to women) in the Cambridge Tripos Examinations of 1889, became – along with her sister Girtonian, Alice Everett – one of the first women to be employed at Greenwich. Annie worked in the solar photography department, and when her marriage to Edward Walter Maunder in 1895 required her retirement from the Observatory, she devoted the next half-century of her life to high-level amateur astronomy, and was a formative influence upon the early B.A.A. Along with Elizabeth Brown, and others, she had travelled to Norway in 1896 – where the obscured sun was 'hissed' by the party [47] – and in 1898 to India.

It was in India that Annie Scott Dill Russell Maunder obtained superb photographs of the solar corona. Using highly sensitive plates, and experimenting with her own photographic techniques, she obtained images of four coronal streamers, one of which was the largest ever recorded, extending for 13.9 solar radii into space [48]. Annie Maunder's photographic work, based as it was upon a professional scientific training and close connections with the Royal Observatory via her husband, was to be of the highest importance in solar studies. It also showed what determined amateurs could really do.

The B.A.A. eclipse expeditions of 1896 and 1898 brought together three young women astronomers. At Talni, some 400 miles from Bombay, Annie and Walter Maunder set up their photographic equipment alongside the observing station of John Evershed [49]. On the 1896 Norwegian expedition, Evershed had met Miss Mary Ackworth Orr, whom he married in 1906; and on this same 1896 expedition Miss Orr met Mrs Maunder, and the two young women formed a friendship that was to last for life, though Mary Orr did not travel to India in 1898. (Elizabeth Brown was by then 68 years old, and also did not travel to India.) In 1898, Annie and Walter Maunder were concerned with recording the solar corona, and John Evershed concentrated on photographing the flash spectrum (indeed, he discovered new features of the hydrogen lines in the ultraviolet). 500 miles away, at Buxar, near Benares, the Revd John Mackenzie Bacon, together with his son Fred and daughter Gertrude, observed the corona, and obtained what was probably the first cinematic, or 'animatographic', record of a solar eclipse [50]. Exactly when Miss Gertrude Bacon first became fascinated with astronomy, photography, cinematography, and ballooning is not clear, but she probably picked them up from her reverend father; in later years she was to lecture on them.

Annie Maunder occupied an important place in the B.A.A. for 56 years, not only as visual astronomer and solar expert who successfully photographed the eclipses of 1900 and 1905 from Mauritius and Canada respectively, but as an editor and an acknowledged authority on astronomical chronology. The B.A.A. eclipse expeditions also included other ladies, not all of whom went to the same observing stations, but whose travels provide us with a glimpse of what comfortably-off women astronomers were doing by 1900. For the 1905 eclipse, indeed, Miss Hart-Davis and Miss Catherine Octavia Stevens went to Burgos in Spain, along with the B.A.A. expedition led by Andrew Claude de la Cherois Crommelin, of the Royal Observatory, Greenwich, and his wife Loetitia.

Indeed, it is interesting to see the emergence of a dedicated band of women eclipse observers in the early B.A.A. who were willing to travel considerable distances in pursuit of their work. Over the first two decades of the twentieth century, Loetitia Crommelin observed from Algiers, Palma (Majorca), and St. Germain-en-Laye [51]. Similarly, Catherine Stevens travelled to Algiers, Majorca, and Quebec, in addition to spending a year in the Shetland Islands to see if she could establish a correlation between auroral phenomena and the weather. Like Elizabeth Brown before her, Miss Stevens owned a private astronomical and a meteorological observatory, and was also a serious communicator of her ideas and researches [52]. Mary Proctor – the daughter of Richard A. Proctor – travelled to observe no less than six eclipses worldwide between 1896 and 1936, and at the age of 65 made the daring innovation of following the zone of totality of the 1927 eclipse from an aeroplane flying across Yorkshire. But more will be said of Mary Proctor when discussing women astronomical writers and lecturers.

In the same way that the calling-up of men into the armed forces opened up many jobs to women during the 1914–1918 War, so the presence of the Revd Martin Davidson at the front as military chaplain left open the B.A.A. Meteor Section. It is true that Miss Stevens had been Director between 1905 and 1911, but two women took over very effectively the running of the Section during the War. Alice Grace Cook was an early pioneer of meteor studies in Britain, as was Mrs Fiametta Wilson [53]. Both women came from well-to-do Suffolk backgrounds, living in Stowmarket and Lowestoft respectively, and as had been the case with Elizabeth Brown, Fiametta Wilson's medical and amateur scientific father encouraged her passion for astronomy and meteorology. Fiametta Wilson in particular seems to have been one of those irrepressible gentlewomen who were so abundant in the early astronomical societies. In her reluctance to leave a clear sky, she once came perilously close to being washed into the sea off a Cornish jetty during a storm, and during the First World War not even Zeppelin raids were enough to make her seek shelter if the sky happened to be good. Shrapnel flew past Mrs Wilson as she, quite unconcerned, stayed on meteor and comet watch with her Zeiss field glasses, and on one occasion she was even suspected of being a spy, for in addition to being able to speak German, her use of a flashlamp to record her meteor observations was mistaken for secret signals intended for German airships. Fiametta Wilson's sudden death at the age of 56, in 1920, prevented her from taking up the Pickering Fellowship for Women offered by Harvard University, although the Fellowship for 1920–21 was accepted by her friend and colleague, Miss Cook [54].

In addition to those interested in the Sun, eclipses, and meteors, the early B.A.A. attracted women lunar observers. Although she never became a member of the Association, the B.A.A. Library has seven volumes of lunar observations made by Miss Mary Ashley between 1878 and 1885 [55]. This Gloucestershire clergyman's daughter, like so many other women astronomers of her time, enjoyed the independence of means to purchase two fine Wray refractors and use them to accomplish a series of detailed selenographical studies. She was especially interested in the rills and clefts of the Hyginus region of the Moon, and being a member of the Liverpool Astronomical Society, she published work in its *Journal*, and in the *Selenographical Journal* and *English Mechanic*; while she possessed sufficient stature to be listed in A. Rebière's *Les Femmes dans la Science* in 1897 [56]. And Mary Adela Blagg, a solicitor's daughter born at Cheadle, Staffordshire, in 1858, was especially concerned with the confusion which existed in the late nineteenth century about the naming of specific features on the lunar surface. Her work in mathematics and in the rationalisation of lunar taxonomy was of great importance, and acknowledged by Professor Herbert Hall Turner [57].

In addition to the above-mentioned women who joined or were otherwise connected with the early B.A.A., one must remember that the Association also enrolled the most distinguished female astrophysicist of the day: Lady Margaret Lindsay Huggins, wife and co-worker of William. For as was discussed in Chapter 7, the Hugginses became a research partnership after their marriage in 1875. Indeed, the 27-year-old Margaret had come to the marriage with an established love of science behind her, combined with a long-standing intellectual admiration for the 51-year-old man who became her husband. Their marriage was to become what was later described as 'one of the most successful husband-and-wife partnerships in the whole of astronomy' [58], and for over fifteen years Margaret was to work in the observatory, where she pioneered the use of the new dry gelatine over against the slower and less sensitive wet collodian plates in the recording of stellar spectra. And after about 1890, when astronomers in other parts of the world who were working under better skies than those above the Hugginses' south London observatory began to advance spectral photography, Margaret and her husband started to concentrate on laboratory spectroscopy [59].

It is further interesting to note how many late-nineteenth-century astronomical women were of Irish descent or connection, as that extraordinary scientific energy which had given rise to Lord Rosse, Romney Robinson, and many other male astronomers in Ireland, also produced several women. There had been Mary Ward, who was mentioned at the beginning of the present chapter. A cousin of Lord Rosse, she had combined the duties of an upper-class wife and mother with a love of both astronomical and microscopical observation. In addition to being the first woman to write a book on the use of the microscope, she was also a keen astronomical observer, who in the 1860s wrote *The Telescope: a familiar sketch*, which was intended to show what useful work could be performed with a 2-inch Dollond refractor. Her career came to a tragic yet curiously technological end when she fell under the wheels of the Fourth Earl of Rosse's experimental steam car in 1869 [60].

And while Alice Everett – who worked at Greenwich, joined the B.A.A., and then worked and travelled in America and Germany – was Glasgow-born, she was brought up in Belfast from the age of 2. The daughter of a Queen's University, Belfast, Professor, she was among the first generation of women to receive a first-class academic education, before becoming an undergraduate at Girton College, Cambridge, in 1886 [61]. Annie Maunder was also Irish, being the daughter of a Belfast minister, while Lady Margaret Huggins was the daughter of a wealthy Dublin solicitor John Murray, and the grand-daughter of a successful banker [62].

Agnes Mary Clerke was not really an observational astronomer, and while she had received some training in practical observatory work from Sir David Gill at the Cape Observatory in 1888, and had refused the offer of a 'Lady Computership' at Greenwich, her great contribution lay in elucidating the historical development of astronomy in the nineteenth century. And like so many other late Victorian astronomical women, she was Irish: the daughter of a Skibereen, County Cork, bank manager. As with all of the other Irish women astronomers, however, with the exception of Mary Ward, her scientific career lay outside Ireland [63].

Although she was born in 1842 and hence a generation too old to benefit from the new university colleges for women, Agnes Clerke had nonetheless received a sound education. Her father was a graduate of Trinity College, Dublin, and a keen amateur astronomer, whose small transit observatory also provided the local time service. On her mother's side, she was descended from the influential Deasy family, who amongst other assets owned a successful brewery at Clonakilty. Agnes and her sister Ellen were educated at home by their scholarly parents, and received a thorough training in mathematics, astronomy, classics, and modern languages, before travelling to Italy, largely on account of Agnes' frail health. Both Agnes and Ellen Clerke became famous as writers: Ellen on Italian history and politics, and Agnes on the history of astronomy. Returning from Italy to Britain in 1877, the Clerke family settled in London, where their brother practised as a barrister [64].

It was the dramatic change through which astronomy had passed in the nineteenth century – from Sir William Herschel's work on the nebulae down to spectroscopy and astrophysics – which seized Agnes' attention, and in 1885 she published her monumental *A Popular History of Astronomy during the Nineteenth Century*. In this hefty and detailed volume, Agnes built upon her earlier published pieces dealing with the rise of astrophysics to produce a work which is still a primary source for the history of the field. Though a naturally shy person, she had come to be acquainted with Sir Joseph Norman Lockyer, Sir David Gill, and the Hugginses, not to mention many other astronomers world-wide, so that this range of contacts, and fluency in all of the main European languages, found her ideally placed to write the history of a science which was still being made. One must also bear in mind that Agnes Clerke used the word 'Popular' in the title of her great work in the same context that Sir John Herschel had used the word 'Outline' in his: not as a thing which, as Lady Huggins later put it, was intended to 'tickle' the ears of dilettantes, but to give the intelligent 'general public ... an indefeasible right of access to [the] lofty halls' of modern science [65].

No other single figure has a better case when it comes to being considered as the founder of the history of astronomy as a serious, scholarly study than does Agnes Clerke. Not only did she have a thorough grasp of the science itself, but very importantly, she possessed the historian's sense of context, and respect for what had gone before, combined with the ability to evaluate complex technical documentary sources. She also practised in that most intellectually dangerous and fluid branch of history: the history of her own time. Her subsequent books, such as *The System of the Stars* (1890) and *Problems in Astrophysics* (1903) were state-of-the-art productions that earned a respect that went across the scientific community, for in the sense of the Greek root-word *historia*, they provided technical yet non-mathematical accounts of where the science stood at the time of writing.

Agnes Clerke was a professional scholar and scientific writer, who used her erudition to make a living by elucidating contemporary astrophysics within its historical context. Her amateur status within astronomy, therefore, derives more from her sex than from anything else. For had she been a man, or had she perhaps been an American, she could well have held some professional appointment in an academical institution. Of all the other figures discussed in this chapter, and perhaps in this book, therefore, she was really a 'stateless' person insofar as she was neither a Grand Amateur, an assistant, a spouse, nor a lady of ample means, but a woman who stood outside the professional scientific community yet who made a living by studying its history and writing about it.

Mary Ward and Agnes Clerke both made their impressions – very different as they might have been in terms of enduring historical impact – as astronomical writers. It was, indeed, to be an avenue whereby several other Victorian women entered into the astronomical world. As early as 1825, for instance, a box of printed, hand-coloured astronomical cards entitled *Urania's Mirror*, and intended to teach elementary astronomy, had been issued 'On a plan perfectly original, designed by a lady', later specified in the text as being 'young'. But when Peter Hingley began to research *Urania's Mirror*, he found that the real author was indeed a man, for when the Revd Dr Richard Rouse Bloxam was proposed Fellow of the R.A.S. in 1830, his election certificate clearly described him as the 'Author of Urania's Mirror' [66]. One wonders, therefore, how far, especially at this early date, a feminine authorship was intended to imply either a simplicity of contents or a hoped-for female readership.

Agnes Giberne joined the B.A.A. in 1890, and already had an established reputation as an astronomical writer behind her. By 1898 her *Sun, Moon and Stars: Astronomy for Beginners* (1879) had already sold over 24,000 copies [67], and it is interesting to see how Miss Giberne's concept of a beginner was very different from that of Webb, Smyth or John Herschel. Her style is clear and sometimes inclined to be sentimental, but she succeeded very well in imparting a great deal of basic astronomy, accompanied by several part-coloured illustrations, in a five-shilling (25p) volume. Agnes Giberne also knew the elderly Revd Charles Pritchard, Savilian Professor of Astronomy at Oxford, who both advised upon and proof-read her *Sun, Moon and Stars*, and with whom she corresponded [68].

It is not clear where Agnes Giberne acquired the astronomical knowledge from which she wrote *Radiant Suns* (1895), *The Oceans of the Air* (1890) and *The World's Foundations* (1881), but it could well have been stimulated by her unnamed father, who was 85 in 1898, and was the dedicatee of *Sun, Moon and Stars*. A strongly religious spirit runs through the book, and we must not forget that the religious motivation which inspired so many male amateurs (and Pritchard himself had been an amateur before becoming an ordained Oxford professor), was also found in the women. Mary Ashley and Catherine Stevens, discussed above, were the daughters of Anglican clergymen, while Agnes Clerke and her sister Ellen were devout Roman Catholics. Gertrude Longbottom – a Girton graduate and lunar dynamicist who in 1916 was among the first women Fellows to be admitted to the R.A.S. – was an active Methodist; and many of these women saw no incongruity in speaking of Providence in an astronomical context [69].

It was mentioned in Chapter 10 that for some reason several British towns produced a peculiar abundance of astronomers in the nineteenth century; and Hester Periam Hawkins came from the same municipality in which Admiral Smyth, Samuel Whitbread, Thomas Maclear, Thomas G.E. Elger and several others had lived and worked: Bedford. Her father had been a prosperous apothecary who had given her a good education, and in 1868 the 22-year-old Hester married the Revd Joshua Hawkins [70]. Joshua Hawkins, in fact, had come to minister at the Wesleyan Methodist chapel of which Hester's father was a leading dignitary, and clearly had astronomical interests, for he joined the B.A.A. in its foundation year of 1890 [71].

Hester Hawkins' astronomy derived from that particular perception of divine Providentialism mentioned above, combined with a commitment to popular education, social reform and good works. It is not clear how far she was engaged in practical observation, or how she learned her astronomy, but her writings on astronomy proper seem to have begun during the Edwardian period when she was in her fifties, and in the wake of an earlier career as a writer of hymns and religious tracts. Her star annual *Almanacs* and *Calendars*, however, indicate a sound familiarity with basic constellation astronomy, although none of her work suggests the regular use of telescopes. On the other hand, her *ABC Guide to Astronomy* (1912) was a good alphabetically-arranged *vade mecum* of astronomical knowledge that was not afraid of coming to grips with technical points, and it was very well received by the popular papers and by the B.A.A. Her subsequent tract *Halley's Comet* (1910) was a largely historical work which presented a good account of cometary astronomy from antiquity onwards, and even corrected the bad Latin of 'Isti Mirant [ut] Stella' which designates Halley's Comet in the Bayeux Tapestry [72].

Hester Hawkins' most significant work, however, was the astronomical novel *Stella Maitland: or Love and the Stars* (1921). This book was significant because she took up the theme first treated by Thomas Hardy in *Two on a Tower* in 1882: namely, the exploration of a romantic relationship as seen against an astronomical backdrop. But where she differed most fundamentally from Hardy was in her reaction against his sense of the hopelessness of the human condition when seen against the vastness of the cosmos. Unlike Hardy's Swithin and Viviette, Hester Hawkins' characters Stella and John are both committed Christians, and there is no emotional conflict or cosmological doubt to add poignancy. As a work of literature it is distinctly dull, partly because of a lack of intellectual drama combined with its rather wooden style and self-consciously didactic tone [73]. And in comparing Hester Hawkins and Thomas Hardy, indeed, one cannot escape the conclusion that the Devil had the best tune.

Even so, as an astronomical educator Hester Hawkins was not without significance, and in 1921, at the age of 74, she joined that first group of women who had been elected into the R.A.S. Her proposer was the leading clerical astronomer of the day, the Secretary and future President of the R.A.S., the Revd Theodore Evelyn Reece Phillips, while one of her seconders was Miss Mary Proctor (the daughter of Richard A. Proctor), who had been elected to the Fellowship in 1916 [74].

By 1900, British women astronomers were well in evidence as observers, writers, and organisers, and by that date a small number of women astronomy lecturers were also emerging within the B.A.A. and provincial societies lecture circuit. Their relative scarcity on the lecture platform at this date, however, may have had more to do with the lack of voice amplification equipment at the beginning of the twentieth century than with anything else, for while women were long since established on the theatrical stage by 1900, an untrained voice could be very hard to hear in even a modest hall. Many of the professional men who were so prominent in the early societies, by contrast, especially clergymen, lawyers and retired army officers, would have been accustomed to public speaking. Yet the problem of inaudibility is still with us today, as speakers of both sexes now assume that modern microphones can unscramble their incoherent mumblings and project lucid speech to the back of a large hall.

Annie Maunder turned down the Presidency of the B.A.A. on the grounds that her voice was not strong enough to carry over a large assembly [75]. On the other hand, she had no difficulty in addressing an audience assembled in the Manchester City Art Gallery in May 1898, when lecturing on her work and that of her friend Miss Gertrude Bacon in photographing the solar corona and spectra during the recent Indian eclipse. Indeed, it is clear that Mrs Maunder knew how to keep the attention of a general audience, for in addition to all the scientific detail, according to the *Manchester Guardian*, she 'put some local touches of a humorous character into her story, and these made an instructive and enjoyable lecture the more entertaining' [76].

Yet one area in which research cries out to be undertaken, but which is beyond the scope of the present study, is the degree to which those women who were active in amateur astronomy, especially in its public dimension, were also active in other areas of contemporary concern to women. It would be interesting to know how many were suffragettes, involved in cycling or motoring – which first gave personal mobility to women in the 1890s – temperance, or the 'Rational Dress' movement. Indeed, this last movement could well have had a strong appeal to women lecturers in particular, in its campaign to get rid of corsets and other constricting garments. It was, after all, during the very years that the amateur astronomical societies were coming into being that the suffragette movement also rose to pre-eminence in national life. And the public speaking skills which a woman developed to address rowdy political assemblies could be easily transferred to the delivery of a scientific lecture. We certainly know that Hester Periam Hawkins was also involved with the temperance movement [77], as drink was seen as one of the main causes of domestic violence; and it is likely that other women astronomers, especially those whom we know to have been Methodists, had similar sympathies. But one suspects that Elizabeth Brown, whose financial independence came from the family wine business, was not a teetotaller.

One woman amateur astronomer and lecturer who quite probably had suffragette associations was Miss Florence Taylor of Leeds. Before becoming Mrs Hildred and emigrating to America in 1898, she delivered those two lectures to the Leeds Astronomical Society on Caroline Herschel and Mary Somerville, which were mentioned above. At the conclusion of the version of her 1897 lecture on Mary Somerville that was published in the *Journal* of the Leeds Society, there was a capitalised peroration on 'THE EMANCIPATION OF WOMEN' [78]. It says a great deal about the genuine intentions of the Leeds Society, which in August 1895 had minuted the need to welcome lady members, when the President, Charles Whitmell, responded to Miss Taylor by saying that Agnes Clerke, Elizabeth Brown, Annie Maunder, Alice Everett and other astronomers were indeed advancing the place of women in society [79].

By the Edwardian period, three women had emerged as lecturers of considerable popularity among the amateur astronomical societies. Sallie Duffield Proctor–Smyth had been the second wife of R.A. Proctor, and step-mother of Mary Proctor. A native of Missouri, U.S.A., she had married Proctor in 1881 at the age of 25, when both were mourning the loss of their first spouses. The Proctors settled in Kew, London, but in 1884 they re-crossed the Atlantic to reside in a rather lonely and insalubrious part of Florida. It was when Richard was in New York *en route* for England, with plans for the family's return to Britain in 1888, that he died from a sudden malarial complication. Although Sallie no doubt enjoyed the royalties from Proctor's best-selling astronomical books, and further obtained a British Civil List pension of £100 a year, she found herself widowed for a third time by the age of 49, when Dr J.C. Smyth died at their home in Altringham, Cheshire [80].

In 1892, however, the newly-remarried Sallie Proctor-Smyth was clearly still active on the commercial lecture circuit, upon which she had entered to fulfil R.A. Proctor's numerous lecture engagements after his sudden death in 1888 [81]. For on 13 October 1892, when she delivered her largely astrophysical lecture to the North-West branch of the B.A.A. in Manchester, the Minute Book recorded that she had been obtained through 'The Lecture Agency', off the Strand, London [82]. Over the years, however, she came to develop a particular relationship with Manchester – no doubt from her home in nearby Altringham – and in 1905 took her seat on the Council of the city's newly-independent Astronomical Society. Between 1905 and 1909 she delivered at least three lectures to the Manchester Astronomical Society alone, including titles such as 'Other worlds than our own' (1905) and 'America and American Observatories' (1906). The former lecture attracted an audience of 200 people, and no doubt the Manchester Society was proud to have such a celebrity in its ranks [83].

Sallie's step-daughter, Miss Mary Proctor, followed in her father's astronomical footsteps. Being only six years younger than her step-mother, Mary Proctor had received professional training in astronomy from Columbia University, New York (where Lewis Rutherfurd's benefaction had established a major school of astronomy), and had already published two astronomical books before she both graduated and joined the B.A.A. at the age of 35 in 1897 [84]. Along with her lecturing, international standing, world-wide astronomical travel, and authorship of sixteen books between 1895 and 1940, she appears to have enjoyed close relations with her step-mother and with Manchester. At the Council meeting of 4 March 1908, for instance, Sallie Proctor–Smyth's proposal that Mary should be invited to deliver the opening lecture for the 1908–1909 session, on 7 October, was unanimously accepted. This 'popular' lecture, on the 'Wonders of the Solar System', was duly presented, and Mary was recorded as being awarded two guineas (£2.10) as a contribution towards her expenses [85]. Although the financial provision for Mary from either her father or stepfather, Dr Smyth, is not known, it is clear that she set about earning her own living as an astronomical lecturer and writer [86]. And like so many other astronomical ladies – such as Catherine Stevens, Mary Blagg, Gertrude Longbottom, her own step-mother Sallie and many others – she lived to a ripe old age, dying in 1957 at the age of 95.

One presumes that Miss Gertrude Bacon received her first scientific encouragement from her aeronautical, astronomical, photographic clergyman father, the Revd John M. Bacon. It will be recalled that she accompanied her father on the B.A.A. eclipse expeditions of 1898 and 1900, and by the spring of 1898 she was sufficiently well known as a lecturer to have the Manchester City Art Gallery made available to her. For when Annie Maunder spoke in that venue on the Indian eclipse in May 1898, it was to deputise for the indisposed Miss Bacon [87]. It is clear that she must have been bitten by the astronomical bug when very young, for she joined the B.A.A. in its foundation year of 1890 when even her father was still only 44 years old [88].

Gertrude was clearly an interesting and perhaps an exciting lecturer, judging from the remarks recorded about her performances. This excitement, moreover, derived from personal experiences when accompanying her father on his adventures. We do not know, unfortunately, how far the 'local touches of a humorous character' in the Manchester lecture of 1898 derived from Gertrude Bacon's written manuscript, and how much came from her deputy, Annie Maunder. But we do know that her ballooning lecture 'Voyages to Cloudland', delivered to the Astronomical Society of Wales in the Cory Memorial Hall in 1901, was described as being 'one of the most delightful lectures delivered in Cardiff this winter'. Gertrude was obviously a professional lecturer insofar as she usually charged a fee for her services, for the *Cambrian Natural Observer* specifically recorded that she had waived it for her 1901 visit [89]. It would indeed be interesting to know what her fee was.

Gertrude Bacon also lectured to the Manchester Astronomical Society on several occasions during the Edwardian decade. The joint photography and ballooning theme was clearly in evidence in her 1904 Manchester lecture 'In odd corners with a Camera', although in the following year her 'Evenings with the Telescope' seems to have been more obviously astronomical. She was back in Manchester to open the 1909-1910 lecture session, in the same slot which Mary Proctor had occupied a year earlier [90].

The Leeds Astronomical Society had three women members who followed in the lecturing footsteps of the emigrated Florence Taylor, although I have no record of them speaking to any other societies. Miss C.A. Barbour was the daughter of William Donald Barbour – who, it will be recalled, had been one of the original 1859 founders of the Leeds Astronomical Society – and was one of four Barbours in the Society including her mother and possibly her brother [91]. On 25 October 1899, Miss Barbour delivered a lecture on the history of astrology, where she attacked its 'dupedom' and 'baneful superstitions' as part of a broader discussion upon ancient science. She followed it up in 1904 by another historical lecture on 'Mythological Astronomy' [92]. Her colleague in the Society, Miss Tranmar, also shared Miss Barbour's historical interests, and her erudition, for both of these women – neither of whose Christian names were recorded – delivered and published several excellent historical lectures, which were acknowledged as carrying on the tradition started by Florence Taylor.

From her critical literary erudition, one suspects that Miss Tranmar might have been a schoolmistress. Her 1905 lecture on 'Chaucer's Astrology' was a piece of considerable scholarship even by modern standards, and not only displayed an assured familiarity with Middle English texts, and their French and Latin cognates, but also with the astrolabe and a wide range of medieval scientific ideas [93]. Miss Tranmar, like Miss Barbour, had pronounced historical interests within astronomy, combined with a love of astrologer-bashing. In 1904 she had lectured on the almanacs of 'Zadkiel, Old Moore, & Co.', in which she demonstrated the contradictions inherent in astrological almanacs, while in 1901 she had written a poem about the astronomical aspects of Egyptology. Her 1907 offering 'Has Science killed Romance?' was a witty and erudite courtroom drama, which concluded with the jury returning the verdict that 'Romance is not murdered but married' to science [94].

Fifteen years later, and as the Leeds Astronomical Society was emerging from the disruption of World War I, one Miss Jagger, who might have been a relative of the Edwardian member Mr E. Jagger of Rotherham, was lecturing on the 'Evolution of the Planetary System' [95]. Miss Jagger possessed an M.Sc. degree, but it is interesting to note how the historical bias was retained, as she looked at the cosmological ideas of Kant and Laplace. One wonders if there was any conscious history of science interest shared between the Misses Jagger, Barbour, and Tranmar within the Leeds Astronomical Society. For not only were all three women still active members in 1921, but the former Miss Taylor had retained her membership since her emigration to Minnesota twenty-three years previously, while a third female member of the Barbour family, Miss M.H., had now joined the Society [96].

But the admission of women to the R.A.S. remained a sticking point, for it is clear that even by the mid-1880s there were women who were both intellectually and financially qualified to become Fellows. After Caroline Herschel and Mary Somerville had been made Honorary Members of the R.A.S. in 1835, the only other woman to be given that status, in 1862, was Miss Anne Sheepshanks, the sister of the Revd Richard Sheepshanks, who in her deceased brother's name became a major benefactress to the Society [97]. Then, in 1886 it was proposed that Elizabeth Isis Pogson Kent be elected to the Fellowship. She was the daughter of Norman Pogson, and had acted for many years as her father's unpaid but highly competent assistant at the Madras Observatory. But for constitutional reasons within the Society, Mrs Pogson Kent's name never reached the balloting stage [98].

Then, as we have seen, efforts were made to elect Agnes Clerke and Margaret Huggins in 1892, and later the American Annie Jump Cannon, without success, although the compromise situation arose where Miss Clerke and Lady Huggins were given cards which enabled them to attend R.A.S. meetings and use its Library [99]. It was not until 1915, however, that the stumbling block of the Society's 1831 foundation charter was amplified by a supplementary charter which provided for women Fellows, and from January 1916 onwards women were admitted. These included amateurs like Grace Cook and Fiametta Wilson, writers and lecturers like Mary Proctor and Gertrude Bacon, and women who had worked professionally in astronomy, like Gertrude Longbottom, all of whom became Fellows in 1916 [100]. In 1926, Dorothea Klumpke Roberts – the widow of Isaac Roberts, who had received a professional scientific training at the Sorbonne and in 1893 became the first woman to head a department in the Paris Observatory before marrying Roberts and becoming a Grand Amateur – was also elected Fellow [101].

While it is usual for feminist historians to condemn the long delays which the R.A.S. allowed to occur before the admission of women to full Fellowship, and to cavil at the still small percentage of women in the world of astronomy even by 1920, it is all too easy to approach the matter from a post-1960s perspective. The undertaking of astronomical research, the power to be economically free and innovative, the membership of privileged corporations, and the exercise of social and political authority had been male prerogatives since pre-Biblical times, and the confusion about the position of women which so many men felt by 1900 was symptomatic of the titanic social changes which had taken place during the Victorian age. That change demanded a fundamental re-think of so many ancestrally-held beliefs; and to condemn the Fellows of the R.A.S. who in 1886 haggled about the male pronoun used in their 1831 Charter as unduly reactionary grossly underestimates the shock that was reverberating through late-nineteenth-century European and American society. Indeed, what is truly remarkable is the speed with which the ancient barriers actually fell once things got going, as women found their voice in British astronomy between 1881 and 1916.

Yet still absent from this world were working-class women astronomers. For not only had institutional barriers to fall before their admission into the science, but fundamental changes regarding wages and leisure had to take place. And for that, one had to wait until the late 1950s.

15

Conclusion and postscript: the amateur astronomer into the twentieth century

The expansion and development of amateur astronomy in nineteenth-century Britain was one of the truly remarkable features in the overall history of the science, encompassing as it did its intellectual, technological and social dimensions. For as we have seen, it was the independently-financed and motivated amateurs who took the first major initiatives in double-star astronomy, nebula cosmology, and the study of the lunar, solar and planetary surfaces.

One of the greatest strengths of the amateur tradition was its ability to respond quickly and decisively to new problems, investigations, or potential technologies in astronomy. It is hard to imagine that spectroscopy and photography would have produced such rapid or such fertile applications if they had depended on official initiatives and patronage. The one ironic exception to this Grand Amateur rule, however, was Father Angelo Secchi – that pioneer of stellar spectroscopy, photography, and photographic photometry [1] – whose observatory in Rome was funded by the Jesuit Order: a religious Society which, according to the more hoary mythologies of the history of science, was supposed to be more concerned with suppressing scientific knowledge than with advancing it.

It is also hard to imagine how the development of the big-aperture reflecting telescopes, with either speculum or silvered glass optics, would have gone but for the obvious amateur initiatives that we have seen, not to mention the costly technology of mounting big reflectors in the equatorial plane. In one branch of the science after another, therefore, it was the amateurs who led the way, leaving the professionals to follow in their beaten tracks once the new research paths were shown to be worth pursuing and investing public or corporate money in.

But the nineteenth-century British (and American) amateur tradition did more than lead the way in the obvious areas of original research and technical innovation. It was also a social pioneer, as bodies like the R.A.S. and the local and national amateur astronomical societies that came into being after 1859, and especially after 1881, showed that there was a creative format for scientific association that was fundamentally different from that of the centralised academy. And inevitably, these more eclectic, non-centralist forms of association that operated through Fellowship, rather than through a chain of command, mirrored the society from which they came, along with the political and economic ends to which the majority of members of that society aspired.

It is all too easy, moreover, to underestimate the purely educational impact of amateurs in an age when the formal provision of basic scientific teaching was virtually non-existent. It is hard to estimate how many young men who were the pupils of the schoolmaster–astronomer, the Revd Charles Pritchard (prior to his receiving an Oxford chair) found their lives and minds enriched by an encounter with this extra-curricular subject. The same can be said for the pupils of Rugby and Stonyhurst Schools where several generations of schoolmaster–astronomers were active from the 1850s onwards. We know for certain that the Revd Thomas Henry Espinall Compton Espin was first bitten by the astronomical bug as a schoolboy at Haileybury when the appearance of Coggia's Comet in 1874 stimulated his form master to giving astronomical talks to his class [2]. A similar, if less eminent, amateur astronomer, J. Willoughby Meares, later traced the genesis of his own interests in astronomy to winning a copy of Richard A. Proctor's *The Orbs Around Us* as a prize at Winchester School in 1885 [3]. And right from the start, the amateur astronomical societies went out of their way to educate the public – as the Leeds Astronomical Society was doing from its Love Lane Observatory from March 1861 – while by the 1890s, women amateur astronomers were lecturing to audiences in London, Wales, and the north.

The Victorian amateur tradition, however, had one overwhelming bias in its approach, and that was, as Airy had pointed out in his 1832 *Report*, towards practical observation. Very few of these men and women – with the possible exception of Sir John Herschel – were concerned with matters of pure theory or with mathematics that did not possess a direct bearing upon an aspect of instrumental investigation. This was, one might say, a major weakness of the amateur approach and was the inevitable consequence of a multiplicity of private initiatives taken by separate persons whose first love lay in building or looking through telescopes. To find an emphasis upon, and an excellence in, the theoretical branches of the physical sciences, on the other hand, it was to Britain's universities that one turned, as one did in France or Germany, to find figures of the calibre of John Couch Adams or James Clerk Maxwell. Yet the interests of these men lay in different directions from those of the Brotherhood of the Big Reflecting Telescope.

Another weakness of the Grand Amateur tradition stemmed from its very individualism, as seen in its inability to sustain any projects demanding meticulously collated teamwork. This was most evident in the ultimate failure of the B.A.A.S. Selenographical Project, for while individual astronomers might be able to work in micrometrically-measured angular units, such as when measuring the coordinates of binary star components, and then compare their results in a scientific or social gathering, it proved impossible to organise a couple of dozen lunar observers working with similar telescopes over a period of years to map their allocated regions of the Moon's surface as part of a larger grid.

Yet in many ways this pragmatism and individualism were symptomatic of so much nineteenth-century British science in both its amateur and professional branches, with figures of the standing of Sir Humphry Davy, Michael Faraday, and Sir George Biddell Airy. Airy himself, who had been an acclaimed First Wrangler and winner of every mathematical prize of his undergraduate generation, was consistently criticising the questions set before the prestigious Smith's Prizemen in Cambridge in the 1850s, 1860s and 1870s because of their stress on purely theoretical topics: 'The use of the Examination is to test the power of candidates to command the application of Mathematics where required', and *not* 'that clever and abtruse algebra, without any reference to its benefit as an application of a tool to purposes' [4]. The young men would be better employed displaying their mathematical creativity to the analysis of engineering structures and the like. Indeed, it shows how close the leading professional astronomer of the age was to that pragmatism which ran through so much British science, and is entirely in keeping with his 1881 retiring statement that pure research was best pursued and funded by private individuals, although, as we have seen, the Grand Amateur tradition was beginning to wane by that date [5].

In the early twentieth century, the examples of the Revd T.H.E.C. Espin and the Revd T.E.R. Phillips epitomised the new type of relatively leisured, serious, and highly-educated astronomical enthusiast that was beginning to emerge in a science which was coming to be increasingly dominated by academic professionals. But both men were by profession clergymen of the Church of England, for whom, like their mentor and inspiration, the Revd Thomas Webb, the sincere discharge of a sacred office came first and foremost. Yet this in no way impeded their involvement in amateur scientific work of the highest order. Espin's Tow Law parsonage in County Durham, the living to which he was appointed in 1888, was not only 'a public house which provided mental stimulant' to all his parishioners, but a serious private scientific institution, containing a 17¼-inch Calver reflector and other telescopes, an experimental X-ray apparatus, a geological museum and a library, and was a point of reference for any question in natural history [6]. Similarly, Phillips's rectory at Headley, in Surrey, was a reference centre and major amateur observatory, also equipped with large telescopes by Cooke and Calver [7].

Thomas Espin conducted the first whole-sky search for double stars to have been undertaken by a single individual with a reflecting telescope since that of Sir John Herschel, while his work on red stars and stellar spectra placed him within the mainstream research tradition of the R.A.S [8]. And as a visual planetary observer and draughtsman, T.E.R. Phillips was perhaps without equal. Both men received the Jackson Gwilt Medal and major awards from the R.A.S. for their work, and Phillips was the last clergyman to serve as President of the R.A.S., although William Herbert Steavenson would be its last amateur astronomer President, serving between 1957 and 1959. Espin and Phillips encapsulate, in so many respects, the ethos of the Victorian Grand Amateurs, extending the tradition right into the fourth and fifth decades of the twentieth century. As securely beneficed clergymen of the Church of England, and comfortably off in their own right (Espin left £12,399 in 1934), neither felt the need to toe departmental or scientific career lines. The urbane and generous-hearted Phillips would ignore the most spectacular observing nights if they fell on a Saturday, because that was time which he set aside for the preparation of his Sunday sermon. Even so, he was recognised by an honorary D.Sc. degree from his old university of Oxford shortly before his death: a rare honour for an amateur to receive by 1942 [9].

And Thomas H.E.C. Espin epitomised the rich, eccentric parson, in a way which makes him at one with figures like William Pearson, Temple Chevallier, Thomas Romney Robinson, John Mackenzie Bacon, and John Lee's friend Joseph B. Reade. His remarkable generosity to his parishioners, which included the provision of an isolation house for consumptives and a chancel screen and spire for his church, was also set in the context of his own personal autocracy: he would never tolerate a Parochial Church Council, nor admit women into the choir. As an accomplished amateur musician and composer he even controlled Ss. Philip and James' music, and acted as his own organist [10]. Espin emerges, therefore, as an ecclesiastical autocrat, magistrate, experimental scientist, academically-honoured astronomer, local benefactor, President of the Newcastle Astronomical Society and confirmed misogynist; and one wonders exactly what his contemporaries made of him. It would, however, be impossible today to imagine a man possessing such independence of action in Church or science – let alone in both at the same time.

Yet while the university- and government-employed professional astronomers and physicists were coming to take over most of the official posts within the R.A.S. by 1940, the independent amateurs continued to be very much within the mainstream of the Society's intellectual and social life for long after that date. And perhaps nowhere was this more evident than in the R.A.S. Dining Club, where figures like Reginald Lawson Waterfield and William Thompson Hay perpetuated the ambience of Grand Amateur conviviality, as dons, clergymen, government scientists, armed forces officers, and successful professional men could regularly come together on equal and easy terms in a way which would have been impossible in an astronomical culture where the professionals and the amateurs occupied different spheres. When talking some time ago to a distinguished Fellow of the R.A.S. who had known the Society and its Club since the 1930s, I was informed that the disappearance of those astronomers who 'would have been at the telescope the night before' had been one of its most conspicuous membership changes over the last sixty years [11].

After the success generated by the B.A.A. and the provincial amateur societies between the 1880s and the First World War, amateur astronomy waned somewhat in its vigour and popular appeal during the inter-war years. The Liverpool Astronomical Society virtually went out of existence until it was revived by Alan Sanderson in the early 1950s [12], while the B.A.A. and other amateur societies experienced lean times. No doubt the 1930s depression played a major part in this malaise, not to mention the broader uncertainties of another war looming in Europe. And as is always the case in such times, the better-off and the securely independent throve best, which is perhaps why the amateur membership of the R.A.S. remained so significant. Indeed, many amateur observers were still active, and the Leeds Astronomical Society continued to prosper, with enthusiastic industrialists like Scriven Bolton, who did significant work with an 18¼-inch reflector by George Calver [13]. And within the B.A.A. itself the year 1930 saw the foundation of the new Historical Section within the Society, which was directed by Mrs Mary A. Evershed, who had also been an eclipse photographer [14].

Successful books were also produced during the inter-war years, and T.E.R. Phillips and W.H. Steavenson's sumptuous two-volume compilation of major astronomical essays, *The Splendour of the Heavens* (1923), gives one a sense of what the leading amateurs were doing. And in the schoolmaster–astronomer tradition of figures like the young Charles Pritchard and the Rugby and Stonyhurst astronomers, Arthur Philip Norton, who taught at the Judd School, Tonbridge, Kent, produced the first edition of that *Star Atlas* in 1910 which became, and remains, a fundamental point of reference in the finding of celestial objects for amateurs [15]. It was intended to be used as an adjunct to Webb's *Celestial Objects*. Similarly, the painstaking selenographical researches of Thomas Gwyn Elger and H. Percy Wilkins produced a lunar chart which, through its subsequent editions and revisions, has acquired an almost classical status among twentieth-century amateur lunar observers [16]. And perhaps nowhere does one more clearly get that sense of the sustained love of observing and recording the sky in a twentieth-century figure than in the astronomical biography of George Alcock, *Under an English Heaven* (1996). The long and productive life of George Alcock connected the world of the late-Victorian observers like Espin and Phillips, who were old when he was young, and that of those amateurs who came to the science after its revival which began in the late 1950s.

What brought amateur astronomy out of its comparative doldrums, and restored it to a popularity which matched, and then exceeded, the expansion of the 1890s, was the Space Race. The Russian and American bid to 'conquer' space which was heralded by the Russian 'Sputnik' earth-orbiting satellite in 1957 created a mass concern with looking at the sky that surpassed anything that had gone before, as even self-confessed scientific ignoramuses left their snug firesides at 10 o'clock in the evening to stand in the garden in the hope of catching a glimpse of some satellite that was due to fly over their house. 'Telstar', the early telecommunications satellite of 1962, so gripped the popular imagination that a successful pop song was written about it. And whilst this sky-fever was sweeping most other countries of the Western world as well as Britain, the people of Britain felt a particular pride, because the newly-completed 250-foot dish of the Jodrell Bank Radio Telescope near Manchester proved to be the most versatile and powerful artificial satellite-tracking instrument in the world. The first American astronauts to land on the Moon in 1969 only added to the popular space-momentum.

Yet in addition to the Space Race, three other factors turned out to be extremely influential in the growth of popular astronomy after 1957. One of these was the appearance of the bright comet Arend–Roland, which was the best naked-eye comet seen for many years. Another was the first screening in April 1957 of the B.B.C. television programme *The Sky at Night*, presented by Patrick Alfred Caldwell-Moore.

The début of Patrick Moore, as one who would become the public voice of British astronomy, was to be of enormous importance in that popular spread of the science which took place after 1957. Combining lucidity with an infectious excitement for the subject, he created a new television science format which would go on to have an unbroken run for over forty years, and make broadcasting history. But in a medium increasingly dominated by professional 'presenters', who were very much anchored to their scripts, Patrick Moore came to television as an accomplished practical astronomer in his own right, with an internationally recognised standing as a lunar observer.

In this respect he stands as an historical successor to the men and women already discussed in this book – Webb, Elizabeth Brown, Espin, Mary Proctor, Phillips, and others – who were both active amateur astronomers and used their expertise to write books and to lecture to wider audiences. For in addition to his use of the new medium of television, Patrick Moore has always worked in the traditional formats of lecturing and writing, and his books, such as *The Amateur Astronomer* (1957) and *Guide to the Moon* (1953), have gone through multiple editions and revisions to become classics in amateur astronomy [17]. Patrick Moore's credentials as an ex-R.A.F. officer also continued that long association of British amateur astronomy with the armed forces, seen in figures like Admiral Smyth and Captain Noble.

The third factor in the growth of post-1957 amateur astronomy was social and economic, and derived from the delayed post-war boom. This was the Britain of 'you've never had it so good' prosperity, where high wages, full employment, and the welfare state meant that ordinary people had a higher real income and more leisure than could ever have been imagined by their parents and grandparents. That same boom which turned television sets, Ford Cortinas and holidays on the Costa del Sol into ordinary parts of British life also gave the wherewithal to buy those modestly-priced Japanese telescopes which were coming on to the market, and those popular astronomical books which one could now purchase from high-street newsagents' shops.

The new opportunities which better economic conditions and a media revolution made possible led to the founding of numerous amateur astronomical societies around Britain in the 1960s. And unlike their Victorian and Edwardian forebears, these new societies, which gave astronomical enfranchisement to new social groups, including children, were not necessarily dominated by middle-class social protocols. It was also within the ranks of these new post-1960s societies that one found the first women amateur astronomers who were neither schoolmistresses, writers, nor ladies of private means.

Although there was a resurgence in B.A.A. membership over this same period, the proliferation of autonomous local societies led to the formation of the Federation of Astronomical Societies in 1974, to act as a coordinating body. The monthly magazine *Astronomy Now*, moreover, began to develop as an unofficial organ of the movement in Britain after its first appearance in 1987, along with the annual 'European AstroFest' conventions, during which it is advertised that 'The Universe Comes to London' [18].

Many amateurs are keen not only to observe the heavens, but also to make their observatories – whether privately or society-owned – open to the public to provide an astronomical educational facility to the community. The return of Halley's Comet in 1985 and the appearance of comet Hale–Bopp in 1997 stand out as occasions where amateur observatories across Britain, Europe, America, Australasia and Japan were besieged by members of the public keen to view these objects with large instruments.

And the optical power of the instruments in private and amateur hands gets higher all the time. It was mentioned in Chapter 12 that personal computers and CCD imaging techniques, along with cheap high-speed photographic film, have given more power at a modest cost to amateur astronomers than anything since the adoption of the silver-on-glass mirror in the 1860s. Members of the late twentieth century amateur astronomical community, therefore, can choose either to work on projects of substantial scientific merit, such as monitoring variable stars or galactic supernovae, or to simply enjoy viewing the planets and nebulae for relaxation, depending upon how their commitment takes them, and can easily find the appropriate equipment and information in the high street.

Societies like that of Salford, Lancashire, formed by William Arthur Taylor around 1967, and equipped with an ex-professional 18-inch Newtonian reflector, or that of Crayford Manor House, Kent, founded in 1960, with a 24-inch built by John Wall, do serious deep-sky and other work which is reported to the wider astronomical world [19]. Similarly, the privately-owned Conder Brow Observatory of Denis G. Buczynski (which has housed both the Revd T.W. Webb's original 9¼-inch mirror and the Revd T.E.R. Phillips' 18-inch, both by With) regularly reports its observations and discoveries to the International Astronomical Union, and in 1995 Lancaster University awarded an honorary M.Sc. degree to Mr Buczynski in recognition of his work [20], in a tradition reminiscent of the honorary degrees awarded to Phillips, Roberts, Common, Lassell, Pearson, and others. The late Jack Ells, also of the Crayford Manor House Society, completed a remarkable private observatory in the 1980s which developed upon the 'snug' observatory concept of James Nasmyth. Jack Ells sat in a warm, insulated wooden room that rotated horizontally around a firm pivot. A large-aperture reflecting telescope moved on the same horizontal pivot outside, sending its light through a hollow vertical trunnion support to an eyepiece that remained fixed within the snug observing room, in a configuration that was similar to Caroline Herschel's 'Comet Sweeper' mentioned in Chapter 14. A sophisticated computer control system could make the altazimuth telescope and its attached room find and track any object visible in the heavens via a system of electric motors [21].

And to emphasise the fact that age is no impediment within the modern amateur astronomical world, not only are junior members warmly encouraged in most societies, but the ex-Somerville College mathematician, Dame Kathleen Mary Ollerenshaw, took up observational work in her mid-seventies. She joined the Manchester Astronomical Society in the late 1980s, and over the next decade, with help from the Society and its members, turned herself into an expert observer and astrophotographer, with publications on computing and CCD astronomy for amateurs based upon work performed in her Lake District observatory 'Lovell II', named in honour of her friend and fellow-Mancunian astronomer, Sir Bernard Lovell, of Jodrell Bank [22]. ('Lovell I' is the official name of the 250-foot Jodrell Bank Radio Telescope, named in honour of its creator.)

One can trace a line of intellectual descent from the heyday of the Victorian Grand Amateurs to the modern practitioners of the science, though as I have indicated elsewhere in this book, amateur astronomy changed fundamentally between 1840 and 1890, and especially in the last couple of decades of the nineteenth century. Most noticeable in this respect are the reversed roles between amateur and professional when it comes to technological innovation.

While modern amateurs use computers, CCD imaging techniques, and telescopes of increasingly large aperture, all of these devices ride on the back of prior-developed professional technologies. While the modern amateur might be able to see and obtain high-quality photographic images of what was only available to the professionals forty years ago [23], the professionals themselves have shot yet further ahead, in a reversal of the situation that existed in 1845. On the other hand, modern instrument-building amateurs, and the optical industry that supplies telescopes to those who prefer to purchase their instruments complete, have come up with some remarkable innovations that are tailored to the needs of the modern observational astronomer.

The changing social conditions of the car, small gardens, and light pollution have placed a high premium on telescopes that combine large aperture and lightweight optics with compact tubes and maximum portability. Modern catadioptric reflectors, with 10-inch mirrors, 2-foot plastic tubes, aluminium stands, and personal computer controls, can be taken from their owner's flat and easily driven to a Pennine ridge or a Kentish field on a clear night. The 'Dobsonian' mount, using lightweight materials and low-friction bearing pads, now makes it possible for two friends to pick up, carry, and use a reflecting telescope of similar power to William Lassell's great cast-iron instrument of 1845. Yet all of these new technologies, from microchip controls to low-friction bearings, are the hand-me-downs from professional astronomical or space-research innovation.

And much of the original work undertaken by dedicated modern observers, alas, is now done in relation to wider professional researches. The search for new minor planets, comets, supernovae, and variable and binary stars is undertaken within the spirit of amplifying the details of the international professional research programmes, rather than leading the way in terms of new cosmological investigations. On the other hand, this in no way diminishes either its scientific value, or the thrill of discovery when an amateur astronomer is the first person in the world to see a new phenomenon. It is not for nothing that Japanese and American amateurs have had the majority of recent comets named after them, or that the regular sky patrols of the British amateurs Brian G.W. Manning of Kidderminster, George Sallit of Newbury Amateur Astronomical Society, and Stephen Laurie of Church Stretton, Shropshire, have the discoveries of asteroids officially accredited to them, whilst supernova discoveries have been accredited to Mark Armstrong of Rolvenden, Kent, Stephen Laurie, Tom Boles of Wellingborough, Northamptonshire, and Ron Arbour of South Wonston, Hampshire. Others, like John Fletcher of Tuffley, Gloucestershire, have a high reputation due to their skills in astrophotography and CCD imaging.

Amateur astronomy has always worked on a variety of levels of appeal, including the roles of patron, armchair enthusiast, and instrument-maker, as well as active observer; and in this respect one can trace the emergence and long-term development of a community of great adaptability. It is not only distinct economically and technically from the professional tradition that grew out of it, but in many ways it is historically more interesting, insofar as it attracted people from a much wider social spectrum, who came in pursuit of a much broader range of objectives.

And because its funding came directly from the pockets of its devotees, and was not filtered through a complex system of state or corporate patronage, amateur astronomy was much more sensitive to changing scientific priorities, and could often seize initiatives in advance of the professionals. Not only was this true across most departments of the science in 1845, but even today, as we saw above, it is often the amateurs who steal the lead when it comes to the 'patrol' work of comets, asteroids, supernovae, meteors, aurorae, and other phenomena. And perhaps nowhere are amateurs more in evidence today than in the field of popular astronomical education at the local level, where local astronomers freely give their time not only to open up their observatories to the general public, but also to work with schools in an attempt to rectify an education system which until recently has failed to acknowledge that celestial phenomena warrant inclusion in the official science curriculum. And even in 1998, the teaching of astronomy in our state schools is often extremely poor when compared with the teaching of physics, chemistry and biology.

In many respects, therefore, one can see a tradition which was not only of fundamental intellectual and technical importance in the nineteenth century, but which produced a line of descent extending down to our own time. For to use the word 'amateur' in that pejorative context engendered and perpetuated by a world becoming increasingly smitten by the formally-organised and paid experts of the early twentieth century, one not only misses the point of so much Victorian astronomical endeavour, but fails to do justice to the society from which it emerged.

ADDENDUM:

FOUR ASTRONOMERS BROUGHT TO THE AUTHOR'S NOTICE SINCE THE 1998 EDITION

John Birmingham, of Tuam, Ireland. Peter Möhr, 'A Star in the Western Sky: John Birmingham, Astronomer and Poet', *The Antiquarian Astronomer* **1** (2004) 23-33.

John Glass, the miller of Tarland, Aberdeenshire, Scotland. Alex Smith, 'A Working Man Astronomer and his Telescope', *The English Mechanic and World of Science*, no. **1810,** (1 December 1899) 360–1.

Moses Holden, of Bolton and Preston, Lancashire. Steve Halliwell, 'Holy Moses! The Story of a City Astronomer, Missionary . . . and Genius', *The Lancashire Evening Post*, Wednesday, 27 November, 2013.

John Wheatley, b. 1812, carpenter of Bluntisham, Norfolk. Amateur astronomer, telescope maker, and achromatic lens maker. I am indebted to David Jackson of the North Norfolk Astronomical Society for sending me unpublished details of John Wheatley, along with a photograph of his house and telescope, from local archival sources.

NEW PUBLICATIONS

The Antiquarian Astronomer and *Bulletin* of the Society for the History of astronomy (S.H.A.)

The Speculum published by the William Herschel Society

Journal of Astronomical History and Heritage

Notes and references

Abbreviations

A.N.	*Astronomische Nachrichten*
B.A.A	British Astronomical Association
B.A.A.S.	British Association for the Advancement of Science
C.U.L.	Cambridge University Library
D.N.B.	*Dictionary of National Biography*
D.S.B.	*Dictionary of Scientific Biography*
J.B.A.A.	*Journal of the British Astronomical Association*
J.H.A.	*Journal for the History of Astronomy*
M.H.S.	Museum of the History of Science, Oxford
M.N.R.A.S	*Monthly Notices of the Royal Astronomical Society*
Mem. [R.]A.S.	*Memoirs of the Astronomical Society of London (later R.A.S.)*
Mem. R.A.S.	*Memoirs of the Royal Astronomical Society*
Notes and Records	*Notes and Records of the Royal Society*
Phil. Trans.	*Philosophical Transactions of the Royal Society*
Proc. R. Soc.	*Proceedings of the Royal Society*
Q.J.R.A.S.	*Quarterly Journal of the Royal Astronomical Society*
R.A.S.	Royal Astronomical Society
R.G.O.	Royal Greenwich Observatory

1 AMATEUR ASTRONOMY IN THE ROMANTIC AGE

1   G.B. Airy, 'Report on the Progress of Astronomy during the present Century', dated 2 May 1832, *Transactions of the B.A.A.S.*, 1833, 125–89.
2   Airy, 'Report', 126–7, 180–9.
3   Airy, 'Report', 182. F.W. Bessel, *Fundamenta Astronomia pro anno MDCCLV deducta ex observationibus... James Bradley...* (Regiomontani, 1818). J.F.W. Herschel, *A Brief Notice of the Life, Researches, and Discoveries of Friedrich Wilhelm Bessel* (London, 1847), 6–7, extracted from its original publication in the *Annual Report* of the R.A.S.
4   Airy to W.V. Harcourt, 5 September 1831 (Harcourt MS), published in Jack Bowes Morrell and Arnold Thackray (eds.), *Gentlemen of Science. Early Correspondence* (Royal Historical Society, 1984), letter 117, p. 150.

5   Airy, 'Report', 146–9. Airy's three-page Section V in the Report deals with the whole of double and variable star astronomy, along with nebulae.
6   J.D. Forbes to W.V. Harcourt, 3 November 1831 (Forbes Papers), published in Morrell and Thackray (ref.4), letter 68, pp. 91–3.
7   Airy to *English Mechanic*, 25 February 1881, 586–7. See heading 'The Endowment of Research'.
8   This move towards professionalisation underpins the arguments implicit in Marie Boas Hall's *All Scientists Now: The Royal Society in the Nineteenth Century* (C.U.P., 1984), and Peter Alter's *The Reluctant Patron. Science and the State in Britain, 1850–1920* (Oxford, Hamburg, New York, 1987). A similar attitude – especially towards the British Grand Amateur tradition – is also present in Donald W. Osterbrock's *Pauper and Prince: Ritchey, Hale and Big American Telescopes* (University of Arizona Press, 1993).
9   The parson–geologists William Buckland and Adam Sedgwick, for example, both came to the science as hobbyists, yet obtained respectively Oxford and Cambridge professorships in geology, held in conjunction with cathedral canonries at Christ Church, Oxford, and in Norwich. Surprisingly, however, no one speaks of these men, or any of their colleagues, as 'amateurs'. See Nicholaas A. Rupke, *The Great Chain of History: William Buckland and the English School of Geology, 1814–1849* (Clarendon Press, Oxford, 1983), 3–28.
10 Characteristic of this attitude was James Nasmyth, who in 1865 refused an F.R.S. and in 1883 a St. Andrews LL.D.: William Pole (Royal Society) to Nasmyth, 16 February 1865; Dr Robert Grant (St. Andrews) to Nasmyth, September 1883: both in a collection of letters to Nasmyth in Edinburgh Public Libraries (Arts Library), WND 478 N 25J (3 vols.). I am indebted to Dr David Gavine for this information.
11 Margaret Gowing, 'Science, Technology and Education: England in 1870', *Notes and Records*, **32**, 1 (July 1977), 71–90. By the latter part of the century, it was 'a sine qua non for British chemists to append the German Ph.D. to their names': Donald S.L. Cardwell, *The Organisation of Science in England* (Heinemann, London, 1957, 1972), 163.
12 This caution about too much central control is evident in two letters by Sir John Herschel, to W.V. Harcourt, 5 September 1831, and to William Whewell, 20 September 1831, printed as letters 37 (pp. 55–7) and 45 (pp. 66–8) in Morrell and Thackray (ref.4).
13 Charles Babbage, in *Reflections on the Decline of Science in England and some of its causes* (London, 1830), set out the perceived 'declinist' view of non–centrally-organised British science, contrasted with its assumed professionalism in France and Germany. See also Roy M. MacLeod, 'Whigs and Savants: reflections on the reform movement in the Royal Society, 1830–1848', in Ian Inkster and Jack Morrell (*eds.*), *Metropolis and Province, Science in British Culture 1780–1850* (London, 1983).
14 Sir David Brewster to W.V. Harcourt, 5 December 1831 (Harcourt MS) in Morrell and Thackray (ref.4), letter 70, p. 95.
15 Cardwell (ref.11) 28–32 argues that the German universities represented a natural

cultural and political expression in a way that Britain's universities did not.
16 This was, of course, the ethos enshrined in Adam Smith's *Inquiry into the meaning of the Wealth of Nations* (1776), which was so profoundly influential upon English social and economic attitudes in the early nineteenth century.
17 Sir John Herschel, 'Presidential Address', *Mem. R.A.S.* **17** (1849), 192.
18 N. Pogson to W.H. Smyth, 22 July 1858, *Aedes Hartwellianae* (London, 1860), 69.
19 Sir William Herschel, in his nebula papers, had often reminded the readers of *Phil. Trans.* that such objects were not visible in 'common telescopes'; and little had changed by the time of his son: A. Chapman, 'William Herschel and the Measurement of Space', *Q.J.R.A.S.* **30** (1989), 407, 408.
20 Anne Secord, 'Science in the Pub: Artisan Botanists in Early Nineteenth-Century Lancashire', *History of Science* **32** (1994), 269–315.
21 Mrs Gordon (Elizabeth Oke), *The Life and Correspondence of William Buckland... by his Daughter* (London, 1894), 11–15, for many examples of Buckland's popularity and enthusiasm.
22 This combination of a love of travel, broad culture, and geology is clearly present in the *Life, Letters, and Journals of Sir Charles Lyell, Bart., edited by his sister-in-law, Mrs Lyell*, 2 vols. (London 1881): see vol. 1, pp. 55–6 for Lyell's romantic poem 'Lines on Staffa'. Darwin inherited a fortune in excess of £45,000, along with Down House (bought for him by his father upon marriage, at a cost of £2,200), and made more money from subsequent investments: Peter J. Bowler, *Charles Darwin, the Man and his Influence* (Basil Blackwell, Oxford, 1990), 90–4. It would distort the overall picture, however, if one failed to acknowledge the existence of a vigorous European amateur tradition in subjects like botany and natural history, where a great deal of important work was done. Yet these were subjects that did not require complex technologies or the kind of prestigious capital outlay which were more likely to attract official patronage in Continental European countries.
23 William Whewell, *Astronomy and General Physics considered with reference to Natural Theology* (London, 1833). The sense of the intellectual primacy of astronomy amongst the sciences is also evident in Sir J.F.W. Herschel's *Preliminary Discourse on the Study of Natural Philosophy* (London 1831).
24 A.J. Meadows, 'Astronomy and Geology, Terrible Muses! Tennyson and 19th-Century Science', *Notes and Records* **46**, 1 (1992), 111–18.
25 L. Pearce Williams, *Michael Faraday* (Chapman and Hall, London, 1965), 102–6.
26 G.B. Airy, 'Family History of G.B. Airy', Enid Airy MS, dated 22 May 1889. I am very grateful to the Airy family for access to this 46-page manuscript which was then in private family hands, but is now lodged in Cambridge University Library, (in process of cataloguing, 1997). G.B. Airy, *Autobiography, ed.* Wilfred Airy (C.U.P., 1896), 65, for pupils in the Lake District: 288, 290, 303 'to verify the localities mentioned in the *Lady of the Lake*'. Airy was visiting Cumberland with his daughter when his wife died in 1875: 310. His social relations with Wordsworth and Southey: 64, 76, 149, 150.
27 See Herbert Hall Turner, *The Records of the R.A.S. Club, 1820–1910* (Oxford, private publication, 1910).
28 Sir John Herschel, 'Presidential Address' 1849 (ref.17) 192.

29 For 'Length, breadth and depth... of space', see Sir William Herschel, 'Astronomical Observations, and Experiments tending to investigate the local arrangement of the celestial bodies in space', *Phil. Trans.* **107** (1817), 302–31; also in *Scientific Papers of Sir William Herschel* II, ed. J.L.E. Dreyer (London, 1912), 575.

2 GENTLEMEN AND PLAYERS: AMATEURS AND PROFESSIONALS IN 1840

1 Thomas Sprat, *A History of the Royal Society of London* (London 1667, 2nd edn., 1702), 67.
2 Robert T. Gunther, *Early Science in Oxford* II (1923) 79, for a list of Greaves' instruments. For my own analyses of these preserved instruments, see A. Chapman, 'The Design and Accuracy of Some Observatory Instruments of the Seventeenth Century', *Annals of Science* **40** (1983), 457–71; also A. Chapman in *Oxford Figures. 800 Years of Mathematics*, eds. J. Fauvel, R. Wilson et al. (O.U.P., forthcoming, 1999).
3 Flamsteed to unnamed recipient, 24 October 1715, Cambridge University Library Royal Greenwich Observatory (C.U.L. RGO) MS, PRO Ref. 35, p. 191, printed in Francis Baily, *An Account of the Revd John Flamsteed* (London 1835) 316.
4 Derek Howse, *Nevil Maskelyne, The Seaman's Astronomer* (C.U.P., 1989), 61.
5 Howse, *Maskelyne* (ref.4), 201–2. North Runcton was a Trinity College, Cambridge, living (*D.N.B.*).
6 A. Chapman, 'Pure research and practical teaching: the astronomical career of James Bradley 1693–1762', *Notes and Records* **47**, 2 (1993), 205–12. For government purchase of Bradley's instruments, Derek Howse, *Greenwich Observatory*, **3**, *The Buildings and Instruments* (Taylor and Francis, London, 1975) 63, 81.
7 The instrument maker Edward Nairne valued them at £375 14s 6d, plus £10 10s for his own valuation fee: 'A Catalogue of the Philosophical Apparatus belonging to The Revd. Dr. Hornsby of Oxford', 10 November 1790, a 12-page MS inventory of the 1790 sale: Bodleian Library MS Top[ographical] c. 236:3–13. R.T. Gunther, *Early Science in Oxford* XI (1937), 399.
8 Gunther, *Early Science in Oxford* II (ref.2), 88–91; Ivor Guest, *Dr. John Radcliffe and his Trust* (Radcliffe Trust, London, 1991), 224–32. A. Chapman, 'Thomas Hornsby and the Radcliffe Observatory', in *Oxford Figures* (ref.2).
9 G.B. Airy to Lord Auckland (First Lord of the Admiralty), 8 October 1832, C.U.L. RGO6 1/151–152.
10 Airy to the Duke of Sussex, 19 May 1834, C.U.L. RGO6 1/145. Airy to Lord Auckland, 10 October 1834, C.U.L. RGO6 1/153. *Autobiography of Sir G.B. Airy*, ed. W. Airy (C.U.P., 1896), 105–8. Warrant for Lady Airy's pension: 10 March 1835, in file 'Private & Personal', Enid Airy Papers, consulted when in family hands, but given to Cambridge University Library in 1997.
11 Airy to T. Spring Rice (Chancellor of the Exchequer), 10 December 1835, in Airy, *Autobiography* (ref.10), 112–13.

12 *Autobiography* (ref.10), 296. G.B. Airy to Otto Struve, 23 July 1872, to inform him of K.C.B.: C.U.L. RGO6 952 f 623. 'Astronomer Royal's Journal', 30 July 1872: C.U.L. RGO6 26.
13 Guest, *Dr. John Radcliffe* (ref.8), 251–2.
14 'In 1813 William Airy [George's father] lost his appointment of Collector of Excise and was in consequence very straitened in his circumstances', Airy, *Autobiography* (ref.10), 17. This embarrassing 'great misfortune', which almost led to George being apprenticed to a local tradesman until his father insisted on his continuing with his education, is also alluded to in the MS autobiography, in 'Family History of G.B. Airy', Enid Airy Papers (ref.10), 18. William Airy could have been the unnamed Excise Officer who absconded with £600 in late August 1813: *Ipswich Journal & Cambridgeshire Advertiser*, 4 September 1813. I am indebted to Miss Clarice Chapman of Colchester for this reference.
15 Sir Charles Wood (Secretary to the Navy) to G.B. Airy, 15 August 1836, C.U.L. RGO6 1/283.
16 Airy to Wilfred Airy, 30 September 1855, in 'Private & Miscellaneous' packet, Enid Airy Papers (ref.10).
17 Richarda Airy to Lady Margaret Herschel, 10 September 1844, in Lady Richarda Airy's correspondence, Enid Airy Papers (ref.10).
18 Airy, *Autobiography* (ref.10), 165–6.
19 J.L.E. Dreyer, article 'Observatory' in *Encyclopaedia Britannica* (8th edn, 1884), 710–11. They were Greenwich, Oxford Radcliffe, Cambridge, Durham, Liverpool, Kew, Edinburgh, Glasgow, Dublin (Dunsink) and Armagh. Dreyer also listed Oxford University and Cork, both of which were founded in the 1870s.
20 Guest, *Dr John Radcliffe* (ref.8), 246.
21 Guest, *Dr John Radcliffe* (ref.8), 253. A.D. Thackray, *The Radcliffe Observatory, 1772–1972* (Radcliffe Trustees, 1972), 8ff.
22 A. Chapman, revision (1997) and expansion of Agnes M. Clerke's article 'Manuel Johnson' for *New D.N.B.* (O.U.P., forthcoming). John Henry Newman, *Apologia pro vita sua* (1864) Part VI (end) (Scott edn., 1913) vol. 2, p. 97. Newman spent his last night in Oxford as Johnson's guest at the Radcliffe Observatory house, prior to leaving for Italy to become a Roman Catholic priest and eventually a Cardinal.
23 Ada Pritchard, *Charles Pritchard... Memoirs of his Life* (London, 1897), 24–81.
24 The most extensive work on the history of the Oxford University Observatory to date has been performed by my friend and sometime research student, Roger Hutchins, of Magdalen College. See his 'John Phillips, 'Geologist–Astronomer' and the Origins of the Oxford University Observatory, 1853–1875', *History of Universities* XIII (O.U.P., 1994), 193–249. Ada Pritchard, *Charles Pritchard* (ref.23), 144–67.
25 Airy, *Autobiography* (ref.10), 86.
26 E.J. Routh, 'G.B. Airy', obituary, *Proc. R. Soc.* **51** (1892), i–xxi. H.H. Turner, 'G.B. Airy', *M.N.R.A.S.* 52, 4 (1892), 212–29. A. Chapman, extensive revision (1997) and expansion of Agnes M. Clerke's article 'G.B. Airy' for *New D.N.B.* (O.U.P., forthcoming).
27 Lettie S. Multhauf, 'Franz Xavier von Zach', *D.S.B.* J.F.W. Herschel, *A Brief*

Notice of the Life, Researches, and Discoveries of Friedrich Wilhelm Bessel (London, 1847), a 16-page pamphlet extracted from *Annual Report of the R.A.S.* A. Chapman, 'The [German] Astronomical Revolution', in *Möbius and his Band, Mathematics and Astronomy in Nineteenth-Century Germany*, eds. J. Fauvel, R. Flood, R. Wilson (O.U.P., 1993), 34–77.

28 J.F.W. Herschel, *Outlines of Astronomy* (2nd edn., London, 1849), 307–8, para. (505). William Sheehan and Richard Baum, 'Observation and inference: Johann Hieronymous Schroeter, 1745–1816', *J.B.A.A.* **105**, 4 (1995), 171–5.

29 Lettie S. Multhauf, 'Heinrich Olbers', *D.S.B.* Kevin Krisciunas, *Astronomical Centers of the World* (C.U.P., 1988), 99–119.

30 David W. Hughes, 'Only the first four asteroids', *J.B.A.A.* **107**, 4 (1997), 211–13.

31 J.H. Mädler and Wilhelm Beer, *Der Mond nach seinen kosmischen und individuellen Verhältnissen*, 2 vols., (Berlin, 1837).

32 Agnes M. Clerke, *A Popular History of Astronomy during the Nineteenth Century* (3rd edn., London, 1893), 155–7.

33 Dreyer, 'Observatory' (ref.19), 712–13.

34 Dreyer, 'Observatory' (ref.19), 713.

35 Alan H. Batten, *Resolute and Undertaking Characters: The Lives of Wilhelm and Otto Struve* (Reidel, Dordrecht, etc., 1988).

36 Dreyer, 'Observatory' (ref.19), 712. Bruno Morando, 'The Golden Age of Celestial Mechanics', in *The General History of Astronomy* II, *Planetary Astronomy* Part B, eds. René Taton and Curtis Wilson (C.U.P., 1995), 214–21. Patrick Moore, *The Planet Neptune: An Historical Survey before Voyager* (2nd edn., Praxis–Wiley, Chichester and New York, 1996), 22–3.

37 I am indebted to Rosa Alonzo and to Dr Amparo Sebastián, Directora, Museo Nacional de Ciencia y Tecnologia, Madrid, for arranging this visit for me in May 1995. The Spanish Royal Naval Observatory, San Fernando, Cadiz, similarly possesses rich collections, by makers such as Jesse Ramsden, Thomas Jones and others: Francisco José Gonzalez Gonzalez and Manuel Berrocosa Domínguez, *Instrumentos Magistrales Para La Astrometria Española en el Siglo XIX: El Observatorio de San Fernando*, Boletin Roa No. 15/93 (Real Instituto y Observatorio de la Armada en San Fernando, 1993).

38 Airy, 'Report on the progress of astronomy during the present century', May 1832 (*Transactions of the B.A.A.S.* 1833), 184–9.

39 R.T. Gunther, *Early Science in Oxford* II (ref.2), 326, 327ff.

40 Dreyer, 'Observatory' (ref.19), 710. An undated handwritten note – perhaps a copy of a note by Sir James South whom it seems to be quoting – is in Dr John Lee's Astronomical 'Album', M.H.S. Gunther 36, 4v. It says that Kew (presumably before its overhaul in 1842) had a salary bill of £776 p. a. for two astronomers, an assistant, and a servant, yet not 'one single astronomical observation... has emanated from this Observatory'.

41 G.D. Rochester, 'The History of Astronomy in the University of Durham from 1835 to 1939', *Q.J.R.A.S.* **21** (1980), 369–78. Original documents for Observatory foundation: 'Observatory Manuscripts, IV, Administrative Records, 1837–1839', v.70–274, Durham University MSS.

42 Temple Chevallier, *First Report of the Curators of the [University of Durham] Observatory* (23 June 1840) 1. 'Senate Minutes' I, 16 March 1841, p. 125, Durham University MSS, for Chevallier's £500 salary as Astronomy and Mathematics Professor. *D.N.B.*, 'Chevallier'.
43 Samuel Lewis, *A Topographical Dictionary of England* II (London, 1837), under 'Esh', for Ushaw College. T. Chevallier to Archdeacon Thorpe, 23 December 1834, insisted upon having an ecclesiastical living upon his taking up his Durham University chair, so as not to 'secularise' himself.
44 Chevallier to George Elwes Corrie, 14 May 1839, Durham University Add. MSS 837/40. Durham University holds 129 Chevallier letters in Add. MSS 837, many of which express alarm about Roman Catholicism. That of 23 February 1839, to Corrie, Add. MSS 837/37, says 'Dr Gilly declares it [Ushaw College] a Jesuitical scheme to blow up the University.'
45 Edgar Jones, *University College, Durham. A Social History* (Edgar Jones, Aberystwyth, 1996) 76–7.
46 'Senate Minutes' II, 27 June 1847, Durham University MS, for use of telescope. Chevallier to G.B. Airy, 13 November 1851, Observatory MS No. 319, Durham University.
47 They were: John S. Browne, Arthur Beanlands, George Rümker, Albert Marth, Edward G. Marshall, John I. Plummer (all Durham); Revd Robert A. Thompson, R.C. Carrington (both Cambridge); William Ellis, F.R.S., and Mondeford R. Dolman. Extracted from Durham University *Calendar*, 1841–1872, by Professor G.D. Rochester, and deposited in typescript in the Observatory archives, Durham University Library.
48 *Calendar* (ref.47) 1841. Owen Chadwick, *The Victorian Church 2, 1860–1901* (S.C.M., London, 1970) 366, for Cathedral stipends. Lewis, *Topographical Dictionary* (ref.43) under 'Esh'. *Crockford's Clerical Directory* says that Esh had been augmented to £200 by 1868.
49 H.A. and M.T. Brück, *The Peripatetic Astronomer, The Life of Charles Piazzi Smyth* (Hilger, Bristol and Philadelphia, 1988), 1–15.
50 John Hartnup (senior), obituary, *M.N.R.A.S.* **46** (1886), 188–91.
John Hartnup (junior), obituary, *M.N.R.A.S.* **53** (1893), 218.
51 'Borough of Liverpool Observatory Committee Minutes', 26 August 1843, 73.
52 'Mersey Dock and Harbour Board, Register of.... Salaried Officers Employed in the Observatory', 1869, for salaries of John and John (jun.) Hartnup. Originals preserved (no catalogue number) in the Liverpool Dock Museum, and copies in the Earth and Physical Sciences Department, Liverpool Museum, William Brown Street, Liverpool.
53 Dreyer, 'Observatory' (ref.19) 716. Krisciunas, *Astronomical Centers* (ref.29) 194–5.
54 Dreyer, 'Observatory' (ref.19) 716.
55 Norman Robert Pogson/G.B. Airy correspondence 1866–1869 regarding the future of the Madras Observatory, C.U.L. RGO6 149–150.
56 David S. Evans, *Lacaille: Astronomer, Traveler, With a New Translation of his Journal* (Pachart, Tucson, 1992), 109–30.

57 Hermann A. Brück, *The Story of Astronomy in Edinburgh from its Beginning until 1975* (Edinburgh U.P., 1983) 10–14. David M. Gavine, 'Astronomy in Scotland, 1745–1900', unpublished Open University Ph.D. thesis 1982, 304ff.
58 Brück and Brück, *Peripatetic Astronomer* (ref.49) 4–15.
59 Brück and Brück, *Peripatetic Astronomer* (ref.49) 4, for Piazzi's salary. It was later increased to £400: list in *Royal Commission on Scientific Instruction and the Advancement of Science* (February 1871), copy in Royal Observatory, Edinburgh. In 1835 Robert Main at Greenwich had received a salary of £300 p. a. (Airy to Sir Charles Wood, C.U.L. RGO6 1/183), plus a £50 house allowance (C.U.L. RGO6 1/241, 9 January 1836).
60 Brück and Brück, *Peripatetic Astronomer* (ref.49) 51: Piazzi borrowed a heliometer from Nasmyth for his 1856 Teneriffe expedition.
61 Hermann A. Brück, 'Lord Crawford's Observatory at Dun Echt', *Vistas in Astronomy* 35 (1992), 81–138: 127.
62 Brück, 'Lord Crawford's Observatory' (ref.61) 126–34.
63 William Ruat to R. Simpson, correspondence regarding instrument purchases, Glasgow University *Senate Archives*, Faculty Minutes, 1754–1755. A. Chapman, *Dividing the Circle. The Development of Critical Angular Measurement in Astronomy 1500–1850* (2nd edn., Praxis–Wiley, Chichester and New York, 1995) 142–4.
64 Dreyer, 'Observatory' (ref.19) 710.
65 Patrick A. Wayman, *Dunsink Observatory 1785–1985: a Bicentennial History* (Dublin Institute for Advanced Studies, Royal Dublin Society, Dublin, 1987).
66 Dreyer, 'Observatory' (ref.19) 710.
67 J.L.E. Dreyer and H.H. Turner, *History of the Royal Astronomical Society 1820–1920* (R.A.S., 1923), 55.
68 W. Valentine Ball, *Reminiscences and Letters of Sir Robert Ball* (London 1915), Chapter VI, provides an illuminating insight into Ball's warm relations with Lord Rosse and the Irish astronomical world of the mid-1860s. Amongst other things he relates (p. 66) how one of Rosse's 'favourite amusements' (like that of W.E. Gladstone) was the felling of trees in his Park.
69 'Thomas Romney Robinson', obituary by Robert S. Ball, *M.N.R.A.S.* 43 (February 1883), 181–3 gives Robinson's appointment date as 1824; James A. Bennett, *Church, State and Astronomy in Ireland: 200 Years of the Armagh Observatory* (Armagh Obs. and Inst. of Irish Studies, Queen's University, Belfast, 1990) 66, says that on 10 November 1823 Robinson was certified competent by Pond, the Astronomer Royal, to Archbishop Beresford, as a preliminary to his Observatory directorship. William Pearson, article 'Observatory' in A. Rees, *A New Cyclopaedia, or Universal Dictionary of Arts, Sciences, and Literature* 25 (London, 1819), claimed that Archbishop Robinson had spent £15,000 on the founding of the Armagh Observatory and its related institutions: see under 'Irish Observatories', Sig. PP3r/v.
70 Bennett, *Church, State, and Astronomy in Ireland* (ref.69) 22, 64. See also Sister Dolores Crowe, 'Thomas Romney Robinson (1792–1882) Director of Armagh Observatory (1823–1882)', *The Irish Astronomical Journal* 10, 3 (September 1971), 93–101.

71 Thomas R. Robinson, *Places of 5,345 stars observed from 1828 to 1854 at the Armagh Observatory* (Dublin, 1859). His subsequent observations were published, under the editorship of J.L.E. Dreyer, as a *Second Armagh Catalogue of 3,300 stars for the Epoch of 1875...* (Dublin, 1886).
72 Romney Robinson corresponded on familiar terms with Lord Rosse (Rosse Archives, Birr Castle), Sir James South and others. Unfortunately Alexander Thom's *Irish Almanac and Official Directory* (Dublin) for 1850 and 1860, which I have checked, does not record the value of Robinson's Irish ecclesiastical benefices, as *Crockford's* does for English livings.
73 Bennett, *Church, State and Astronomy* (ref.69).
74 Bennett (ref.69) 146, citing from Robinson's 1869 *Report* of the Armagh Observatory.
75 Bennett (ref.69) 146–7, citing Robinson's *Report*.
76 Dreyer, 'Observatory' (ref.19) 711. Ian Elliott, 'An Irish Galaxy', *Irish Studies Review* **4** (Autumn 1993), 19–20. Charles Mollan, William Davis and Brendan Finucane (*eds.*), *Some People and Places in Irish Science and Technology* (Royal Irish Academy 1985) 104–5.
77 Howard Grubb, 'On the Equatorial Telescope and on the New Observatory of the Queen's College, Cork', *Scientific Proceedings of the Royal Dublin Society* **2** (1880), 347–69. For the church-like Observatory, which cost £10,000, *The Irish Builder*, 1 May 1879, 130, and 1 December 1879, 376. I am indebted to Tony Ryan and Dr Ian Elliott of Dublin for much of the above information.
78 Dreyer, 'Observatory' (ref.19) 711–12.
79 'Observatory' in Sir David Brewster (*ed.*), *Edinburgh Encyclopaedia* **15** (1830) 439–48. I am indebted to Kevin Johnson of the Science Museum, South Kensington, for drawing my attention to a loose single-sheet manuscript list of 24 'Private Observatories mentioned in the *Edinburgh Encyclopaedia* 1830'. William Pearson, in 'Observatory', Rees's *Cyclopaedia* (ref.69) also gives a total list of over twenty substantial private observatories operating in England in *c*.1810, including some that were not on Brewster's later list.
80 John Weale, *Description of Public and Private Observatories in London and its Vicinity* (London, 1851).
81 *Astronomical Register* **4**, 1866, 91 reported a census of private observatories, which it suggested should be repeated each decade.
82 Sir George Shuckburgh, 'An Account of the Equatorial Instrument', *Phil. Trans.* **83** (1793), 67–128.
83 Weale, *Description of... Observatories* (ref.80) 62–4.
84 *James Nasmyth, Engineer. An Autobiography*, ed. Samuel Smiles (London 1889) 338. S. Smiles, *Men of Invention and Industry* (London, 1884), 334, for Nasmyth's 8-inch Cooke refractor. The Science Museum, South Kensington, possesses photographs of Nasmyth's Penshurst observatory.
85 T.W. Backhouse's results were issued privately under *Publications of the West Hendon House Observatory, Sunderland*, I–IV (Sunderland, 1891–1915) and in other works.
86 'The Craig Telescope at Wandsworth Common', *Illustrated London News*, 28

August 1852, 168. Henry Mayhew, 'The Telescope Exhibitor', in *London Labour and the London Poor; A Cyclopaedia of the Conditions and Earnings of Those that Will Work, Those That Cannot Work, and Those That Will Not Work* III (London 1861) 79–83:82–3. The gossip about 'the Revd. Mr. Craggs' [*sic*], or Craig, was supplied by one Mr Tregent, a street telescope exhibitor, without any source being cited.

87 Thomas Woods, *The Monster Telescope* (Parsonstown [Birr], Ireland, 1845) 4.
88 Joseph Ashbrook, 'John Tebbutt, his observatory, and a probable nova', in *The Astronomical Scrapbook, Skywatchers, Pioneers, and Seekers in Astronomy*, ed. Leif J. Robinson (C.U.P., and Sky Publishing Corporation, Cambridge, Mass., 1984) 66–71.
89 Wilhelm Gotthelf Lohrmann, *Mondcharte in 25 Sectionen und 2 Erlä uterungstafeln* (Leipzig, 1878).
90 Benjamin Scott to *The Times*, undated cutting 'The new interior planet', probably autumn 1859, in Dr J. Lee's 'Album', M.H.S. Oxford Gunther 38, vol. 5, 23v. For the most thorough modern treatment, see Richard Baum and William Sheehan, *In Search of Planet Vulcan: The Ghost in Newton's Clockwork Universe* (Plenum Trade, New York and London, 1997).
91 Dreyer, 'Observatory' (ref.19) 715.
92 The main source for the Harvard Observatory foundation is W.C. Bond, 'History and Description of the Astronomical Observatory of Harvard College', in *Annals of the Astronomical Observatory of Harvard College* I, 1 (Cambridge, Mass., 1856), vi ff.: xliii–xlvi, 'The Great Refracting Telescope'. Pp.lxiii–lxvi list the Observatory's subscribers from 1839 to 1855, including David Sears, and Edward David Blomfield's 1848 $100,000 donation. Edward S. Holden, *Memorials of William Cranch Bond... and of his son George Phillips Bond* (San Francisco and New York 1897) 28. Charles Dickens on his visit to America in 1842 reckoned the U.S. Dollar to exchange at 4 shillings (20p) Sterling: *American Notes* (Chapman and Hall edn., 1907) Chapter III 'Boston', 28. I have used this exchange rate throughout the earlier sections of the present book.
93 F.G. Wilhelm Struve, *Description de l'Observatoire Astronomique Central de Pulkova*, 2 vols. (St. Petersburg, 1845). Krisciunas, *Astronomical Centers* (ref.29) 108.
94 Krisciunas, *Astronomical Centers* (ref.29) 124.
95 Deborah Jean Warner, 'Astronomy in Antibellum America', in *The Sciences in the American Context: New Perspectives*, ed. Nathan Reingold (Smithsonian, Washington, 1979) 55–75. D.J. Walker, *Alvan Clark and Sons. Artists in Optics* (Smithsonian, Washington, 1968), 39–112, (new edition, 1996), indicates how widespread were Clark's instruments among American colleges.
96 Robert A. McCutcheon, 'Amateurs in Astronomy', in *History of Astronomy. An Encyclopaedia*, ed. John Lankford (Garland, New York, 1997) 6–10. Prof. Lankford himself has published several works on American amateur astronomy, such as 'Amateurs and Astrophysics: A Neglected Aspect in the Development of a Scientific Specialty', *Social Studies of Science* **11** (1981), 275–303, and 'Astronomy's Enduring Resource', *Sky and Telescope* **76** (1988), 482–3.

97 Dreyer and Turner, *History of the R.A.S.* (ref.67) 2. David P. Miller, 'Method and the 'Micropolitics' of Science: The Early Years of the Geological and Astronomical Societies of London', in *Politics and Rhetoric of Scientific Method, Historical Studies*, ed. J.A. Schuster and R.R. Yeo (Reidel, Dordrecht, 1986) 227–57; for Astronomical Society, 245–50.
98 Dreyer and Turner, *History of the R.A.S.* (ref.67) 16, quoted from Augustus De Morgan's *Memoirs*, Section III, 41.
99 J.F.W. Herschel, *Memoir of Francis Baily* (London 1845), published as a pamphlet from *M.N.R.A.S.* 6, November 1844. This deals with Baily's work as a stockbroker, and (pp. 9–10) founder of the R.A.S.
100 Dreyer and Turner, *History of the R.A.S.* (ref.67) 5, cited from *Mem. [R.]A.S.* 1, 4.
101 Dreyer and Turner, *History of the R.A.S.* (ref.67) 11.
102 Friedrich Max Müller, *Chips from a German Workshop* 4 (London, 1875). Müller uses this phrase twice in his 'Life of Colebrooke', *ibid.*, pp. 338, 409.
103 Weale, *Description of... Observatories* (ref.80) 57.
104 J.B. Morrell and A. Thackray, *Gentlemen of Science. Early Years of the British Association for the Advancement of Science* (Clarendon Press, Oxford, 1981) 328–9, 509, 510.

3 AN INHERITENCE, A WIFE, A BENEFICE, OR A BREWERY: FINANCING FUNDAMENTAL RESEARCH

1 Jack Bowes Morrell and Arnold Thackray, *Gentlemen of Science. Early Years of the British Association for the Advancement of Science* (Clarendon Press, Oxford, 1981).
2 J.F.W. Herschel to William Whewell, 20 September 1831. J.B. Morrell and A. Thackray, *Gentlemen of Science: Early Correspondence* (Royal Historical Society 1984), letter 45, 66–8.
3 G.B. Airy, 'Report on the Progress of Astronomy during the Present Century', *Transactions of the B.A.A.S.* 1833, 180–6.
4 Airy to Sir Charles Wood, 22 March 1847, C.U.L. RGO6 2/259–264.
5 Airy to W.V. Harcourt, 5 September 1832: Morrell and Thackray, *Gentlemen of Science* (ref.2) letter 117, 150.
6 A. Chapman, 'The accuracy of angular measuring instruments used in astronomy between 1500 and 1850', *J.H.A.* 14 (1983), 133–7, reprinted in Chapman, *Astronomical Instruments and their Users: Tycho Brahe to William Lassell* (Variorum, Aldershot, 1996).
7 James Bradley, 'A Letter to Dr. Halley, giving an Account of a New Discovered Motion of the Fixed Stars', *Phil. Trans.* 35 (1728), 637–61. A. Chapman, *Dividing the Circle: the Development of Critical Angular Measurement in Astronomy, 1500–1850* (Praxis–Wiley, Chichester and New York, 1990; 2nd edn., 1995) 98–102.
8 Chapman, *Dividing the Circle* (ref.7) 146–52.
9 J.A. Bennett, *Church, State, and Astronomy in Ireland. 200 Years of the Armagh Observatory* (Armagh Observatory and Institute of Irish Studies, Queen's University, Belfast, 1990) 22, 59–66.

10 William Pearson, *Introduction to Practical Astronomy* II (London, 1829) 413–17.
11 Henry C. King, *The History of the Telescope* (Griffin, London, 1955) 230–2.
12 William Henry Smyth, *Cycle of Celestial Objects* 1 (1844) 335. Gavin Lowe (1744–1815) of Islington does not seem to have been an F.R.S., although he, and some of his astronomical work, is recorded in the Royal Society archives: 'Letters and Papers' Sequence, No. 10, 96, R. Soc. MS. He also wrote in the *Philosophical Magazine* **30** (1808), 67–9 (on the 1807 comet), and **45** (1815), 21–2 (on Polaris). He may have been related to the Miss Lowe (of Islington) who married Dr William Pearson, as mentioned later in this chapter (ref.24).
13 A.R. Martin, *No. 6 Eliot Place Blackheath, The House and its Occupants, 1797–1972* (Greenwich and Lewisham Antiquarian Society 1974) 11–24. I am indebted to Dick Chambers, of the Crayford Manor House Astronomical Society, Kent, for supplying me with details of Groombridge's house, and for searching, without success, for the northern meridian mark which Groombridge engraved on the south wall of Greenwich Park.
14 *A Catalogue of Circumpolar Stars, Deduced from the Observations of Stephen Groombridge*, ed. G.B. Airy (London 1838) vii ff., xxvii. Pearson, *Practical Astronomy* II (ref.10) 402–10.
15 Groombridge, *Catalogue* (ref.14) iii.
16 Groombridge, *Catalogue* (ref.14) xxx.
17 Groombridge, *Catalogue* (ref.14) iii.
18 Groombridge, *Catalogue* (ref.14) xxv.
19 John Pond, 'On the Declinations of some of the principal fixed Stars; with a Description of an Astronomical Circle; and some Remarks on the Construction of Circular Instruments', *Phil. Trans.* **96** (1806), 420–54.
20 Pond, 'Astronomical Circle' (ref.19) 423–4.
21 John Pond, obituary, *Gentleman's Magazine* (November 1836) 546–8. It is stated, p. 547, that it was Pond's work 'while residing in the country' that had convinced the Royal Society that he was 'the fittest man to succeed' the aged Maskelyne. I am indebted to Andrew Murray for his insights into the life and work of John Pond, as part of his preparation of the *New D.N.B.* article.
22 S.J. Gurman and S.R. Harratt, 'Revd. Dr. William Pearson (1767–1847): a Founder of the Royal Astronomical Society', *Q.J.R.A.S.* **35** (1994) 271–92: 272. This article is the most complete scholarly study of Pearson currently in print.
23 W. Pearson, obituary by Richard Sheepshanks, *Mem. R.A.S.* **17** (1848) 128–33 (reprinted in *M.N.R.A.S.* **8** (February 1848) 69–73), says that Pearson produced 63 articles, though Gurman and Harratt (ref.22), 290, say he only produced 62. Some of the horological articles were reprinted in Rees' *Clocks, Watches and Chronometers 1819–1820* (David and Charles, Newton Abbot, 1970), being extracted from Abraham Rees, *The New Cyclopaedia, or Universal Dictionary of Arts, Sciences, and Literature* (London, 1819).
24 Gurman and Harratt (ref.22) 281, ref. 29, state that Frances Low (Lowe), Pearson's first wife, was sister to Gavin Lowe, but give no source. Pearson, in his article 'Equatorial' in Rees, *Cyclopaedia* (ref.23), unpaginated, column 17, sig. 3E6, speaks of 'Mr. Lowe of Islington', while specifying no degree of relationship

to him; in his article 'Observatory' in Rees, unpaginated, sig. PP2v, he lists Gavin Lowe's Islington establishment as one of twenty English amateur observatories operational around 1810.
25 Gurman and Harratt (ref.22) 274, 286.
26 Pearson, 'Roof', sub-section 'Roof, *Rotative*, or *Rotatory* in Astronomy', in Rees, *Cyclopaedia* (ref.23), unpaginated, sig. 3^R3v. For a detailed account of past and present (*c*.1810) observatories in Britain, Europe, and elsewhere, see article 'Observatory' in Rees (nos. 23, 24).
27 Pearson, *Practical Astronomy* II (ref.10) 434–46 for a detailed account of the St. Petersburg Circle and its acquisition by an 'English purchaser' (435). For 500 guineas reference see W. Pearson to G.B. Airy, 30 October 1843, C.U.L. RGO6, cited in Gurman and Harratt (ref.22) 282.
28 For an account of the tests carried out on this Guinand–Tulley lens (not of Pearson's ownership) see 'Report of the Committee appointed by the Council of the Astronomical Society of London [later R.A.S.], for the purpose of examining the Telescope constructed by Mr. Tulley, by Order of the Council', read 12 May 1826, *Mem. [R]A.S.* **2** (London, 1826) 507–11. No money is mentioned in this Report, however; on that matter see J.L.E. Dreyer and H.H. Turner, *History of the Royal Astronomical Society, 1820–1920* (R.A.S., London, 1923) 16. Pearson's obituary, *M.N.R.A.S.* (ref.23) 73, says that he paid £250 for the Tulley lens.
29 Pearson's book earned him the R.A.S. Gold Medal in 1829: *History of the R.A.S.* (ref.28) 252.
30 Smyth, *Cycle* I (ref.12) 366–9.
31 Wrottesley, obituary, *M.N.R.A.S.* **28** (1868) 64–8.
32 R.C. Carrington, *Results of Astronomical Observations made at the Observatory of the University, Durham, from October, 1849, to April, 1852, under the General Direction of the Revd. Temple Chevallier...* (Durham 1855).
33 R.C. Carrington, *A Catalogue of 3,735 Circumpolar Stars Observed at Redhill, In the Years 1854, 1855 and 1856... for 1855.0* (London 1857), titlepage.
34 Carrington, *Redhill Catalogue* (ref.33), 'Introduction' v–x. Carrington's 5-inch aperture transit circle later passed to the Radcliffe Observatory, Oxford: E. Dunkin, *The Midnight Sky. Familiar Notes on the Stars and Planets* (Religious Tract Society, London, 1891) 177. It is now in the Museum of the History of Science, Oxford.
35 Carrington, *Redhill Catalogue* (ref.33) IV. For Simmonds's salary, Carrington to G.B. Airy, 7 November 1854, C.U.L. RGO6 /212, 158. Carrington's handwriting is rather cryptic, though I interpret it as saying that Simmonds received 'ten pounds, ten shillings' per month, or £126 p. a. This is a salary similar to that of a junior Greenwich Assistant for the period.
36 Carrington, *Redhill Catalogue* (ref.33) IV for zone technique.
37 For the clock case, see T.R. Robinson, 'On Mr. Carrington's note...', *M.N.R.A.S.*, January 1873, 121–3. R.C. Carrington, *Observations of the Spots on the Sun, from November 9, 1853 to March 24, 1861, made at Redhill* (London, 1863). Sir Norman Lindop, 'R.C. Carrington and his solar work', *Astronomy*

Now, May 1997, 41–4. E.T.H. Teague, 'Carrington's method of determining sunspot positions', *J.B.A.A.* **106**, 2 (1996) 82–6.

38 R.C. Carrington, 'Description of a Singular Appearance seen in the Sun on September 1, 1859', *M.N.R.A.S.* **20** (November 1859). Carrington's friend R. Hodgson of Highgate made a confirming observation, and noted its magnetic disturbance, *ibid.* 15–16. Agnes M. Clerke, *A Popular History of Astronomy during the Nineteenth Century* (3rd edn., London, 1893) 198–200, says that a magnetically-related storm was raging on the Sun from 28 August to 4 September 1859.

39 Clerke, *A Popular History of Astronomy* (ref.38) 179–84.

40 Account of resulting trial in *The Surrey Advertiser*, 30 March 1872, 'The Tragedy at the Devil's Jumps, Farnham'. Norman C. Keer, *The Life and Times of Richard Christopher Carrington B.A., F.R.S., F.R.A.S.* (1826–1875) (N.C. Keer, Heathfield, East Sussex, 1996) 44–7. I am indebted to Norman Keer and Sir Norman Lindop for information on Carrington.

41 William Herschel, 'On the Direction and Velocity of the Motion of the Sun, and Solar System', *Phil. Trans.* **95** (1805), 233–56. The following year Herschel published further computations in 'On the Quantity and Velocity of the Solar Motion', *Phil. Trans.* **96** (1806), 205–37.

42 William Herschel, 'Catalogue of a second thousand of new nebulae and clusters of stars; with a few introductory remarks on the construction of the heavens', *Phil. Trans.* **79** (1789), 212–55; reprinted in *The Scientific Papers of Sir William Herschel*, ed. J.L.E Dreyer, 2 vols. (Royal Society and R.A.S., London, 1912): vol. I, 336. A. Chapman, 'William Herschel and the Measurement of Space', *Q.J.R.A.S.* **30** (1989), 399–418: 402.

43 Joseph Ashbrook, 'The story of Groombridge 1830', in Ashbrook's *The Astronomical Scrapbook, Skywatchers, Pioneers, and Seekers in Astronomy* (Sky Publishing Corporation, Cambridge, Mass., and C.U.P., Cambridge, England, 1984) 352–9.

44 Robert Grant, *History of Physical Astronomy from the Earliest Ages to the Middle of the Nineteenth Century* (London, 1852) 551.

45 William Herschel, 'Account of the changes that have happened, during the last Twenty-five years, in the relative Situation of Double Stars; with an investigation of the causes to which they are owing', *Phil. Trans.* **93** (1803), 339–82; also in *Scientific Papers* II (ref.42) 277–96.

46 Pearson, *Practical Astronomy* II (ref.10) 57–9, for the excellence of the clock drive of Fraunhofer's Dorpat refractor, and the difficulty in performing micrometric work without a clock drive.

47 King, *The History of the Telescope* (ref.11) 188–9.

48 John Weale, *Description of the Public and Private Observatories in London and its Vicinity* (London, 1851) 57.

49 James South, obituary, *M.N.R.A.S.* **28** (1868), 69–72.

50 Spoken by Baily, President of the R.A.S., when bestowing the Gold Medal of the R.A.S. upon South, *Mem. [R.]A.S.* **2** (1826) 547; cited in M.A. Hoskin, 'Astronomers at War: South v. Sheepshanks', *J.H.A.* **20** (1989) 75–212: 75.

51 South, obituary (ref.49) 71.
52 James South and John Herschel, 'Observations of the Apparent Distances and Positions of 380 Double and Triple Stars, made in the years 1821, 1822, and 1823, and compared with those of other Astronomers', *Phil. Trans.* **114** Part III (1824) 1–412.
53 James South, Letter to *The Times*, 20 November 1838.
54 C.U.L. RGO6 803/1–803/136. Airy and South were on intimate terms from at least October 1823 to May 1830.
55 This story is related, without references, in King, *The History of the Telescope* (ref.11) 236.
56 Weale, *Description of... Observatories* (ref.48) 57.
57 South, obituary by T.R. Robinson, *Proc. R. Soc.* **16** (1867–8) xliv–xlvii. The around £1,000 cost of the Cauchoix lens is mentioned by Agnes M. Clerke in her article on South in *D.N.B.*
58 Hoskin, 'Astronomers at War' (ref.50).
59 South Papers 3, R.A.S. Archives: 'Appendix', facing p. 93, for α Lyrae; cited in Hoskin, 'Astronomers at War' (ref.50) 187, 209 (n. 75). For details of the Equatorial's design, Dreyer and Turner, *History of the R.A.S.* (ref.28) 52–5. For the original £1,235 cost and eventual £1,470 award, Hoskin, *ibid.*, 184, 193.
60 R. Sheepshanks, *A Letter to the Board of Visitors of the Greenwich Royal Observatory in reply to the calumnies of Mr. Babbage at their Meeting in June, 1853* (London, 1854) 17. As shown by Michael Hoskin, 'Astronomers at War' (ref.50) 178, 206, by 1853 Sheepshanks' 'loathing for South knew no bounds'.
61 Dreyer and Turner, *History of the R.A.S.* (ref.28) 54.
62 South's printed poster (1839) and engraving (1842?) of the smashed–up Troughton Equatorial, and his totalled scrap profits, are in the R.A.S. Archives, and are reproduced by Hoskin, 'Astronomers at War' (ref.50) 194–5, 196, 198.
63 Airy to W.H. Smyth, 29 July 1846, C.U.L. RGO6 2/296, for 'ignorant Blackguard' quote. South launched a full-scale attack on Airy in 1846 and 1847, including a critique of his management of the Royal Observatory, South to the Admiralty, 12 March 1847, RGO6 2/251-258, culminating in the accusation that in his failure to pay attention to Adams in 1845 Airy had failed to secure for Greenwich the 'Glory of the Greatest Astronomical Discovery of Modern Times' – the discovery of Neptune! Needless to say, Airy defended himself vigorously: RGO6 2/259–267. See also my own, and other scholars', contribution to the James South debate in Michael Hoskin, 'More on 'South v. Sheepshanks'', *J.H.A.* **22** (1991), 174–9, which cites further contemporary documents expressing opinions on Sir James.
64 The tone of South's many letters to Lord Rosse and T.R. Robinson in the Birr Castle Archives conveys his warm regard for his friends in Birr and Dublin. See also Robinson's obituary of South, *Proc. R. Soc.* (ref.57).
65 Félix Savary, 'Sur la détermination des orbites que décrivent autour de leur centre de gravité deux étoiles très rapproché es l'une de l'autre', *Connaissance de Temps* 1830, 56–69, 163–79. Sir John Herschel, *Outlines of Astronomy* (London, 1849) 572.
66 W.R. Dawes, obituary, *M.N.R.A.S.* **29** (1869) 116–21.

67 Agnes M. Clerke, 'W.R. Dawes', in *D.N.B.*, and R.A. Marriottt, 'W.R. Dawes', in *New D.N.B.* (O.U.P., forthcoming). Edward Crossley, Joseph Gledhill and James M. Wilson, *A Handbook of Double Stars* (London 1879) 35–6.
68 R.A.S. Lassell MSS 1:4, 5, 9:2, 9:8. Also A. Chapman, 'William Lassell (1799–1880): Practitioner, Patron and 'Grand Amateur' of Victorian Astronomy', *Vistas in Astronomy* 32 (1989) 341–70: 345; reprinted in Chapman, *Astronomical Instruments and Their Users* (ref.6), XVII.
69 Obituary (ref.66) 117.
70 Obituary (ref.66) 117.
71 For a detailed scholarly treatment of this discovery, see R.A. Marriott, 'Dawes, Lassell and Saturn's dusky ring', *J.B.A.A.* 96, 5 (1986), 270–7. By September 1851 Dawes was visiting Lassell in Liverpool to test, on the 20-foot reflector, the solar eyepiece that was to bear Dawes' name: R.A. Marriott, 'Dawes' Solar Eyepiece', *Bulletin of the Scientific Instrument Society* 14 (1987), 2–3.
72 Deborah Jean Warner, *Alvan Clark & Sons. Artists in Optics* (Smithsonian Institution, Washington DC, 1968), 50–4, (new edition, 1996).
73 Warner, *Alvan Clark* (ref.72) 52.
74 Samuel Smiles, *Men of Invention and Industry* (London, 1884), 344. Dawes used his 8-inch Cooke to obtain his important drawings of Mars: R.J. McKim and R.A. Marriott, 'Dawes' observations of Mars, 1864–65', *J.B.A.A.* 98, 6 (1988) 294–300.
75 Warner, *Alvan Clark* (ref.72) 52, 62. Also 'Biographical and Historical Notes on the Pioneers of Photometry in Ireland', in C.J. Butler and Ian Elliott, *Stellar Photometry. Current Techniques and Future Developments, Proceedings of the I.A.U. Colloquium* No. 136, 1992 (C.U.P., 1993), 3–12: 9. I am indebted to Dr. I. Elliott for this reference.
76 John Herschel Ms. 'Diary', transcript, 19 June 1846, R. Soc. Lib. MS.
77 George Bishop, obituary, *M.N.R.A.S.* 22 (February 1862) 104–6.
78 Bishop, obituary (ref.77) 105.
79 J.R. Hind, obituary, *M.N.R.A.S.* 56 (1895) 200. Agnes M. Clerke, 'John Russell Hind', *D.N.B.* On 30 September 1846, following his Neptune sighting, Hind wrote to *The Times*, Sir John Herschel (R. Soc. MS H.S. 9:333), James Challis (University of Cambridge Institute of Astronomy 'Neptune file'), and others.
80 G. Bishop, Royal Society 'Election Certificate', 16 February 1848, R. Soc. MS IX.211.
81 W.R. Dawes to *The Astronomical Register* 2 (1865) 43–4.
82 *Burke's Irish Family Records* (Burke's Peerage, London, 1976) 273.
83 R.A. Cauchoix, 'Nachrichten über Herrn *Cauchoix*'s Fernröhre mit Objectiven aus Glas und Cristall', *A.N.* No. 212 (Juillet 5, 1831) 349–52. Cauchoix lists several large lenses for sale, including that of 1 foot 1.86 pouces (English), which Cooper bought. Agnes M. Clerke, 'Cooper', *D.N.B.* I.S. Glass, *Victorian Telescope Makers, The Lives and Letters of Thomas and Howard Grubb* (Institute of Physics, London, 1997) 13–16. King, *The History of the Telescope* (ref.11) 180–1.
84 'Report of Council' of R.A.S., *M.N.R.A.S.* 11 (1851) 104. E.J. Cooper, obituary, *Proc. R. Soc.* 13 (1864) i.
85 E.J. Cooper, *Catalogue of Stars near the Ecliptic observed at Markree during the*

Years 1848, 1849, & 1850 and whose places are supposed to be hitherto unpublished, 4 vols. (Dublin, 1851–6): for Graham reference, see vol. II (1853) 'Preface'. Cooper's recognition of the Grand Amateur context in which he worked was expressed in his *Cometic Orbits, with Copious notes and addenda* (Dublin 1852), 'Dedication' and 'Preface'.

86 Airy seems to have first heard of Astraea in a letter from Johann Encke in Berlin: 'Astronomer Royal's Journal', 22 December 1845, C.U.L. RGO6 24. D.W. Hughes, 'Only the first four asteroids', *J.B.A.A.* **107**, 4 (1997) 211–13.
87 Cooper, *Catalogue of Stars* I (ref.85) V–XII.
88 Cooper, *Catalogue of Stars* I (ref.85) VI.
89 Cooper, *Catalogue of Stars* I (ref.85) XII. For surviving correspondence of the Markree Observatory, see Michael A. Hoskin, 'Archives of Dunsink and Markree Observatories', *J.H.A.* **13** (1982) 146–52, and Susan McKenna-Lawlor and M.A. Hoskin, 'Correspondence of Markree Observatory', *J.H.A.* **15** (1984) 64–8.
90 Glass, *Victorian Telescope Makers* (ref.83) 15.
91 I am indebted to Roger Hutchins for access to his forthcoming Oxford University D.Phil. thesis section which discusses Graham and Markree: 'A Mismatch of Ideals and Resources: British University Observatories *c.* 1820–1939'. See also W. Doberck, 'Markree Observatory', *The Observatory* **7**, 90 (1884) 283–8.
92 T.R. Williams, 'Amateurs in Astronomy', in *History of Astronomy. An Encyclopedia*, ed. John Lankford (Garland, New York, 1997). John Lankford, 'Amateurs and Astrophysics. A neglected Aspect in the Development of a scientific Specialty', *Social Studies of Science* **11**, 3 (August 1981) 275–305. In his recent comprehensive and masterly study, John Lankford, assisted by Ricky L. Slavings, *American Astronomy. Community, Careers, and Power 1859–1940* (University of Chicago, Chicago and London, 1997), is overwhelmingly concerned with employed, career astronomers. The benchmark of excellence, of course, is membership of the National Academy of Sciences; when Grand Amateurs like Henry Draper and Lewis Rutherfurd become Academicians, Prof. Lankford weighs them in as part of the professional establishment. In his references to amateurs, such as on pp. 5 and 19, he invariably assumes that they belonged to a lesser, disadvantaged breed that 'rarely had access to resources, power or the reward system' (p. 5).
93 J. Ashbrook, 'The legacy of S.W. Burnham', in Ashbrook, *Astronomical Scrapbook* (ref.43) 84–92. Espin, obituary by T.E.R. Phillips, *M.N.R.A.S.* **95**, 4 (1935) 319–22.
94 Ashbrook, 'Burnham' (ref.93) 84.
95 S.W. Burnham, *A General Catalogue of Double Stars Within 121° of the North Pole* (Carnegie Institution Publications, no. 5, Washington D.C., 1906). This had been preceded by Burnham's monumental paper 'Double Star Observations made at Chicago with the 18½-inch Refractor of the Dearborn Observatory, comprising: I. A Catalogue of 251 New Double Stars with Measures; II. Micrometrical Measures of 500 Double Stars', *Mem. R.A.S.* **44** (1877–9) 141–305.
96 Ashbrook, 'Burnham' (ref.93) 91. D.J. Warner, 'Burnham' in *D.S.B.* Lankford, *American Astronomy* (ref.92) 52ff. discusses Burnham's quasi-professional career,

as a man who 'never quite lost his amateur status'. It is true, however, that by the 1890s it was becoming harder for men from Burnham's background to carry out serious research work, and he was very much an exception. But this had not been the case in 1870.
97 An outstanding exception to this rule was Dr W.H. Steavenson, a medical man who, at the end of World War I, was permitted to use the Greenwich 28-inch: D.W. Dewhirst, 'Steavenson' obituary, *Q.J.R.A.S.* **18** (1977) 147–54: 149.

4 SIR JOHN HERSCHEL: A MODEL FOR THE INDEPENDENT SCIENTIST

1 W. Pearson, 'Dr. Herschel's Observatory at Slough' (*c*.1810), a sub-section (unpaginated, Sig. PP2) in the article 'Observatory', Abraham Rees, *A New Cyclopaedia, or Universal Dictionary of Arts, Sciences, and Literature* (London 1819). Günther Buttmann, *The Shadow of the Telescope: A Biography of John Herschel* (Lutterworth, Guildford and London, 1974), 7–20. Constance Anne Lubbock (*ed.*), *The Herschel Chronicle: The Life-story of Sir William Herschel and his Sister Caroline* (Cambridge, 1933). Professor Michael J. Crowe, of Notre Dame University, Illinois, U.S.A., has completed a CD–ROM *Calendar of the Correspondence of Sir John Herschel*, which will provide a synopsis and index of some 14,817 letters. I am much indebted to Professor Crowe for his assistance in locating particular Herschel letters.
2 Cited by Agnes M. Clerke, *The Herschels and Modern Astronomy* (London, 1895), 142–3.
3 Reproduced in the *Harvard Library Bulletin* 32 (1984), 73–82. Sections of it are also included in Owen Gingerich, *The Great Copernicus Chase and other Adventures in Astronomical History* (Sky Publishing Corp. and C.U.P., 1992), 160–4: 162.
4 *Scientific Papers of Sir William Herschel* I (Royal Society and R.A.S., 1912), Herschel's note for 1771, p. xxi. This £400, moreover, was not an exceptional sum but an indication of a growing prosperity: in December 1769 he reckoned his year's receipts at £316, and in 1770, £352: *ibid.* p. xxi. Herschel's notes, and those of Caroline, recorded in *Scientific Papers* xix–xxiii, *Herschel Chronicle* (ref.1), and elsewhere, indicate a busy yet prosperous lifestyle.
5 The 7-foot reflector of *c*.1795, formerly owned by the Revd Nathaniel Jennings F.R.A.S., and presented to the Museum of the History of Science, Oxford, in 1927 by Dr H.N. Evans, bears the label that William Herschel 'charged 100 guineas [£105] for this size of instrument'. No source is cited. William Herschel, obituary, in Edwin Dunkin, *Obituary Notices of Astronomers* (London, 1879) 112–13.
6 These cover the period 1800–1806, and, with a dividend of £34 11*s* 6*d* (£34.55), clearly represent only a fraction of his investments: M.H.S. MS Museum 58.
7 Will of Sir William Herschel, 17 December 1818, proved 3 October 1822, transcript, PRO PROB. 10/4640 and PRO PROB. 11/1662, Quire 527. I thank Peter Hingley, Librarian of the R.A.S., for locating this document. The R.A.S. Library, Add. MS 88/74, has an unidentified newspaper cutting of *c*.1822 which

provides a synopsis of Herschel's Will, and which led to the search for the original documents. We know that 'The Crown' Inn, a fine coaching inn with ten large windows overlooking the street, was Baldwin property that Lady Mary brought into her marriage with Sir William (a splendid Edwardian photograph survives, in local Slough records; however, 'The Crown' fell victim to 'planners' blight' in the 1960s); I am indebted to Brian Colthorpe of Slough for local information regarding the Baldwin and Pitt families in the district. The total extent of Sir William Herschel's estate at the time of his death in 1822 has not yet been computed. It was valued at under £60,000, but it is still not yet known how much under. Land Registry documents could give the worth of the named but unvalued properties included in his Will, while the as yet untraced Will of Lady Mary Herschel could be revealing as to the eventual estate which Sir John finally inherited.

8   Dunkin, *Obituary Notices* (ref.5) 112, speaks of Lady Mary Herschel, Sir William's wife, as 'a lady of considerable property' and daughter of a 'City Merchant'. For Lady Margaret, Sir John's wife's background, see *Lady Herschel: Letters from the Cape 1834–1838*, ed. Brian Warner (Friends of the South African Library, Cape Town, 1991), 9–10.

9   Francis J. Ring, 'John Herschel and his heritage', in *John Herschel, 1792–1871: A Bicentennial Commemoration. Proceedings of a Royal Society Meeting held on 13 May 1992*, ed. Desmond King–Hele (Royal Society, 1992), 3–15: 6.

10 Ring, 'John Herschel' (ref.9) 9. Herschel's Ms. transcript 'Diary' in the Royal Society Library contains many references to ill–health, while in a letter of 27 February 1854 Herschel was reminded by his Cambridge friend of forty years' standing, Adam Sedgwick, to guard his 'tender lungs', for 'when a cough settles in them you do not easily serve the fiend with an ejectment': *Life and Letters of Adam Sedgwick* II, ed. J.W. Clark and T.M. Hughes (C.U.P., 1890), 268.

11 G. Melvyn Howe, *Man, Environment and Disease in Britain, A Medical Geography* (Pelican, 1976), 176.

12 *Memoirs and Correspondence of Caroline Herschel by Mrs. John Herschel* (Murray, London, 1879), 25 January 1813, 120; 17 March 1813, 120.

13 John Herschel to Charles Babbage, 18 December 1815, Royal Society MS H.S. 20:30. See also *Collections of the Royal Society, Letters and Papers of Sir John Herschel: A Guide to the Manuscripts and Microfilm*, project editor Paul Kersaris (Royal Society, 1990), xxix.

14 Lubbock, *Herschel Chronicle* (ref.1) 351. Buttmann, *Shadow of the Telescope* (ref.1) 17.

15 John Herschel's first research paper was published anonymously when he was 20: 'Analytical formulae for the tangent, cotangent & C.', *Nicholson's Journal* **31** (1812), 133. His first specifically optical paper was 'On the Optical phenomena of Mother–of–Pearl, depending on its internal structure', *Edinburgh Philosophical Journal* **2** (1820) 114–21.

16 Lubbock, *Herschel Chronicle* (ref.1) 352.

17 Herschel to Margaret Herschel (wife), date uncertain, 10 August 1841 or 1843. The original letter is in the Harry Ransom Humanities Research Center, The

University of Texas, Austin, and is cited by Larry J. Schaaf, 'The poetry of light: Herschel, Art, and Photography', in King–Hele (*ed.*), *John Herschel* (ref.9) 77–99: 94, 99. It is clear that the Herschel family's cosmological quest had always struck sympathetic chords in romantic culture: see the anonymous 'Sonnet to the Late Dr. Herschel', dated September 1822, preserved as a cutting from an unspecified journal in the 'Albums' of Dr John Lee, M.H.S. Oxford Gunther 37, vol. 4, p. 89.

18 Herschel gives a good account of his instrumental procedures in 'Account of some Observations made with a 20-feet Reflecting Telescope', *Mem. [R.]A.S.* **2** (1826) 459–97: 460. A. Chapman, 'An Occupation for an Independent Gentleman: Astronomy in the Life of John Herschel', *Vistas in Astronomy* **36** (1994), 1–25; published in corrected form in Chapman, *Astronomical Instruments and their Users* (Variorum, 1996), V.

19 Several prints from Herschel's 1839 photograph of his father's now decaying 40-foot telescope survive. I reproduce the one in the M.H.S. Oxford collection in my 'Occupation for an Independent Gentleman' (ref.18).

20 South to *The Times*, 20 November 1838. In this letter South says Sir John Herschel is no longer welcome at his house. He also states that his greatly-admired deceased friend Sir William Herschel 'could place his 20-foot telescope on any star in the heavens... within five minutes of time'. (South recorded further reminiscences of Sir William Herschel and their joint observations of nebulous objects during the 1810s: *A.N.* 536 (1845–6), 115.) On 31 January 1846 South again wrote to *The Times*, recording Sir John Herschel's early efforts in observational astronomy, and provoked an exasperated reaction from Herschel when he read the letter in the next morning's paper: Herschel, 'Diary', 1 February 1846, R. Soc. MS.

21 Sir John Herschel's range of interests, and the wide spectrum of people who wrote to him, are evident from a perusal of the large collection of his letters and diaries, preserved in the Royal Society Library and the R.A.S. Library. A CD–ROM *Calendar of the Correspondence of Sir John Herschel* has been prepared by Professor Michael Crowe, of Notre Dame University, Illinois (ref.1).

22 Herschel's *Preliminary Discourse on the Study of Natural Philosophy* (London 1831) was immensely influential in forming the concept of scientific induction. Darwin told Herschel, 11 November 1859, when presenting him with a copy of *On the Origin of Species*, that 'Scarcely anything in my life made so deep an impression on me.' Original document in Harry Ransom Humanities Research Center, University of Texas, reproduced in Larry J. Schaaf, *Out of the Shadows. Herschel, Talbot, and the Invention of Photography* (Yale University Press, New Haven and London, 1992), 16.

23 The text of R.A. Locke's 'The Great Astronomical Discoveries Lately made by Sir John Herschel at the Cape of Good Hope', from the New York *Sun*, 1835, is reproduced in *The Man in the Moon*, ed. Faith K. Pizor and T. Allan Camp (Sidgwick and Jackson, London, 1971) 190–216.

24 Mary T. Brück, 'Mary Somerville, mathematician and astronomer of underused talents', *J.B.A.A.* **106**, 4 (1996), 201–6: 203.

25 This independence of approach also comes over in Edwin Dunkin's life of Sir William in his *Obituary Notices* (ref.5) 112–13. Dunkin, though only an infant

Plate 72. Liverpool Astronomical Society crest. The Latin motto reads 'Thus it is gone to the stars'.

Plate 73. Astronomical Society of Wales crest, 1896. Courtesy of David Powell and the Cardiff Astronomical Society.

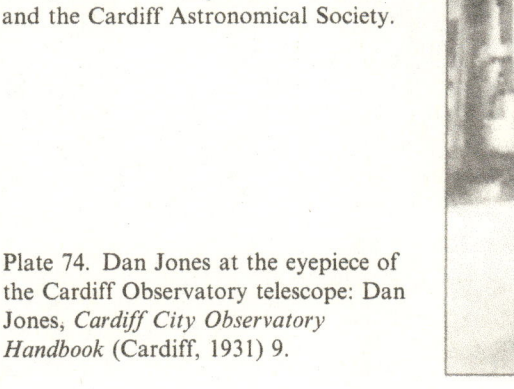

Plate 74. Dan Jones at the eyepiece of the Cardiff Observatory telescope: Dan Jones, *Cardiff City Observatory Handbook* (Cardiff, 1931) 9.

Plate 75. An early experiment in aperture synthesis? Mr J. Alun Lloyd of the Astronomical Society of Wales and his telescope, *c*.1905. There is no explanation for the railway-line telescope mount (in old Welsh colliery railway gauge). One wonders if it was an ingenious attempt to provide a stable mount that gave maximum sky access to an awkward site. Evans; *Seryddiaeth* (pl.43) 306.

Plate 76. John Willoughby Meares (1871–*c*.1946) with his 3¼-inch aperture Wray refractor on the roof of the Hove Electric Lighting Company, *c*.1893. Meares, 'Astronomical Notes and Drawings, Vol. 1, Planets' (volume dated 24 August 1894) 158, kindly loaned to the author by the Meares family.

Plate 77. J.W. Meares' naked-eye watercolour of the 22 January 1898 total solar eclipse observed from Buxar, India. Note the extensive coronal streamers. Meares, 'Astronomical Notes and Drawings, Vol. II' (pl.76) 8.

Plate 78. H.J. Townsend, late President, Leeds Astronomical Society, with his 9½-inch Herschel-type reflecting telescope in his garden, *c*.1900. *Leeds Astronomical Society Journal and Transactions* 1902, frontispiece.

Plate 79. Washington Teasdale, patriarch of the Leeds Astronomical Society, and President 1893–97. *Leeds Astronomical Society Journal and Transactions* **5** (1897–1898), frontispiece.

Plate 80. Saturn, drawn by H.J. Townsend between 26 May and 10 June 1895, with powers ×200, 300 and 500, with his 9½-inch aperture reflector. *Leeds Astronomical Society Journal and Transactions* **3** (March 1896), frontispiece.

when Sir William died in 1822, came to know and work with Sir John in the R.A.S., and gained thereby a great deal of first-hand insight into the life of Sir William. The same private independence shines through Caroline Herschel's *Memoirs* (ref.12).

26 John Herschel to W.V. Harcourt, 5 September 1831, and to William Whewell, 20 September 1831, in J.B. Morrell and A. Thackray, *Gentlemen of Science, Early Correspondence* (Royal Historical Society, London, 1984) letters no. 37, pp. 55–7, no. 46, pp. 66–8.

27 This quotation of Sir John Herschel was recorded in the *Memoirs* (sect. iii, 41) of Augustus De Morgan, and is cited in J.L.E. Dreyer and H.H. Turner, *History of the Royal Astronomical Society 1820–1920* (R.A.S., 1923), 16.

28 For British Association, see Herschel to Whewell, 20 September 1831 (ref.26), no. 46, pp. 66–8. The Royal Society Presidency, which a campaign in *The Times* and amongst many scientists had been hoping to secure for Herschel, was firmly refused by Sir John himself, who only wanted that 'retirement which is indispensable' for original scientific research: Herschel to C. Babbage, 26 November 1830, Royal Soc. MS H.S. 9:257. The same letter collection contains other documents pertaining to efforts made to persuade the reluctant Herschel to contest the R.S. Presidency. Throughout his life, John Herschel was watchful of his independence from faction, and concerned with his privacy and intellectual freedom, showing no wish for office of any kind. The controversy within the Royal Society, however, had been sparked off by C. Babbage's *Reflections on the Decline of Science in England, and Some of its Causes* (London, 1830). See also Roy M. MacLeod, 'Whigs and savants: reflections on the reform movement in the Royal Society, 1830–1848', in *Metropolis and Province, Science in British Culture 1780–1850, ed.* Ian Inkster and Jack Morrell (Hutchinson, London, 1983), 55–90.

29 'Sir John Herschel on the choice of a Standard Length', *The Leeds Intelligencer*, 31 October 1863, 7.

30 H.C. King, *The History of the Telescope* (Griffin, London, 1955), 236. Anita McConnell, *Instrument Makers to the World: A History of Cooke, Troughton and Simms* (Sessions, York, 1992), 29.

31 William Pearson, *Introduction to Practical Astronomy* II (London 1829), 57–9.

32 John Herschel, *A Treatise on Astronomy* (London, 1833), 388–95.

33 Félix Savary, 'Sur la détermination des orbites que décrivent autour de leur centre de gravité deux étoiles très rapproché es l'une de l'autre', *Connaissance de Temps* (Paris, 1830), 56–69, 163–72. J. Herschel, *Outlines of Astronomy* (London 1849), 564–80.

34 W.H. Smyth, *A Cycle of Celestial Objects*, 1 (London, 1844), 292.

35 John Herschel, *Results of Astronomical Observations made during the years 1834, 5, 6, 7, 8, at the Cape of Good Hope* (London, 1847): see total of his double stars, 171–242.

36 T.H.E.C. Espin, obituary by T.E.R. Phillips, *M.N.R.A.S.* **95**, 4 (February 1935), 319–22: 320.

37 J. Herschel, 'Account of some Observations' (ref.18) 459–61.

38 *Life and Letters of Adam Sedgwick*, II (ref.10) 107.

39 Maria Edgeworth to Harriet Butler, reprinted in Christina Colvin (*ed.*), *Maria Edgeworth, Letters from England 1813–1844* (Clarendon Press, Oxford, 1971), 506. See also Mary T. Brück, 'Maria Edgeworth: Scientific 'Literary Lady' ', *Irish Astronomical Journal* **23**, 1 (1996), 49–54.
40 This accident is reported in King, *History of the Telescope* (ref.30) 199, and ascribed to Sir David Brewster's *A Treatise on Optics* (London, 1831), 357. Though this page refers to Brewster's account of Ramage's telescope it says nothing whatsoever about Brewster's lucky fall. I have failed to trace it in *The Home Life of Sir David Brewster, by his daughter Mrs. Gordon* (Edinburgh, 1869). I cannot therefore offer a source for this story which, considering the acquaintance of Ramage and Brewster, has a ring of truth about it.
41 *Memoirs... of Caroline Herschel* (ref.12), 31 December 1783, 54–5.
42 The final Herschel statement on the nebulae came in John Herschel's *Cape Results* (ref.35) 137ff., 146ff. See also his *Outlines* (ref.33) 598.
43 This was the mirror of the 20-foot telescope, described in his 'Account' of 1826 (ref.18) 460. William Herschel, 'On Nebulous Stars properly so called', *Phil. Trans.* **81** (1791), 71–8 (reprinted in Herschel, *Scientific Papers* I, 422 (ref.4)) discusses the taxonomic problems of nebula classification. The most complete study of William Herschel's cosmological ideas, with extensive quotations of Herschel's original papers, is M.A. Hoskin's *William Herschel and the Construction of the Heavens* (Oldbourne, London, 1963). See also A. Chapman, 'William Herschel and the Measurement of Space', *Q.J.R.A.S.* **39** (1989), 399–418: 404ff., reprinted in Chapman, *Astronomical Instruments and their Users* (ref.18).
44 This provision of observational benchmarks of nebulae underpinned J. Herschel's observations of the Orion and Andromeda nebulae, described in detail in the 'Account' (ref.18) 487ff. and 495, and in *Cape Results* (ref.35) 31.
45 Herschel, *Cape Results* (ref.35) 25–32.
46 Herschel, *Outlines* (ref.33) 601–2. John Herschel was very cautious (more so, indeed, than his father) when it came to cosmological speculation from limited evidence.
47 John Herschel first discussed the 'hollow shell' idea of planetary nebulae in 'Observations of Nebulae and Clusters of Stars, made at Slough, with a Twenty-feet Reflector between the years 1825 and 1833', *Phil. Trans.* **123** (1833), 359–505: 500. Herschel, *Outlines* (ref.33) 602–4, for planetary nebulae. Chapman, 'An Occupation' (ref.18) 10–13.
48 William Herschel, 'On Nebulous Stars' (ref.43) 119.
49 J. Herschel discusses his 'particulate' idea of nebulous matter in 'Observations of Nebulae' (ref.47) 501.
50 Herschel, 'Observations of Nebulae' (ref.47), which totalled 2,307 nebulae and clusters. In his definitive 'Catalogue of Nebulae and Clusters of Stars', *Phil. Trans.* **154** (1865), 1–137, 7, Herschel compiled a grand total of 5,079 objects. (The paper was received by the Royal Society on 16 October 1863, though the *Phil. Trans.* volume bears the titlepage date 1865.)
51 *Herschel at the Cape: Diaries and Correspondence of Sir John Herschel*, eds. D.S. Evans, T.J. Deeming, B.H. Evans, S. Goldfarb (University of Texas, Austin and London, 1969), xx.

Notes and references

52 Herschel occupied the eighteen–roomed 'Feldhausen' from 23 April 1834 at a rent of £225 per annum, and purchased it 'out and out' for £3,000 on 18 November 1834: Diary entries in *Herschel at the Cape* (ref.51) 65, 107.
53 Herschel, *Outlines* (ref.33) 596.
54 Herschel, *Cape Results* (ref.35) Chapter IV, esp. 389.
55 Lady Margaret Herschel (Cape) to Thomas and Mary Baldwin, 9 April 1834, in *Lady Herschel, Letters from the Cape, 1834-1838*, ed. Brian Warner (Friends of the South African Library, Cape Town, 1991), 32-33; John Herschel (Cape) to Caroline Herschel, 6 June 1834, in *Herschel at the Cape* (ref.51) 71-3.
56 Lady Margaret Herschel (Cape) to Mrs Duncan Stewart, 15 July 1835, in *Lady Herschel* (ref.55) 79.
57 Agnes M. Clerke, *The System of the Stars* (London, 1905), 66, 246, 274-5.
58 Herschel suggested to Sir William Hamilton in June 1836 that the 'Milky Way is not a stratum but a ring': cited, without source, by Agnes M. Clerke, *The Herschels and Modern Astronomy* (Cassell, London, 1895) 173. Sir John Herschel, always very cautious about cosmological speculation (ref.46), usually kept to straightforward descriptions of objects, as in *Outlines* (ref.33) Chapter XVII. Herschel's definitive statement on the structure of the Milky Way is to be found in *Cape Results* (ref.35) Chapter IV.
59 For the 'Sequences' and 'Astrometer' techniques, see *Cape Results* (ref.35) 353-72, and *Outlines* (ref.33) 524-6.
60 William Herschel's classic statement on the similar generic size of the stars, and their original equidistance from one another in space – his 'distance luminosity' ratio – is given in 'Catalogue of a Second Thousand new Nebulae or Clusters of Stars; with a few introductory remarks on the Construction of the Heavens', *Phil. Trans.* **79** (1789) 212-56: 113-14, reprinted in Herschel, *Scientific Papers* I (ref.4) 336. For John's revision of this principle, derived from his 'astrometric' results at the Cape, see *Cape Results* (ref.35) 353-72: 371; also *Outlines* (ref.33) 522-3.
61 Lord Rosse used Sir John Herschel's 1833 nebula Catalogue (ref.50) as the benchmark against which he began his own surveys with the 72-inch, as is clear from Rosse's statements in 'Observations of the Nebulae', *Phil. Trans.* 140 (1850) 499-514.
62 Herschel, transcript of Ms. 'Diary' (ref.10), 10 November 1838.
63 Herschel, 'Diary' (ref.10), 29 November 1838 and 14 January 1839.
64 Herschel, 'Diary' (ref.10), 20 July 1839, for 'Collingwood' repair costs; 29 August 1839 for purchase price.
65 *Lady Herschel* (ref.55) 156-7 for accounts. It is also stated (p 156) that the Herschels' expenditure for 1832 had been £829, which was a greater sum than his 3% consols could have yielded, tending to confirm the presence of further funds. It is sobering to realise, however, that Sir John was troubled by the tax man: 'Attended Appeal Income Tax at Cranbrook expecting (like a blockhead) plain common justice. Got None.': 'Diary' (ref.10) 10 January 1843. At this time, Sir Thomas Maclear, as Astronomer Royal at the Cape, received a salary of £600 p. a.: *Herschel at the Cape* (ref.51) xx.
66 The Herschels clearly lived well, both at the Cape and in England. According to

Lady Margaret's Accounts, on 1 December 1834 she paid 'Month's wages to 1st. inst. as formerly to 8 servants £13-10-0' with an extra £1 for the coachman. On 5 December 1834, they also paid £120 'Transfer tax on £3000', presumably for the purchase of 'Feldhausen'. Sir John's 'Diary' (ref.10) when he was back in England clearly indicates that he lived very well as a Kent country gentleman, with a fine house, 'Collingwood', and a coachman who, on 12 March 1844, for instance, drove him to 'Staplehurst Station in our Carriage'. On 15 June 1850, he recorded the comings and goings of servants: 'Crick left – Jeffrey came at 16s a week & house & garden.' These were very good wages for an early Victorian servant. It is possible, however, that the Cape Expedition, followed by the purchase of 'Collingwood', had taken something of a toll on Sir John's finances, for Herschel documents from 1842 in The Humanities Research Center, University of Texas, Austin, indicate that a lack of funds to the extent of £50–£60 hampered his development of the 'Actinoscope' for photographing solar spectrum colours. Herschel even obtained a Royal Society grant for his work, but when the 'Actinoscope' proved unworkable in 1843, he dutifully returned the £100 to the Society, and presumably paid Dollond's firm for their wasted efforts out of his own pocket: Larry Schaaf, *Out of the Shadows. Herschel, Talbot, and the Invention of Photography* (Yale University Press, New Haven and London, 1992) 132.

67 Herschel, 'Diary' (ref.10), 23 July 1844. Mauvais' Comet had been discovered in Paris, 7 July 1844: J.R. Hind, *The Comets: A Descriptive Treatise upon these Bodies...* (London 1852) 165. And on 11 June 1845, he observed 'Gibsons Comet' with the 7-foot reflector: 'Diary' (ref.10).

68 Herschel's post-Cape 'Diary' (ref.10) is that of a celebrity fleeing to a country house, 'Collingwood', 12 miles from the nearest railway station (Staplehurst) in the vain hope of finding the peace in which to work: 14 March 1845 'Finished Magellanic Clouds & began collating the double Star Catalogue' for the forthcoming *Cape Results* (ref.35). On the other hand, his status demanded regular visits to London (see 'Diary'), such as on 15 June 1844, when Sir Robert Peel, the Prime Minister, introduced him to the boorish visiting King of Saxony. Herschel objected to being referred to simply as 'un mathématicien', and to the King turning his back to him. Herschel had also been attending a ball at Buckingham Palace on 31 July 1843.

69 Even from 8 to 11 January 1839 'this vile sciatica' caused him to be covered 'at Robert's recommendation with "Poor man's Plaister", a composition of Cobler's [*sic*] Wax spread thin on paper – from hip to ancle and across the loins – which makes one feel like an armadillo.' On 15, 21 and 25 July 1855 his medical brother-in-law and other surgeons operated on his knee joint. However, his 'Twelfth Child & ninth daughter' was born to the 63-year-old Herschel on 29 May 1855, leaving one to assume that his vigour was still far from exhausted. See 'Diary' (ref.10) under dates.

70 For an authoritative treatment of his Mint work, see G.P. Dyer, 'One of the best men of business: Master of the Royal Mint', in *John Herschel, 1792–1871* (ref.9) 105–13.

71 The 'Diary' (ref.10) charts the breakdown of Herschel's health during his last

months at the Mint, under great pressure from the currency reform, and from the toll taken by the Crimean War upon the government of which he was now part: *e.g.* 16 December 1854, 19 February 1855 (where he reported 'have lost half an inch'). The portrait of him by J. Dugden, 1853, reproduced in Buttmann, *Shadow of the Telescope* (ref.1) 86, shows the strain. Whether financial considerations (a £1,000 p.a. salary) played any part in Sir John's acceptance of the Mint Mastership is unclear; at his death in 1871 he was still a very wealthy man. From Sir John's Estate Accounts, 20 November 1885, produced after the death of Lady Margaret Herschel, the grand total came to £29,554: PRO Ref. IR 59/86. For this information I am indebted to Peter Hingley, who is researching the Wills of the Herschel family.

72 Herschel, 'Diary' (ref.10), 23 June 1847. 'Struve, Le Verrier, Adams, Airy & Co. Arranged for these four to meet at Hawkhurst after the [British] Association [meeting] is over.' They were all together at Collingwood on 10 July 1847.
73 Herschel, 'Diary' (ref.10), 25–27 May 1864. For Nasmyth's side of the relationship with Herschel, which he greatly valued, see *James Nasmyth, Engineer. An Autobiography, ed.* Samuel Smiles (Murray, London, 1889) 374, 379–80.
74 Herschel, 'Diary' (ref.10), 1 February 1846.

5 AN ASTRONOMICAL HOUSE-PARTY: THE BEDFORD–AYLESBURY AXIS

1 Hartwell House, on the western outskirts of Aylesbury, is now a hotel: Gervase Jackson–Stops, 'Hartwell House, Bucks.', *Country Life*, 22 November 1990, 68–73.
2 Harry Wilmot Buxton, 'John Lee', obituary, *Records of Buckinghamshire* III (1863–9) 215–36: 234. (The original Ms. of this obituary is in the M.H.S., Oxford: MS Buxton 1.)
3 I.F., 'Admiral William Henry Smyth', *Astronomical Register* 4 (1866) 117–24. This beautiful piece of biographical writing brings Smyth to life as a genial, learned, and widely loved man. Its author had clearly known him well for a long time. I suspect that 'I.F.' was John Fiott, the name with which John Lee was christened, and in which he followed the classical Latin usage of scribing the letter J as I. He took the name Lee to qualify for part of his inheritance. Admiral Smyth was grandfather to Lord Baden–Powell, the First Scout, and is amusingly described in E.E. Reynolds, *Baden–Powell, A Biography* (O.U.P., 1942) 12–13.
4 W.H. Smyth, *Aedes Hartwellianae, or Notices of the Manor and Mansion of Hartwell* (London 1851) 223–4.
5 Revd G.D. Reade, 'Account of the observatory at Hartwell House', read by Lee in the first person (as though dictated to Reade) to the R.A.S., in 'Portfolio' in the *Aylesbury News*, 12 August 1854. The pasted cutting of this article is in Lee's 'Scrapbook', 71v, para. 2: M.H.S. Oxford Gunther 37. This article describes Lee's Observatories, and gives a flavour of his convivial sharing of his astronomy with a circle of friends in their 'synod'.

6   Lee, 'Album' 1 (on fly sheet) 'Autographs, Colworth 1824 to Hartwell 1844', M.H.S. Oxford Gunther 9, 57 (loose sheet). Colworth was another of Lee's estates.
7   This was the Hartwell 'synod', over which Admiral Smyth presided as the guiding spirit: 'Account of the Observatory' (ref.5) para. 3. Thomas Dell was a relatively poor man, whose Aylesbury transit observatory had been set up by Lee, so that they could work together: Smyth, *Aedes* (ref.4) 227.
8   William Pearson, article 'Observatory' in A. Rees's *The New Cyclopaedia, or Universal Dictionary of Arts, Sciences and Literature* **25** (London 1819). W. Pearson, *An Introduction to Practical Astronomy* II (London 1829), does not describe the observatory as a building, but does describe virtually every type of instrument, and its usage, that could be found in a research observatory in 1829.
9   Brewster's *Edinburgh Encyclopaedia* (1830) contained major articles on 'Astronomy' (Brewster, Wallace, Pond), 'Astronomical Instruments' (Troughton), 'Optics' and 'Graduation' (Brewster), 'Observatory' (Pond), etc.
10 H.H. Turner, *Records of the R.A.S. Club, 1820–1910* (Oxford 1910). Smyth was one of the mainstays of the R.A.S. Club for 40 years. Turner's *Records*, 'Preface', mentions that in 1856 Smyth agreed to edit a book of 'anecdotal and biographical matter' relating to the Club, but sadly never got around to doing so – 'alas! gone beyond recall'. On 10 December 1849, Smyth wrote to Airy from his London address, 3 Cheyne Walk, Chelsea, regretfully excusing himself from the imminent Club dinner because of lumbago. Like many of Smyth's letters to friends, this one also carries one of the Admiral's cartoons: a sketch of his sick–room, with its bottles, lozenges, and 'fissick': C.U.L. RGO6 231/265.
11 Lee, 'Album' 1 (ref.6) M.H.S. Oxford Gunther 9, 60. 'Account of a private Observatory recently erected at Bedford. By Captain W.H. Smyth, R.N., Foreign Secretary to the Society', 10 December 1830, *Mem. [R.]A.S.* **4** (1831) 545–68.
12 W.H. Smyth, *A Cycle of Celestial Objects*, I (London, 1844) 326, 335.
13 See Smyth's(?) drawing entitled 'A Half–pay Officer's Work–shop, not 1,000 miles from London, 1829', Lee 'Album', Gunther 9 (ref.6) 60 (*V.A.A.* Plate 3). This sounds surprisingly similar to the 3-inch aperture Tulley equatorial, though with an 8-foot tube, which was at Hartwell, though named 'the Smythean' in 1829: Lee, 'Album' 1 (ref.6) 57.
14 Smyth, *Cycle* (ref.12) 338.
15 Smyth, *Aedes* (ref.4) 243, where Smyth provides further information about his acquisition of the ex-Sir James South 5.9-inch lens, of which South said 'I believe it to be Tulley's chef d'oeuvre.' In an undated but sequentially 1829 entry in Lee's 'Album' 1 (ref.6) 61v, are drawn as circles the clear diameters of several object glasses: Bedford (Smyth's) 6 inches [5.9 clear]; Hartwell, 3, etc.
16 See Chapter 3, Sir James South. I.S. Glass, *Victorian Telescope Makers, The Lives and Letters of Thomas and Howard Grubb* (Institute of Physics, Bristol, 1997) 13–16 for Markree and Grubb's mounting of Cauchoix's lens.
17 Smyth, *Aedes* (ref.4) 227.
18 Buxton's obituary of Lee (ref.2) brings out the eclectic range of his interests, as does H.A. Hanley, *Dr. John Lee of Hartwell* (Buckinghamshire Record Office, 1983) 15–18. Smyth's *Aedes* (ref.4) contains a ground plan of the Mansion,

showing the observatory leading off from the library, which is reprinted in Jackson-Stops' 'Hartwell House, Bucks.' (ref.1) 68.
19 Lee, 'Album' 1 (ref.6) 76, 81, 89r, for brick-laying certificates, 1831.
20 Lee, 'Album' 1 (ref.6) 65 for Smyth's jocular challenge for the completion of the mount for a 5-foot equatorial at Hartwell within the next six years from 18 December 1829. Lee accepted.
21 Lee, 'Album' 1 (ref.6) 102, for Davies' Brick Certificate.
22 The exact chronology is not clear. Smyth, in *Aedes* (ref.4) 238, says the equatorial tower was started three years after the transit room, which dated from 1831. Yet Smyth's 5.9-inch Tulley refractor was later acquired by Lee in 1839, when Smyth went to Cardiff to superintend the construction of the new floating dock: Smyth, obituary by 'I.F.' (ref.3) 117–24. It is likely that a smaller equatorial building of c.1834, for the wagered 5-foot, 3-inch aperture equatorial, was later extended to accommodate the much larger 5.9-inch aperture instrument.
23 John Lee, [Manifesto] 'To the Electors of the County of Buckingham', Hartwell, 4 December 1863: C.U.L. RGO6 381/481. In this printed address, to support Lee's unsuccessful stand for Parliament, he advocated anti-tobacco laws, the abolition of the game preservations or hunting laws, the extension of the suffrage to *all* intelligent people (such as members of Mechanics' Institutions) *including women over 21*, and mentioned the Peace and Temperance Festivals held in the grounds of his mansion. A more detailed Manifesto was issued on 10 December 1863: RGO6 381/482. Lee was 80 years old in 1863, though still full of vigour, maintaining his legal practice in London, and had just completed his term as President of the R.A.S. On 22–23 July 1845 Lee hosted one of his 'Peace and Brotherhood' rallies in the park surrounding Hartwell House, complete with the 'Thame Teetotal Brass Band' to entertain the crowd. He also hosted anti-hunting and anti-duelling meetings: John Lee, 'Album' 2, M.H.S. Oxford Gunther 10, 76v–77v.
24 Hanley, *John Lee* (ref.18) 11, cited without reference.
25 Hanley, *John Lee* (ref.18) 27, cited without reference. Cecelia Lee's poor penmanship is visible in several places: Lee, 'Album' 2 (ref.23), 4 May 1848, 111v–113.
26 Smyth, *Cycle* I (ref.12) 335.
27 'I.F.' (John Lee?), obituary 'Smyth', *Astronomical Register* (ref.3) 117–24.
28 This is the observatory described in Smyth's *Aedes* (ref.4) Chapter IV. His residence at St. John's Lodge is mentioned in an obituary previous to 'I.F.'s (ref.3): see obituary 'Admiral Smyth', *Astronomical Register* 3 (1866) 241–3.
29 All Smyth's obituaries, listed in his *D.N.B.* entry, describe his naval career. 'I.F.', *Astronomical Register* (ref.3) 117, however, is especially detailed, giving dates of service in specific ships.
30 'I.F.', *Astronomical Register* (ref.3) 117. Hermann A. and Mary T. Brück, *The Peripatetic Astronomer, The Life of Charles Piazzi Smyth* (Hilger, Bristol, 1988) 1–3.
31 W.H. Smyth, *The Cycle of Celestial Objects, continued at the Hartwell Observatory to 1859... containing details of the Aedes Hartwellianae* (London, 1860) 63.
32 Richard Sheepshanks, *A letter to the Board of Visitors of the Greenwich Royal*

Observatory in reply to the calumnies of Mr. Babbage at their Meeting in June, 1853, and in his book entitled 'The Exposition of 1851' (London, 1854), 2nd edn., 1860, 'with correspondence prefixed', 16–17. Cited by Michael Hoskin, 'Astronomers at War: South v. Sheepshanks', *J.H.A.* **20** (1989) 178–80, 206.

33   Smyth, *Cycle* 1 (ref.12) 366: 'Hints to Amateur Astronomers'.
34   Smyth, *Cycle* 1 (ref.12) 370–8.
35   Smyth, *A Cycle of Celestial Objects*, 2: *The Bedford Catalogue* (London, 1844).
36   Smyth, *Aedes* (ref.4) 312–41.
37   John Lee, 'Album' 2 (ref.23) 57.
38   Smyth, *Aedes* (ref.4) 303.
39   Herbert Sadler, 'Notes on the late Admiral Smyth's "Cycle of Celestial Objects", Volume the Second, commonly known as the "Bedford Catalogue"', *M.N.R.A.S.* **39** (January 1879) 183–95.
40   Joseph Ashbrook, 'The Sadler-Smyth scandal', reprinted from *Sky and Telescope*, in *The Astronomical Scrapbook, Skywatchers, Pioneers, and Seekers in Astronomy*, ed. Leif J. Robinson (Sky Publishing Corporation and C.U.P., 1984) 51–6, provides an elegant and well-researched account of the 1879 'scandal'.
41   Ashbrook, 'Sadler–Smyth scandal' (ref.40) 55.
42   *Astronomical Register* **3** (1865) 186–7.
43   Roger Hutchins, 'John Phillips 'Geologist-Astronomer', and the Origins of the Oxford University Observatory, 1853–1873', *History of Universities* XIII (O.U.P., 1994) 193–249.
44   The Revd T.W. Webb, *Celestial Objects for Common Telescopes* (London, 1859) 47, was also discussing the British Association's proposed mapping of the Moon to a large scale with powers of 1,000. Hutchins, 'John Phillips' (ref.43) 206–10. Phillips also experimented with lunar photography, and on 1 February 1853 wrote to Lord Rosse that as such 'momentary an exposure is sufficient to affect the plate, it may *easily* get lunar pictures from your mirror': Rosse Archives, Birr Castle, K17: 35. Walter Goodacre, 'The Moon', Chapter V in *Splendour of the Heavens* I, ed. T.E.R. Phillips and W.H. Steavenson (Hutchinson, London, 1923) 257–8.
45   'W.R. Birt', obituary, *M.N.R.A.S.* **42**, 4 (February 1882) 142–4. I am also very much obliged to Roger Hutchins for the typescript copy of his life of W.R. Birt (August 1996), soon to be published in the *New D.N.B*. This is the best scholarly assessment of Birt to have been written in recent times, and clearly defines Birt's relationship with Lee and Phillips.
46   Many Victorian amateurs were keen lunar draughtsmen in their own right, and the putative alteration of the crater Linné in 1866 directed a great deal of attention to the lunar surface: in *M.N.R.A.S.* **28** (1868) 185–8, no less than three eminent amateurs – Revd T.W. Webb, Edward Crossley, and Capt. William Noble – supplied *independent* observations as part of the same article. English amateurs like H.P. Wilkins and Walter Goodacre, in the early British Astronomical Association – working in the independent tradition – produced fine lunar studies, and Wilkins' map is still sometimes used by modern amateurs. The present author, when a schoolboy around 1961 purchased a new edition of Wilkins' map, revised by Patrick Moore.

47 Thomas Maclear, 'Description of a small Observatory at Biggleswade', *Mem. R.A.S.* **6** (1833) 147–8. Maclear joined the R.A.S. as member 308 on 9 November 1828, with Smyth, Stratford, South, Beaufort and Lee as the signatories on his certificate.
48 Maclear, 'Description' (ref.47) 148.
49 Maclear, 'Description' (ref.47) 148.
50 Agnes M. Clerk, obituary, 'Thomas Maclear', *D.N.B.*
51 In spite of his relatively early withdrawal, Bedford School ranked Charles Piazzi Smyth as one of its worthies: see his obituary in *The Ousel* (Bedford School Magazine), 1900, 22. Also Brück and Brück, *Peripatetic Astronomer* (ref.30) 4, 16.
52 Lee, 'Album' 1 (ref.6) 108, 22 August 1833.
53 Lady Richarda Airy to Lady Margaret Herschel, 25 July 1845, discussing the imminent arrival of Lady Maclear's thirteenth child, in Cape Town. It is clear that the three astronomical wives were in close contact. Enid Airy Papers: I am indebted to the Airy family for access to Lady Airy's papers in their hands. In 1997, the collection was donated to Cambridge University Library, and is in process of classification.
54 Lady Richarda Airy's letters to Lady Herschel, 1834–1852, contain many such references to scientific-*cum*-social visits: Cambridge, 19 June 1845; Edinburgh, 15 January 1845; Hawkhurst, 13 December 1845: Enid Airy Papers (ref.53).
55 Obituary, 'Samuel Whitbread', *M.N.R.A.S.* **40** (February 1880) 207. J.M. Walker and W.S. Pike, 'Pen Portraits of Presidents – Samuel Whitbread, F.R.S.', *Weather* **52**, 12 (December 1997).
56 William Simms, *The Achromatic Telescope and its Mountings, Especially the Equatorial to which are added Some Hints on Private Observatories* (London 1852) 44, 45.
57 Charles Whitbread Observing Journal, 'Observatory, Cardington, 1850–[1852]', MS volume in Bedford Record Office, W. 4035. I am indebted to Michael J. Watt of the Bedford Astronomical Society for drawing my attention to this volume, and to Miss Louisa Watt for providing me with a complete transcript of the volume.
58 Whitbread, 'Observatory, Cardington' (ref.57), 31 October 1851. The Journal suggests that the move to the new Observatory extended over several weeks: on 10 December 1851, for instance, Whitbread 'fixed the Stone for the Transit for good in Portland Cement'.
59 Whitbread, Observing Journal (ref.57), 9 September 1851.
60 De La Rue was at Cardington observing over 28–29 February 1852. Indeed, Whitbread's Observing Journal, like Lee's Hartwell 'Albums', is full of social activity, especially as Whitbread invited local gentry and friends from London to see the heavens: 10 February 1852, 'I took Mr. Harvey to see the Transits of Three stars, and Mars, and I showed him the Nebula in Orion… and the Galaxy in Perseus'; 19 February 1852, 'Emma Lefevrer [?] went down from London with me, in order to see my Observatory.'
61 Whitbread, Observing Journal (ref.57), 20 August 1851. Warren De La Rue went on to build his own speculum polishing machine for 13-inch mirrors, and openly thanked William Lassell and James Nasmyth for their help: De la Rue, 'On

figuring Specula', *M.N.R.A.S.* **13** (1852) 44–51; S. Smiles (*ed.*), *James Nasmyth, Engineer: An Autobiography* (Murray, London, 1889) 314.
62 J. Weale, *Description of the Public and Private Observatories in London and its Vicinity* (London 1851) 62.
63 Weale, *Description of... Observatories* (ref.62) 62.
64 Weale, *Description* (ref.62) 73–5 provides a detailed account of this amateur–professional cooperation in meteorology.
65 Lee, 'Album' 2 (ref.23), 3 April 1850, 134v–135. This MS entry states that a previous Meteorological Society had been founded by Dr Birkbeck in 1823, but it had ceased after publishing a single volume of *Transactions*.
66 Smyth, *Cycle... continued... to 1859* (ref.31) 383ff.
67 Smyth, *Cycle... continued... to 1859* (ref.31) 394.
68 Weale, *Description* (ref.62) 75 for an account of how the national network of volunteer meteorological observers despatched their readings to London for collation and publication in the *Daily News*.
69 Various subsequent meetings of the 'British Meteorological Society' are recorded in Lee's 'Album' 2 (ref.23): on 16 September 1856, 215v, it met, and included Glaisher's friend Henry Perigal, and James' precocious 8-year-old son, James Whitbread Lee Glaisher. See also John L. Hunt, 'James Glaisher FRS (1809–1903) Astronomer, Meteorologist and Pioneer of Weather Forecasting: 'A Venturesome Victorian'', *Q.J.R.A.S.* **37**, 3 (1996) 315–47: 322–3; also R. Corless, 'A Brief History of the Royal Meteorological Society', *Weather* **5** (1950) 78.
70 Norman Pogson to W.H. Smyth, 22 July 1858, reproduced in Smyth's *Aedes* (ref.4) 69.
71 For Glaisher's salary, see G.B. Airy, 'Scheme for the payment of salaries of the personnel establishment of the Royal Observatory, Greenwich', C.U.L. RGO6 4/132, 24 January 1859; also Cecilia Glaisher, 5 August 1854, Lee 'Album' 2 (ref.23) 189.
72 J.W.L. Glaisher (aged 8), Lewisham, 29 September 1856: Lee 'Album' 2 (ref.23) 162. Letter glued into 'Album'. J.W.L. Glaisher went on to a brilliant Cambridge career in his own right; see John L. Hunt, 'J.W.L. Glaisher, F.R.S., Sc.D. (1848–1928)', *Q.J.R.A.S.* **37**, 4 (1996) 743–57.
73 Richarda Airy, 22 December 1854: Lee 'Album' 2 (ref.23) 194v. On 29 October 1834, Richarda Airy had confided to Lady Herschel that George Airy 'dreads and detests crowded assemblies and public dinners: they make him ill'. Similarly, Lady Airy again, 15 February 1845, commented upon the fact that her husband had 'borne two successive *dinners* without a headache': Enid Airy Papers (ref.53).
74 Airy, 'Astronomer Royal's Journal', C.U.L. RGO6 25, 14–15 December 1854. Hubert, G.B. Airy's physician son, produced a major clinical study on migraine, which was communicated to the Royal Society by his father. It is clear that Hubert, his father, Sir John Herschel and several other scientists of the age were plagued by this disease: Hubert Airy, 'On a distinct form of Transient Hemiopsia', *Phil. Trans.* **155** (1870) 247–64, with plates depicting the classic 'fortification' flashes of acute migraine.
75 Lee, 'Album' 2 (ref.23) 122.

76 Lee, 'Album' 2 (ref.23) 140.
77 Lee, 'Album' 2 (ref.23) 175.
78 Smyth, *Cycle... continued... to 1859* (ref.31) 123.
79 Lee, 'Album' 2 (ref.23) 213v.
80 Lee, 'Album' 2 (ref.23) 115.
81 Lee, 'Album' 2 (ref.23) 168.
82 Lee, 'Album' 2 (ref.23) 168.
83 Lee, 'Album' 2 (ref.23) 169.
84 Lee, 'Album' 2 (ref.23) 63, continued on 74.
85 Lee, 'Album' 2 (ref.23) 104.
86 Lee, 'Album' 2 (ref.23) 223v–224.
87 Lee, 'Album' 1 (ref.6) 65.
88 Lee, 'Album' 1 (ref.6) 45.
89 Lee, 'Album' 1 (ref.6) 77, 83.
90 Lee, 'Album' 2 (ref.23) 115.
91 Lee, 'Album' 2 (ref.23) 147.
92 Smyth, *Aedes* (ref.4) 247, 254–83.
93 Lee, 'Scrapbook' 4 (ref.5).
94 Smyth, *Aedes* (ref.4) 247.
95 Samuel Horton, lunar drawing and printed text from an unspecified publication, 1858, in Lee, 'Scrapbook', M.H.S. Oxford Gunther 38, 7.
96 Smyth, *Cycle... continued... to 1859* (ref.31) 127.
97 Smyth, *Cycle... continued... to 1859* (ref.31) 129.
98 Smyth, *Cycle... continued... to 1859* (ref.31), illustration, 129.
99 Weale, *Description* (ref.62) 68. The Revd G.D. Reade in his 'Hartwell House' (ref.5) 71v mentions that after 1827 some 'ten or eleven observatories' were erected in Bedfordshire, following Smyth's and Lee's example. The Revd G.D. Reade was no doubt related to the Revd Joseph B. Reade, F.R.S., who was a good friend of Lee, and wrote to him familiarly from Stone Vicarage, as on 5 April 1843, in Lee, 'Scrapbook', M.H.S. Oxford Gunther 36, 109.
100 Weale, *Description* (ref.62) 69. It is clear that John Lee was Dell's astronomical patron, and had helped him set up his modest observatory: Smyth, *Aedes* (ref.4) 227.
101 Lee, 'Scrapbook', Gunther 36 (ref.99) 97–104 for Dell's observations.
102 Dell to Lee (?), 25 August 1844 (?) (undated, but year from context on same page) in Lee, 'Scrapbook', Gunther 36 (ref.99) 97. John Russell Hind, *Comets, A Descriptive Treatise upon these Bodies...* (London, 1852), makes no mention of Melhaps, nor does Donald K. Yeomans, *Comets. A Chronological History...* (Wiley, New York, 1991). Hind, 165, however, says that Mauvais first sighted his comet in Paris, 7 July 1844.
103 Dell to Lee, 12 April 1852: Lee, 'Scrapbook', Gunther 37 (ref.5) 74v. Dell suggested that the light might be a cometary tail.
104 Weale, *Description* (ref.62) 69.
105 Joseph Turnbull to Lee, Aylesbury, 18 January 1858: Lee, 'Scrapbook', Gunther 38 (ref.95) 35v–36.

106 Charles Shea to Smyth, Connaught Square, Hyde Park, 2 August 1861: Lee, 'Scrapbook', Gunther 38 (ref.95) 48.
107 Account of observation of eclipse by Charles Howell and Charles Shea [Shay] from Howell's Hove, Sussex, observatory, in *Morning Herald*, 18 October 1856, in Lee, 'Scrapbook', Gunther 37 (ref.5) 75v. I am also indebted to Michael Feist, of Portslade, for information from his own local researches into the location of the Victorian Hove Observatory.
108 Dillwyn Llewelyn to Smyth, 3 January 1859: Lee, 'Scrapbook', Gunther 38 (ref.95) 17v–18.
109 J. Fletcher and Captain Jacob to Smyth, Clifton & Crossbarrow Collieries, nr. Workington, 15 March 1860, in Lee, 'Scrapbook', Gunther 38 (ref.95) 38, r and v. This was probably the same 9¼-inch (perhaps exaggerated to 9½ by its owner) refractor which Thomas Cooke supplied to Isaac Fletcher, of nearby Cockermouth, in 1857: S. Smiles, *Men of Invention and Industry* (London, 1884) 344.
110 Joseph Conter to Samuel Horton, 29 October 1859: Lee, 'Scrapbook', Gunther 38 (ref.95) 19.
111 Weale, *Description* (ref.62) 73–5, for description of Glaisher's system. As was seen in ref.95, however, Samuel Horton performed astronomical as well as meteorological observations.
112 Weale, *Description* (ref.62) 57.
113 The 'Uranian Society' was advanced by J.M. Cavalier, Hon. Sec., 1 Belle Vue Place, Mile End Road, London, on 6 June 1839 (handbill), and notice given in the *Argus* newspaper, 14 July 1839, by Mr W. White. A meeting was proposed 'at the rooms of the Meteorological Society'. Lee, 'Scrapbook', Gunther 36 (ref.99) 55–6.
114 Several are preserved in Lee's 'Scrapbook', Gunther 38 (ref.95) 15v, 16, 19.

6 THE BROTHERHOOD OF THE BIG REFLECTING TELESCOPE

1 For William Herschel's description of his original 20-foot telescope with an 18.7-inch mirror, see his 'Catalogue of One Thousand new Nebulae and Clusters of Stars', *Phil. Trans.* **76** (1786) 457–99: 457–63, reprinted in *The Scientific Papers of Sir William Herschel* (2 vols., R. Soc. and R.A.S., London, 1912) I, 260–303. For John Smeaton's detailed first-hand account of the operating procedures of Herschel's 20-foot, 4 November 1785, see William Pearson's article 'Telescopes' in Abraham Rees, *The New Cyclopaedia, or Universal Dictionary of Arts, Sciences, and Literature* (London 1819) fol. Nn 2. John F.W. Herschel, 'Account of some Observations made wih a 20-feet Reflecting Telescope', *Mem. R.A.S.* **2** (1826) 459–97: 459–61 for operating procedures and quality.
2 William Herschel, 'Description of a Forty-Feet Reflecting Telescope', *Phil. Trans.* **85** (1795) 347–409; 485–6 for an account of his earlier telescopes, including the 20-foot (reprinted in Herschel, *Scientific Papers* I (ref.1) 485–527). J.A. Bennett, 'On the power of penetrating space: the telescopes of William Herschel', *J.H.A.* **7** (1976) 75–108.
3 W. Herschel, 'Account of a Comet' [discovery of Uranus], *Phil. Trans.* **71** (1781)

492–501. W. Herschel, 'An Account of the Discovery of Two Satellites revolving around the Georgian Planet', *Phil. Trans.* **77** (1787) 125–9 and 'On the Georgian Planet and its Satellites', *Phil. Trans.* **78** (1788) 364–78 (reprinted in Herschel, *Scientific Papers* I (ref.1) 30–8; 312–14; 317–26). His discovery of the satellites in 1787 was facilitated by the extra light grasp made possible by tilting the mirror, to observe the incident light at the 'Herschellian' focus.

4 John Hadley developed the 6-inch aperture Newtonian reflector, on an altazimuth stand, in 'An Account of a Catadioptrick Telescope, made by John Hadley Esq., F.R.S. With the Description of a Machine contriv'd by him for the applying it to use', *Phil. Trans.* **32** (1723) 303–12. The leading manufacturer of reflecting telescopes (usually Gregorians) before Herschel was James Short: see David Bryden, *James Short and his Telescopes* (Edinburgh, 1968).

5 David S. Evans, 'John F.W. Herschel', *D.S.B.* VI 324, claims without reference that his father, Sir William Herschel, sold £16,000 worth of telescopes during his career, though this sum also included manufacturing costs. Nasmyth supplied Warren De La Rue with a 13-inch diameter speculum casting in 1840, though whether it was sold or given is not clear: *James Nasmyth, Engineer, An Autobiography*, ed. Samuel Smiles (London, 1889) 314. De la Rue subsequently mentions his indebtedness to Nasmyth and William Lassell for assisting him with his early 13-inch mirrors: 'On figuring Specula', *M.N.R.A.S.* **13** (1852) 44–51.

6 Sir Howard Grubb, 'On the Great Telescopes of the Future', *Scientific Transactions of The Royal Dublin Society* 1 (New Series) Memoir No. 1 (Dublin 1877) 1–12: 3. S.C.B. Gascoigne, 'The Great Melbourne Telescope and other 19th Century Reflectors', *Q.J.R.A.S.* **37** (1996) 101–28: 108.

7 William Sheehan and Richard Baum, 'Observation and Inference: Johann Hieronymous Schroeter, 1745–1816', *J.B.A.A.* **105** (1995) 4.

8 John Herschel's 'Account of some Observations...' (ref.1) 459–61 brings home the high degree of familiar and idiosyncratic personal management which these big reflectors needed. William Pearson's *An Introduction to Practical Astronomy* II (London, 1829) 82 reinforces the same, when describing John Ramage's 25-foot instrument, which was elevated 'into any altitude, [for] the *meridional sweeps*, formerly practised by Sir William Herschel, and continued by his son Mr. Herschel with great success.' Ramage's 25-foot reflector was brought from Aberdeen and set up at Greenwich, where Pearson saw it in the late 1820s, though it was removed in 1835 when Ramage died and Airy became Astronomer Royal. There is no evidence that it was ever used with any seriousness at Greenwich. Some smaller Herschel reflectors were used for comet-watching and similar activities in professional observatories, however, such as the 7-foot supplied to Maskelyne at Greenwich, and the 10-foot focus, 8-inch aperture Newtonian instrument commissioned for the Duke of Marlborough in 1812, and later passed on to the Radcliffe Observatory, Oxford: R.T. Gunther, *Early Science in Oxford* II (London 1923) 317.

9 William Lassell, 'Description of a Machine for Polishing Specula... with Directions for its Use; together with Remarks upon the Art of Casting and Grinding Specula, and a Description of a Twenty-foot Newtonian Telescope

equatoreally mounted at Starfield near Liverpool', *Mem. R.A.S.* **18** (1849) 1–20.
10 John Edwards, 'Directions for making the best composition for metals of reflecting telescopes; and the Method of casting, grinding, polishing, and giving them the true parabolic figure', *Nautical Almanac* 1787, 1–48, 49–60.
11 R.A.S. Lassell MS 2; see fols. 21–22 for date.
12 W. Herschel, 'Description of a Forty-feet Reflecting Telescope' (ref.2). By a 'common telescope' Herschel meant a normal, good-quality instrument, such as those used to observe clusters of stars in the '77th of the Connaissance de Temps': Herschel, 'Astronomical Observations relating to the siderial part of the Heavens, and its Connection with the nebulous part; arranged for the purpose of a critical Examination', *Phil. Trans.* **104** (1814) 248–84: 281, and *Scientific Papers* (ref.1) II, 539.
13 See his 'On the Power of penetrating into Space by Telescopes: with a comparative Determination of the Extent of that Power in natural Vision, and in Telescopes of various Sizes and Constructions; illustrated by Select Observations', *Phil. Trans.* **90** (1800) 49–85, reprinted in Herschel, *Scientific Papers* (ref.1) II, 31–52. Such telescopes, however, lacked the great 'Space Penetrating Power' of his 20- and 40-foot focus instruments. Herschel mentions this 'Space Penetrating' or resolving power (as opposed to mere magnification) in most of his cosmological papers after 1783.
14 Sir William Herschel left a fortune in excess of £25,000 at his death in 1822: Will, 17 December 1818, proved 3 October 1822, transcript PRO PROB. 10/4640 and PRO PROB. 11/1662, Quire 527. For fuller details see Chapter 4, ref. 7. I thank Peter Hingley for drawing this document to my attention.
15 This remarkable instrument, with its 18-inch mirror and octagonal wooden tube, 12 feet long, survives in almost complete condition. It was made by James Short for the Duke in 1742, and rebuilt by Thomas Short, his brother, in 1770. There is no evidence that it ever did any serious work, and it was neglected by the Radcliffe Observatory, Oxford, to which it was presented in 1812. Its mirror still has a decent reflectivity, and would probably produce an image. M.H.S. No. 34/7.
16 William Kitchener, *The Economy of the Eyes – Part II. Of Telescopes; Being the Result of Thirty Years' Experiments, with Fifty-One Telescopes of from one to nine inches in Diameter* (London, 1825) 217. Kitchener was a telescope enthusiast, and his rather rambling book is a major primary source on the London instrument-making community *c.*1790–1820.
17 David Brewster, *A Treatise on Optics* (London, 1831) 357.
18 'Mr. John Ramage's Description of his large Reflecting Telescope', *Mem. R.A.S.* **2** (1826) 413–18. Pearson, *Practical Astronomy* II (ref.8) 79–82.
19 William Parsons, Third Earl of Rosse, obituary, *M.N.R.A.S.* **29** (1869) 123–30. Daubeny gave the toast and thanks when Lord Rosse entertained the British Association visit to Birr in 1857: 'Excursion of the British Association to Parsonstown – The Earl of Rosse and the Monster Telescope', *The King's County Chronicle*, Wednesday 9 September 1857.
20 Lord Oxmantown (the title of Lord Rosse before he inherited his Earldom), 'An Account of Experiments on the Reflecting Telescope', *Phil. Trans.* **130** (1840) 503–27: 503.

21 William Herschel presented his 'classic' view on the 'particulate matter' that composed the stellar universe in 'On the Construction of the Heavens', *Phil. Trans.* **75** (1785) 213–66, and *Scientific Papers* I (ref.1) 223–59: 224–5. John Herschel, *Results of Astronomical Observations made during the years 1834, 5, 6, 7, 8 at the Cape of Good Hope* (Smith, Elder and Co., London, 1847) 138–9, where John argues for the 'carbonaceous', particulate composition of nebulae. No one in 1850, moreover, would have disagreed with William Herschel's 1814 statement that 'We have no reason to expect that an increase of light and distinctness of our telescopes would free us from ambiguous objects', in 'Astronomical Observations relating to the Sidereal Part of the Heavens, and its connection with the nebulous part: arranged for the purpose of a critical examination', *Phil. Trans.* **104** (1814) 248–84: 262, and *Scientific Papers* (ref.1) II, 520–41: 528. For the best modern scholarly treatment of Herschel's cosmology, see Michael Hoskin, *William Herschel and the Construction of the Heavens* (Oldbourne, London, 1963).

22 Bindon B. Stoney to Lady Rosse, 24 December 1864, Rosse Archives, Birr Castle J/19/1, thanking her for her photographs of drawings, and also discussing the Orion and Hercules nebulae and clusters. Also *The Photographic Journal*, 15 December 1856, 96, reporting from The Photographic Society of Ireland, announced that 'The model for the best paper negative was obtained by the Countess of Rosse.' The journal *Photographic Notes*, 15 September 1859, spoke of the Rosses as '... so distinguished a man of science, and so clever a photographer as his lady has the reputation of being'. Both publications in Rosse Archives, Birr Castle. Also 'Mary Wilmer Field, The girl from Heaton', in Horace Hird, *Bradford Remembrancer* (Bradford, 1972) 44–53.

23 'Death of an old Parsonstonian', obituary for William Coghlan (1815–1896), *King's County Chronicle*, 12 March 1896. I am indebted to Miss Bridgid Roden, Project Director, Birr Castle Demesne, for this information.

24 *King's County Chronicle*, 9 September 1857 (ref.19). 'Excursion', cutting preserved in Rosse Archives, Birr Castle.

25 As early as 1827, Lord Oxmantown (Rosse) had been experimenting with mirrors: see Lord Oxmantown, 'An Account of an Apparatus for grinding and polishing the Specula of Reflecting Telescopes', in D. Brewster (*ed.*), *The Edinburgh Journal of Science* 9 (1828) 213–17. Oxmantown, 'An Account of Experiments...' (ref.20) 505–6: 512 for solid 36-inch disc.

26 Mary Somerville, Rome, to Lord Rosse, 11 November 1843, Rosse Archives, Birr Castle K17: 16.

27 Lord Rosse to Mary Somerville, 12 June 1844, Rosse Archives, Birr Castle K17 Additional. For Sir William Herschel's (90, or) 100 hours, see 'On the Power of penetrating into Space' (ref.13) 84 (*Scientific Papers* (ref.1) II, 31–52: 51). Sir James South wrote to his friend Lord Rosse to apprise him of the superiority of his new 36-inch reflector to Sir John Herschel's 20-foot, 18¼-inch aperture: South to Oxmantown, 11 November 1840, Rosse Archives, Birr Castle K1/2(1).

28 Lord Rosse, 'On the Construction of a Specula of Six-feet Aperture; and a selection of Observations of Nebulae made with them', *Phil. Trans.* **151** (1861) 681–747: 681.

29 Rosse to Mary Somerville, 12 June 1844 (ref.27).
30 Lord Rosse, 'On the Construction of Specula of Six-feet...' (ref.28). This was Lord Rosse's *magnum opus* on large speculum mirror casting, wherein he discussed the earlier techniques used to produce the 36-inch mirror of 1840 (ref.20) as well as the casting and figuring sequences, pp. 682–6, needed to produce the 72-inch. Also, D.W. Dewhirst and Michael Hoskin, 'The Rosse Spirals', *J.H.A.* **22** (1991) 258–66.
31 Rosse, 'On the Construction of Specula of Six-feet...' (ref.28). For the 16 strokes per minute of the Rosse polishing machine (for the 36-inch), see Thomas Romney Robinson to Rosse, 15 December 1841, Rosse Archives, Birr Castle K5.5.
32 Lord Rosse, 'Observations on the Nebulae', *Phil. Trans.* **140** (1850) 499–514: 505. Sir James South related his enthusiastic impression of objects seen through the 72-inch in *Astronomische Nachrichten* No. 536 (1846) 113–18. G.B. Airy presented his own account to the R.A.S. in 'Substance of the Lecture delivered by the Astronomer Royal on the large Reflecting Telescopes of the Earl of Rosse and Mr. Lassell at the last November [1848] Meeting', *M.N.R.A.S.* **9** (March 1849) 110–21; 120 for description of Jupiter as resembling a coach lamp.
33 For an excellent semi-primary source assessment of the significance of the 72-inch, see Agnes M. Clerke, *A Popular History of Astronomy during the Nineteenth Century* (London, 1893) 143–8. Also Michael Hoskin, 'The first drawing of a Spiral Nebula', *J.H.A.* **13** (1982) 97–101. Dr Hoskin examines Sir John Herschel's 1833 drawing of M51, depicted as a bifurcated ring with an appendage, in contrast to the true spiral observed by Rosse in April 1845.
34 Thomas Woods, *The Monster Telescope* (1845) 4. I am indebted to Michael Turbridy, the Dublin engineer in charge of the 1996–7 restoration of the 72-inch telescope, for the primary source of the oft-quoted but never referenced £12,000 original construction cost. John Pringle Nichol also gave the same sum, £12,000, in an undated c.1850 article 'The Wonders of the Telescope', preserved as a cutting in John Lee's 'Scrapbook' 4, M.H.S. Oxford Gunther 37, vol.4, 42.
35 *Letters on the State of Ireland, by the Earl of Rosse* (London, 1847). A 36-page pamphlet containing Rosse's anonymous and rejected letters to *The Times*: Rosse Archives, Birr Castle, pamphlet J/11. Also J.T. (or F.) Burgoyne, Dublin Castle, to Lord Rosse, 8 March 1847, Rosse Archives J/11, thanking Rosse for 'unravelling the *mysteries* of Irish affairs', saying 'that your principles are contrary to some favourite dogma that they [the Famine Relief Commissioners] wish to inculcate'.
36 Lady Richarda Airy to Lady Herschel, 16 October 1848, relaying the contents of G.B. Airy's letter from Ireland: Enid Airy Papers, in Airy family possession until deposited in Cambridge University Library in 1997.
37 *King's County Chronicle*, 9 September 1857 (ref.19).
38 *Illustrated London News*, 3 February 1851.
39 For the Grubb 15-inch Cassegrain at Armagh, see I.S. Glass, *Victorian Telescope Makers. The Lives and Letters of Thomas and Howard Grubb* (Institute of Physics, Bristol and Philadelphia, 1997) 17–19. Robinson suggested that Rosse should 'Kyanize' the wood (a form of creosote treatment) to preserve it, as well as employ Fairbairn's for the castings: Robinson to Rosse, 10 January 1841, Rosse Archives,

Birr Castle K5.1b. There is no record of who made the great iron castings for the 72-inch, such as the massive iron pivot upon which the 52-foot tube moved. Tradition has it that the entire telescope was the product of local resources, and while the pivot casting appears to be the product of a professional foundry, neither it nor any other of the original ironwork for the 72-inch bears any maker's stamp. Lord Rosse did have dealings with Fairbairns of Manchester, however, when in February 1845 he purchased a high-pressure 4 horse-power steam engine from them for £163 10s 0d: Lord Rosse's Scrapbook, Rosse Archives, Birr Castle. The engine was probably used to power the 72-inch mirror polishing machine.

40   Robinson to Rosse, 7 April 1841, Rosse Archives, Birr Castle K5:4.
41   Robinson to Rosse, 1 December 1840, Rosse Archives, Birr Castle K5:1a.
42   William Huggins, obituary for William Lassell, *M.N.R.A.S.* **41**, 4 (February 1881) 188–91: 181; Margaret Lindsay Huggins, obituary, 'William Lassell', *The Observatory* **3** (November 1880) 587–90 (William Huggins and his wife Margaret were great admirers of Lassell and came to know him well). Also A. Chapman, 'William Lassell (1799–1880): practitioner, patron, and 'Grand Amateur' of Victorian astronomy', *Vistas in Astronomy* **32** (1989) 341–70, reprinted in Chapman, *Astronomical Instruments and their Users: Tycho Brahe to William Lassell* (Variorum, Aldershot, 1996).
43   Edwards, 'Directions for making... reflecting telescopes' (ref.10). See R.A.S. MS Lassell 1.4 for Lassell's 23-page v and r transcription of Edwards' article 'extracted from the *Nautical Almanac* for 1787, W.L. 18 Sep. 1821'.
44   Lassell to Otto Struve, 25 February 1848, in R.A.S. MS Lassell notebook 8:2 'Correspondence & Co., 1840 to 1855', fols. 34–35: 'I regret that owing to my necessary occupations in business my leisure rarely affords opportunity for me to do more than make observations.'
45   These incidents are recorded in R.A.S. MS Lassell notebooks 1, 5 and 6. MS 6 15v records the trials between Lassell's Dollond refractor and Alfred King's Tulley refractors on 11 August 1828.
46   R.A.S. MS Lassell 6, 'Astronomical Observations 1821–1847', 8v–12 for 1821–1822 references to the 7½-inch Gregorian. For 9-inch mirrors, see R.A.S. MS Lassell III, 3–4 September 1833, 87–8; 6–12 September 1833, 91–5.
47   R.A.S. MS Lassell III, 79–81. Joseph N. Lockyer, *Stargazing: Past and Present* (London, 1878) 128 also discusses the damage done to a mirror by over-polishing.
48   Chapman, 'William Lassell' (ref.42) 345–6.
49   William Lassell, 'Description of an Observatory erected at Starfield, near Liverpool' (read to the R.A.S. on 7 April 1841), *Mem. R.A.S.* **12** (London, 1842) 265–72: 270.
50   Lassell (ref.49). The optics of this telescope dated from the summer of 1833, R.A.S. MS Lassell III, 87–8, when they were being tested, though the iron equatorial mount probably dates from 1838–40.
51   Glass, *Victorian Telescope Makers* (ref.39): Markree 13–16; Armagh 17–19.
52   Glass (ref.39) 18. I am indebted to Professor Patrick A. Wayman for kindly showing me a copy of Chapter 2, 'Five Early Grubb Telescopes', from his forthcoming study on Thomas Grubb.

53 *James Nasmyth* (ref.5) 121–34, 231–42. A. Chapman, 'James Nasmyth, Astronomer of Fire', *1997 Yearbook of Astronomy*, ed. Patrick Moore (Macmillan, 1996) 143–67: 145.
54 *James Nasmyth* (ref.5) 392–6. In a letter to Thomas Romney Robinson, 11 October 1849, Nasmyth claims to have been using the chill plate and iron hoop method of casting mirrors (as opposed to sand casting) since 1829, when he was 21 years old, although it was not until he had read Lord Rosse's description of the technique in 1840 that its real value was brought home to him. In the same letter to Robinson, Nasmyth claims that he had first taught William Lassell how to cast mirrors with the chill plate, and thereby produce more uniform speculae than was possible with sand castings. Nasmyth to Robinson, 11 October 1849: Lord Rosse's Scrapbook, Rosse Archives, Birr Castle.
55 *James Nasmyth* (ref.5) 314. Warren De La Rue, later in his own career often paid his debts to Nasmyth (and Lassell), as in 'On figuring Specula' (ref.5) 24: 'I have also to express acknowledgements to Mr. James Nasmyth for his most valuable hints in polishing and grinding.'
56 *James Nasmyth* (ref.5) 313 for Lassell's remark recorded by Nasmyth.
57 J. Nasmyth to Richard Harris, 11 June 1844: R.A.S. MS Letters, 1844, under date order, 'Nasmyth'.
58 Lassell, 'Description of a Machine' (ref.9) thanks Rosse, 1–2, and Nasmyth, 4, for guidance on 'chill' casting techniques; for Nasmyth and the polishing machine, 5.
59 Lassell to Lord Rosse, undated, sequentially between December 1844 and April 1845, copy in R.A.S. MS Lassell 8:2 fol. 24.
60 Lassell, 'Description of a Machine' (ref.9) 4–5.
61 Lassell, 'Description of a Machine' (ref.9) 5–20.
62 Richard Sheepshanks to G.B. Airy, 14 September 1849, C.U.L. RGO6 231/504–507. For some reason Sheepshanks felt that Lassell 'cannot help having a little prejudice against me', in spite of his own regard for Lassell, and asks Airy, who knew Lassell better, to help sweeten him.
63 Lassell, 'Description of a Machine' (ref.9) 17, says that it took nine or ten minutes to find any object in the sky with the 24-inch when aided by an assistant. This time included leaving his house, walking down the garden, and opening up the dome shutter. On his own, unaided, the same procedure took 'three or four minutes longer'. When a full-sized working duplicate of Lassell's 24-inch was completed by Liverpool Museum in October 1996, Martin Suggett, Alan Bowden (of Liverpool Museum) and myself suspected that Lassell's times for finding a single object in the sky sounded optimistic, especially when working alone. But, we were raw beginners with the instrument, and by 1849 Lassell was an expert user. Even so, the widely-separated adjustments on this vast iron instrument would have made one-man operation very awkward.
64 When Lassell was showing Edward Holden (later of the Lick Observatory) his 24-inch in the late 1870s, Speculum 'A' was still deemed 'faultless': astronomical observations July 1869 – May 1878: 25 July 1876, R.A.S. Lassell MS 17:4. Lassell had first commented upon the 24-inch polishing machine, which was 'in my

judgement a perfect engine for figuring specula', in 'Extract of a Letter... to Mr. Dawes', 8 November 1852, *M.N.R.A.S.* **13** (1853) 14.
65 See J.R. Hind's letter, 30 September 1846, to *The Times* (pasted into Observing Book), 2 and 3 October 1846 for first sighting of Neptune, 10 October for Triton: all in R.A.S. Lassell MS 9:7. See also Robert W. Smith, 'William Lassell and the Discovery of Neptune', *J.H.A.* **14**, 1 (February 1983) 30–2. Chapman, 'William Lassell' (ref.42) 360.
66 Robert Grant, *History of Physical Astronomy* (London, 1852) 236. Also John F.W. Herschel, 'An Address Delivered to the Annual Meeting of the Royal Astronomical Society', *Mem. R.A.S.* **27** (1849) 192–200: 199.
67 W. Lassell to *The Times*, 29 September 1847, for further observations of Triton. R.A. Marriott, 'Dawes, Lassell and Saturn's dusky ring', *J.B.A.A.* **96**, 5 (1986) 270–7. Lassell saw Saturn's Hyperion, 18 September 1848: *M.N.R.A.S.* **8** (1848) 195–7. He saw Uranus' Ariel and Umbriel on 24 October 1851: *M.N.R.A.S.* **11**, Supp. 9 (1851) 248. Bond's simultaneous discoveries were recorded at virtually the same time: see also Chapman, 'William Lassell' (ref.42) 368–9.
68 W. Lassell, 'Observations of the Nebula of Orion, made at Valletta [*sic*], with the Twenty-foot Equatoreal', *Mem. R.A.S.* **23** (1854) 53–7.
69 Lassell, 'Nebula of Orion' (ref.68) 54–5.
70 Lassell, 'A Catalogue of New Nebulae discovered with the four-foot Equatoreal in 1863–65', *Mem. R.A.S.* **36** (1867) 53.
71 W. Lassell, 'Observations of the Planet Saturn with 20-foot Equatoreal at Valletta', *Mem. R.A.S.* **22** (1854) 151–65: 151.
72 W. Lassell, 'Observations of Planets and Nebulae at Malta', *Mem. R.A.S.* **36** (1867) 1–32, for a detailed account of the 48-inch reflector and its operations. R.A.S. Lassell MS 7 contains Lassell's working notebooks for the construction of the 48-inch reflector. It was most likely for use with the 48-inch at Malta that Lassell commissioned a divided lens micrometer of advanced design for William Simms. This is probably the same beautifully made and complex instrument – ideal for measuring planetary diameters and satellites – that is now owned by Mr Christopher J.R. Lord, who kindly demonstrated it to me and permitted me to examine it in 1996. See C.J.R. Lord, 'Divided Lens Micrometer by Troughton & Simms circa 1860', unpublished 10-page paper presented by Mr Lord to the author, January 1998.
73 Nasmyth to Rosse, 15 December 1852, Royal Society Misc. MS 12:66, and published in *Correspondence concerning the Great Melbourne Telescope in Three Parts, 1852–1870* (R. Soc., London, 1871), date order. Nasmyth proposed an iron equatorial telescope with a 48-inch mirror of 32 feet focal length.
74 Lassell to Rosse, 6 May 1852, Rosse Archives, Birr Castle K10.1.
75 'Letter from Mr. Lassell, Valetta, 29 December, 1852', *M.N.R.A.S.* **13** (1853) 34–6.
76 J. Nasmyth to John Phillips, 2 June 1851: Oxford University Museum of Natural History, Letters to Phillips, 1851/3.
77 Nasmyth, *Autobiography* (ref.53) 339.
78 Nasmyth, *Autobiography* (ref.53) 324.

348 Notes and references [Ch. 6

79 Airy observed the 1842 eclipse in Turin, G.B. Airy, *Autobiography*, ed. W. Airy (C.U.P., 1896) 153–4, and Lassell that of 1852 at Trollhätten, Sweden, 'Trollhätten Falls, Observations by W. Lassell Esq.', *Mem. R.A.S.* **21**, 1 (1852) 44–50, so both astronomers were familiar at first hand with the newly-discovered phenomenon of solar prominences. I am not aware that Nasmyth ever saw prominences at first hand, however. The account of his solar-disk-obscuring apparatus is in correspondence with Airy in C.U.L. RGO6 113/123–67, under 'Eclipses'. See especially Nasmyth to Airy, 18 August 1851, and 30 August 1851, RGO6 113, 123 and 129, for sketches of Nasmyth's unsuccessful prominence projector.

80 He originally communicated his results to the *Memoirs of the Literary and Philosophical Society of Manchester*, 3rd series, vol. I, 407. See also Nasmyth, *Autobiography* (ref.53) 370, 378. On 9 January 1864, Nasmyth despatched a case of 'my drawings in relation to my discovery of the Willow Leaf shaped objects' to the R.A.S.: see R.A.S. Letters, 1864: date order.

81 Nasmyth to John Phillips, 20 January (?) 1863. Royal Society MS A.P. 45:10.

82 The purpose of the 'Nasmyth' design was to 'render the use of Gigantic Telescopes at once Easy and *comfortable*'; Nasmyth to Isaac Williams, 26 June 1849; R.A.S. MS Letters, 1849, date order. On 14 May 1849, Nasmyth had written to J.D. Forbes at St. Andrews, describing the optical arrangement for his telescope, and mentioning a construction cost of £1,000, though it is not clear whether this applies to the 20-inch, or a projected upgrading to 60 inches: Forbes Papers, St. Andrews University Library. The 20-inch was clearly a major subject of discussion 1848–51, and on 13 October 1848 Nasmyth promised to invite his Sheffield friend, Clifton Sorby, over to Manchester for a 'scamper over the sky when my 20" is finished': Sorby Papers, Sheffield Literary and Philosophical Society Library. In 1849, Nasmyth sent a working model of the 20-inch to the Royal Institution, complete with a dressed doll with clothes by Mrs Nasmyth sat in the observing seat: *Autobiography* (ref.53) 340. By 22 February 1852, however, Nasmyth was writing to John Williams, R.A.S. Secretary, for the return of the model to Manchester, as it seems to have been rejected by the Great Exhibition Committee and subsequently neglected: R.A.S. Letters, 1852, date order.

83 Glass, *Victorian Telescope Makers* (ref.51) 17–19.

84 It is preserved at Armagh Observatory: see Glass (ref.51) Fig. 1:4 for photograph.

85 Lady Airy to Lady Herschel, 16 October 1848, Enid Airy Papers, Cambridge University Library, (in process of classification; I examined these papers in 1995 when they were still in the possession of the Airy family). Lady Airy reported the visit to Lord Rosse communicated by her husband G.B. Airy, from a letter that no longer survives; however, it was probably similar to the report Airy presented to the R.A.S. in November 1848 (ref.32). Lord Rosse gave an account of his telescope and its mirror, supported in a 'jimmal'-mounted hoop, in 'Drawings to illustrate Recent Observations on Nebulae. By the Earl of Rosse. With Remarks by the Rev. Dr. Robinson', *Report of the Twenty-Second Meeting of the B.A.A.S.* [at Belfast] *1852* (London, 1853) 22–3.

86 Lassell seems to have seen the 'ring' on the night that he first saw Neptune, on 2 and

3 October 1846: R.A.S. Lassell MS 9:7. Various reports by him to *M.N.R.A.S.* **7** and **8** (1847–8). See Robert Smith and Richard Baum, 'William Lassell and the Ring of Neptune: a case study in instrument failure', *J.H.A.* **15** (1984) 1–17. Lassell, 'On a Method of Supporting a large Speculum, free from sensible Flexure in all positions', *Report of the B.A.A.S.* 1850, 180–3. Chapman, 'William Lassell' (ref.42) 169–70.

87 Noticed by the author and other members of the Royal Insurance Lassell Telescope Project when the mirror was examined in 1995. They were used to remount the mirror within the astatic cell of the replica instrument in 1996.
88 John F.W. Herschel, *Results of Astronomical Observations made during the years 1834, 5, 6, 7, 8, at the Cape of Good Hope* (London, 1847) 135. J. Herschel, *Outlines of Astronomy*, 2nd edn. (London, 1849) 595–6 (art. 868).
89 Nasmyth to Rosse, 15 December 1852; Royal Society Misc. MS 12:66, also printed in *Correspondence... Melbourne Telescope* (ref.73), date order.
90 Nasmyth to Rosse, 15 December 1852 (ref.73).
91 Nasmyth to Rosse, 30 December 1852, Royal Society Misc. MS 12:74, printed in *Correspondence... Melbourne Telescope* (ref.73) 15–16, for roller support. Nasmyth to Thomas Bell, 2 July 1853, Royal Society Misc. MS 19:109, printed in *Correspondence... Melbourne Telescope* (ref.73) 37–40. Nasmyth submitted two beautifully executed 35 × 38-inch engineering drawings to the Royal Society as possible designs for the 'Great Southern Equatorial', almost certainly the products of his Patricroft Works drawing office: Royal Society Misc. MS 12:66 and 19:117.
92 Nasmyth to Rosse, 30 December 1852, R. Soc. Misc. MS 12:74, printed in *Correspondence... Melbourne Telescope* (ref.73) 15–16. A copy of this letter, complete with drawings for a sprung railway trolley to convey the mirror for repolishing, is also preserved in the Rosse Archives, Birr Castle, K17:20, along with other Nasmyth, Lassell and Rosse items pertaining to the Great Southern Equatorial.
93 These drawings were beautifully conserved by the Royal Society in 1996: *Bibliotheca: News from the Royal Society Library and Archives* **8** (January 1997) 1–111.
94 This seminally-designed instrument, with its skeleton tube, equatorial mount, and still remarkably luminous speculum, is still preserved in the M.H.S., Oxford. It was the same instrument for which Warren De La Rue wrote: 'On Figuring Specula' (ref.5); 24. It was at De La Rue's Canonbury observatory, as discussed in Chapter 5.
95 Lassell, 'Observations... at Malta' (ref.72), illustrations.
96 Nasmyth to Rosse, 30 December 1852, Royal Soc. Misc. MS 12:74, and *Correspondence... Melbourne Telescope* (ref.73) 38 for 'snug box'. Royal Society Misc. MSS 12:66 and 19:117 (ref.91) depict the 'snug box'.
97 Nasmyth to Rosse, 15 December 1852, Royal Society Misc. MS 12:66, reprinted in *Correspondence... Melbourne Telescope* (ref.73).
98 William Pole (on behalf of the Royal Society Council) to Nasmyth, 16 February 1865, urging Nasmyth to 'make a martyr' of himself and agree to become an F.R.S. along with the other men of science: 'Let me know at once, then, a good "*Fellow*".'

350 Notes and references [Ch. 6]

  Dr Robert Graham, Errol Manse, to Nasmyth, September 1883 about St. Andrews University's wish to confer an honorary LL.D. upon him. Nasmyth refused both honours. Documents in a collection of letters to James Nasmyth, Edinburgh Public Library (Art Library) WND 478 N 25 J, 3 vols. I am indebted to Dr David Gavine for informing me about this Edinburgh letter collection, and transcribing them for me.

99 Nasmyth's *Autobiography* (ref.51) 444 ends with an engraved 'James Nasmyth Hys [thumbprint] Marke', often found on his letters to friends. Nasmyth's leisure activities also extended to the maintaining of at least one mistress, Flossie (Russell?), with whom he had a daughter, Minnie, in 1859. His letters to Flossie, and then to Minnie, were never signed by his name, but were invariably full of jokes, banter, sketches, and 'Hys Markes'. Nasmyth's very idiosyncratic handwriting and jokes make the letters unmistakably his. His generally undated letters to Flossie were sometimes signed 'Yours Ever affectionately *The Old File* [or *Hold File*] hys * marke.' Minnie was left £1,000 in 3% Consols by Nasmyth, but a dishonest lawyer later swindled her out of the money. She died in 1940. I am indebted to Mr C.J.R. Abbott of Essex, the great-grandson of Mrs Minnie Abbott, who very kindly sent me copies of the manuscript letters and photographs in April 1997. These letters, indeed, open up an aspect of the Grand Amateur world which never got into *M.N.R.A.S.*

100 Nasmyth to J. Phillips, 20 January (?) 1863 (ref.81) for Nasmyth's observations of Mars with his 8-inch Cooke refractor. Thomas Cooke, who was a friend of Nasmyth, had supplied him with the 8-inch refractor in 1861: S. Smiles, *Men of Invention and Industry* (London, 1884) 344. The Science Museum, South Kensington, has photographs of Nasmyth's refractors, on the terrace at Penshurst, Kent, *c*.1870.

101 Nasmyth to Richard Hodgson, 26 July 1863, R.A.S. Add. MSS 69:127–128.

102 Nasmyth to Hodgson, 26 July 1863 (ref.101).

103 Lassell, 'Observations... at Malta' (ref.72), plate.

104 Lassell to Lord Rosse, Valletta [*sic*], Malta, 30 December 1852, Royal Society Misc. MSS, printed in *Correspondence... Melbourne Telescope* (ref.73) 15–18: 16, for Lassell's belief that Nasmyth's 'snug box' would be too unsteady and vibration-prone.

105 Nasmyth to Bell, 2 July 1853, Royal Society Misc. MS 19:109, in *Correspondence... Melbourne Telescope* (ref.73) 37–40: 39 for proposed skeleton tube.

106 Cited in H.A. and M.T. Brück, *The Peripatetic Astronomer, The Life of Charles Piazzi Smyth* (Adam Hilger, Bristol and Philadelphia, 1988) 107.

107 Lassell, letter 27 August 1877 to *The Observatory* **6** (1877) 178–9 for the dismantling of the 48-inch with its superb mirrors, which he could not even give away. Thomas Grubb to T.R. Robinson, 3 December 1862, cited in Glass, *Victorian Telescope Makers* (ref.51) 48, for original estimate of £4,600. The finished cost was £5,000; cited in Gascoigne, 'The Great Melbourne Telescope' (ref.6) 126, n. 22.

108 Gascoigne, 'The Great Melbourne Telescope' (ref.6) 119–22.

109 This stretch of railway seems to have been only a dozen or so yards long,

however, running from the mirror-cell of the telescope (when pointing to the zenith), and across a turntable, so that the mirror could then be placed on a heavy flat-wheeled trolley (which still survives at Birr Castle) to be drawn several hundred yards over the lawns to the re-polishing shop. I am indebted to Mr Michael Turbridy for this information and for the 1879 Ordnance Survey map which clearly shows the telescope and its short railway. The railway can also be seen very clearly on the detailed nineteenth century model of the 72-inch telescope in the Science Museum, South Kensington. The railway trolley for the mirror is also depicted in Rosse, 'On the construction of a Specula of Six-feet Aperture' (ref.28) plate XXIV; 1 and 2.
110 R.A.S. Lassell MS 16:1, Astronomical Observations September 1849–September 1850, p. 83. Lassell and Struve were trying the 72-inch reflector at Birr on Monday 26 August 1850, from whence they returned to Liverpool. But the weather was poor in both places.
111 The necessity for, and great weight of, two speculum mirrors in large reflecting telescopes always made them prohibitive for professional observatories: Thomas Grubb costed in an additional £800 to provide a second mirror and its necessary polishing machinery in his 1862 estimate for the Melbourne Telescope: Glass, *Victorian Telescope Makers* (ref.51) 48. The 'B' mirror increased the original estimate from £3,800 to £4,600, although the completion cost was £5,000.
112 Lassell, 'Description of a Machine' (ref.9) 1–4. Lassell's notebooks also contain abundant information about mirror-making: R.A.S. Lassell MS 3, 6, etc.
113 It is clear from these fragments in the Science Museum store that after casting they were deliberately broken in two, so that the internal consistency and luminosity of the metal could be ascertained.
114 Though Lassell's 24-inch telescope was given to the Royal Observatory, Greenwich in 1883, after his death in 1880 – see Derek Howse, *Greenwich Observatory* 3: *The Buildings and Instruments* (Taylor and Francis, London, 1975) 118 – and eventually broken up, its 4-inch aperture finder survived intact. After many years in America, it was given to Liverpool Museum in 1989: *Liverpool Daily Post*, 20 November 1989. I am indebted to Martin Suggett and Alan Bowden of Liverpool Museum for these accounts of Saturn. Unfortunately, during my own visits to Liverpool the sky was cloudy.
115 Lassell to Thomas Bell, 2 July 1853, Royal Society Misc. MS 19:110 (ref.105), printed in *Correspondence... Melbourne Telescope* (ref.73) 41.
116 Lassell to Bell, 2 July 1853 (ref.115). Lassell, 'Description of a Machine' (ref.9) 3. The speculum alloy contained 136 pounds of tin, 295 pounds of copper, and 18 pounds of arsenic, though a small fraction was probably lost in the heating process.
117 H.C. King, *The History of the Telescope* (London, 1955) 261–81.

7 THE NEW SCIENCES OF LIGHT: SPECTROSCOPY, PHOTOGRAPHY AND THE GRAND AMATEURS

1. Auguste Comte, *Cours de Philosophie Positive* II (Paris, 1835), 'Astronomie' 8–9. Several authors writing on the history of astrophysics cite this statement by Comte, but invariably fail to provide a precise source.
2. John William Draper, 'On the Process of Daguerreotype, and its application to taking Portraits from the Life', *Philosophical Magazine and Journal of Science* 17 (1840) 217–25: 222, for his use of a heliostat and 3-inch lens to obtain a 'very strong impression' of the Moon with a half-hour exposure. Agnes M. Clerke, *A Popular History of Astronomy during the Nineteenth Century*, 3rd edn., (London, 1893) 190.
3. Joseph Norman Lockyer, *Stargazing: Past and Present* (London, 1878) 392 for Fraunhofer, and reproduction of his 1814–1815 spectrum. G.A. Kirchhoff, 'Researches on the Solar Spectrum and the Spectra of the Chemical Elements', 1861, translated by Henry E. Roscoe, 1862, and reprinted in Harlow Shapley and Helen E. Howarth, *A Source Book in Astronomy* (McGraw-Hill, New York and London, 1929) 279–82.
4. Kirchhoff (ref.3) 279. Agnes M. Clerke, *History of Astronomy* (ref.2) 164–7.
5. Clerke, *History of Astronomy* (ref.2) 168–70. Albrecht Unsöld, *The New Cosmos* (Springer Verlag, Berlin, 1969) 96.
6. Charles E. Mills and C.F. Brooke, *A Sketch of the Life of Sir William Huggins* (London, 1936). This 63-page essay, produced indirectly at the behest of Lady Huggins, is the most complete source for Huggins' early life: see pp. 7–15.
7. Mills and Brooke, *Huggins* (ref.6) 21. William and Margaret Huggins, *An Atlas of Representative Stellar Spectra*, being *Publications of Sir William Huggins' Observatory* I (London, 1899) 6. Pp.3–24 provide the primary source for the early work of Huggins' Tulse Hill Observatory. Deborah Jean Warner, *Alvan Clark & Sons, Artists in Optics* (Smithsonian Institution, Washington, 1968) 52 (new edition, 1996). Also R.A. Marriott, 'The 8¼-inch Clark refractor of the Temple Observatory, Rugby', *J.B.A.A.* 101, 6 (1991) 343–50.
8. Mills and Brooke (ref.6) 23.
9. W. Huggins, 'On the Spectra of some of the Chemical Elements', *Phil. Trans.* 154 (1864) 139–60. Huggins' Tables, 152–9, list 24 elements, plus solar and air spectra.
10. W. Huggins and W.A. Miller, 'On the Spectra of some of the Fixed Stars', *Phil. Trans.* 154 (1864) 413–35.
11. W. Huggins and W.A. Miller, 'On the Spectra of some of the Nebulae... A Supplement to the Paper "On the Spectra of some of the Fixed Stars"', *Phil. Trans.* 154 (1864) 437–44. Also Huggins, 'On the Spectrum of the Great Nebula in the Sword-handle of Orion', *Procs. of the R. Soc.* 14 (1865) 39–42. Also Huggins and Huggins, *Atlas of... Spectra* (ref.7) 11–12.
12. W. Huggins, 'Further Observations on the Spectra of some of the Stars and Nebulae, with an Attempt to determine therefrom whether these Bodies are moving towards or from the Earth, also Observations on the Spectra of the Sun and of Comet II 1868', *Phil. Trans.* 158 (1868) 529–64: 549 for Sirius' recession velocity.

13 S.C.B. Gascoigne, 'The Great Melbourne Telescope and other 19th-century Reflectors', *Q.J.R.A.S.* **37** (1996) 101–28.
14 Mary T. Brück, 'The Family Background of Lady Huggins (Margaret Lindsay Murray)', *The Irish Astronomical Journal* **20**, 3 (March 1992) 210–11. M.T. Brück, 'Companions in Astronomy, Margaret Lindsay Huggins and Agnes Mary Clerke', *The Irish Ast. J.* **20**, 2 (September 1991) 70–7. Barbara J. Becker, 'Eclecticism, Opportunism, and the Evolution of a New Research Agenda: William and Margaret Huggins and the Origins of Astrophysics' (unpublished Ph.D. thesis, Johns Hopkins University, Maryland, U.S.A., 1993) 290–2, for an analysis of the working relationship between the Hugginses.
15 Angelo Secchi, 'Note sur les spectres prismatiques des corps célestes', *Comptes Rendus* **57** (Paris, 1863) 71–5: 74. This was the first of Secchi's many papers which laid the foundation for stellar spectral classification, beginning, p. 75, with his original groups of red and yellow stars. J.B. Hearnshaw, *The Analysis of Starlight, One hundred and fifty years of Astronomical Spectroscopy* (C.U.P., 1986) 57–66.
16 *Astronomical Register* **1** (1863) 69–70. G.B. Airy, 'Apparatus for the Observation of the Spectra of Stars', *M.N.R.A.S.* **23** (1863) 188–91. I am indebted to Dr Barbara Becker for the above references. For Carpenter and the Orion Nebula, see G.B. Airy to Edmund Stone, 4 January 1864, C.U.L. RGO6 31/237. *G.B. Airy, Autobiography*, ed. Wilfred Airy (C.U.P., 1896) 257, mentions that 'the Prismatic-Spectrum-Apparatus had been completed in 1863'.
17 J.L.E. Dreyer, 'Observatory', *Encyclopaedia Britannica*, 8th edn (1884) 713. Kevin Krisciunas, *Astronomical Centers of the World* (C.U.P., 1988) 122.
18 Secchi's work on the photography of the planets – such as his ability to record the Cassini Division with an 8-minute exposure, and the different exposure times needed for the Moon and Saturn, calculated from their distance from the Sun – was reported in *The Leeds Intelligencer* newspaper, 23 October 1858.
19 *James Nasmyth, Engineer, An Autobiography*, ed. S. Smiles (London, 1889) 314.
20 Lockyer, *Stargazing* (ref.3) 464. As early as 1853, however, De La Rue had obtained 'A good collodian picture of the Moon ... taken by him in 30 seconds', *M.N.R.A.S.* **14** (1853–4) 134. For further reports of De La Rue's work with his silver-on-glass reflector, and its application to photography, see *M.N.R.A.S.* **19** (1859) 138, 353–8. On pp. 171–2 De La Rue describes his silvering process, based upon Justus von Liebig's account in *Annalen der Chimie und Pharmacie* **98**, 132.
21 De La Rue, account of 'Mr De la Rue on Celestial Photography', *M.N.R.A.S.* **19** (1859) 353–8. The photoheliograph aperture is clearly stated by De La Rue, p. 357, to be 3 inches, though many sources give it as 3½ inches. Lockyer (ref.3) 461–2.
22 The 13-inch reflector, with its equatorial mount and skeleton tube, is preserved in the store of the Museum of the History of Science, Oxford. Its tightly-lidded speculum still possessed a remarkably good luminosity in 1995 when Tony Simcock and I uncased it and washed it with distilled water. Had the telescope been mounted to receive it, I am sure that the mirror would still have yielded a good image of the Moon and planets. See also Roger Hutchins, 'British University Observatories c.1820–1939: A Mismatch of Ideals and Resources' (forthcoming, unpublished Oxford University D.Phil. thesis, 1997) Chapter 2, for De La Rue,

Phillips and Pritchard in Oxford. Many references to De La Rue are to be found in *The Life and Work of Charles Pritchard... compiled by his daughter, Ada Pritchard* (London, 1897).

23 'Lewis Morris Rutherfurd', obituary, *M.N.R.A.S.* **53** (February 1893) 229–31: 229. Edward S. Holden, *Memorials of William Cranch Bond... and of his son George Phillips Bond* (San Francisco, 1897) 240. Deborah J. Warner, 'Lewis M. Rutherfurd: Pioneer Astronomical Photographer and Spectroscopist', *Technology and Culture* **12**, 2 (1971) 190–216.

24 Rutherfurd subsequently obtained a solar spectrograph – made up from 15 overlapping sections – that was 2.1 metres long: Warner (ref.23) 206. L.M. Rutherfurd, 'On the construction of the Spectroscope', *The American Journal of Science*, new series, **39** (May 1865), 129–32. Rutherfurd, 'Observations on Stellar Spectra', *Am.J. Sc.*, second series, **36** (November 1863) 154–7.

25 'Rutherfurd', obituary (ref.23) 231. Warner (ref.23) 206 says more precisely that the speculum metal grating had 17,296 lines to the inch.

26 Draper, 'On the Process of Daguerreotype' (ref.2).

27 Henry Draper, 'On a Reflecting Telescope for Celestial Photography, erected at Hastings, near New York', *Report... of the B.A.A.S. 1860* (London, 1861), Sections' Report, 63–4. In this paper, the 23-year-old Draper describes the as yet unfinished 'first observatory that has been erected in America expressly for celestial photography', before the mirror's faults had become apparent. H. Draper, 'On the Construction of a Silvered Glass Telescope Fifteen and a half inches in Aperture, and its use in Celestial Photography', *Smithsonian Contributions to Knowledge* **14**, art. 4 (1864) 1–15: 2–3. 'Prof. Henry Draper', obituary by W.H.M. Christie, *The Observatory* **6** (January 1883) 23–4.

28 H. Draper, '... Silvered Glass Telescope' (ref.27) 3. Draper's article was considered so significant that it was reprinted in a French translation 41 years later: 'Construction d'un Telescope à Miroir Argenté de 15 pouces et demi d'ouverture et son emploi en Photographie Céleste', *Mémoires publies par la Société d'Encouragement pour l'Industrie Nationale* (Paris, 1905). Draper's 'Construction', after all, was ground-breaking insofar as it provided the first step-by-step account of how to make a substantial silvered glass reflecting telescope intended for photography.

29 H. Draper, '... Silvered Glass Telescope' (ref.27) 3.

30 'Professor Henry Draper', obituary, *Nature* **27** (30 November 1882) 108.

31 Simon Newcomb, *Popular Astronomy* (London, 1898) 135 says that Draper is 'the only American who has ever successfully undertaken the contruction of large reflecting telescopes'. H. Draper, 'Photographs of the Spectra of Venus and α Lyrae', *The Americal Journal of Arts and Sciences*, 3rd series (1877) 95, and in *Philosophical Magazine*, 5th series, **3** (1877) 238, and *Nature* **15** (1877) 218.

32 H. Draper, 'Photographs of the Nebula in Orion', *The American Journal of Science and Arts*, 3rd series, **20** (1880) 433. Draper, 'Photographs of the Nebula in Orion', *Philosophical Magazine*, 5th series, **10** (1880) 388.

33 Henry Draper, 'On Photographs of the Nebula in Orion, and of its Spectrum', *M.N.R.A.S.* **42**, 8 (June 1882) 367–8. Draper, 'Présentation d'une épreuve

photographique de la nebuleuse d'Orion', *Comptes Rendus* **92** (1881) 173: a letter from Draper to M. Cornu, 11 December 1880. His habit of publishing his major discoveries in triplicate – in American, English, and French journals – clearly indicates Draper's awareness of his international standing as an astronomer, while at the same time being self-financed.

34 An excellent account of the range and character of Draper's researches can be gained from H. Draper, 'On Photographing the Spectra of the Stars and Planets', *American Journal of Science and Arts* **18** (1879), and his other major papers, published together as *Researches on Astronomical Spectrum-Photography by the late Professor Henry Draper, M.D., LL.D.* (C.U.P., 1884). Also Hearnshaw, *Analysis of Starlight* (ref.15) 74–6.

35 John Lankford, 'Gender and Science: Women in American Astronomy, 1859–1940', *Physics Today*, March 1990, 58–65, where Lankford argues that the American 'great factory observatories' brought women into professional astronomy, but only as second-class citizens, and without prospects.

36 Annie J. Cannon, *The Henry Draper Memorial*, booklet reprinted from *Journal of the R.A.S. of Canada*, May–June 1915, 202–15. Mrs Anna Mary Draper also created a $6,000 Trust, for the award of a $200 Gold 'Henry Draper Medal': *The American Journal of Science*, 3rd series, **25** (1883) 428.

37 David Dewhirst and Michael Hoskin, 'The Message of Starlight: The Rise of Astrophysics', in *The Cambridge Illustrated History of Astronomy*, ed. M. Hoskin (C.U.P., 1997) 299. Howard N. Plotkin, 'Harvard College Observatory', in *History of Astronomy, An Encyclopaedia*, ed. John Lankford (Garland, New York, 1997) 252–3.

38 A.J. Meadows, *Science and Controversy: A Biography of Sir Norman Lockyer* (Macmillan, London, 1972). Mary T. Lockyer and Winifred L. Lockyer, *The Life and Work of Sir Norman Lockyer* (London, 1928) 7, 8–44.

39 'Sir Joseph Norman Lockyer', obituary, 'A.S.E.' [Eddington], *M.N.R.A.S.* **81**, 4 (February 1921) 261–6. 'Memorial Tribute to Sir Norman Lockyer', Archibald Geike, *Nature* **106** (2 September 1920) 20–1.

40 J.N. Lockyer, 'Memorandum on the Solar Researches Carried on by Sir Norman Lockyer, 1863–1906' (private publication, London, 1906), cited in Father A.L. Cortie, 'Sir Norman Lockyer 1836–1920', *The Astrophysical Journal* **53**, 4 (May 1921) 233–48.

41 J.N. Lockyer, *The Chemistry of the Sun* (London, 1887) 106–7 for 'Artificial Eclipses', and discussions with Balfour Stewart in 1866.

42 'Sir Joseph Norman Lockyer', obituary by 'A.S.E.' [Eddington] (ref.39) 262–6: 262.

43 Lockyer's discovery only became possible when a Royal Society grant enabled him to equip his 6¼-inch refractor with a new high-dispersion spectroscope: Lockyer, 'Memorandum', cited in Cortie, 'Sir Joseph Norman Lockyer' (ref.40) 235. Lockyer gave an account of Janssen's 1868 post-eclipse Indian observation in *Contributions to Solar Physics* (London, 1874) 213–14.

44 Lockyer had the greatest respect for Jules Janssen, dedicating *Contributions to Solar Physics* (ref.43) to him, and Balfour Stewart, and calling Janssen 'a spectroscopist second to none', 213.

45 Cortie, 'Sir Norman Lockyer' (ref.40) 236.
46 'Sir Joseph Norman Lockyer', 'A.S.E.' [Eddington] (ref.42) 263.
47 Kevin Johnson, 'The Sun Spotteries: The South Kensington Solar Physics Observatory, 1879–1913' (unpublished M.Sc. thesis, Imperial College, London, 1996) 1–5.
48 Johnson (ref.47) 9. Cortie, 'Sir Norman Lockyer' (ref.45) 239. Lockyer was on the 1879 Solar Physics Committee, lecturing at South Kensington 1880–1881, and full Professor in 1882.
49 J.N. Lockyer, *The Meteoritic Hypothesis. A Statement of the Results of a Spectroscopic Inquiry into the Origins of Cosmical Systems* (London, 1890).
50 William J.S. Lockyer, *Handbook to the Norman Lockyer Observatory* (N.L.O. Corporation, 1935). I have visited the observatory several times in recent years. It is currently operated by the Norman Lockyer Observatory Society (a union of the Sidmouth and District Astronomical Society and the Sidmouth Amateur Radio Society), which publishes a Newsletter. See also George A. Wilkins, 'The Early Years of the Norman Lockyer Observatory at Sidmouth, Devon', *Transactions of the Torquay Natural History Society* 1997. I am indebted to Dr Wilkins and other members of the N.L.O. Society for their kind assistance.
51 'James Ludovic Lindsay, Lord Crawford', obituary, *M.N.R.A.S.* **74** (February 1914) 271–3.
52 Hermann A. Brück, 'Lord Crawford's Observatory at Dun Echt, 1872–1892', *Vistas in Astronomy* **35** (1992) 81–138: 81, 85 for Pulkowa associations. On p. 104, Prof. Brück reproduces part of a set of accounts (now in the R.O.E. Library, Edinburgh) which give figures for £10,185, along with a note mentioning the *c.* £20,000 total valuation.
53 George Forbes, *David Gill, Man and Astronomer. Memories of Sir David Gill, K.C.B., H.M. Astronomer (1879–1907) at the Cape of Good Hope* (London, 1916) 80–1, 233.
54 J.L.E. Dreyer, 'Ralph Copeland', obituary, *M.N.R.A.S.* **66** (February 1906) 164–74: 164–6. J. Gerhardt Lohse, a Göttingen graduate, was taken on as a second Assistant in 1877: Brück, 'Lord Crawford's Observatory' (ref.52) 118.
55 George Forbes, *The Transit of Venus* (Macmillan Nature Series, London and New York, 1874) 17.
56 R. Copeland, 'On Hartwig's Nova Andromedae', *M.N.R.A.S.* **47**, 2 (December 1886) 49–50. The nova was discovered on 1 September 1885 in Germany, and studied by Copeland and Lord Crawford on 1–2 September. Copeland, 'On the visible Spectrum of the Great Nebula in Orion', *M.N.R.A.S.* **48**, 8 (June 1888) 360–2.
57 Hermann A. Brück, *The Story of Astronomy in Edinburgh from its Beginnings until 1975* (Edinburgh, 1983) 53–65. Brück, 'Lord Crawford's Observatory' (ref.52) 126–34.
58 H.H. Turner, 'Frank McClean', obituary, *M.N.R.A.S.* **65** (February 1905) 338–42. H.F.N., 'Frank McClean', obituary, *Procs. of the R. Soc.* **78** (January 1907) xix–xxiii.
59 One of the finest accounts of McClean's work is to be found in Sir Robert Ball's

Address, when presenting him with the R.A.S. Gold Medal, *M.N.R.A.S.* **59** (February 1899) 315–24. F. McClean, 'Comparative Photographic Spectra of Stars to the 3½ magnitude', *Phil. Trans.* Series A, **191** (1898) 127–38, plus 17 spectra plates.
60 McClean, obituary, *M.N.R.A.S.* (ref.58) 340. McClean, obituary, *Procs. of the R. Soc.* (ref.58) xx. McClean left £2,000 to the R.A.S. in 1905: J.L.E. Dreyer and H.H. Turner, *History of the R.A.S. 1820–1920* (London, 1923, 1987) 246. The Museum of the History of Science, Oxford, has a small collection of photographs and documents pertaining to McClean.
61 Sir F.W. Dyson, 'Andrew Ainslie Common', obituary, *M.N.R.A.S.* **64**, 4 (February 1904) 274–8: 274. 'A.A. Common', obituary by H.H. Turner, *Procs. of the R. Soc.* **75** (1905) 313–18: 313.
62 Edmund Stone, President, 'Address, Delivered by the President on Presenting the Gold Medal of the Society to Mr. Common', *M.N.R.A.S.* **44** (February 1884) 221–3: 222. Common, obituary, *M.N.R.A.S.* (ref.61) 275. For an account of the 36-inch, see A.A. Common, 'Particulars of the Mounting of a Three-Foot Reflector', *Mems. of the R.A.S.* **46** (1880–81) 173–82.
63 Stone, 'Address' (ref.62) 221.
64 Stone, 'Address' (ref.62) 222.
65 Stone, 'Address' (ref.62) 222. A.A. Common, 'Note on a Photograph of the Great Nebula in Orion and Some New Stars Near θ Orionis', *M.N.R.A.S.* **43** (March 1883) 255–7. Common, obituary, *M.N.R.A.S.* (ref.61) 275. Dreyer and Turner, *History of the R.A.S.* (ref.60) 213.
66 Common, 'Note on an Apparatus for correcting the Driving of the Motor Clocks of large Equatorials for long Photographic Exposures', *M.N.R.A.S.* **49** (March 1889) 297–300. Common provides detailed engineering drawings of his plate corrector.
67 For an account of Crossley's observatory–mansion prior to the acquisition of the Common 36-inch, see E. Crossley, J. Gledhill and J.A. Wilson, *A Handbook of Double Stars* (London, 1879) 30–2. For Crossley's use and donation of the 9-inch and 36-inch telescopes, see Frank P. Andrews and Edwin Budding, 'Carter Observatory's 9-inch Refractor: The Crossley Connection', and Frank Dyson, 'A Brief History of Crossley's 36-inch Reflector', both in *Southern Stars: J.R.A.S. of New Zealand (Inc.)* **34**, 6 (March 1992) 358–66, 367–72.
68 Common, 'On the Construction of a Five-foot Equatorial Reflecting Telescope', *Mem. R.A.S.* **50** (1891) 112–204. His first mirror, finished in 1888–9, was not satisfactory, however, and a new glass blank was ordered from Paris and delivered to Ealing at the end of 1890 (p. 119), to be finished by 3 July 1891 (p. 147). Like his 36-inch, Common's 61-inch telescope had a similar photographic plate tracking mechanism, to override any motor irregularities (p. 194). See also ref.66 above.
69 Common, obituary, *M.N.R.A.S.* (ref.61) 277.
70 Isaac Roberts, obituary by Sir R.S. Ball, *Procs. of the R. Soc.* **75** (1905) 356–363: 358. Isaac Roberts, *A Selection of Photographs of Stars, Star-Clusters and Nebulae* I (London, 1893) 21–5, where Roberts gives an account of his 20-inch photographic reflector, which he completed in April 1885, and of his photographic

technique. Stephen H.G. James, 'Dr Isaac Roberts (1829–1904) and his Observatories', *J.B.A.A.* **103**, 3 (1993) 120–3. I am indebted to the late Stephen James for the materials on Roberts which he kindly sent to me. I. Roberts, 'Photographic Maps of the Stars', *M.N.R.A.S.* **46**, 3 (January 1886) 99–103.

71 Agnes M. Clerke, in her *The System of the Stars* (London, 1890), Chapter XXVI, 'Status of the Nebulae', and Chapter XX, 'The Great Nebula [Andromeda]', was firmly of the opinion that 'The question whether nebulae are external galaxies hardly any longer needs discussion' (p. 349), so firmly was she of the view that 'recent research', such as the photographs of Isaac Roberts, had confirmed that 'No competent thinker, with the whole of the available evidence before him can now, it is safe to say, maintain any single nebula to be a star system of co-ordinate rank with the Milky Way.' Galactic novae and nebulous structures all argued for the relative closeness of these objects, otherwise such details would not be visible. In short, there were no island universes, but everything was contained within the Milky Way. Agnes Clerke maintained this view in the 1905 edition of her *System*. Not until Vesto Slipher's work on the spiral nebulae was published, a decade later, did arguments once more swing in favour of island universes. Roberts, *Photographs of Stars* I (ref.70) 21, where he draws attention to the superiority of photography over visual techniques when it comes to detecting long-term changes.

72 Roberts, *Photographs of Stars* I (ref.70) 31. In the late 1880s Roberts was firmly of the opinion that he had photographed the dark bands in the Andromeda Nebula recorded by George Phillips Bond at Harvard: 'An Account of the Nebula in Andromeda', *Memoirs of the American Academy of Arts and Sciences*, new series, **III** (1848). Roberts argued, p. 32, that his photograph would act as a permanent record by which the assumed changes within the Andromeda Nebula would go, in accordance with the Nebular Hypothesis. Very clearly, Roberts saw the Andromeda Nebula not as an 'island universe' but as a nascent solar-system-like structure in the process of condensation.

73 Roberts believed that he saw 'symmetrical rings of nebulous matter surrounding the large diffuse centre of the nebula', *Photographs of Stars* I (ref.70) 31–2.

74 Roberts, *Photographs of Stars* I (ref.70) 31–2 for photograph, 29 December 1888, and its original interpretation. Also Roberts, *Photographs of Stars* II (1899), 63. By 1899 Roberts had come to identify M31, the Andromeda Nebula, to be 'a left-hand spiral, and not annular as I at first suspected'.

75 Because 15 years of nebular photography, especially of the Orion (M42), Andromeda (M31) and Triangulum (M33) nebulae, had failed to reveal any structural changes over time, Roberts concluded by 1899 that 'this therefore indicates that its [M33's] distance from the earth is very great': *Photographs of Stars* II (ref.74) 66.

76 Roberts, *Photographs of Stars* I (ref.70) 47 for Pleiadean nebulosity. Roberts' Pleiades and nebulosity photograph, 8 December 1888, was published as the frontispiece to Lockyer's *Meteoritic Hypothesis* (ref.49).

77 Robert Burnham, Jnr., *Burnham's Celestial Handbook. An Observer's Guide to the Universe Beyond the Solar System* (Dover, New York, 1978) p. 1880, provides a

concise account of the early study of the Pleiadean nebulosity. Isaac Roberts, *Photographs of Stars* I (ref.70) 47, however, says that a 'concise history' of the Pleiadean nebulosity was published by W.H. Wesley (Assistant Secretary to the R.A.S.) in the *Journal of the Liverpool Astronomical Society* **5**, 148–50; however, I have not been able to trace a copy.

78   'Dorothea Klumpke Roberts', obituary by J.H. Reynolds, *M.N.R.A.S.* **104**, 2 (1904) 92–3. Kenneth Weitzenhoffer, 'The Triumph of Dorothea Klumpke Roberts', *Sky and Telescope*, August 1986, 109–10. Roberts left £40,000 on his death in 1904: obituary, *Procs. of the R. Soc.* 1905 (ref.70) 362.

79   G.M. Minchin, 'The Photo-Electric Cells', *Astronomy and Astro-Physics* **11** (Northfield, Minnesota, USA, 1892) 702–5, and W.H.S. Monck. 'The Photo-Electric Effect of Star-light', *ibid.* 843–4. Stephen M. Dixon, 'The Photo-Electric Effects of Star-light', *ibid.* 844.

80   W.H.S. Monck drew attention to his 'excellent' 7½-inch Clark refractor, formerly owned by Dawes, in 'Photo-Electric Effects...' (ref.79) 843. Deborah J. Warner, *Alvan Clark & Sons, Artists in Optics* (Smithsonian Institution, Washington, 1968) 52 (new edition, 1996). Wentworth Erck, obituary, *M.N.R.A.S.* **51** (February 1891) 194–6: 194. Erck had also used this same 7½-inch Clark refractor to make important observations of the newly-discovered satellites of Mars in 1877: Erck, 'On the Satellites of Mars', *Scientific Transactions of the Royal Dublin Society* 1877, 29–37.

81   C.J. Butler, 'Early Photoelectric Photometry in Dublin and Daramona', *The Irish Astronomical Journal* **17**, 3 (1986) 373–7.

82   William Edward Wilson, obituary, *Procs. of the R. Soc.* Series A, **83** (1910) iii–vii. William Edward Wilson, obituary, *M.N.R.A.S.* **69** (February 1909) 254–5. C.J. Butler and Ian Elliot, 'Biographical and Historical Notes on the Pioneers of Photometry in Ireland', in *Stellar Photometry – Current Techniques and Future Developments*, Proceedings of the I.A.U. Colloquium No. 136, Dublin, Ireland, 4–7 August 1892 (C.U.P., 1993) 8.

83   W.E. Wilson, P.L. Gray (communicated by G. Johnstone Stoney), 'On the Temperature of the Carbons of the Electric Arc; with a note on the Temperature of the Sun, Experiments made at Daramona, Streete, Co. Westmeath', *Procs. of the R. Soc.* **58** (1895) 24–38. W.E. Wilson, 'The Effective Temperature of the Sun', *Procs. of the R. Soc.* **69** (1902) 312–20. The firm of Yeates, in its various partnership titles, was one of Dublin's leading and most enduring precision instrument-making firms: J.E. Burnett and A.D. Morrison-Low, *Vulgar and Mechanick. The Scientific Instrument Trade in Ireland, 1650–1921* (Royal Dublin Society, 1989) 140.

84   G.M. Minchin, 'The Electrical Measurement of Starlight. Observations made at the Observatory of Daramona House, co. Westmeath, in April 1895. Preliminary Report', *Procs. of the R. Soc.* **58** (1895) 142–54. See 142–6 for photometric apparatus and drawings.

85   A.A. Rambaut, 'W.H.S. Monck', obituary, *M.N.R.A.S.* **76** (February 1916) 264–6. Joel Stebbins of Illinois University, and a group of professional scientists in Germany, rather than the Irish researchers, are generally acknowledged to be the

founders of photoelectric photometry after 1912: see J.B. Hearnshaw, 'Photometry, Astronomical', in *History of Astronomy, An Encyclopedia*, ed. J. Lankford (Garland, New York, 1997) 395–401: 400.

86   Laurie G. Brock, 'George Higgs, 1841–1914' (unpublished University of Liverpool M.Sc. thesis, 1996). Dr. Brock, who was a retired physician, died shortly after the acceptance of his fine and scholarly thesis. I am indebted to Alan Bowden of Liverpool Museum for drawing it to my attention. The Museum holds a copy of the thesis.

87   Brock, 'George Higgs' (ref.86) 58.

88   H.A. and M.T. Brück, *The Peripatetic Astronomer. The Life of Charles Piazzi Smyth* (Adam Hilger, Bristol and Philadelphia, 1988) 231.

89   Higgs was not working in total isolation or without recognition, however, for he received several small Royal Society grants amounting to the sum of £200: Brock (ref.86) 53. He was also active in the Liverpool Astronomical Society, was known to other societies, and was the author of some eighteen publications: Brock 6–7.

90   Brück and Brück, *The Peripatetic Astronomer* (ref.88) 47–68. The Astronomer Royal, G.B. Airy, was also active in organising mountain-top observations at Teneriffe: Airy Papers, C.U.L. RGO6 217–221.

91   Charles A. Whitney, 'Henry Draper', *D.S.B.*

92   Frank Dyson, 'Brief History of Crossley's 36-inch Reflector' (ref.67).

93   *Sir Henry Bessemer F.R.S. An Autobiography* (The Offices of 'Engineering', London, 1905) 349–54; plates XLVII–XLIX. 'The Bessemer Observatory', *J.B.A.A.* 8 (1898) 384.

94   Richard Anthony Proctor, *Wages and Wants of Science Workers* (London 1876), reprinting Proctor's letter: *English Mechanic*, June 1873, 98–105: 99.

95   The first of these American millionaire-founded observatories was that of James Lick: 'Lick Observatory was built with money from the fortune of a wealthy man, a pattern that was followed later by Yerkes Observatory, Mount Wilson Observatory, Palomar Observatory, and McDonald Observatory, the great American research observatories built up to the time of World War II', Donald Osterbrock, *James E. Keeler, Pioneer American Astrophysicist and the Early Development of American Astrophysics* (C.U.P., 1984) 36.

8 THE ASTRONOMERS' GENTLEMEN: THE GRAND AMATEURS' PROFESSIONAL ASSISTANTS

1   John Weale, *Description of the Public and Private Observatories in London and its Vicinity* (London, 1851) 57, 62.

2   W.H. Smyth, *Aedes Hartwellianae, or Notices of the Manor and Mansion of Hartwell* (London, 1851) 247–83 for James Epps' observations and work. Prior to working for Lee at Hartwell, 'Mr. Epps spent some years in giving Mathematical Lessons, not only to Youth, but to Gentlemen upon their entering the University...' and had a large teaching collection of books and instruments: Epps' card is preserved in M.H.S. Oxford Gunther 37, 45. Similarly, Gunther 9,

17v contains a transcription of James Epps' (30 July 1773–10 August 1839) tombstone inscription, yet without specifying the place of burial.
3. The brief life histories of Thomas Phelps (born Chalford, Oxfordshire, in January 1694 and still active when engraved in 1776) and John Bartlett (born Stoke Talmage, Oxfordshire, 22 August 1721 and still active in 1776) are copied from a large engraving depicting the two men in the Blenheim Observatory. The information, 'Inscription of another [superimposed upon 'the'] Engraving of Phelps and Bartlett' has been copied, typed out, and pasted upon the back of the wooden frame in which the picture hangs, Museum of the History of Science, Oxford.
4. Airy to Algernon West, 22 December 1869, C.U.L. RGO6 6/365.
5. Airy to Benjamin Templar (unsuccessful candidate), 11 September 1850, C.U.L. RGO6 3.10/62 '. . . I may [take] leave to say that the extent of qualifications which you mention is far greater than we commonly look for in Assistants beginning their course here.'
6. Airy to Charles Wood, 15 July 1837, C.U.L. RGO6 2/213.
7. I am indebted to Roger Hutchins for this information, obtained from Cambridge Observatory Archives, which will appear in the Cambridge Observatory section of his 'A Mismatch of Ideals and Resources: British University Observatories c.1820–1939' (Oxford University D.Phil. thesis, forthcoming, 1998), Chapter II. As I have not seen a final text of this thesis, it is not possible to give page numbers. Also 'Andrew Graham' by R. Hutchins, *New D.N.B.* (O.U.P., forthcoming).
8. 'Auto-Biographical Notes. By Edwin Dunkin F.R.S., F.R.A.S., formerly Chief Assistant at the Royal Observatory, Greenwich, and Past President of the Royal Astronomical Society, and of the Royal Institution of Cornwall' (1894), R.A.S. Additional MS 55.4, R.A.S. Library no. 22837, p. 118. This autobiography is due to be published by the R.A.S. and Royal Institution, Cornwall, in 1999.
9. This is exactly what Dunkin did, at the age of 24, in 1845: 'Auto-biographical Notes' (ref.8) 129ff.
10. Airy often referred to the Library, its role in the intellectual life of the Royal Observatory, and the education of its Assistants: *Astronomer Royal's Report* to Board of Visitors (3 June 1837) 2; for the education of the young Assistants, *Report* (6 June 1857) 17; and *Report* (5 June 1869) 8, where Airy writes 'I attach much importance to the Library', and its sustained 'intellectual character', which 'has contributed materially to raise the [intellectual] tone of the Observatory'. Such an access to books and training would have been unlikely in most private observatories.
11. Airy, 'Scheme proposed for the payment of Salaries to the personal establishment of the Royal Observatory, Greenwich', 24 January 1859, C.U.L. RGO6 4/132.
12. Airy, 'Regulations for Assistants' Holidays for 1848', C.U.L. RGO6 35/191. The four senior assistants got 35 days a year, their five lower colleagues 28 days, not including the no-Moon days. James Glaisher claimed, by his employment agreement, a full six weeks' holiday a year: Glaisher to Airy, 4 December 1843, RGO6 33/286.
13. Airy to Robert Main, 23 August 1835, C.U.L. RGO6 1/321–322. Main, like the

rest of the Warrant Assistants, had five hours' office 'Confinement', from 9 a.m. to 2 p. m. But Main was called upon to observe at night only 'occasionally but not frequently'.

14   Airy to Charles Wood, 30 August 1835, C.U.L. RGO6 1/183, for Main's starting salary of £300 p. a. Main received an extra £50 p. a. for rent allowance: 9 January 1836, RGO6 1/241.
15   Airy to James Challis, 30 June 1846, Cambridge University Observatory MS. Airy described 'James Breen, a rough genius, but who can go through any quantity of work'. Also James Breen to Challis, 8 August 1846, *ibid*.
16   Catherine Wicks to Airy, 29 January 1859, C.U.L. RGO6 4/135, and subsequent letters to Airy, RGO6 4/136. For Breen's prospectus for his failed Catholic academy, see 'November 27' [1858], RGO6 4/122. Breen was offering classics, mathematics, German and 'commercial' courses, at £35 p. a., for 14-year-old boarders, which gives some insight into the breadth of his own education. His seventh Warrant Assistant salary at Greenwich in 1852 had totalled £130: Airy to A. O'Brien Stafford (Secretary of the Admiralty), 17 July 1852, C.U.L. RGO6 3/74–75.
17   Main to Airy, 7 January 1860. The exact identity of this 'Mr. Breen' is unclear however; it could have been James' and Hugh's brother, John William. For further amplification of the Greenwich–Cambridge Observatory 'exchange' system, see D.W. Dewhirst, 'The Greenwich–Cambridge Axis', *Vistas in Astronomy* **20** (1976) 109–11.
18   'William Ellis', obituary, *M.N.R.A.S.* **77**, 4 (February 1917) 295–9. For Ellis at Durham (and his successors R.C. Carrington, Albert Marth, Gabriel Goldney, etc.) see G.D. Rochester, 'The History of Astronomy at the University of Durham from 1835 to 1839', *Q.J.R.A.S.* **21** (1980) 369–78: 370, 378.
19   *Morning Chronicle*, Wednesday 10 September 1856, copy preserved in C.U.L. RGO6 4/566.
20   Manuel Johnson (Radcliffe Observatory, Oxford) to Airy, 31 October 1854, C.U.L. RGO6 212/382, for Pogson's salary.
21   Temple Chevallier (Durham) to Airy, 1 November 1854, C.U.L. RGO6 212/214, for Rümker's salary.
22   Challis (Cambridge) to Airy, 31 October 1854, C.U.L. RGO6 212/178.
23   R.C. Carrington, *A Catalogue of 3735 Circumpolar Stars Observed at Redhill In the Years 1854, 1855 and 1856...* (London 1857), Section 2, IV for friendship with Simmonds. I am not wholly certain of Simmonds' salary, due to Carrington's scrawled and abbreviated handwriting: see Carrington to Airy, 7 November 1854, C.U.L. RGO6 212/158, for 'ten [?] pounds, ten [?] shillings, & no pence' for Simmonds' monthly salary at Redhill.
24   T.R. Robinson to Lord Rosse, 6 November 1847, Rosse Archives, Birr Castle, K.5:9. For more about Edmondson, whose salary was increased *ad hominem* after 23 years' service to £105 p. a. in 1860, see J.A. Bennett, *Church, State, and Astronomy in Ireland. 200 Years of the Armagh Observatory* (Armagh Observatory and Institute of Irish Studies, Queen's University, Belfast, 1990) 87–8.
25   Robinson to Rosse, 6 November 1847 (ref.24). Andrew Graham became Assistant

at the Cambridge University Observatory in 1864, on a salary of £150, rising to £230 by the time of his retirement at the incredible age of 88. Information supplied by Roger Hutchins from Cambridge Observatory records, and to appear in 'A Mismatch of Ideals and Resources...' (ref.7).
26 Wallace was awarded an extra £50 p. a. for merit. For salaries, see Royal Observatory, Edinburgh, archives, especially the *Royal Commission on Scientific Instruction and the Advancement of Science* (February 1871), R.O.E. Library. I am indebted to Dr David Gavine for Edinburgh salaries information.
27 R.A. Proctor, *Wages and Wants of Science Workers* (London, 1876; Frank Cass facsimile reprint, 1976) 89.
28 Airy to Edmund James Stone (Chief Assistant), 8 January 1864, C.U.L. RGO6 31/249–250, for payment of 12s to one Johnson, for a six-day working week. But James Stride, an observatory labourer, received 18s per week: 21 September 1863, RGO6 5/121–122. Michael Sheeky, however, in September (?) 1863, received only 7s per week as a Gate Porter: RGO6 5/111.
29 John Henry Belville, known as 'Mr. Henry', a recently deceased Senior Warrant Assistant, was styled 'gentlemanly': Airy to Osborne, 4 September 1856, C.U.L. RGO6 4/91. And in his description of the qualities of an Assistant Astronomer, Airy pointed out the need to 'acquire the knowledge and bearing which fits them to associate with gentlemen': Airy to the Secretary of the Admiralty, 15 January 1872, RGO6 6/158–163.
30 Charles Dickens, *Nicholas Nickleby* (1838–9), Chapter XXXV in The New Oxford Illustrated Dickens series (O.U.P., 1950), 457.
31 Airy, 'Scheme proposed for the payment of salaries to the present establishment at the Royal Observatory, Greenwich', 1 September 1856, C.U.L. RGO6 4/93.
32 Patrick Moore, *The Astronomy of Birr Castle* (Birr, Ireland, 1971), Appendix 1, 'The Astronomers of Birr Castle', 73.
33 'Supper allowed to the Men when employed at the Telescope' from 8 January 1848: Rosse Archives, Birr Castle, unnumbered slip of paper between K17:42 and 43. These items of bread, butter, tea and coffee, totalling 1s 10d for four men, were probably intended for the workmen who operated the winches that controlled the 72-inch.
34 *Reminiscences and Letters of Sir Robert Ball*, ed. W. Valentine Ball (London 1915), Chapter VI, 'Parsonstown, 1865–1867'. Ball also records of Lord Rosse, who died, probably of heart disease, in 1867, that 'Felling trees was a favourite amusement of Lord Rosse, and we frequently spent an afternoon so employed', 66–7. Ball, in *Great Astronomers* (London, 1906), also reminisced about his happy time at Birr, 'my esteemed friend, Dr. Johnstone Stoney', 287, and Lord Rosse who 'socially... endeared himself to all with whom he came in contact', 288. He also recorded that 'the value of the mere copper and tin entering into the composition of each of the mirrors was about £500', 283.
35 Moore, *Astronomy of Birr Castle* (ref.32) 61–6, 75, 78.
36 'Sir Charles Todd', obituary by G.F. Dodwell, *M.N.R.A.S.* **71**, 4 (February 1911) 272–5. Also Todd, obituary by W.H.B., *Procs of the R. Soc.* Series **A**, **85** (1911) xiii–xvii: xv has a good account of the laying of the overland telegraph. I have

found no written source for the naming of Alice Springs after Todd's wife, Alice Gillam Bell, though I have heard the same in verbal anecdotes.
37 George Forbes, *David Gill, Man and Astronomer, Memories of Sir David Gill, K.C.B., H.M. Astronomer (1879–1897) at the Cape of Good Hope* (London, 1916). Also Sir David Gill, obituary by 'A.S.E.' [Eddington], *M.N.R.A.S.* **75**, 4 (February 1915) 236–47: 238 for Dun Echt and Cape appointment.
38 'Norman Robert Pogson', obituary by J.L.E. Dreyer, *M.N.R.A.S.* **52**, 4 (February 1892) 235–8.
39 Airy to Manuel Johnson, 30 October 1854, C.U.L. RGO6 212/380, for Airy's commendation of Pogson's 'intelligence and Spirit' displayed at Harton. For Pogson at Hartwell, see W.H. Smyth, *The Cycle of Celestial Objects, continued at the Hartwell Observatory to 1859... including details from Aedes Hartwellianae* (London, 1860), 128–9 for Pogson's duties.
40 J.L.E. Dreyer and H.H. Turner, *History of the Royal Astronomical Society* (R.A.S., London, 1923) 233–4.
41 *Public Home Department Proceedings, December 18 1869. Preparation of a Code of Instructions for conduct of work in the Madras Observatory*, C.U.L. RGO6 150/2/11–17. Although Pogson feared that the new 1869 Code would reduce his Observatory to a mere Magnetic Station, he was gratified when it retained its astronomical functions, and assisted with the drawing up of the Code. Article 8 stated 'The [Madras] Astronomer will be placed in official communication with the Astronomer Royal.' See also Pogson (Madras) to Airy, 5 August 1870, RGO6 150/2/22–23.
42 'Pogson', obituary (ref.38) 237. In spite of his unrelieved service, he was created a Companion of the Indian Empire.
43 'James Glaisher', obituary by W.E., *M.N.R.A.S.* **64**, 4 (February 1904) 281–7.
44 For Glaisher's social position within the Grand Amateur world, see Chapter 5, pp. 84–90. Also J.L. Hunt, 'James Glaisher, F.R.S. (1809–1903) Astronomer, Meteorologist, and Pioneer of Weather Forecasting: "A Venturesome Victorian"', *Q.J.R.A.S.* **37** (1996) 315–47.
45 See the early account of Hind's work in *The Illustrated London News*, 28 August 1852, 168. By that date he had discovered three comets and six asteroids.
46 Hind to J.F.W. Herschel, 30 September 1846, R. Soc. MS H.S. 9:333, where he specifies the receipt of Brünnow's letter.
47 'Discovery of Le Verrier's Planet: To the Editor of The Times', reporting Brünnow's communication from Berlin, 25 September 1846, and his own observation and letter date, 30 September 1846. Cutting of letter preserved in R.A.S. Lassell MS 9:7.
48 'John Russell Hind', obituary by T. Almond Hind, *M.N.R.A.S.* **56**, 5 (February 1896) 200–5. This obituary, 200, politely omits reference to Hind defeating Adams for the Nautical Almanac Superintendentship.
49 Fanny Copeland, 'Memoirs', photocopy of an original MS, in private hands in Edinburgh, placed in the Royal Observatory Edinburgh Library by Michael and Mary Smyth. The original document is held by a family friend of Fanny Copeland, who died in Yugoslavia around 1970. I am indebted to Dr Mary Brück for this

information, and for sending me a photocopy of the R.O.E. copy. Also 'Ralph Copeland', obituary by J.L.E. Dreyer, *M.N.R.A.S.* **66**, 4 (February 1906) 164–74: 167 for Dun Echt. Hermann A. Brück, 'Lord Crawford's Observatory at Dun Echt, 1872–1892', *Vistas in Astronomy* **35** (1992) 81–138: 113–19 for Copeland.

50 'Albert Marth', obituary by E.B. Knobel., *M.N.R.A.S.* **58**, 4 (February 1898) 139–43: 140 for South Villa. A significant and perceptive biography of Albert Marth has been written by Roger Hutchins for the *New D.N.B.* (O.U.P., forthcoming), and I am indebted to the author for a copy of his typescript.

51 The nebula work, said Lassell, was 'entirely the work of Mr. Marth on these occasions when the telescope was not otherwise engaged [on solar system work]'. W. Lassell, 'A Catalogue of New Nebulae discovered with the Four-Foot Equatorial in 1863–65', *Mem. R.A.S.* **36** (1867) 53.

52 Marth to the Fourth Earl of Rosse, 29 January 1874, Rosse Archives, Birr Castle, K. 25:1, where Marth applies for a job at Birr, being at the same time highly critical of Newall. Even so, Marth seems to have been offered the Birr job, only to turn it down: Marth to Rosse, 18 February 1874, Rosse Archives, Birr Castle, K. 25:2. Marth claims to have had no facility for practical mechanics. In the same letter, Marth claimed to have turned down a post with the new Melbourne Telescope, before adding a further swipe at persons involved with that instrument. These letters give one an insight into what it was about Marth that angered so many people.

53 A. Marth, 'On the Polar Distances of the Greenwich Transit Circle', *A.N.*, nos. 1260–1263, 177–230. This 54-column paper, analysing the flexures and errors of the Greenwich Transit Circle of 1850 (which Airy regarded as a triumph of applied engineering), certainly displays Marth's gifts as a scientist, if not as a diplomat.

54 Airy to the Fourth Earl of Rosse, 18 April 1874, Rosse Archives, Birr Castle, K. 3:12.

55 'Thomas William Backhouse', obituary, *M.N.R.A.S.* **81**, 4 (February 1921) 254–5. Election certificate to R.A.S., 11 April 1873, for occupation as 'Coal-Owner'.

56 Backhouse's Observation Books, 1 January 1858 to 10 February 1920, are in Durham University Library: 36 vols., v65/1–36, which I first looked at around 1988. I am indebted to Don Simpson for the page count: Simpson to Chapman, 6 September 1994, personal correspondence. Backhouse also issued the *Publications of West Hendon House Observatory, Sunderland* I–IV (Sunderland 1891–1915), and *Catalogue of 9842 Stars, or All Stars Very Conspicuous to the Naked Eye for the Epoch 1900* (Sunderland, 1911): both private publications, with copies in R.A.S. Library. He also regularly contributed to *M.N.R.A.S.* following his election as F.R.A.S. in 1873.

57 For Backhouse's Astronomical Secretaries see W.H.M. Christie papers, C.U.L. RGO7, 247.

58 A.J. Edmunds, personal information communicated from Joan Newiss to Don Simpson, 2 June 1994, and relayed, Simpson to Chapman, 27 June 1996. For Sharp see *The Friend*, 15 March 1940, p. 158.

59 Personal communication, Simpson to Chapman, 9 March 1994, and many subsequent letters.

366   Notes and references   [Ch. 8

60 Simpson to Chapman, 16 May 1994: a 5-page private communication detailing the life of Charles Paton (1849–1938), the writer's grandfather, from Simpson family records.
61 Simpson to Chapman (ref.60). Simpson to Chapman, 31 May 1994, states that many of the recollections of Charles Paton relayed in the above and other private communications were primarily based upon family memory, and not written record.
62 Simpson to Chapman (ref.60).
63 Joseph Foster, *Alumni Oxonienses: the members of the University of Oxford, 1715–1886*, II (London, 1888), 535 for Goldney. *Durham Directory and Almanack* (1906), 30 October 1905, p. 70, Durham University Department of Palaeography Collection. The *Almanack* states that Goldney was born in Bristol, and was the nephew of Sir Gabriel Goldney M.P., and had been Head Scholar at Lincoln College.
64 Goldney to Prof. Pearce, 15 May 1874. Goldney stated that he was 26 years old: Durham University Observatory MS, item 332. William Christie (Greenwich Chief Assistant) to Pearce, 11 May 1874, specifies that Goldney was a Roman Catholic, Durham University Obs. MS item 331.
65 Airy to Pearce, 15 May 1874, Durham University Obs. MS item 333. Pearce to Goldney, 28 April 1885, *ibid.* item 340 (B6) and Goldney to Pearce, 17 July 1885, *ibid.* item 349.
66 Goldney, obituary, *Durham County Advertiser*, 3 November 1905, and *Almanack* (ref.63).
67 Goldney, obituary, *Durham County Advertiser* (ref.66).
68 G.B. Airy, *Report of the Astronomer Royal to the Board of Visitors of the Royal Observatory, Greenwich*, 6 June 1846; on p. 7 Airy names the nine Warrant Assistants on the Observatory Staff.
69 Airy, *Report* (ref.68), 1 June 1872, p. 20.
70 'Salaries of Lady Computers, 1890, April', in Christie's hand, C.U.L. RGO7, 140, cited in Mary T. Brück, 'Lady Computers at Greenwich in the early 1890s', *Q.J.R.A.S.* **36** (1995) 83–95: 86, 93.
71 Mary T. Brück, 'Alice Everett and Annie Russell Maunder, torch-bearing women astronomers', *The Irish Astronomical Journal* **21** (March and September 1994) 281–90: 282 for £4 per month salary. Brück, 'Lady Computers' (ref.70) 90–1 for Alice Everett as a photographer.
72 Mary T. Brück, 'Agnes Mary Clerke, Chronicler of Astronomy', *Q.J.R.A.S.* **35** (1994) 59–79: 66–7.
73 Cambridge University Observatory Syndicate *Report*, 1882, 3. See R. Hutchins, 'A Mismatch of Ideals and Resources . . .' (ref.7), section on Cambridge Assistants.
74 Cambridge University Obs. Syndicate *Report*, May 1885, 1.
75 Hutchins, 'A Mismatch of Ideals and Resources . . .' (ref.7), section on Cambridge Assistants.
76 On 25 September 1996, the National Maritime Museum held an 'Old Observers' reunion and public discussion session, for a group of around ten people who had worked at the Royal Observatory, Greenwich, prior to its vacating the old

Greenwich site for Herstmonceux in 1954. I was delighted to be invited to act as Chairman of the meeting, and to begin the historical questions. What soon became very clear is how the old junior Computership system of entry had survived virtually intact down into the 1950s. My friend Gilbert Satterthwaite was the last of the breed; he joined the staff in 1952 from school, and made the last published observation with the Airy Transit Circle of 1850 on 30 March 1954: the last 'official' observation made at Greenwich. Like other Greenwich Assistants before him, such as John Henry Belville, and Edward Walter Maunder, Mr Satterthwaite, who now works in the physical laboratories of Imperial College, London, became fascinated with the history of the Royal Observatory, and in 1995 completed a thesis which in itself forms the definitive historical study of the Greenwich instrument which he last used officially in 1954: 'The History of the Airy Transit Circle at the Royal Observatory, Greenwich', M.Sc. Dissertation, University of London, September 1996.

77   Charles Dickens, 'The Planet Watchers of Greenwich', *Household Words* **9** (25 May 1850) 200–4: 200. Airy, however, objected to much of the routine work, especially chronometer regulation, which the Admiralty forced upon Greenwich, and threatened to reduce the Assistants 'into a mere bureau of clerks': *Astronomer Royal's Report* to the Board of Visitors, 3 June 1837, 2.

78   Isaac Roberts, who died suddenly in 1904 (see Chapter 7, pp. 137–139), was probably the last British Grand Amateur to pioneer true fundamental research at his own expense. His own photographic assistant had been William Sadler Franks, who first worked for Roberts in 1892 'and was there engaged in photographing nebulae and star clusters with the 20-inch reflector, until the time of his employer's death in 1904'. Franks, who was the last Professional Assistant to the last major Grand Amateur, later worked at John Franklin-Adams' Mervel Hill Observatory, and then Frederick J. Hanbury's East Grinstead Observatory, though doing much less significant work than he had done with Roberts: 'William Sadler Franks', obituary by W.H. Steavenson, *M.N.R.A.S.* **96**, 4 (February 1936) 291–2.

9 A PENNY A PEEP: THE ASTRONOMICAL LECTURERS OF THE PEOPLE

1   *Autobiography of Sir George Biddell Airy*, ed. Wilfred Airy (C.U.P., 1896) 66–70. Also 'Family History of G.B. Airy', and 'Personal & Private', in Enid Airy Papers, C.U.L. MS, which record his early career at Cambridge. Airy's personal financial accounts (1820–1824) preserved in these papers show that even while still an undergraduate he was receiving fees from other students for tuition.

2   *Autobiography* (ref.1) 220–1. For the dimensions of the Central Hall, South Shields, measuring 46 × 30 feet with a stage 15 × 30, rated to hold 400–500 people, see 'Journal kept by Edwin Dunkin', Mon. 16 October 1854, C.U.L. RGO6 213/275–305. Airy lectured to a packed hall.

3   Temple Chevallier to Airy, 22 October 1854, C.U.L. RGO6 212/210. According to the 'Auto-Biographical Notes' of Edwin Dunkin (1894), R.A.S. Additional MSS 55.4, Library No. 22837, Airy lectured to 'several hundred working men', 257.

4   *Illustrated London News*, 4 November 1854. Also *North and South Shields Gazette*, 27 October 1854. 'Airy and the Coalhole', *Punch* **27**, 11 November 1854, 199. For Airy's amused response, *Autobiography* (ref.1) 221.
5   Airy to Richarda Airy, 14 March 1848, in Airy, *Autobiography* (ref.1) 196–7. See also 'Astronomer Royal's Journal', C.U.L. RGO6 25.
6   'Sir George Biddell Airy', obituary by E.J. Routh, *Procs. of the R. Soc.* **51** (1892) i–xxi: xv.
7   J.F.W. Herschel lectured on 'Sensorial Vision' at the Leeds Philosophical and Literary Society, 30 September 1858: Herschel, 'Diary', Royal Society MSS 583–6, date order. Documents preserved in the 'Album' of the Leeds Astronomical Society, including a handbill of 1861, show the presence of an Astronomical Society in Leeds, with Sir John Herschel as Honorary President. Sir John wrote a lecture which was read to the Society in late October 1863: *The Leeds Intelligencer*, **31** October 1863, 17.
8   The lecture was delivered at the Salford 'Lyceum': Thomas Coates, *Report on the State of Literary, Scientific and Mechanics' Institutions in England: with a list of such Institutions and a list of Lectures* (Society for the Diffusion of Useful Knowledge, London, 1841) 11.
9   *Sheffield Iris*, 15 April 1847, 22; cited in Ian Inkster, 'Advocates and Audience – Aspects of Popular Astronomy in England, 1750–1850', *J.B.A.A.* **92** (1982) 119–23: 121. This excellent article gives an important insight into popular astronomy lecturing, especially in northern Britain.
10 W.R. Birt was associated with the Hackney Scientific Association (founded c.1867), giving the lecture 'A Method of Ascertaining the Existence of Lunar Changes' on 24 January 1871: *Astronomical Register* **9**, 1871 (1872) 91, 184. In 1884 Elger was lecturing on his lunar work to the Liverpool Astronomical Society: *Astron. Reg.* **22**, 1884 (1884) 112.
11 'T. Dell's Lecture on Eclipses of the Sun', undated, but as the forthcoming eclipse being discussed was due to take place on 18 July, it seems to have been 1860. Printed account from an unspecified journal, preserved in Lee's 'Album', M.H.S. Oxford Gunther 38, 45.
12 'London Lecturers – The Astronomical Mania', cutting from an unspecified newspaper(?) dated sequentially to c.1839, in the Lee 'Album', M.H.S. Oxford Gunther 36, 1–3.
13 'Astronomical Mania' (ref.12).
14 G.B. Airy, *Popular Astronomy: A series of lectures delivered at Ipswich* (6th edn., London, 1868). Airy does not specify the year in which he delivered these lectures, though from his *Autobiography* (ref.1) 196–7 it was 1848.
15 Quoted by Inkster, 'Advocates and Audience' (ref.9) 122.
16 Newspaper cutting, unspecified and undated, c.1850, in Lee, 'Album', M.H.S. Oxford Gunther 37, 41.
17 'Astronomical Mania' (ref.12) 3.
18 'Recollections from an Astronomical Lecture for Promoting Emigration', signed G. Rosetta Smyth and Caroline Mary Smyth, 11 September 1848, Lee, 'Album', M.H.S. Oxford Gunther 37, 98.

19 New and punitive measures against the Irish destitute after 1846 led them to flee to England, where in a workhouse they would at least receive food. Cardiff to Dublin coal ships gave free passages on the empty run home, the human ballast being cheaper to load and unload than stone and shingle: Cecil Woodham-Smith, *The Great Hunger, Ireland, 1845–9* (Hamish Hamilton, London, 1962) 270ff. The English magistrates did everything they could to encourage these unfortunates to re-emigrate to Australia or America. The lecture attended by the Misses Smyth might have been in aid of a charity that tried to provide assisted passages.
20 Thomas Chalmers, *A Series of Discourses on the Christian Revelation, viewed with its connection with The Modern Astronomy* (Glasgow, 1817). This influential work built seven Discourses upon a foundation of modern astronomical knowledge, to show its harmony with Christian truth. Thomas Dick, *Celestial Scenery: or Wonders of the Planetary System displayed: Illustrating the Perfections of Deity and a Plurality of Worlds* (London, 1838). See also David Gavine, 'Astronomy in Scotland, 1745–1900', unpublished Open University Ph.D. thesis, 1982; also William Astore, 'Observing God: Thomas Dick (1774–1857), Evangelicalism and Popular Science in Victorian Britain and Antebellum America', unpublished Oxford University D.Phil. thesis, 1995. I am indebted to Drs Gavine and Astore for their assistance and conversations about Thomas Dick.
21 'Lecture on Science and Religion', upon which Lee has written 'Castle Street Lectures. Aylesbury', unspecified Aylesbury (?) newspaper cutting, early 1854, Lee, 'Album', M.H.S. Oxford Gunther 37, 100.
22 Doxology: an ancient prayer or hymn of thanksgiving, best known in English in the hymn by the seventeenth-century Bishop Thomas Ken, 'Praise God from Whom all blessings flow'.
23 *Aylesbury News*, 12 January 1861: cutting in Lee, 'Album', M.H.S. Oxford Gunther 38, 51.
24 *Aylesbury News*, 12 January 1861 (ref.23). Even in the early 1880s, farm labourers at Juniper Hill, to the east of Banbury and about 25 miles from Aylesbury, were earning only 10*s* a week, from which they had to pay cottage-rent: Flora Thompson, *Lark Rise to Candleford* (1947; Penguin edn., 1979), Chapter 1, 20–1.
25 Handbill, for lecture 8 June 1842, in Lee, 'Album', M.H.S. Oxford Gunther 36, 91.
26 *Manchester Weekly Advertiser*, 10 December 1853, cutting in Lee, 'Album', M.H.S. Oxford Gunther 37, 98.
27 See *Punch*, 6 November 1858, 190, cartoon 'The Comet as Seen from our Area', and 'Canzonet on the Comet' (sung to the tune 'Little Bo-peep'): 'The Comet has flown/Where? – it is not quite known;/But in two thousand years we may find him,/Says AIRY, he'll then/come and see us again,/And bring his tail behind him.'
28 Jabez Inwards, 'THE COMET. IS THE WORLD IN DANGER?', handbill in Lee, 'Album', M.H.S. Oxford Gunther 37, 111. The year 1843 was rich in comets, and three, including the 'Great Comet' of February–April 1843, are described by John Russell Hind, *The Comets: A Descriptive Treatise upon these bodies...* (London, 1852) 164. For the 'Millerites' and the Great Comet of 1843, see Donald K. Yeomans, *Comets, A Chronological History of Observation, Science, Myth and Folklore* (John Wiley, Chichester and New York, 1991) 178.

29 Inwards, 'THE COMET' (ref.28). For Cratchit at Camden Town see Charles Dickens, *A Christmas Carol* (1843; Thomas Nelson & Sons contemp. edn., London, 1912) 15.
30 Inwards, 'THE COMET' (ref.28).
31 'T. Dell's Lecture on Eclipses of The Sun' (ref.11).
32 'T. Dell's Lecture' (ref.11). This total eclipse, 8 July 1842, was seen across northern Italy and southern France. Airy and his wife Richarda were among those who saw the 'pink flames' at totality. See also A. Pannekoek, *A History of Astronomy* (London, 1961) 406.
33 Undated cutting from the *Gateshead Observer*, in Lee, 'Album', M.H.S. Oxford Gunther 4, 41.
34 Inkster, 'Advocates and Audience' (ref.9) 122. For Edward Mills (1803–1865) of Llanidloes with his 66-foot diameter orrery, see John Silas Evans, *Seryddiaeth a Seryddwyr* [Astronomy and Astronomers] (Cardiff, 1923) 271. I am indebted to Rhys Morris of the University of Wales in Cardiff for translating sections of *Seryddiaeth* for me. A 'primitive-looking orrery' and other badly-taught simple science as part of a Twelfth-Night Show was recalled by the elderly James Bonwick when a pupil at the Borough Road School, Southwark, London, in 1823: J. Bonwick, *An Octogenarian's Reminiscences* (London, 1902), reprinted in John Burnett (*ed.*), *Destiny Obscure: Autobiographies of Childhood, Education, and Family from the 1820s to the 1920s* (Allen Lane, London, 1982).
35 Henry Solly, *Working Men; a Glance at some of their Wants, With Reasons and Suggestions for Helping Them to Help Themselves* (London, 1863) 5.
36 The above narrative is from Henry Mayhew, *London Labour and the London Poor: A Cyclopaedia of the Conditions and Earnings of Those That Will Work, Those That Cannot Work, and Those That Will Not Work*, 4 vols., III (London, 1861) 'Exhibitor of the Microscope' 83–8: 84.
37 Mayhew, *London Labour* III (ref.36) 84.
38 Mayhew, *London Labour* III (ref.36) 81.
39 A good example of these pyrotechnics is presented in Arnold Bennett, *Anna of the Five Towns* (1902; O.U.P., 1995), Chapter V, 52–60.
40 Elizabeth Gaskell, *Mary Barton, A Tale of Manchester Life* (1848; Penguin edn., 1970), Chapter V, 75.
41 Mayhew, *London Labour* III (ref.36), 'The Telescope Exhibitor', 79–83: 79–80. Mayhew seems to have interviewed Tregent on 26 October 1856.
42 Mayhew, *London Labour* III (ref.36) 80. Samuel Smiles, *Men of Invention and Industry* (London, 1884) 336–7, discusses the intellectual lives of shoemakers, who, often working alone, tended to be thinkers; whereas tailors, who worked with others, were more likely to be communicators.
43 Mayhew, *London Labour* III (ref.36) 80.
44 Mayhew, *London Labour* III (ref.36) 81.
45 Mayhew, *London Labour* III (ref.36) 80, 82–3, for Tregent's discussion of object glasses. He does not specify, 79, which of Dr William Kitchener's works he is referring to, but it was most probably his *Practical Observations of Telescopes, Opera-Glasses and Spectacles* (London, 1818). The Revd Edward Craig's (whose

name Mayhew wrote down as 'Cragg') enormous refractor on Wandsworth Common was illustrated and described in *The Illustrated London News*, 28 August 1852, 168, but the financial and other details were omitted. It would be interesting to know the sources from which Tregent came by this information, although the telescope was the focus of much general discussion.

46 Mayhew, *London Labour* III (ref.36) 83.
47 Mayhew, *London Labour* III (ref.36) 81.
48 Mayhew, *London Labour* III (ref.36) 81.
49 Mayhew, *London Labour* III (ref.36) 81–2.
50 Mayhew, *London Labour* III (ref.36) 81.
51 Mayhew, *London Labour* III (ref.36) 81.
52 Mayhew, *London Labour* III (ref.36) 82. William Wordsworth, 'Star-Gazers', in *The Poetical Works of William Wordsworth*, ed. E. de Selincourt (Clarendon Press, Oxford, 1952) 219–20 (composed 1806, published 1807). Ben Weinreb and Christopher Hibbert (eds.), *The London Encyclopaedia* (Macmillan, Book Club Associates, London, 1983) 840, reproduces a photograph of one of Tregent's professional descendants, posed with a fine brass refractor (fitted with a solar eyepiece) by Boudicca's statue in Parliament Square c.1912.
53 Inkster, 'Advocates and Audience' (ref.9) 119–20.
54 Mayhew, *London Labour* III (ref.36) 84.
55 Mechanics' Institution lending libraries, however, proved enormously popular with artisan members, and by 1841 the Ancoats, Manchester Lyceum Library 'circulated' over 100 books a day, including loans to women: Coates, *Report* (ref.8) 49. See also Coates, Chapter 7, on Institution Libraries.
56 Gavine, 'Astronomy in Scotland' and Astore, 'Observing God' (both ref.20).
57 Works such as Benjamin Martin's *General Magazine* (London, 1755) and *The Young Gentleman's and Lady's Philosophy* (London, 1772). See also John Millburn, *Benjamin Martin: Author, Instrument-Maker and 'County Showman'* (Leiden 1976).
58 Richard Lovell Edgeworth, *Practical Education, or the History of Harry and Lucy* (London, 1780). These simple conversational books, '... Written upon the Principles of Association... Inscribed to Dr. Priestley', taught genteel young children how to observe nature, and take an interest in useful skills such as dairy-craft, brick-making, and blacksmithing.
59 Mayhew, *London Labour* III (ref.36) 79.
60 Mayhew, *London Labour* III (ref.36) 85.
61 Dionysius Lardner's *Cabinet Cyclopaedia* series ran to 133 volumes between 1830 and 1851. The prices of these volumes are printed on the paper title labels glued on to the spines of the original cloth bindings. Herschel's *Treatise on Astronomy*, for instance, is labelled 6s. For the price of nineteenth century science books, see A. Chapman, 'Science for the working man: the failure of a Victorian dream', *Transactions of the Eccles and District Historical Society* (Eccles, Manchester, Supplement 1987–8) 1–16:11–13.
62 Denison Olmsted, *The Mechanism of the Heavens; or, Familiar Illustrations of Astronomy. With Historical and Biographical Sketches* (Nelson, London,

372   Notes and references [Ch. 9

Edinburgh and New York, 1856). Also Ormsby McKnight Michell, *The Orbs of Heaven, or, The Planetary and Stellar Worlds. A Popular Exposition of The Great Discoveries and Theories of Modern Astronomy* (4th edn., London 1853). Both Olmsted and Michell were professional astronomers who were also involved in public exposition: the former was a Yale professor and the latter the Director of Cincinnati Observatory.

63   Richard Anthony Proctor, *Wages and Wants of Science-Workers* (London, Smith, Elder, 1876; Frank Cass reprint, 1976) 8.

64   *A Familiar Introduction to the Science of Astronomy* (1851–2), an anonymous and undated booklet to accompany a box of 38 hand-painted astronomical slides, now deposited in the Physical Sciences department of Liverpool Museum. Though the booklet is undated, it mentions the fourth of J.R. Hind's asteroids, Irene, discovered 19 May 1851, but not his fifth, Melpomene, discovered 24 June 1852, or his subsequent ones, so that the text seems to have been composed somewhere between these two dates. See p. 12.

65   *A Familiar Introduction* (ref.64) 29–30.

66   George Orwell, *Down and Out in Paris and London* (1933; Penguin edn., 1989), Chapter 30, 166–9.

10 ASTRONOMY AND THE MODEST MASTER-CRAFTSMAN

1   John Weale, *Description of the Public and Private Observatories in London and its Vicinity* (London 1851) 57. The early numbers of the *Astronomical Register* (1863–1886), with their discussions of telescopes, observing projects, obituaries, accounts of lectures, and book reviews, convey a lively impression of the cultivation of astronomy amongst the middle classes, though one assumes from its tone that its readers were far from hard-up.

2   Anne Secord, 'Science in the Pub: the Artisan Botanists in early Nineteenth-Century Lancashire', *History of Science* 32 (1994) 269–315; also 'Corresponding interests: artisans and gentlemen in nineteenth-century natural history', *British Journal for the History of Science* 27 (1994) 383–408.

3   John Horsefield to *Manchester Guardian*, 24 April 1850, 6. I am indebted to Dr Secord for this reference.

4   Henry Mayhew, *London Labour and the London Poor: a Cyclopaedia of the Conditions and Earnings of Those That Will Work, Those That Cannot Work, and Those That Will Not Work* III (London, 1861) 82.

5   Samuel Bamford, *Passages in the Life of a Radical*, 2 (Manchester[?], 1842–1844) 177. The Bodleian Library copy of Bamford's book carries neither a date nor a place of publication. It was, however, printed by Heywood (probably of Manchester), and vol. II, p. 240 terminates with the date 27 July 1842. I am indebted to Dr Secord for drawing my attention to this reference.

6   'Johnson Jex, A Study for the Million', *Norfolk Advertiser*, 17 January 1852; cutting in Dr John Lee's 'Album', M.H.S. Oxford Gunther 37, 82–83.

7 Arthur Young, *General View of the Agriculture of the County of Norfolk* (London, 1804) 74.
8 'Johnson Jex' (ref.6). I have been unable to trace any further references to this head in Norfolk, or to Bianchi's work with him.
9 'Johnson Jex' (ref.6). I am indebted to David Todd, who has kindly, yet unsuccessfully, enquired into members of the Jex family still living around Letheringsett. Mr J.A. Jex, Managing Director of Jex Engineering Company, Ltd., of Grimsby, is not familiar with Johnson as a relative, although he is of Norfolk descent, with a great-grandfather who had been Mayor of Norwich. Johnson, of course, never married; however, further research in Norfolk archives is necessary to ascertain how extensive the wider Jex family might have been.
10 One of the most complete primary sources on Veitch is *The Home-Life of Sir David Brewster, by his daughter Mrs. Gordon* [Margaret Maria] (Edinburgh, 1869) 22–50, 76–9, 139–40, etc.
11 *Home-Life of Brewster* (ref.10) 25.
12 J.N. McKie, 'James Veitch, 1771–1838', *J.B.A.A.* **87**, 1 (1976) 44–50: 45.
13 *Home-Life of Brewster* (ref.10) 30 for Brewster's account of his own boyhood visits to Veitch. Veitch used to walk with the young Brewster back into Jedburgh because the road was supposedly haunted.
14 Sir Walter Scott to James Ellis, April 1818, a letter printed in Willis's *Current Notes*, 25 January 1856, and cited in *Home-Life of Brewster* (ref.10) 26.
15 G. Watson, *The Border Magazine* V (III) (March 1900) 45, 47, cited in McKie, 'James Veitch' (ref.12) 46. Unfortunately I have not been able to locate actual copies of *The Border Magazine* V (1900), and rely upon Mr McKie's transcription, and other information about Veitch with which he has generously supplied me. The Herschel 100-guinea price I derive from a label pertaining to a 7-foot reflector of c.1795 formerly owned by the Revd Nathaniel Jennings, F.R.A.S., and presented to the Museum of the History of Science, Oxford, by Dr H.N. Lewis in 1927: see the file copy of the label, M.H.S. Oxford.
16 Veitch to Brewster, 5 August 1812, cited in *Home-Life of Brewster* (ref.10) 78–9, for Veitch's first sighting of the comet, 27 August 1811. It had, however, been first seen by Flaugergues at Viviers, France, in the Rhone valley (42° 28' N; 4°41'E) on 26 March 1811: J.R. Hind, *The Comets. A Descriptive Treatise upon those Bodies* (London, 1852) 110.
17 Veitch to Mary Somerville, 12 October 1836, reproduced in Mary Somerville, *Personal Recollections from Early Life to Old Age of Mary Somerville. With Selections from her Correspondence by her Daughter, Martha Somerville* (London, 1873) 101–2. Mrs Somerville discusses her recollections of Veitch between pp. 99 and 103. In this letter, Veitch also cites Thomas Dick's authority for his discovery – made with a home-made telescope in his sawpit – of the 1811 comet.
18 Somerville, *Personal Recollections* (ref.17) 110.
19 *Home-Life of Brewster* (ref.10) 24, 27.
20 *Home-Life of Brewster* (ref.10) 139.
21 J.N. McKie, 'Gideon Scott (1765–1833)', *J.B.A.A.* **76**, 1 (December 1965) 53–8. For the reference to the surviving Scott telescope in 1965, see McKie, 56–7.

374   Notes and references   [Ch. 10

22   'John Pringle Nichol' (1804–1859), obituary, *M.N.R.A.S.* **20**, 1 (1859) 131.
23   Sir David Brewster, *A Treatise on Optics* (London, 1831) 356–7: 'Mr Ramage's Telescope'.
24   Robinson to W.E. Gladstone: several letters, late 1869, explaining Leach's activities and worth, C.U.L. RGO6 6/362ff. Also G.B. Airy, 'Astronomer Royal's Journal', 22 December 1869, RGO6 26– '... letter about an applicant to Mr Gladstone'.
25   Robinson to Airy, 25 February 1870, C.U.L. RGO6 6/388, for Robinson's absence in the south of France. Gladstone does not appear to have been wintering in the south of France in the first weeks of 1870, though some years later he started to visit Cannes: Roy Jenkins, *Gladstone* (Macmillan, London, 1995) 466.
26   Robinson to Airy, 25 January 1870, C.U.L. RGO6 6/383. Airy also told Gladstone, 14 January 1870, RGO6 6/376–380, that Isaac Todhunter (whom Airy knew well in Cambridge) had no recollection of Leach.
27   Airy to Gladstone, 14 January 1870, C.U.L. RGO6 6/376–380.
28   Robinson to Airy, 11 March 1870, C.U.L. RGO6 6/391. Airy to Leach, 26 March 1870, RGO6 6/404–405.
29   Leach's list, expressing his desire for a good achromatic telescope, was enclosed in a letter from Robinson to Airy: see C.U.L. RGO6 6/389–390.
30   Airy to Gladstone, 14 January 1870, C.U.L. RGO6 6/376–380. For the achromatic telescope prices, see Airy to Robinson, 1 February 1870, RGO6 6/383–386.
31   Leach's list (ref.29).
32   Airy to Messrs. Horne and Thornthwaite, 12 March 1870, C.U.L. RGO6 6/394. Airy to Horne and Thornthwaite, 14 March 1870 and 19 April 1870, RGO6 6/395 and 413, regarding the larger object glasses for the Star Finders. Horne and Thornthwaite's booklet, *The Star-Finder and how to use it*, is preserved in RGO6 6/396–399.
33   For 'catch penny', see Leach's list (ref.29). Airy to Horne and Thornthwaite, 19 April 1870, recommending the fitting of larger object glasses, C.U.L. RGO6 6/413. Leach to Airy, 7 June 1870, RGO6 6/415, thanking Airy for his attention to the matter, and his instruction to Horne and Thornthwaite to replace the smaller diameter object glass with a larger one of 1¼ inches.
34   Malby's Globe Catalogue & Prices, in C.U.L. RGO6 6/407. Airy to Leach, 26 March 1870, RGO6 6/404–405, is a remarkably detailed and instructive letter, in which the Astronomer Royal went into considerable detail to instruct the shoemaker about equatorials, globes, telescopes, and their various merits. It is evident that Airy had examined the 'Star-Finder' personally. Clearly, Airy took Leach seriously, and one sees the public educator side of Airy's personality coming to the fore.
35   Leach to Airy, 7 June 1870, C.U.L. RGO6 6/415.
36   Census Return, Frodsham, Cheshire, 1871, PRO Ref. 3693 fo 82; microfilm 24/21, Cheshire Record Office. I am indebted to Gerard Gilligan, of the Liverpool Astronomical Society, for locating and photocopying this document.
37   On several occasions in the correspondence regarding Leach, Airy repeats that

injunction which he invariably made to those who threatened to presume upon his time, that no *continuing* correspondence could develop.
38 T. Chevallier to Airy, 22 October 1854, C.U.L. RGO6 212/210.
39 The urban working-man got the vote in 1867, when the Tory Prime Minister, Benjamin Disraeli, slipped through the Reform Act. It was the Liberal, Gladstone, who benefited from it, however, when he came to office in 1868. Gladstone's 'People's William' populism first swept through British politics in the early 1860s, and grew to hero-worshipping proportions by the 1880s: Jenkins, *Gladstone* (ref.25) 230ff.
40 Airy, 'Astronomer Royal's Journal', 30 March 1870, C.U.L. RGO6 26.
41 Article on the proposed Nottingham Observatory in *Lloyd's Weekly News*, 28 December 1851; cutting in John Lee, 'Album', M.H.S. Oxford Gunther 37, 76v.
42 'Mr. J.R. Hind', *The Illustrated London News*, 28 August 1852, 168.
43 'The Working Men's International Exhibition', *The Times*, 12 July 1870, 12. The bibliographical sources for some of the foregoing Bush material are complex. In addition to clearly-marked copies of *The Times* and *Daily Telegraph*, many references to Bush's achievements around 1870 derive from a collection of cuttings, often from unspecified local newspapers, now held by Nottingham University Library. When these cuttings are cited, I give their University of Nottingham archive numbers. I am indebted for assistance on Bush to Patrick Fleckney of Nottingham, whose 'The Story of Thomas William Bush (1839–1928) F.R.A.S., Secretary of Nottingham General Hospital 1873–1888', *The Nottingham Historian* (1995) is the most complete published life of Bush. Mr Fleckney very generously supplied me with photocopies of many local documents.
44 'Thomas William Bush' (1839–1928), obituary by W.S.F. [Franks], *M.N.R.A.S.* **89**, 4 (February 1929) 298.
45 This was presumably J.F.W. Herschel's *The Telescope* (Edinburgh, 1861), the 190-page book version of Herschel's major article 'From the *Encyclopaedia Britannica*': see the long letter headed 'The Nottingham Telescope', dated 19 July 1870, by S. Alex Renshaw to the editor of *The Nottingham Journal*; cutting in Nottingham University Library, UXW 1/1/9.
46 *The Times*, 12 July 1870 (ref.43). A picture of the 13-inch telescope was also published in *The Nottingham Daily Guardian*, 13 July 1870.
47 *The Times*, 12 July 1870 (ref.43). This statement by the journalist of *The Times* is ambiguous insofar as it is not entirely clear whether these condemned items of working-class handicraft were present in the same exhibition as Bush's telescope, or elsewhere.
48 *The Times*, 12 July 1870 (ref.43). The smoky air in the part of Nottingham where Bush lived was mentioned by several newspapers: 'Nottingham Genius and Industry at the Working Men's Exhibition', an unspecified (probably local) and undated newspaper cutting, *c.*1870, Nottingham University Library, UXW 1/1/4.
49 S. Alex Renshaw, 'The Nottingham Telescope' (ref.45) for 'Browning Equatorial' and details of Bush's mount.
50 See ref.45.
51 See *The Times*, 12 July 1870 (ref.43), and Renshaw's article (ref.45) for the weight,

construction features, and optical characteristics of Bush's 13-inch aperture telescope.

52 *The Times*, 12 July 1870 (ref.43). *The Daily Telegraph*, 12 July 1870, 5, also gives an account of the Working Men's Exhibition and its visiting dignitaries, but makes only a one-sentence reference to Bush.

53 Several of the Midlands local newspapers picked up on the Exhibition and the notice received by Bush. For Henry Milward's demonstration of Bush's telescope to Gladstone and the other dignitaries see 'Visit of the Premier and Mrs. Gladstone to the Working Men's Industrial Exhibition', unspecified newspaper cutting pertaining to the 1870 Exhibition in Nottingham University Library, UXW 1/1/2. Also G.B. Airy, 'Astronomer Royal's Journal', 9 and 19 September 1870, C.U.L. RGO6 6/26.

54 Bush, obituary, *M.N.R.A.S.* (ref.44) 298. The Gold Medal instruments presented by Gladstone and co., and the sidereal clock presented in 1879 by Nottingham Friends, are itemised by Percy Sharman ('Beckford, East Grinstead, Sussex') to J.E. Shimeld, Esq., University College, Nottingham, in three letters, 1, 7, and 13 May 1928 respectively: University of Nottingham Library UXW 1/4/1/1–3. Fleckney, *Thomas Bush* (ref.43) 4, 8 (unpaginated; counting from first page).

55 His 1861 Census address seems to have been 39 Canal Street, St. Mary's Parish, Nottingham; preserved in Nottinghamshire Archives Office, Schedule 39, p. 6. By the time of the Exhibition of 1870, however, he was trading from 102 Canal Street, for this was the smoky observing site referred to in *The Times* (ref.43) and other newspaper reports cited above: see the *Catalogue* of the Working Men's Exhibition, p. 49, preserved in Nottingham University Library; also unspecified 1870 newspaper cutting, Nottingham University Library, UXW 1/1/3. For Bush's F.R.A.S. election date see obituary, *M.N.R.A.S.* (ref.44) 298. For 2-guinea R.A.S. subscription fee, R.J. Tayler et al., *History of the Royal Astronomical Society Vol. 2, 1920–1980* (R.A.S. and Blackwell's Scientific Publications, Oxford and London, 1987) 129.

56 On 3 May 1876, Thomas Bush, then living at 102 Canal Street, signed two Borough of Nottingham certificates seeking permission to build a new house 'off Thyra Grove Alexandra Park Estate a Cottage Residence'. Plans of this house and its large observatory, complete with 'Transit Room', 'Equatorial Rooms' and 'Computing Room', are preserved in the Nottinghamshire Archives Office, ref. CA/PL2: Thyra Grove Cottage/Observatory, 5 May 1876, Vol. 1/1113, p. 161. The buildings are also clearly indicated on the 1881 Ordnance Survey Map of Nottingham, with the observatory marked and named. None of them survive.

57 This was certainly the comet better known as '1881 III, Tebbutt', visible in June 1881. Bush refers to it as '1881 b' in several of his private papers, such as on the undated sheet of 'General Hospital, Nottingham' notepaper, where he speaks of it as displaying three spectral lines 'identical to those of some previous comets' and mentions the work of Huggins: Nottingham University Library UXW 1/3/5/1. See also the note, 21 August 1881 (in which he also mentions 'Comet 'C''), UXW 1/3/5/15. In his note of 27 June 1881 – shortly after his first observation of the 1881 comet – he left a sketch of the nucleus and an account of his observation of yellow,

green, and violet spectral lines: UXW 1/3/5/5. Bush's undated letter to *The Nottingham Guardian*, describing the comet, his observations on 25, 26 and 27 June 1881, and his early spectroscopic observations are preserved in a separate newspaper cutting in UXW 1/1/6/1. His lecture on 'The Distance and Magnitude of the Sun', delivered to the Nottingham Literary and Philosophical Society, was reported in detail by a local(?) newspaper: undated and unspecified newspaper cutting, UXW 1/1/1. The lecture must have been post-1873, as he appends the letters F.R.A.S. to his name.

58 John Burnett, *A History of the Cost of Living* (Penguin, 1969) 203–5. Sir Robert C.K. Ensor, *England, 1870–1914* (Clarendon Press, Oxford, 1936) 111.
59 Bush, obituary, *M.N.R.A.S.* (ref.44) 298.
60 Fleckney, 'The Story of Thomas William Bush' (ref.43) 10–11.
61 Bush's connection with his *M.N.R.A.S.* obituarist (ref.44), William Sadler Franks (1851–1935), is unclear, for while Franks' Newark, Nottinghamshire, origins could well indicate a long-standing acquaintance between the two men, Franks was a 'Professional Assistant' whose scientific life from 1892 onwards had been spent in the observatories of Grand Amateurs. Franks had been Isaac Roberts' photographer, and from 1910 to his death was in charge of Frederick J. Hanbury's private observatory at East Grinstead, Sussex (see 'W.S. Franks', obituary, *M.N.R.A.S.* **96** (February 1936) 291–2). I am not aware of Franks possessing an observatory of his own at East Grinstead, where he worked in that of Hanbury; it is likely that Bush's 24-inch was actually in Hanbury's observatory. When Percy Sharman, of East Grinstead, was corresponding with J.E. Shimeld about the transfer of the recently deceased Bush's scientific instruments to the University of Nottingham in May 1928, he specified Franks as a knowledgeable man to consult. Even if the 24-inch was used in Hanbury's observatory, however, it was clearly still Bush's legal property, for it headed Sharman's inventory of the instruments contained within the donation to the University: Nottingham University Library UXW 1/4/1/2; 1/4/2/2/1; 1/4/3; 1/4/21/1.
62 Fleckney, 'The Story of Thomas William Bush' (ref.43) 12–14. Also Sharman to Shimeld, 1 May 1928, urging the University of Nottingham to act promptly 'as all Mr. Bush's belongings have to be removed from his room at the [Sackville] College within a fortnight', Nottingham University Library UXW 1/4/1/2.
63 *The Times*, 12 July 1870, 12 (ref.43).

11 THE DAY-LABOURER ASTRONOMER

1 Samuel Smiles, *Men of Invention and Industry* (London, 1884) 328.
2 Smiles (ref.1) 329.
3 Smiles (ref.1) 326, 330. At least two optical instrument-making firms seem to have brought out 'five-pounders' around 1876. They appear to have had 2¼-inch aperture achromatic object glasses and to have performed well in double-star tests. See W.S. Franks' letter, *English Mechanic*, 3 November 1876, and subsequent correspondence. The *Leeds Mercury* printed several letters on £5 telescopes, 14

and 21 October 1876; and the folio 'Scrapbook' of the Leeds Astronomical Society (which is still in the Society's possession) includes numerous cuttings from 1876 on these 'cheap' telescopes.

4 G.B. Airy to Mr Robinson, 1 February 1870, C.U.L. RGO6 6/385–386.
5 Smiles (ref.1) 328.
6 Smiles (ref.1) 331. None other than William Lassell had observed the total eclipse of 1851 in the eyepiece of his Merz refractor with dark glasses, which broke 'successively with most alarming rapidity. I was driven to my wit's end, fearing I should not have a single bit of dark glass left with which to defend my eye.' 'Trollhättan Falls. Observations by W. Lassell Esq.', *Mem. R.A.S.* **21**, Part 1 (1852) 44–50: 45.
7 Smiles (ref.1) 332.
8 Smiles (ref.1) 333.
9 Smiles (ref.1) 332–3, 335. Part of one of his letters to the Dundee *Evening Telegraph* is reprinted by Smiles, 332–3. Also Robertson, 'Solar Activity', *The Observatory* **5**, 61 (1882) 146; and 'Solar Activity', *ibid.* 5, 68 (1882) 371–2: Robertson says, amongst other things, how a magnetic storm on 17 November 1882 caused railway telegraph bells to ring of their own accord 'as if by the operator'.
10 Smiles (ref.1) 335.
11 Smiles (ref.1) 335. James Symons (1838–1900) had originally worked on storm prediction for the Board of Trade, but by 1883 was coordinating weather data supplied from some 168 observers around Britain, most of whom were amateur volunteers, like Robertson. See Agnes M. Clerke, 'J. Symons', *D.N.B.*, and his entry in *Encyclopaedia Britannica* (11th edn.).
12 Salaries printed in the *Royal Commission on Scientific Instruction and the Advancement of Science* (February 1871). Copy in the Library of the Royal Observatory, Edinburgh. I am indebted to Dr David Gavine for this information.
13 Smiles (ref.1) 336.
14 Samuel Smiles also narrated Jones' career in *Men of Invention and Industry* (ref.1) 361–9. Jones enjoyed a much greater cultural celebrity than any other of the astronomers in this section, however. In addition to Smiles' account, he was the subject of an article by Arthur Mee in *Young Wales* **4**, 48 (December 1898), as well as in John Silas Evans' *Seryddiaeth a Seryddwyr* [*Astronomy and Astronomers*] (Cardiff, 1923) 273: I am indebted to Rhys Morris, Deptartment of Astronomy, University of Wales, Cardiff, for kindly translating parts of the latter work into English for me, as I do not read Welsh. I am also especially indebted to Dr Gwawr Jones of Bangor, a relative of John Jones with whom I was placed in contact by Rhys Morris, for supplying me with family, local, and anecdotal material about Jones. John Jones lived at 32 Albert Street, Upper Bangor, a small terraced cottage that was demolished some years ago. This is a pity, for being the house of Wales' first modern astronomer, it would have warranted a commemorative plaque.
15 Bryngwyn Bach is in south-east Anglesey, near to the coast of the Menai Strait, and only about three miles across the strait from Caernafon. John Jones (or 'Ioan', as he styled himself in Welsh) also wrote, in a clear, firm hand, a two-page MS autobiography in Welsh for Miss Ellis at some date in the 1880s or 1890s: Bangor

University Archives MS no. 479. It is the same in its details as the account related, in the company of Miss Ellis, to Smiles.
16 Smiles (ref.1) 363.
17 Smiles (ref.1) 365–6.
18 Since completing the main text of this chapter I have received a letter with transcribed and photocopied documents from Dr Gwawr Jones (ref.14), which relate a version of John Jones' introduction to astronomy very different from the more canonical one narrated to Smiles. It seems that as a young man 'John Jones loved a maid who, alas, jilted him. He was so maddened against his rival that, for the purpose of working him harm, he visited one who had some knowledge of astrology & a reputation for practising black art. In the event, John Jones, instead of annihilating his rival, became so interested in the methods of astrology that both maid & rival were forgotten & the astronomer in him sprang into being': Miss Grace Ellis to the Revd Thomas Shankland (Librarian of Bangor College), 8 June 1915, Bangor University Archives MS 479.
19 Smiles (ref.1) 365, for wages and 'fair Eve'. The 1881 Census Return for Bangor is problematical, however. It gives Jones' age as 58 (not 63) and that of 'Fa[ ]' as 49. But from her and John's gravestone inscription in Glanadda Cemetery, Bangor, it would appear that Fanny Jones (1823–1906) was only 48 in 1881. It is surprising that neither Smiles nor Miss Ellis makes any reference whatsoever to Fanny Jones. Smiles speaks of Jones as a man who lived alone, although in the cases of other working-class astronomers he visited, he invariably made reference to the spouse. (John Robertson, for instance (ref.1, 328) had a 'clever wife'.) I am indebted to Steffan ab Owain of Gwynedd Council Department of Education and Culture for the 1881 Census transcription, and to Dr Gwawr Jones for the transcription of the Jones tombstone in Glanadda Cemetery, Bangor.
20 Smiles (ref.1) 367.
21 Smiles (ref.1) 366–7. See *ibid.* (ref.3) for reference to £5 and Solomon telescopes.
22 Smiles (ref.1) 367. Jones' 6-inch and his other silver-on-glass reflectors were noted by Parry Jenkins in 1911. Jenkins 'had the honor of knowing John Jones intimately as I once lived in Anglesey, and many were the pleasant evenings we spent together at my observatory and afterwards at his home', using Jones' home-made reflectors: G. Parry Jenkins, 'A Plea for the Reflecting Telescope', *The Journal of the Royal Astronomical Society of Canada*, V (1911) 59–75: 68.
23 Smiles (ref.1) 367–8. Though the word 'Jumbo' has a provenance dating back to 1823, it was the elephant sold to P.T. Barnum's Circus in 1882 which popularised the word: *O.E.D.* Jones' 'Jumbo' was almost certainly the large (10-foot?) Newtonian reflector reproduced in *Seryddiaeth* and *Young Wales* (both ref.14): see my pl. 60. Dr Gwawr Jones (ref.14) when a girl in the 1950s saw 'Jumbo', 'Jumbo Bach' (perhaps his smaller, 6-inch reflector), 'Alice' and other Jones instruments in Bangor Museum, which was then housed in the Old Girls' Grammar School, Bangor. Around 1968 this Museum was moved, and the exact whereabouts of the 'Jumbos' became unknown. Dr Jones is currently trying to locate them. Dr Gwawr Jones to A. Chapman, private communications, 24 March and 16 June 1998.
24 Smiles (ref.1) 368.

25 Smiles (ref.1) 368. It is possible that some of these smaller instruments were amongst those on display in Bangor Museum in the 1950s, and seen by Dr Gwawr Jones.
26 Smiles (ref.1) 368 says that Jones' work 'is perhaps too technical to be illustrated in words, which are full of focuses, parabolas, and convexities'.
27 Smiles (ref.1) 364.
28 Robert Roberts (1777–1836) produced a popular Welsh almanac, *Cyfaill Glandeg* (1805–37) from Holyhead, but with a false Dublin imprint to avoid tax: *The Dictionary of Welsh Biography down to 1940* (London, 1959) 876–7.
29 'Ioan, of Bryngwy Bach', Anglesey, was the Bardic name under which Jones wrote his Welsh poetry: see Smiles (ref.1) 368–9.
30 Arthur Mee, the Scottish journalist who came to work on the Cardiff *Western Mail* newspaper in 1892, and became thereafter something of an 'honorary Welshman', was an amateur astronomer who actively promoted Jones and his memory, writing about him in *Young Wales* in 1898 (ref.14), and probably getting him photographed. It is not known if Mee ever met Jones personally. Silas Evans, in *Seryddiaeth* (ref.14) 281, gives an account of Mee's career up to 1923.
31 *The Cambrian Natural Observer, And Quarterly Journal of the Astronomical Society of Wales*, **3**, 3 (January 1901), List of Members, 49. John Jones, even three years after his death, is still proudly recorded by the Society as 'Seryddwyr, 32, Albert Street, Bangor (deceased)'.
32 Roger Langdon, *The Life of Roger Langdon told by himself, with Additions by his Daughter Ellen* (London 1909) 51.
33 Langdon, *Life* (ref.32) 23. This booklet was probably William Pinnock's *A Catechism of Astronomy* (8th edn? 1820), or a subsequent copy. It cost 9*d*.
34 Langdon, *Life* (ref.32) 37.
35 Langdon, *Life* (ref.32) 64. He soon left Bristol station to work a 12-hour day, at 16*s* per week, as a signalman at Stoke Canon, Devonshire.
36 His 30*s* wage is cited by H. Clifton Lambert of the G.W.R. in his 'Preface' to Langdon's *Life* (ref.32), 5–8.
37 Langdon, *Life* (ref.32) 65–6, 68.
38 Langdon, *Life* (ref.32) 70.
39 Langdon, *Life* (ref.32) 65–6. In the summer of 1997 I visited Silverton. The station itself has now gone, and though it was over two miles from the village by road, it was probably a shorter distance across the fields.
40 Langdon, *Life* (ref.32) 71–2. Ellen Langdon also spoke, 71, of a lens telescope (a refractor) with 'a 1½-inch reflector'. One suspects that she may have been confusing refracting and reflecting telescopes.
41 Langdon, *Life* (ref.32) 65: he had earned £1 per week as signalman at Durston, Somerset, in 1857.
42 Langdon, *Life* (ref.32) 73, 76.
43 Langdon, *Life* (ref.32) 73, 76.
44 Langdon, *Life* (ref.32) 74.
45 Langdon, *Life* (ref.32) 75. Another keen working-man astronomical photographer was Samuel Cooper, a bricklayer of Charminster, Dorset. On 6 December 1882, Cooper – who sometimes wrote to the *English Mechanic* under the *nom-de-*

plume 'The Optical Bricklayer' – secured three excellent photographs of the largely clouded-out transit of Venus of that year, using his home-made 9-inch reflector: Parry Jenkins, 'A Plea for the Reflecting Telescope' (ref.22) 67–8.
46 Langdon, *Life* (ref.32) 79.
47 Langdon, *Life* (ref.32) 82, 84.
48 Langdon, *Life* (ref.32) 85.
49 Langdon, *Life* (ref.32) 8.
50 G.W.R. *Magazine* September 1889 for Lambert's visit, *Wit and Wisdom* June 1889, *Western Weekly News* (1889?). I have been unable to trace any of these primary sources, which are quoted in Langdon, *Life* (ref.32) 84, 85.
51 Langdon, *Life* (ref.32) 74. Langdon recorded, 74, that none other than the Revd T.W. Webb had commended his Venus paper to the R.A.S. I do not know if he corresponded with Webb.
52 R. Langdon, 'Observations of the Planet Venus, with a 6-inch silvered glass reflector', *M.N.R.A.S.* **31–32** (June 1872) 88–91. Langdon was not an F.R.A.S., though his paper was communicated by none other than Joseph Norman Lockyer. He addressed his paper from 'Silverton Station, near Collompton, Devon'. See Langdon's other article, of the same name, *M.N.R.A.S.* **33** (June 1873) 500. Both of these papers are reprinted in Langdon, *Life* (ref.32) 88–91, 92. The Revd T.W. Webb refereed Langdon's papers for the R.A.S.
53 Langdon, *M.N.R.A.S.* **33** (June 1873). Langdon noted that during his observation of 19 April 1873 he saw Venus through an aurora.
54 R. Langdon, 'A Letter from the Man in the Moon', *Exe Valley Magazine*, and 'A Journey with Coggia's Comet', *Home Words*, reprinted without precise location details in Langdon, *Life* (ref.32), as Appendices IV and V: 93–97 and 98–100.
55 Langdon, *Life* (ref.32) 82.
56 T.W. Webb, *Celestial Objects for Common Telescopes* (London, 1859) viii.
57 Burnerd's advertisement is printed in *The Cambrian Natural Observer, And Quarterly Journal of the Astronomical Society of Wales* **10**, Supplementary Number (1908). Copy in the Radcliffe Science Library, Oxford.

12 A GOODLY PURSUIT FOR A GODLY MIND: THOMAS WEBB AND HIS INFLUENCE

1 T.H.E.C. Espin, 'T.W. Webb – A Reminiscence', in Thomas W. Webb, *Celestial Objects for Common Telescopes* 1, (5th edn., revised by T.H.E.C. Espin, London, 1893) xv.
2 T.W. Webb, *Celestial Objects for Common Telescopes* (London, 1859) ix. While *Celestial Objects* was by far his most famous and influential book, he also wrote other significant amateur works, such as *Optics without Mathematics* (London, 1883) and *The Sun, A familiar description of his Phaenomena* (London, 1885).
3 The need for a new edition of Smyth's *Cycle of Celestial Objects* was already being voiced in 1863: *Astronomical Register* **1**, January–December 1863 (London, 1864) 112.

4   Espin, 'T.W. Webb – A Reminiscence', in Webb, *Celestial Objects*, 5th edn. (ref.1) xv. Also Webb, *Celestial Objects* 1859 (ref.2) 2ff.
5   Webb, *Celestial Objects* (ref.2) vii.
6   Webb, *Celestial Objects* (ref.2) viii.
7   A.C.R., obituary, 'Thomas William Webb', *M.N.R.A.S.* **46** (February 1886) 198–201. One of the finest studies on Webb, along with that of Espin (ref.1), was Arthur Mee's 'The Rev. T.W. Webb, in Memoriam', *Observational Astronomy* (1897) 109–13. See also 'Death of the Rev. Prebendary Webb', *Monmouthshire Beacon*, Saturday 30 May 1885; *Burke's Landed Gentry* (1952), 'Thomas William Webb', 2668. Webb's will, dated 12 March 1885 and proved at Hereford, 17 June 1885, bequeathed a gross estate of £16,986 5s1d. His *D.N.B.* entry, however, written by 'A.M-E' shortly after his death, specifies that Webb 'left a sum of over £20,000 to Herefordshire charities'. I am indebted to Alan Dowdell for the most generous loan of his large collection of photocopies of primary documents and MSS pertaining to Webb, whose *Life* he has been researching for several years. Webb's financial independence, moreover, is made very clear when one realises that his Hardwick living was worth only a modest £150 p.a.: *Crockford's Clerical Directory* (London, 1868), see name. By 1882 the living seems to have been augmented to £233 gross: *Crockford's* (1883) 1220. Even so, it is evident that Webb enjoyed ample private means.
8   *Kilvert's Diary. Selections from the Diary of the Rev. Francis Kilvert 1870–1879*, 3 vols. (Jonathan Cape, London, 1938–40): 27 February 1870; March Eve, 1870; March Day, 1870; 7 July 1870; 29 August 1870; 21 May 1871; 29 May 1871; 11 August 1871; 26 September 1871; 15 January 1877; 27 September 1878; 21 November 1878, etc. Some of these references mention astronomy, but most are about Webb in the social context of a conscientious clergyman and country gentleman.
9   Espin, 'T.W. Webb – A Reminiscence' (ref.1) xvi.
10 Webb, *Celestial Objects* (ref.2) 221, footnote.
11 Webb, *Celestial Objects* (ref.2) 25.
12 Webb, *Celestial Objects* (ref.2) 129. I am not aware of Webb's source for this discovery: it does not seem to be reported in *Phil. Trans.* (1665–1666).
13 Espin, 'T.W. Webb – A Reminiscence' (ref.1) xiv. Also Deborah J. Warner, *Alvan Clark & Sons, Artists in Optics* (Smithsonian Institution Press, Washington, 1968) 107 (new edition, 1996).
14 Espin, 'T.W. Webb – A Reminiscence' (ref.1) xiv for Webb's telescopes.
15 Espin, 'T.W. Webb – A Reminiscence' (ref.1) xv.
16 Espin, 'T.W. Webb – A Reminiscence' (ref.1) xiv. According to G. Parry Jenkins, 'A Plea for the Reflecting Telescope', *Journal of the Royal Astronomical Society of Canada* V (1911) 59–75: 62, this $9\frac{1}{3}$-inch With–Berthon was presented to Webb around 1865 by his 'venerable father'. This is most likely the same instrument, on its 'Berthon' stand, as that now in the B.A.A. instrument collection (no.83). The exact diameter of an historical mirror, when one is attempting a positive identification, can often be the subject of conjecture: does one mean the entire diameter of the glass – in this instance let us say $9\frac{1}{3}$ inches – or inside the bevelled edge, which forms the reflecting area – perhaps $9\frac{1}{4}$ inches?

17 Revd E.L. Berthon, 'On Observatories', *English Mechanic* 342 (13 October 1871) 83–4. Also Berthon's letter mentioning Webb's satisfaction with his own 'Romsey' Observatory: *English Mechanic* 343 (20 October 1871) 119. 'Telescope Houses' of the Berthon type were also being discussed in *English Mechanic* 779 (27 February 1880) 600.
18 Webb's Observing Books, R.A.S. MS Webb, 5 volumes of lunar, planetary, and stellar observations, are almost art-works in their own right, with their italic, roman and bold capital handwriting that is sometimes as uniform as letter press. And inserted into these pages are Webb's beautiful sketches of the objects under observation. One can see from the way in which Webb describes long-familiar celestial objects in his notebooks that a peaceful and awesome contemplation of the Divine Creation was probably a stronger motive behind his astronomy than was physical investigative research.
19 T.W. Webb, Observing Books (ref.18) vol. 'Stars and Nebulae' 21 August 1857, 1 May 1869; 'Moon', 1 May 1871.
20 Webb, Observing Book (ref.18), 'Stars & Nebulae, 1871–1885', R.A.S. MS Webb 3.
21 *Astronomical Register* **4** (1866) 21, 91 for private observatories census.
22 'The construction of Berthon collapsible boats is a special industry' (1911): see article 'Romsey, Hants.' in *Nelson's Encyclopaedia* XIX (undated, 1911–12). See also Adam Hart-Davis and Paul Bader, 'Edward Lyon Berthon and his folding Lifeboats', in *The Local Heroes Book of British Ingenuity* (Sutton Publishing, Stroud, 1997) 40–2, including photographs. For Berthon's 'Dynamometer', used for measuring the diameter of the exit pupil of an eyepiece, see *English Mechanic* 339 (22 September 1871) 25, and 341 (6 October 1871) 77.
23 The Revd Edward Lyon Berthon's stable and compact equatorial mount for short-tubed, glass-mirror reflecting telescopes was one of the first telescope mounts that was specifically designed to meet the optical and physical needs of the new 'Leisured Enthusiast' community of amateurs.
24 Revd E.L. Berthon, 'On Observatories' (ref.17) 83–4.
25 Berthon, 'On Observatories' (ref.17). Berthon was a well-beneficed parson, with his Romsey living valued at £430 p. a., or nearly three times the value of Webb's (ref.7): *Crockford's Clerical Directory* (London, 1868), see name.
26 Thomas Maclear, 'Description of a small Observatory at Biggleswade', *Mem. R.A.S.* 1833, 147–8.
27 Roger Langdon, 'Observations of the Planet Venus, with a 6-inch silvered glass reflector', *M.N.R.A.S.* **33**, 8 (June 1873) 500–1. Langdon, in *The Life of Roger Langdon told by himself, with Additions by his Daughter Ellen* (London 1909) 74, mentions that T.W. Webb was the referee of his papers to the R.A.S.
28 Numerous advertisements for instruments and parts, both new and second-hand, appear throughout the two decades that the *Astronomical Register* ran: see offering for sale of an 'excellent' reflector, 7½ inches aperture, 7 feet focal length, with four eyepieces and equatorial amount, for £20: *Astronomical Register* **2** (1864) 307. Solomon's of London, on the same page, were offering 2-inch aperture refractors, 44 inches focal length, for £5 new, and £4 4s second-hand.

29 S. Smiles, *Men of Invention and Industry* (London, 1884) 367 for Jones' mirrors.
30 Mr G.F. Lemmon is recommended in the broadsheet *Sale of Mr With's Choicest Reserves of Silvered Glass Specula for Newtonian Telescopes*, reproduced in facsimile from the collection of R.A. Marriott: R.A. Marriott, 'The Life and Legacy of G.H. With, 1827–1904', *J.B.A.A.* **106**, 5 (1996) 257–64: 258, 260.
31 John Browning, *A Plea for Reflectors, Being a Description of the New Astronomical Telescopes with Silvered-Glass Specula and instructions for adjusting and using them* [price one shilling] (London, 1867) 4.
32 Lord Oxmantown (Rosse), 'An Account of Experiments on the Reflecting Telescope', *Phil. Trans.* **130** (1840) 503–27. Henry Cooper Key, 'On a mode of Figuring Glass Specula for the Newtonian Telescope', *M.N.R.A.S.* **23** (1863) 199–202. Cooper Key's Stretton Sugwas living had an annual revenue of £222: *Crockford's Clerical Directory* (London 1868), see under name and parish.
33 Marriott, 'The Life and Legacy of G.H. With' (ref.30) 258.
34 Marriott, 'The Life and Legacy of G.H. With' (ref.30) 257.
35 Marriott, 'The Life and Legacy of G.H. With' (ref.30) 257.
36 Marriott, 'The Life and Legacy of G.H. With' (ref.30) 258. At least, he possessed sufficient knowledge of Greek, Latin and Hebrew to sign his mirrors in these languages: Marriott, 259.
37 Marriott, 'The Life and Legacy of G.H. With' (ref.30) 258.
38 'Mr. With's Letter', G.H. With, 7 Bridge St., Hereford, to George Calver, 12 October 1893, later published in *English Mechanic* 2327 (29 October 1909) 307–8.
39 Marriott, 'The Life and Legacy of G.H. With' (ref.30) 260.
40 See A. Chapman, 'Scientific Instruments and Industrial Innovation: the achievement of Jesse Ramsden', in *Making Instruments Count*, ed. R.G.W. Anderson, J.A. Bennett and W.F. Ryan (Variorum, Aldershot, 1993) 418–30; reprinted in Chapman, *Astronomical Intruments and Their Users: Tycho Brahe to William Lassell* (Variorum, Aldershot, 1996).
41 Cooper Key, 'On a mode of Figuring Glass Specula' (ref.32) 202, for praise of With's mirrors. W.R. Dawes, 'On the Performance of Object Glasses on γ Andromedae and δ Cygni', *Astronomical Register* **5** (March 1867) 49–51. (See also the following letters on object glasses by George Knott and H.E. Lowe, *Astronomical Register* **5**, 51–3; John Browning, letter in response to Dawes' article on refractor resolution, *ibid.*, 111; Browning not only praises a 6½-inch With mirror's ability to divide γ^2 Andromedae, but even mentions a 6-inch speculum mirror at Kenyon College, Ohio, that can also divide it, thus concluding 'that the superior dividing power of reflectors over refractors must be admitted'. Parts of With's grinding and polishing procedures, relayed to Calver in 1893 (ref.38), are reprinted by Marriott in 'The Life and Legacy of G.H. With' (ref.30) as an Appendix.
43 Marriott, 'The Life and Legacy of G.H. With' (ref.30) 259. William Lassell had also tested his speculae on watch dials in the 1840s: R.A.S. Lassell MSS 3, fol. 137.
44 For 1922 discovery, see P.M. Ryves, letter to *J.B.A.A.* **51**, 9 (October 1941) 332. For '5 feet 7 inches... wonderful perfection', see James L. Cummergen, *J.B.A.A.* **52**, 2 (March 1942) 71.

45 For With's *Choicest Reserves*, and prices, see Marriott, 'The Life and Legacy of G.H. With' (ref.30) 258, 260.
46 Marriott, 'The Life and Legacy of G.H. With' (ref.30) 261–3.
47 Marriott, 'The Life and Legacy of G.H. With' (ref.30) 262. The Leeds Astronomical Society 'Album' of documents, cuttings, and pictures, which is still in the possession of the Society, contains photographs and sources for Scriven Bolton: date order, 1920s. The best modern study of Scriven Bolton and his gifts is Roger Hutchins, 'Yorkshiremen with Clout: The Leeds Astronomical Society of 1859, and the Pursuit of Astronomy at Leeds University after 1904', unpublished, copy presented to me by Dr Hutchins; a copy is also deposited in the Museum of the History of Science, Oxford.
48 Marriott, 'The Life and Legacy of G.H. With' (ref.30) 262. Muschamp Perry is a clerical amateur who warrants more research, both as an astronomer and as a social type. I am indebted to Joe McKie of Newcastle, and to Don Simpson of Sunderland, for their local inquiries; however, he seems to have left surprisingly little local record behind him.
49 In spite of regrets about his inability to afford a 37-inch mirror (Marriott, 'The Life and Legacy of G.H. With' (ref.30) 262), one suspects that Muschamp Perry enjoyed private means and good social connections. In 1879 the living of St Paul's, Alnwick, was worth £250 p. a. This was a decent income, but not one big enough to support his lavish telescope-collecting. He was, moreover, still a young man, having only taken his B.A. at Sidney Sussex College, Cambridge, in 1866, and been ordained priest in 1868: *Crockford's Clerical Directory* (London, 1879), by name and parish. He was clearly an experienced observer as well as telescope collector: see Muschamp Perry, 'Nova Andromedae – A Solar Prominence', *English Mechanic* 1097 (2 April 1886) 103. The 37-inch mirror by Calver that Muschamp Perry so wished he could have afforded was that described in detail by George Calver in 'On the working of the Speculum for Mr. Common's 37-inch silver-on-glass Reflector', *M.N.R.A.S.* **40** (November 1879) 17–23.
50 Nathaniel E. Green, 'Observations of Mars, at Madeira, in August and September 1877', *Mem. R.A.S.* **49** (1877–9) 123–38: 123.
51 See ref.16.
52 Marriott, 'The Life and Legacy of G.H. With' (ref.30) 259. I have not been able to trace a Revd Mr Matthews in the vicinity of Yarmouth in the Anglican *Crockford's Clerical Directory* for the 1860s. Most likely he was a Methodist or a Dissenting Minister, and hence not included in *Crockford's*. I am not aware of George Calver's religious opinions, or of the ecclesiastical circles in which he might have moved. Far less is known about the scientific activities of Dissenting clergy than of those of their Anglican colleagues. On the whole, however, they were poorer, less leisured, and less well recorded.
53 Smiles, *Men of Invention* (ref.29) 367.
54 Warner, *Alvan Clark & Sons* (ref.13) 15–19. Anita McConnell, *Instrument Makers to the World. A History of Cooke, Troughton & Simms* (William Sessions, York, 1992) 50–6.
55 Smiles, *Men of Invention* (ref.29); Mrs Cooke's narrative to Smiles, 339–40.

56 Smiles, *Men of Invention* (ref.29); James Nasmyth's narrative to Smiles, 337.
57 Smiles, *Men of Invention* (ref.29) 337.
58 Smiles, *Men of Invention* (ref.29); Mrs Hannah Cooke's narrative, 340. McConnell, *Instrument Makers to the World* (ref.54) 50.
59 Smiles, *Men of Invention* (ref.29) 344 for list of Cooke's leading clients. For Robertson, Smiles says (p. 330) that he had a fine 3-inch aperture Cooke refractor, and on p. 326 he states that Robertson was using an instrument for which he had paid £30, which I presume was the Cooke.
60 Smiles, *Men of Invention* (ref.29) 343–4.
61 'Telescope' in *Encyclopaedia Britannica* **23** (9th edn., Edinburgh, 1888) 149.
62 McConnell, *Instrument Makers to the World* (ref.54) 57–68, gives a sense of the range of Cooke's manufacturing at the Buckingham Street Works, especially for capital observatory instruments. *The Splendour of the Heavens, A Popular Authoritative Astronomy* II, ed. T.E.R. Phillips and W.H. Steavenson (Hutchinson, London, 1923) 725–34 illustrates a variety of more modest Cooke refractors (and several Calver reflectors).
63 Smiles, *Men of Invention* (ref.29) 344.
64 Smiles, *Men of Invention* (ref.29) 346–7. H.C. King, *The History of the Telescope* (Griffin, London, 1955) 250–4.
65 Lancaster, along with 'Oderic Vitalis', was being encouraged by 'Nova Scotian' to produce his 'long-promised' letter on object glass grinding, for instance, in *English Mechanic* 776 (6 February 1880) 531. Smiles, *Men of Invention* (ref.29) 349.
66 Smiles, *Men of Invention* (ref.29) 349–50: 354 for the furnace.
67 Smiles, *Men of Invention* (ref.29) 354. There is no record of Lancaster becoming an F.R.A.S.
68 Smiles, *Men of Invention* (ref.29) 356.
69 I am not aware of a book by Airy entitled 'Geometrical Optics'. Lancaster was probably referring to Airy's *On the Undulatory Theory of Optics, designed for the use of students in the University* (London and Cambridge, 1866 and 1879).
70 Smiles, *Men of Invention* (ref.29). Smiles added a supplement, p.360, which recorded his second interview with Lancaster conducted in September 1884, where he examined his instruments and home-made optical figuring apparatus, and was informed of the problems experienced by amateurs in making achromatic lenses.
71 *English Mechanic* 776 (ref.65) 531.
72 Smiles, *Men of Invention* (ref.29) 360–1.
73 Smiles, *Men of Invention* (ref.29) 361. Lancaster was by no means the only amateur achromatic lens maker, according to *English Mechanic*. There was 'Oderic Vitalis', W. Bradbury, Dr Hastings and others: *English Mechanic* 779 (28 February 1880) 602–3; 974 (23 November 1883) 258; 1025 (14 November 1884) 232, for the commencement of Dr Hastings' long, subsequently serialised treatise on lensmaking. I am not aware of anything of further significance by Lancaster, however.
74 Thomas Hardy, *Two on a Tower* (1882), Wessex edition, ed. F.B. Pinion (Macmillan, London, 1975). Hardy to Edmund Gosse for quotation, cited in Introduction, 15. Also R.L. Purdy, *Thomas Hardy: A Biographical Study* (O.U.P., 1954, re-issued 1968) 44n.

75   Hardy, *Two on a Tower* (ref.74) 35.
76   Hardy, *Two on a Tower* (ref.74) 62–6.
77   Hardy, *Two on a Tower* (ref.74) 78, for Hilton & Pimms. St Cleve had told her that 'an equatorial such as I describe costs as much as two grand pianos', 70. It would probably have cost much more.
78   Pinion, 'Introduction' to *Two on a Tower* (ref.74) 15, for Proctor reference. Whether or not Hardy visited the Royal Observatory is a matter of scholarly dispute: F.B. Pinion, *Thomas Hardy: His Life and Friends* (1992) does not believe that he did. Although admitting that Hardy never signed the Observatory Visitors' Book, Martin Beech has been through the Greenwich records, and found Hardy's correspondence with Dunkin, the Chief Assistant, which might perhaps have led up to an unrecorded visit: M. Beech, 'Thomas Hardy: Far from the Royal Observatory, Greenwich', *The Observatory* **110** (December 1990) 185–7.
79   Pinion, 'Introduction' to *Two on a Tower* (ref.74) 15.
80   Hardy, *Two on a Tower* (ref.74) 55.
81   R.S. Ball, *The Story of the Heavens*, new revised edn. (London, 1905) 28.

13 THAT CLUBABBLE PASSION: THE AMATEUR ASTRONOMICAL SOCIETY

1   Jack Morrell and Arnold Thackray, *Gentlemen of Science. Early Years of the British Association for the Advancement of Science* (Clarendon Press, Oxford, 1981) 58–9.
2   A. Chapman, 'Science for the working man: the failure of a Victorian dream', *Eccles and District Historical Society*, Supplement (Eccles, Lancashire, 1987–8) 5.1–5.16.
3   David Gavine, 'Astronomy in Scotland, 1745–1900', unpublished Open University Ph.D. thesis, 1982, 304–14. Gavine, 'Thomas Dick (1774–1857) and the Plurality of Worlds', *Astronomical Society of Edinburgh Journal* **28** (September 1992) 4–10. Gavine, 'Navigation and Astronomy teachers in Scotland outside the Universities', *The Mariner's Mirror*, **76**, 1 (February 1990) 5–12. I am indebted to Dr Gavine for information about Dick passed on to me by letter and in conversation. See also Capt. William J. Astore, 'Observing God: Thomas Dick (1774–1857), Evangelicalism and Popular Science in Victorian England and Antebellum America', unpublished Oxford University D.Phil. thesis, 1995: Chapter 6. I wish to thank Capt. Astore for his conversations, and for permission to see his thesis.
4   Hermann A. Brück, *The Story of Astronomy in Edinburgh from its beginnings until 1975* (Edinburgh University Press 1983) 10–12.
5   Uranian Society reported in the *Argus* newspaper, 14 July 1839. Cutting preserved in John Lee's 'Album', M.H.S. Oxford Gunther 36, 55–6.
6   Alan Bowden discussed the efforts of Ebenezer Henderson to establish an observatory in Liverpool in 1836 in his 1991 address to the Scientific Instrument Society, 'The Early Years of the Liverpool Observatory' *Bulletin* 33 (1992) 18–20: 18.

388 Notes and references [Ch. 13

7   Cutting from *Lloyds Weekly Newspaper*, 28 December 1851, in John Lee, 'Album', M.H.S. Oxford Gunther 37, 76v.
8   Written by 'An Associate' of the British Association: cutting from the *Liverpool Mercury*, 6 October 1859, in John Lee, 'Album', M.H.S. Oxford Gunther 37, 4.
9   The Mills Observatory in Dundee, a Victorian philanthropist's gift to the town, is one of the oldest and grandest. I am indebted to Duncan Lunan for showing me the restored Airdrie Observatory in June 1997: *Airdrie Public Observatory* (Astra Society Pamphlet, Troon, 1996). A public observatory was set up at Darlington, Co. Durham, in 1904: Barry Hetherington, *The Darlington Telescope* (Darlington, 1986). I am indebted to Barry Hetherington for information about astronomy in the Darlington area.
10 Sir John Herschel gives a detailed account of his activities in Leeds, 21 September to 1 October 1858, in his Diary: transcription in the Royal Society Library, MS 583-6. There had been an unsuccessful attempt to found an astronomical society in Leeds as early as 1825, however: *The Leeds Intelligencer*, 'Local Intelligence' section, 27 January 1825; while in early March 1823, one Mr Lloyd was offering public lectures demonstrated with his 'Dioastrodoxon' or transparent orrery. I am indebted to Ray Emery M.A. of the Leeds Astronomical Society for this information, and for his generosity and patience in combing local Leeds newspapers of the period to locate early Reports of the Society. For the social composition of the Victorian Leeds Phil. and Lit.: E. Kitson-Clark, *The History of 100 years of Life of the Leeds Philosophical and Literary Institution* (Leeds, 1924). From the 1860s to the 1890s the Institution had a membership of around 700.
11 Although Herschel's lecture to the B.A.A.S. at the Leeds Phil. and Lit. on 30 September 1858 provided the spark for the founding of the Leeds Astronomical Society, it would appear that the momentum to establish a society proper came – as would be the case with hundreds of subsequent societies – from a course of public lectures: *The Leeds Intelligencer*, 21 May 1859, 5, states that such a course was being offered at the Young Men's Christian Institute, Park Row, Leeds. Also, Bill of Sale to the Leeds Astronomical Society of 'a Galvanised Iron House', 21 April 1860; it seems to have been paid for, £5, 5 November 1863, and apparently re-sold, 17 April 1867. Documents pasted onto same sheet in the 'Album' of the Leeds A.S., 1859–1930, and still in the possession of the Society. I am indebted to the Leeds A.S. for the loan of this precious volume, and especially to Ray Emery for providing me with computer scans of many of its pages.
12 Printed advertisement notice, 7 March 1861, preserved in the Leeds A.S. 'Album' (ref.11), date order.
13 'Leeds Astronomical Society', Established 1859, Prospectus and Rules. Printed sheet, with hand-written date, 31 March 1863: Leeds A.S. 'Album' (ref.11) 102v and date order.
14 'Leeds Astronomical Society' (ref.13), in Leeds A.S. 'Album' (ref.11).
15 Printed advertisement notice, 7 March 1861 (ref.12) in Leeds A.S. 'Album' (ref.11) for List of Patrons; also sheet dated 31 March 1863.
16 G.B. Airy to William Trant, 18 March 1863, in Leeds A.S. 'Album' (ref.11), date order.

17 This telescope is still preserved by the Society in excellent condition (see Plate 71), though mounted on what is probably a later stand. Though Herschel recommended the purchase of this instrument, he did not pay for it, and its purchase-cost caused financial problems for the early Leeds Astronomical Society.
18 Reported at length in *The Leeds Intelligencer*, 31 October 1863. Herschel also published it as a separate pamphlet, *The Yard, the Pendulum, and the Metre Considered in Reference to the Choice of a Standard of Length* (Leeds, 1863), for which the Leeds A.S. had to pay £4 16s publication costs for 500 copies, 100 of which went to Sir John. See Printer's Bills, 24 October 1863, in Leeds A.S. 'Album' (ref.11), date order. The Lecture was republished in Herschel's *Familiar Lectures on Scientific Subjects* (London, 1867), Lecture X, 419, where it followed 'On Sensorial Vision', which he had delivered in Leeds on 30 September 1858.
19 John Herschel, 'Diary': Royal Society Library MS 583–6, date order.
20 William Trant to Sir John Herschel, 24 August 1864, University of Texas Library. I am indebted for this Trant reference to Michael J. Crowe of Notre Dame University, Illinois, U.S.A., who includes it in his *Calendar of the Correspondence of Sir John Herschel* (forthcoming, 1998). From the transcription copy of Herschel's 'Diary' made in 1910, now in the Royal Society Library, I had been searching in the Royal Society's Herschel letter collection for a Mr *Frank*, instead of *Trant*, due to an error in the original penmanship, deriving, no doubt, from Herschel's own cursive handwriting. (Herschel's original diary is also in the University of Texas Library.) My 'Frank' for 'Trant' error was pointed out to me in private correspondence (following his reading of my article in *J.B.A.A.* **106**, 5 (1996) 252) by Professor Crowe, who had encountered some 30 Trant letters to Herschel during the compilation of his *Calendar*, some 13 of which are in the Royal Society Library. See also Michael J. Crowe, 'Sir John Herschel at the Leeds Astronomical Society', *J.B.A.A.* **108**, 1 (1998) 33–4.
21 Trant to Herschel, 1 September 1864, in Crowe, *Calendar of Correspondence* (ref.20); communicated to me privately by Professor Crowe.
22 William Trant seems to have been the son of a pastry-cook, for a Mrs Trant (late Scott), 23 North Street, Leeds (the same street in which William Trant lived), 'respectfully solicits a share in the Gentry, Clergy, and the Public's patronage' of her pies, pastries, teas, wines, etc.: see her advertisements in *The Leeds Intelligencer*, second half of 1859 (I am indebted to Ray Emery for this information). For William's subsequent career, see *The Journalist* XVIII, 13 (New York, 9 December 1893); copy in Leeds A.S. 'Album' (ref.11), date order. Also W. Stewart Wallace, *Dictionary of Canadian Biography* (Toronto 1945).
23 'First Circular issued by the re-organised Society', Leeds A.S. 'Album' (ref.11), 1893, date order.
24 Miss Tranmar, obituary, 'William Donald Barbour' (1833–1903), *Leeds Astronomical Society Journal and Transactions* 11 (1903) 9–10; poem 10–11.
25 See ref.8.
26 T.[H.]E.[C.] Espin, 'Astronomical – Fine Objects', *English Mechanic* 774 (23 January 1880) 480. Espin spoke of the 'isolation' of not having an amateur society to which he could belong, though he had 'a recollection of hearing of an Amateur

Astronomical Society years ago'. Frank C. Dennett responded to Espin's plea in the next number of *English Mechanic*, 775 (30 January 1880) 507. In the intervening years, since 1859, it should be noted, other societies had been attempted, such as the Observing Astronomical Society, with its 5s annual subscription: Richard Baum, 'The Observing Astronomical Society – birth of a legend', *J.B.A.A.* **108**, 1 (1998) 42–3.

27 W.H. Davies to *The Observatory* **6** (1883) 268–9 recounts the origins of the L. [Liverpool?] A.S. in 1881. I am especially indebted to Gerard Gilligan, Philip Pennington, Jeff Hall and other members of the present-day Liverpool Astronomical Society for giving me access to the Society's archives, and supplying me with an abundance of documents pertaining to its early history.
28 Davies, *The Observatory* (ref.27) 269.
29 Davies, *The Observatory* (ref.27) 269.
30 Cited from the Liverpool A.S. Rules and Constitution, 1881, still in force today: see Gerard Gilligan *et al.*, *The History of the Liverpool Astronomical Society* (L.A.S., 1996) 13. A similar aspiration, in different words, was expressed by W.H. Davies to *The Observatory* (ref.27) 268–9.
31 *The Observatory* **9** (November 1886) 371. Copies of the early *Journal of the Liverpool Astronomical Society* are now very scarce, beyond the Society's own library, though a broken set (for 1884–1888) is on deposit in the Royal Society Library: Tracts 845/11–13, 774e, and a bound but incomplete set is in the library of the Science Museum, South Kensington.
32 *The Observatory* (ref.31) 371.
33 Robert H. Wigmore, Queen Street, Melbourne, Australia, was attempting to enlist the necessary 50 people to form a Colonial Branch: *Journal of the Liverpool A.S.*, V, pt. 6 (1 April 1887) 191. By February 1887, moreover, it had been reported to *English Mechanic*, 1144 (25 February 1887) 559, that 'the Pernambuco branch [of the Liverpool A.S.] has now become so extended as almost to form a Society in itself'.
34 The present author, who has enjoyed a long and warm relationship with the Society, also had the pleasure of being elected an Honorary Member in 1996.
35 Liverpool A.S. *List of Members, 1st March 1887* (Liverpool, 1887), Royal Society Library, Tracts, 845/11. The *List* included Professor Theodor Bredichin of Moscow, and S.W. Burnham and Asaph Hall in the U.S.A.
36 Gilligan, *History of the Liverpool A.S.* (ref.30) 2.
37 Liverpool A.S. *List of Members* (ref.35).
38 I am indebted to Gerard Gilligan for forwarding this list to me, abstracted from the copies preserved in the present Liverpool A.S. library housed in William Brown Street, Liverpool.
39 'A Fellow of the Royal Astronomical Society' (Capt. William H. Noble), 'A Lady's Observations of the Eclipse of 1884', *English Mechanic* 1324 (8 August 1890) 511 mentions Elizabeth Brown's travels. Liverpool A.S. *List of Members* (ref.35) 14–16 for the Emperor of Brazil's election to the Liverpool A.S. on 14 February 1887. Also *English Mechanic* 1144 (25 February 1887) 559.
40 'Liverpool Astronomical Society', Report of meeting of 11 May 1886, and Address by the President, Dr Isaac Roberts: *English Mechanic* 21 May 1886, 252–

3. See also response by 'F.R.A.S.' (William Noble) in *English Mechanic* 4 June 1886, 297. The affairs of the Liverpool A.S. were the subject of extensive reportage in *English Mechanic* during the 1880s.
41 Davies to *The Observatory* **9** (June 1886) 226–7.
42 Gilligan, *History of the Liverpool A.S.* (ref.30) 2.
43 W.H. Monck, 'An Amateur Astronomers' Association', *English Mechanic* 1321 (18 July 1890) 445–6.
44 Elizabeth Brown to *English Mechanic*, 1322 (25 July 1890), 363.
45 Maunder to *English Mechanic* 1324 (8 August 1890) 513. During 1890 a great deal of correspondence about the Liverpool A.S. and the case for a London-based Society (B.A.A.) was being published in *English Mechanic*.
46 For original Council and Section Directors, see *J.B.A.A.* **1**, 1 (1890) 2–3. 'The British Astronomical Association. The First Fifty Years', ed. Howard L. Kelly, *Memoirs of the B.A.A.* **36**, pt. 2 (December 1948), reprinted in *Memoirs of the B.A.A.* **42**, pt. 1 (1989) 1–132: 7.
47 Obituary, 'Mrs. Walter Maunder', by Mary A. Evershed, *J.B.A.A.* **57** (December 1947) 238. Also 'Annie Scott Dill Maunder', obituary by Mary Evershed, *M.N.R.A.S.* **108**, 1 (1948) 48–9.
48 Report of 24 October 1890 meeting, *J.B.A.A.* **1**, 1 (1890) 1. 'The British Astronomical Association' (ref.46) 7–8.
49 W.H. Noble, obituaries in *J.B.A.A.* **14** (1903–4) 351–5 and *M.N.R.A.S.* **65** (February 1905) 342–3. For the 'gun-carrying' portrait, preserved in the B.A.A. archives, see 'The British Astronomical Association' (ref.46) pl. 6. Also A. Chapman, 'William Henry Noble', *New D.N.B.* (O.U.P., forthcoming).
50 Barnard's Inn was one of the Inns of Chancery in Fetter Lane, though by 1892 it was past its prime: see 'Barnard's Inn', *J.B.A.A.* **1**, 1 (1890) iii. Charles Dickens mentions it in *Great Expectations* (1861).
51 'The British Astronomical Association' (ref.46) 60–5.
52 'The British Astronomical Association' (ref.46) 86.
53 'The British Astronomical Association' (ref.46) 52. Also 'John Mackenzie Bacon', obituary, *M.N.R.A.S.* **65**, 4 (1905) 334. Two of Gertrude Bacon's eclipse expeditions to India and America, on which she travelled with her father, were the subjects of her lectures to the B.A.A. North West (Manchester) Branch in 1898 and 1900: *J.B.A.A.* **10**, 1 (1900) 18 and *J.B.A.A.* **11**, 4 (1901) 149.
54 It is clear from the lives of many of these figures, however, that they enjoyed affluent circumstances. The Revd J.M. Bacon (1846–1904), for instance, was a balloonist, cinematographer, astronomer, and traveller. Though a Trinity, Cambridge, graduate, and ordained in 1870, he never held any living or appointment. He lived for most of his adult life at Coldash, Newbury, as a gentleman of means, who spent his time and money on philanthropic and learned projects. He was also a popular lecturer: see obituary (ref.53).
55 John Willoughby Meares, 'Astronomical Notes and Drawings' II, August 1894, 7–12. One of two volumes of MS notes, drawings and photographs by J.W. Meares loaned to the author many years ago by a member of the Meares family, and not reclaimed.

392 Notes and references [Ch. 13

56 Joseph A. Ashbrook, *The Astronomical Scrapbook. Skywatchers, Pioneers and Seekers in Astronomy* (C.U.P., and Sky Publishing, Cambridge, Mass., 1984) 70. Tebbutt is quoted, but without indication of the primary source.
57 Ashbrook, *Astronomical Scrapbook* (ref.56) 87. The Society came into being under the aegis of Edward S. Holden.
58 R.D. Radcliffe, *A Memoir of Thomas Glazebrook Rylands, of Highfields, Thelwall, Cheshire* (private publication 1901). Copy in the Liverpool A.S. library.
59 Gilligan, *History of the Liverpool A.S.* (ref.30) 4.
60 Leeds A.S. 'Album' (ref.11) 3 and 5 December 1892, date order.
61 Washington Teasdale's notices to *The Yorkshire Post*, *Leeds Mercury*, etc., newspapers, 5–6 December 1892: cuttings in Leeds A.S. 'Album' (ref.11), date order.
62 W.D. Barbour, etc., to *English Mechanic* and other papers: cuttings pasted into the Leeds A.S. 'Album' (ref.11).
63 W.D. Barbour to *Leeds Mercury*, April 1884: cutting preserved in Leeds A.S. 'Album' (ref.11), date order. See also Barbour's MS note about his 1887 visit to Davies in Liverpool, *ibid*.
64 Florence Taylor's 'Miss Caroline Herschel, the Astronomer' is included in the printed lecture syllabus, October 1896, preserved in the Leeds A.S. 'Album' (ref.11), date order. Her lecture 'Mary Somerville, the great Woman Astronomer and Mathematician', dated 28 July 1897, was published in the *Leeds Astronomical Society Journal and Transactions* 5 (Leeds 1897) 33–7. Florence Taylor's Caroline Herschel lecture may also have been printed, but I have not seen a volume of the *Journal and Transactions* for 1896. I wish to express my thanks to the Leeds A.S. for presenting me, in 1996, with an early set of the Society's *Journal and Transactions* (1895, 1897–1922) which has proved invaluable in reconstructing the early history of the Society.
65 P. Lowell to W.D. Barbour, 2 September 1897. Typed letter, preserved in the Leeds A.S. 'Album' (ref.11), date order.
66 Astronomical and Physical Society of Toronto to Leeds A.S., 24 May 1898, in Leeds A.S. 'Album' (ref.11) 103v, date order.
67 William Trant to W.D. Barbour, 25 January 1893, in Leeds A.S. 'Album' (ref.11), date order. Trant went on to distinction in Canada: see article in *The Journalist* (ref.22).
68 Obituary, 'Scriven Bolton', *Yorkshire Evening Post*, 27 December 1929; obituary, 'Greevz Fysher', *Yorkshire Post*, 19 May 1931: both in Leeds A.S. 'Album' (ref.11), 108v and 110v., date order.
69 'Seventieth Anniversary Dinner, 1929, March 11th at the Griffin Hotel, Leeds', menu and tabled speeches by Prof. S. Brodetsky and Prof. H.H. Turner: Leeds A.S. 'Album' (ref.11), date order.
70 Arthur Mee to W.D. Barbour, 5 January 1898, with information about the Astronomical Society of Wales: Leeds A.S. 'Album' (ref.11), date order. The receipt of copies of *The Cambrian Natural Observer* was minuted in the *Leeds A.S. Journal and Transactions* 6 (1898) (ref.64) 10. A set of *The Cambrian Natural Observer* for the years 1900–1911 was sent by Henry Jones of the Cardiff Central

Library to the Bodleian Library, Oxford, 1 May 1912; Jones mistakenly referred to it as the 'Astrological Society of Wales': Radcliffe Science Library, Oxford, Per. 1842 f. 43.
71 John Silas Evans, *Seryddiaeth a Seryddwyr* [*Astronomy and Astronomers*] (Cardiff 1923) 279–80.
72 *The Cambrian Natural Observer, And Quarterly Journal of the Astronomical Society of Wales* III, 1 (May 1900), inside cover. It is also stated that the Society was founded in 1894–5. Bibliographically *The Cambrian Natural Observer* can sometimes be awkward to cite because of its Series and New Series changes.
73 *The Cambrian Natural Observer* IV, 3 (August 1901), 89, for Markwick's observatory.
74 *The Cambrian Natural Observer* published an annual list of members. The greatest membership year seems to have been 1908.
75 *The Cambrian Natural Observer* III, 3 (January 1901): list of members, and for John Jones, see p. 49. It is clear that they also regarded the 16-year-deceased Revd T.W. Webb as a member of the Welsh diaspora – especially as he had apparent Welsh connections by marriage, lived within walking distance of the Welsh border, and was buried at Mitchel Troy, Monmouthshire: Arthur Mee, 'The Grave of Prebendary Webb', *The Cambrian Natural Observer* IV, new series 2 (May 1901) 17–18. Mee acknowledged that 'no man ever did so much for the advancement of observational astronomy in the English-speaking world as Thomas William Webb'. (For more on Webb's burial place, see 'Death of the Rev. Prebendary Webb', *Monmouthshire Beacon*, Saturday 30 May 1885.)
76 Miss E. Graham Hagerty, A.R.C.Sc., seems to have been listed in *The Cambrian Natural Observer* as an original Council Member.
77 Evans, *Seryddwyr* (ref.71) 294 for Harries. I am indebted to Rhys Morris of the Astronomy Department, Cardiff University, for generously translating for me large sections of Evans' book from the original Welsh.
78 Evans, *Seryddwyr* (ref.71) 270–306.
79 Evans, *Seryddwyr* (ref.71) 280. Also Dan Jones, *Cardiff City Observatory Handbook* (Cardiff 1931).
80 Evans, *Seryddwyr* (ref.71) 287.
81 Of these 119 members, 45 had some title or letters after their name: John H. McElderry, 'Historical notes on Irish Astronomy: Members of the Ulster Astronomical Society, from list dated 1st May, 1893', *The Irish Astronomical Journal* 3, 3 (September 1954) 77–79. Unfortunately, the archival location of the original printed or manuscript Membership List is not mentioned. See also J.H. McElderry, 'Historical Notes on Irish Astronomy: Glimpses of popular astronomy in Belfast', *ibid.* 75–7.
82 'Scrapbook' of the Manchester Astronomical Society, compiled by Thomas Weir, January 1892–May 1899, deposited in Manchester City Reference Library, M298/3; items in date order. I am indebted to the Council of the Manchester A.S. for the loan of the 'Scrapbook', along with other early documents of the Society. See also *The History of the Manchester Astronomical Society (The First Hundred Years)* (U.M.I.S.T., Manchester, 1992) 1.

83 Manchester A.S. 'Scrapbook', date order; also *History of the Manchester A.S.* 4 (both ref.82).
84 *History of the Manchester A.S.* (ref.82) 6. There was also a Mr Wilde who demonstrated his 'Magnetarium' to the Manchester Lit. and Phil. in their 36 George St. premises – '(Dalton's Old residence – near the Infirmary)' – in early March 1895: carbon copy of the Society's circular notice, 2 March 1895, in Manchester A.S. 'Scrapbook' (ref.82).
85 On 2 February Father Sidgreaves of Stonyhurst, the first President of the Manchester A.S. (then the North West Branch of the B.A.A.), received a telegram requesting him to photograph the spectrum of the nova which had appeared 2° south of χ Aurigae: Manchester A.S. 'Scrapbook' (ref.82), see date. The *City News*, 9 April 1898, was reporting that Thomas Thorp and Father Sidgreaves were photographing stellar spectra and experimenting with a celluloid diffraction grating ruled with 14,438 lines to the inch: Manchester A.S. 'Scrapbook' (ref.82), date order.
86 Manchester A.S. 'Scrapbook' (ref.82), 1892–3, date order.
87 Manchester A.S. 'Scrapbook' (ref.82), date order, for 1892–3 Council.
88 Cutting from *City News*, 9 November 1895, in Manchester A.S. 'Scrapbook' (ref.82), date order.
89 Society circular notice, 2 January 1897, Manchester A.S. 'Scrapbook' (ref.82), date order.
90 Society circular by Thomas Weir, 27 April 1896, Manchester A.S. 'Scrapbook' (ref.82), date order.
91 I am very much indebted to Kevin Kilburn, formerly Chairman of the Manchester Astronomical Society, for passing his computer printouts of the early membership analyses on to me. Copies are also on deposit in the Manchester A.S. Godlee Observatory, U.M.I.S.T., Sackville Street, Manchester.
92 As a native of Greater Manchester, I have long had an interest in the historical and economic geography of the region, and upon analysing Kevin Kilburn's membership lists of the B.A.A. North West Branch, was not surprised to find that the early members lived in the better-off districts. I am also informed by Ray Emery of the Leeds A.S. that the majority of that Society's early members resided in the prosperous districts of Leeds.
93 This correspondence with the B.A.A. by Thomas Weir is reproduced in *History of the Manchester A.S.* (ref.82) 8–12.
94 Philip Duke to Thomas Weir, 9 June 1894, reproduced in *History of the Manchester A.S.* (ref.82) 9–10.
95 Duke to Weir, 9 June 1894, and Weir to Duke, 22 June 1894, reproduced in *History of the Manchester A.S.* (ref.82) 9–12.
96 Kevin Kilburn to A. Chapman, 20 January 1992, private communication, accompanying Mr Kilburn's printouts of the 1892–1902 membership analysis.
97 'Minute Book, 1903–1909' of the Manchester A.S., deposited in the Manchester City Library, As. 472. Account of meeting, 18 September 1903; date order.
98 Manchester A.S. 'Minute Book' (ref.97), 7 December 1904, for Hoyle as projectionist. By 4 March 1908, he had been replaced by Mr Buss, who also received 3s. See 7 February 1906 for projector electrification.

99 *History of the Manchester A.S.* (ref.82) 38–41.
100 'Midlands Branch (Birmingham)', Report, *J.B.A.A.* **13** (1903) 381.
101 The Revd Alfred Adolphus Cole (1821–1893), a Baptist minister, and William Henry Robinson (1847–1926), a businessman, were leading promoters of science and astronomy around Walsall and elsewhere in the Midlands. Robinson wrote regular astronomical columns for *The Walsall Observer and South Staffordshire Chronicle*, especially in 1916–17. See also Stuart Williams, *William Henry Robinson. A Man ahead of his Time* (Walsall Local History Centre, 1992). I am indebted to Stuart Williams of the Walsall A.S. for information and photocopied documents pertaining to late Victorian and Edwardian astronomy. The original documents for this period are preserved in the Local History Centre, Walsall Public Libraries.
102 T.E.R. Phillips, obituary, 'Thomas Henry Espinall Compton Espin', *M.N.R.A.S.* **95**, 4 (February 1935) 319–22. Also W.H. Davies, obituary, 'T.H.E.C. Espin', *J.B.A.A.* **45**, 3 (January 1935) 128.
103 Espin, for instance, was in the Presidential Chair in January 1911, when the Newcastle Astronomical Society held its A.G.M. in the fashionable Crosby's Café, in Northumberland Street, Newcastle. I am indebted to Dr David Turnbull of the Newcastle A.S. for this and other information. Dr Turnbull has worked on the minute books and early documents of the Newcastle A.S. and has sent me a copy of his unpublished 'History of the Newcastle Upon Tyne Astronomical Society'.
104 *Walsall Science and Art Institute... Programmes of Classes... for the Session 1889–1890.* Walsall Public Library P 378 5WC 5273.
105 'A Grand Popular Astronomical Entertainment, by B.J. Malden Esq., F.R.G.S., & C. (Ten years Lecturer to the Royal Polytechnic)', on 'The Glories of the Heavens': advertisement sheet reproduced in Williams, *William Henry Robinson* (ref.101) 13.
106 *The Cambrian Natural Observer* IV, new series 2 (May 1901) 39–40. Although it is said that Ball had recently delivered *two* lectures in Cardiff, there is no indication of the organisations involved or their venues. He returned to London afterwards on the night train.
107 Manchester A.S. 'Minute Book' (ref.97), Council record, 23 September 1908.
108 Manchester A.S. 'Minute Book' (ref.97), 3 October 1906, for Bellamy. A previous minute, 23 August 1906, recorded that the Society would meet Bellamy's expenses. The 23 August 1906 minute also recorded that the Manchester A.S. operated on about £20 per year, with a £3 –£4 annual profit.
109 Eugene M. Antoniadi to Midlands Branch, B.A.A.; the lecture seems to have been given on 18 December 1902, but was reported in *J.B.A.A.* **13**, 4 (1903) 160–1.
110 Dr Oliver Lodge to Midlands Branch, B.A.A., lecture on 17 October 1901: *J.B.A.A.* **12**, 1 (1901) 17–18.
111 The *City News*, 15 April 1893, reported a lecture to the Manchester A.S. by Dr Orchard of Pendleton on Jeremiah Horrocks and William Crabtree. Mr A. Dodgson also lectured on 'Jeremiah Horrocks and the Transit of Venus' to the Leeds A.S. on 12 April 1899; this lecture was subsequently published in the *Leeds A.S. Journal and Transactions* 7 (1900, for 1899) 25–31. There would obviously

have been considerable local interest in this topic. William Crabtree (1610–1644) had lived nearby in Broughton, Salford, while his friend Jeremiah Horrocks (1619–1641) had lived at Much Hoole, near Preston, Lancashire. In November 1639 they had observed and recorded the first transit of Venus across the Sun's disk, from coordinates calculated by Horrocks. There was also William Gascoigne (1612–1644) of Leeds, who, while he did not observe the transit of Venus, nonetheless invented the telescopic sight, and was a friend of Horrocks and Crabtree. Victorian scientific historians had been deeply taken with these three figures: see J.E. Bailey, *Palatine Notebook* II, III (Manchester 1882–3) 253–266: 17–22, and A.B. Whatton, 'Memoir' of Jeremiah Horrocks, which included an English translation of his Latin *Venus Sub Sole Visa 1639* (London 1859). See also A. Chapman, *William Crabtree. Manchester's First Mathematician* (Manchester Statistical Society 1995).

112 Florence Taylor, 'Miss Caroline Herschel, the Astronomer'; forthcoming lecture, to be given on 23 September 1896, in the 1896 Syllabus: *Leeds Astronomical Society Journal and Transactions* **3** (1896) 7.

113 Thomas Core's lecture 'Sixty Years' Astronomical Progress, 1837–1897' was reported in the *City News*, 9 October 1897: cutting in Manchester A.S. 'Scrapbook' (ref.82), date order. One suspects that in his choice of dates, Core was making the Manchester Astronomical Society's loyal gesture to Queen Victoria's Diamond Jubilee which was also being celebrated in 1897.

14 NOW LADIES AS WELL AS GENTLEMEN

1 Joanna Johnston, *The Life, Manners and Travels of Fanny Trollope. A Biography* (Quartet Books, London, 1980). Isabella Bird, *A Lady's Life in the Rocky Mountains* (London, 1879; Virago, 1982). Mary Kingsley, *Travels in West Africa* (London, 1897, Virago, 1982). Also Eric Newby, *A Book of Travellers' Tales* (Collins, London, 1985).

2 Johnston, *Fanny Trollope* (ref.1) 42–69.

3 *D.N.B.* **22**, Suppl., 'Anning'. 'Mary Anning', *Larousse Dictionary of Scientists*, ed. Hazel Muir (Larousse, Edinburgh–New York, 1994).

4 See Chapter 13, pp. 248, 256, 265–8 for amateur society memberships by 1905.

5 Florence Taylor (later Hildred), 'Miss Caroline Herschel, the Great Astronomer', a lecture due to be delivered to the Leeds A.S., 23 September 1896, in the syllabus of lectures for 1896: *Leeds Astronomical Society Journal and Transactions* **3** (1895–6) 7. Also F. Taylor, 'Mary Somerville, The Great Woman Astronomer and Mathematician', published lecture of 28 July 1897, in *Leeds A.S. Journal and Transactions* **5** (1897) 33–7.

6 H.J. Mozans, *Women in Science* (M.I.T. Press, Cambridge Mass. and England, 1913) 182–90. In spite of her achievements, however, Caroline Herschel always had a touching humility, and regarded herself as no more than 'a well-trained puppy-dog' in the business of astronomical research: quotation in Agnes M. Clerke, *The Herschels and Modern Astronomy* (London 1895) 140. See also the

reassessment of the role of Caroline Herschel made by Dr Michael Hoskin in a paper read to the R.A.S. Special Meeting, 13 March 1998, reported in *The Observatory* and *Astronomy and Geophysics* (both forthcoming). No full biographical study of Caroline has yet been made, though at the time of writing Dr Hoskin was working on a joint volume devoted to her and to her brother William. For a good brief life see Patrick Moore, *Caroline Herschel, Reflected Glory* (William Herschel Society, Bath, 1988).

7 Mrs John Herschel, *Memoirs and Correspondence of Caroline Herschel* (London, 1876), 8 July 1783: p. 54. J.L.E. Dreyer and H.H. Turner et al., *A History of the Royal Astronomical Society, 1820–1920* (R.A.S., 1923; Blackwell's reprint, 1987) 87 for presentation of Caroline's telescope to the R.A.S. in the 1840s. The Science Museum, South Kensington, store contains a black-painted wooden Herschel-type reflecting telescope which is traditionally believed to have been used by Caroline. Also M.A. Hoskin and B. Warner, 'Caroline Herschel's Comet Sweepers', *J.H.A.* **12** (1981) 26–34.

8 Caroline Herschel, *Catalogue of Stars from Mr Flamsteed's Observations contained in the second volume of the Historia Coelestis and not inserted in the British Catalogue* (Royal Society, London, 1798).

9 Flamsteed's unwitting observing and recording of Uranus as the star 34 Tauri has long been one of the legends of astronomy. I am indebted to Kevin Kilburn of the Manchester A.S. for calculating the position of Uranus in December 1690 and January 1691, from the 'Dance of the Planets' computer program. Depending on the brightness of the Moon, the clarity of the sky, and the computed motions of Uranus within Taurus, Mr Kilburn says that Flamsteed could have observed the astronomically well-placed Uranus between the true fixed stars 33 and 35 Tauri on 16 December 1690 (5 December Old Style), 4 January 1691 (24 December 1690 Old Style), and 3 February 1691 (22 January Old Style). Private correspondence, Kevin Kilburn to A. Chapman, 23 February, 5 March and 13 March 1997.

10 *Memoirs and Correspondence of Caroline Herschel* (ref.7) 37.

11 *Memoirs and Correspondence of Caroline Herschel* (ref.7), 31 December 1783: p. 55.

12 Dreyer and Turner, *History of the R.A.S.* (ref.7) 81 for honorary membership; 252 for Gold Medal. See also refs.5 and 6 above.

13 Ellen Mary Clerke, 'Mary Somerville' entry, *D.N.B.* For a double obituary for the astronomical Clerke sisters, see Lady Margaret Huggins, *Agnes Mary Clerke and Ellen Mary Clerke, an Appreciation* (private publication 1907); copy in R.A.S. library.

14 R.A. Proctor, obituary, 'Mary Somerville', *M.N.R.A.S.* **33** (1873) 190. Mary T. Brück, 'Mary Somerville, mathematician and astronomer of underused talents', *J.B.A.A.* **106**, 4 (1996) 201–6.

15 Elizabeth C. Patterson, *Mary Somerville, 1780–1872* (Oxford, 1979) 24. Lord Brougham to Dr Somerville, 27 March 1827, reprinted in *Personal Recollections from Early Life to Old Age. With Selections from her Correspondence by Her Daughter Martha Somerville* (London, 1873) 161. Brougham had written to Dr Somerville in the hope that he would become Brougham's advocate in persuading her to undertake the project.

16 William Whewell to Dr Somerville, 2 November 1831; and William Peacock to Mrs Somerville (no date), emphasising the importance of *Mechanism of the Heavens* to Cambridge mathematics teaching: *Personal Recollections* (ref.15) 170-2: 172. Sir John Herschel was one of Mary Somerville's staunchest supporters: *ibid.* 168-70, etc.
17 Brück, 'Mary Somerville' (ref.14) 204. The bust was completed in 1824.
18 Mary Somerville (Rome) to Lord Rosse, 11 November 1843, Rosse MS, Birr Castle Archives, K17:16. Lord Rosse replied, 12 June 1844, mentioning the current use of the 36-inch and the progress of the Birr Castle 72-inch: Somerville, *Personal Recollections* (ref.15) 215-16.
19 Brück, 'Mary Somerville' (ref.14): title.
20 Proctor, obituary, 'Mary Somerville' (ref.14) 190.
21 The ease with which Mary Somerville moved in the Grand Amateur world is made clear in her *Personal Recollections* (ref.15): pp. 218-19, for instance, which describe her visit, with her husband, to Sir James South's Campden Hill Observatory, and their measurement of the distance of Mercury from the Sun. Also Maboth Moseley, *Irascible Genius: A Life of Charles Babbage, Inventor* (Hutchinson, London, 1964) 155-75.
22 Mozans, *Women in Science* (ref.6) 192-5. Peggy A. Kidwell, 'Mitchell, Maria (1818-1889)', in *History of Astronomy. An Encyclopedia*, ed. John Lankford (Garland, New York, 1997) 332-3.
23 Kidwell, 'Women in Astronomy', in *History of Astronomy* (ref.22) 564-7: 566. Also Deborah J. Warner, *Alvan Clark and Sons. Artists in Optics* (Smithsonian Institution, Washington DC, 1968) 104-5 (new edition, 1996) for Vassar College Observatory.
24 Deborah J. Warner, 'Astronomy in Antebellum America', in *The Sciences in the American Context: New Perspective*, ed. Nathan Reingold (Smithsonian Institution, Washington DC, 1979) 55-75: 62-3.
25 'The Royal Cork Institution', from *Some People and Places in Irish Science and Technology*, ed. C. Mollan, W. Davis and B. Finucane (Royal Irish Academy, Dublin, 1985, 1990) 104-5. See also the *History of the Gallery*, Crawford Municipal Art Gallery, Emmet Place, Cork. I am indebted to Dr Colm O'Sullivan and Dr Ian Elliott for information.
26 Mary Creese, 'Elizabeth Brown (1830-1899), solar astronomer', *J.B.A.A.* **108**, 4 (1998) 193-7. I am indebted to Mary Creese for a pre-publication typescript.
27 Liverpool A.S. *List of Members, 1st March 1887* (Liverpool, 1887) (Royal Society Tracts, 845/11) 1-11: of 300 members in 1887, there were eight women. The Pernambuco Branch of the Liverpool A.S. (pp. 14-16) had 77 members, including one woman.
28 It is not unlikely that Miss E. Graham Hagerty, A.R.C.Sc., of the Higher Grade School, Cardiff, had been a pupil of Sir J.N. Lockyer, who held the Solar Physics Chair at the Royal College (Imperial College) in the 1880s and 1890s: see *The Cambrian Natural Observer* III, I (1900), front inside cover. Miss Graham Hagerty was a Council Member and Secretary of the Astronomical Society of Wales. A set

of *The Cambrian Natural Observer*, 1900–1911, is deposited in the Radcliffe Science Library, Oxford, Per. 1842 f. 43.

29 *Leeds Astronomical Society Journal and Transactions* (ref.5), published membership lists for cited years. Florence Taylor was presented with a 'very valuable book' upon her departure to America in 1898: F. Taylor, 12 Springfield Mount, Leeds, to W.D. Barbour, 13 June 1898, letter preserved in the 'Album' of the Leeds Astronomical Society, and still in the possession of the Society; date order. Florence Hildred – as she became upon marriage – remained an ordinary member of the Leeds A.S. (in spite of living in Minnesota from 1898) until 1909, when she was elected an Honorary Member. She later gave £100 to the Society and Leeds University, and still appeared on the Membership List in the Society's *Journal and Transactions* 29–30 (1921–2) (p. 43), the last date for which I have seen a copy of the *Journal*. I have been unable to trace any reference to Florence Hildred, however, in Minnesota archives. (I am indebted to Dr Rowland Barden, of Moorhead State University, for his generous continuing search in local archives.)

30 *The Cambrian Natural Observer* (ref.28), published membership lists for cited years.

31 John H. McElderry, 'Historical notes on Irish Astronomy: Members of the Ulster Astronomical Society, from list dated 1st May 1893', *The Irish Astronomical Journal* **3**, 3 (September 1954) 77–9. The ten women members included Miss Ellen M. Clerke (Agnes M. Clerke's sister) and Mrs Proctor–Smyth.

32 Mary Brück, 'Lady Computers at Greenwich in the early 1890s', *Q.J.R.A.S.* **36** (1995) 83–95. Also Brück, 'Alice Everett and Annie Russell Maunder, torch-bearing women astronomers', *The Irish Astronomical Journal*, **21**, 3 and 4 (March and September 1994) 281–90.

33 Peggy Kidwell, 'Women Astronomers in Britain, 1780–1930', *Isis* **74** (1984) 534–46: 540.

34 Creese, 'Elizabeth Brown' (ref.26). 'In Memoriam. Elizabeth Brown, F.R.Met.-Soc.', obituary in *J.B.A.A.* **9**, 5 (March 1899) 214–15: 215.

35 'In Memoriam. Elizabeth Brown' (ref.34) 214. Her first paper was 'Observations of proper motions of Sunspots', *Liverpool Astronomical Society. Abstract of Proceedings* **2** (1883–4) 18–19.

36 Creese, 'Elizabeth Brown' (ref.26), 193.

37 Creese, 'Elizabeth Brown' (ref.26), 195–6.

38 Creese, 'Elizabeth Brown' (ref.26), 195.

39 Elizabeth Brown, 'Programmes of the Directors of the Observing Sections. Solar Section', *J.B.A.A.* **I**, 2 (November 1890) 58–60: 59.

40 T.W. Webb, *Celestial Objects for Common Telescopes* (London, 1859) 14–15.

41 'In Memoriam. Elizabeth Brown' (ref.34) 215 mentions sudden and fatal complications following upon 'an apparently slight catarrhal illness'.

42 See E. Brown (written anonymously), *In Pursuit of a Shadow, by a Lady Astronomer* (London 1887) 90–1 for visit to 'Professor B.' (Bredichin). Chapter 6, 89–112, describes the visit of Elizabeth Brown, her unnamed lady companion 'L.', Father Perry and Dr Copeland to Bredichin's estate at Pogost; though sadly the eclipse was clouded out, 106–110. In addition to its astronomical content, *In Pursuit of a Shadow* contains a fascinating account of a journey through Russia in 1887.

43 Creese, 'Elizabeth Brown' (ref.26)194. I have not been able to locate a copy of Elizabeth Brown's *Caught in the Tropics*, with its account of her voyage to Trinidad in 1889. Father Stephen Perry did sail out from Southampton with a party of astronomers including Miss Brown on board the Royal Mail SS *Tagus*. He was conveyed from Barbados (no longer in the group with Miss Brown) to the Isles du Salut (Devil's Island) on HMS *Comus*, to observe the eclipse on 22 December 1889. He died of dysentery, contracted on the island, on 27 December, soon after the *Comus* put back to sea: E.B. Knobel, 'Stephen Joseph Perry', obituary, *M.N.R.A.S.* **50**, 4 (February 1890) 168–75: 172. See also George Bishop, 'Stephen Perry (1833–1889): Forgotten Jesuit Scientist and Educator', *J.B.A.A.* **89**, 5 (1979) 473–84: 483.

44 The best account of the 1896 expedition is R.A. Marriott, 'Norway 1896: The B.A.A.'s first organised eclipse expedition', *J.B.A.A.* **103**, 3 (1991) 162–9: 164.

45 'In Memoriam. Elizabeth Brown' (ref.34) 215.

46 Creese, 'Elizabeth Brown' (ref.26)197, reference 31. The gross value of the estate in 1907 was £63,182, or £29,014 net.

47 'The History of the British Astronomical Association. The First Fifty Years', ed. Howard L. Kelly *et al.*, *Memoirs of the B.A.A.* **36**, pt. II (December 1948, reprinted 1989) 51. One presumes that the obscured sun was hissed in the same manner as a disappointing music-hall turn.

48 M.A. Evershed, 'Mrs Walter [Annie Scott Dill Russell] Maunder' (1868–1947), obituary, *J.B.A.A.* **57**, 6 (December 1947) 238. 'History of the British Astronomical Association' (ref.47) 51. Also E.W. Maunder, 'The Total Eclipse, January 22 1898', *Knowledge*, 1 March 1898, 49–50, and 'The Recent Eclipse', *ibid.*, 2 May 1898, 107–9: for streamers, 108–9. It would appear that Annie Maunder had a medical sister or other female relative who pressed medical advice on those about to embark for India to observe the 1898 eclipse: Hester Dill Russell, M.B., B.Ch., ('communicated by Mrs. A.S.D.R. Maunder'), 'Hints for Board Ship and Tent Life in India during the Eclipse Expedition', *J.B.A.A.* **8**, 1 (17 November 1897) 38–9, reprinted in B.A.A. *Occasional Reprints* 1 (1995). The detailed character of her advice suggests that Dr Hester Russell had personal experience of living in the Far East.

49 A.D. Thackeray, 'Mary Acworth Evershed' (1867–1949), obituary, *M.N.R.A.S.* **110**, 2 (1950) 128–9. See also Mary T. Brück, 'Mary Acworth Evershed née Orr (1867–1949), solar physicist and Dante scholar', *Journal of Astronomical Heritage and History* **1**, 1 (June 1998) 45–59: 46–7.

50 'History of the British Astronomical Association' (ref.47) 52. I wish to express my indebtedness to Anthony J. Kinder, Director of the Historical Section of the B.A.A., for his generous assistance, and for supplying me with photocopies of the Association's early membership records.

51 Obituary, 'Loetitia Crommelin', *J.B.A.A.* **32**, 1 (October 1921) 28–9.

52 J.H.D., obituary notice, 'Catherine Octavia Stevens', *J.B.A.A.* **70** (1960) 103–4.

53 Surprisingly, Mary Proctor (1862–1957) had no obituary either in *J.B.A.A.* or in *M.N.R.A.S.* A brief account of her life to date, including her aeroplane observation of the total eclipse of 1927, is given in 'Who's Who in the Moon.

First Memoir of the Historical Section', *ed.* Mrs John [Mary Acworth] Evershed, *Memoirs of the B.A.A.* **34**, 1 (1938) 1–130: 100. A. Hunter, Notification of the death of Alice Grace Cook to the B.A.A. Meeting, 25 June 1958: *J.B.A.A.* **68** (1958) 302. W.F.D., obituary, 'Mrs Fiametta Wilson', *M.N.R.A.S.* **81** (February 1921) 266–9; A. Grace Cook, obituary, 'Mrs Fiametta Wilson', *J.B.A.A.* **30**, 10 (September 1920) 330–1.

54 'Mrs Fiametta Wilson', obituary, *M.N.R.A.S.* (ref.53) 268.
55 'Mary Ashley', biographical computer database, B.A.A. I am indebted to Anthony J. Kinder of the B.A.A. for this initial information. I have examined Miss Ashley's volumes of 'Lunar Drawings', 1880–1885, preserved in the B.A.A. library. She was then working with a 4-inch aperture Wray equatorial refractor, from 16 New King Street, Bath, only a few doors away from 19 New King Street, from where William Herschel had discovered Uranus in 1781.
56 Adolphus Rebière, *Les Femmes dans la Science* (2nd edn., Paris, 1897).
57 P.M. Ryves, obituary, 'Mary Adela Blagg', *M.N.R.A.S.* **102**, 2 (1945) 65.
58 John B. Hearnshaw, *The Analysis of Starlight. One Hundred and Fifty Years of Astronomical Spectroscopy* (C.U.P., 1986) 74, for quotation. Also Barbara J. Becker, 'Eclecticism, Opportunism, and the Evolution of a New Research Agenda: William and Margaret Huggins and the Origins of Astrophysics', vol.II (unpublished Ph.D. thesis, Johns Hopkins University, Baltimore, 1993) 290.
59 Mary T. Brück, 'Companions in Astronomy: Margaret Lindsay Huggins and Agnes Mary Clerke', *Irish Astronomical Journal* **20**, 2 (September 1991) 70–7: 73.
60 The Hon. Mrs [Mary] Ward, *The Telescope, A Familiar Sketch...* (London, undated, *c.*1860). Also Timothy Collins, 'Some Irish Women Scientists', U.C.G. Women's Studies Centre *Review* **1** (1992) 39–53: 44.
61 Brück, 'Alice Everett and Annie Russell Maunder' (ref.32) 281–4.
62 Brück, 'Alice Everett and Annie Russell Maunder' (ref.32) 281, 287–9.
63 See M. Huggins, *Agnes Mary... and Ellen Mary Clerke* (ref.13) for the Clerke sisters' careers.
64 I am indebted to Dr Mary T. Brück for her extensive private correspondence with me on the Clerke sisters, for information about the backgrounds of the Clerke and Deasy families, and for insights into the social and religious climate of early Victorian Ireland.
65 For 'tickling' the ears, see Lady M.L. Huggins' review of Agnes M. Clerke's *The System of the Stars* (London 1890), in *The Observatory* **13**, 169 (December 1890) 382–6: 384. For access to astronomy's 'lofty halls', see Clerke, *System of the Stars*, Preface, vii.
66 Peter D. Hingley, 'Urania's Mirror – a 170-year-old mystery solved?', *J.B.A.A.* **104**, 5 (1994) 238–40. Bloxam's R.A.S. Election Certificate is reproduced on p. 239.
67 Agnes Giberne, *Sun, Moon and Stars: Astronomy for Beginners* (London and Guildford, 1879). See title page for sales figures.
68 Giberne, *Sun, Moon and Stars*, new and revised edn. (London, 1898): 'Twenty-fourth Thousand', on titlepage.

402   Notes and references   [Ch. 14

69   Obituary, 'Gertrude Longbottom', *M.N.R.A.S.* **96** (February 1936) 295 specifies her Methodist activities.
70   Mary McKeown, 'Hester Periam Hawkins: Fellow of the Royal Astronomical Society, 1846–1928', *Bedfordshire Magazine* **25**, 197 (Summer 1996) 208–14: 208.
71   For Joshua Hawkins and the B.A.A., see biographical computer database, B.A.A., under 'Hawkins, Hester Periam'.
72   Hester Periam Hawkins, *Halley's Comet, With a Plan of its Pathway in the Heavens, and some notes on Comets and Meteors* (London; undated, but from internal evidence, 1910) 6; the bad Latin of the Bayeux Tapestry reads 'Isti Mirant Stella', whereas H.P. Hawkins correctly points out that it should read 'Isti Mirantur Stellam'.
73   Hester P. Hawkins, *Stella Maitland: or Love and the Stars* (London; undated, but 1921 in Bodleian Library Catalogue).
74   H.P. Hawkins, R.A.S. Election Certificate, proposed 12 November 1920 and elected 14 January 1921: facsimile reproduced in McKeown, 'Hester Periam Hawkins' (ref.70) 213.
75   M.A. Evershed, 'Mrs Walter Maunder' (ref.48) 238.
76   *The Manchester Guardian*, 12 May 1898; date order. Cutting preserved in the Manchester Astronomical Society (the North West Branch of the B.A.A.) 'Scrapbook', January 1892–May 1899, compiled by Thomas Weir, Manchester City Central Reference Library, MS M298/3.
77   Obituary, 'Hester Periam Hawkins', *M.N.R.A.S.* **89** (February 1929) 308–9.
78   Florence Taylor, 'Mary Somerville' (ref.5) 37.
79   At a 'Conversational Meeting' of the Leeds A.S., 14 August 1895, Washington Teasdale, the President, stressed the Society's wish to encourage lady members: *Leeds A.S. Journal and Transactions* **3** (March 1896) 34. Following Florence Taylor's 'Mary Somerville' lecture, 28 July 1897, the then President, Charles Whitmell, responded, praising the work of women astronomers: *Leeds A.S. Journ. & Trans.* n.s. 5 and 78, 37.
80   Mary Proctor, 'Sallie Duffield Proctor–Smyth' (1856–1941), obituary, *M.N.R.A.S.* **102**, 2 (1942) 73–4.
81   'Sallie Duffield Proctor–Smyth' (ref.80) 73: 'On learning of her husband's death Mrs Proctor hurried to New York... determined to take his place for the lectures which he had engaged to give in England.' Also A.C.R., 'Richard Anthony Proctor' (1837–1888), obituary, *M.N.R.A.S.* **49**, 4 (February 1889) 164–7.
82   Manchester A.S. 'Scrapbook' (ref.76), 13 October 1892.
83   Manchester A.S. 'Minute Book', 5 April 1905, 3 January 1906, etc., date order. Manchester City Central Reference Library MS AS.472.
84   Mary Proctor's lecture circuit, indeed, was truly international, 'and has done much to rouse interest in [astronomy] by lectures in the U.S.A., Canada, England, Australia, and New Zealand': 'Who's Who in the Moon' (ref.53) 100. See her popular books, *Stories of Starland* (1895) and *Giant Sun and his Family* (1896).
85   Manchester A.S. 'Minute Book' (ref.83), 7 October 1908, date order.
86   Catherine Stevens died aged 95, Gertrude Longbottom 81, Mary Blagg 84, Sallie

Proctor–Smyth 86, and Mary Proctor 95. Their longevity paralleled that of their predecessors, Caroline Herschel, 98, and Mary Somerville, 92.
87 Manchester A.S. 'Scrapbook' (ref.76), 13 October 1892; date order.
88 'Revd John Mackenzie Bacon' (1846–1904), obituary, *M.N.R.A.S.* **65**, 4 (February 1905) 334. Bacon was a Trinity College, Cambridge, graduate who lived at Coldash, Newbury. He held no Church living or any apparent paid job, and seems to have been a gentleman of ample means. As he married Miss Gertrude Myers in 1871, their daughter Gertrude, the astronomical lecturer, can have been no more than 18 at most when she joined the B.A.A. in 1890.
89 'Lecture by Miss Bacon', *The Cambrian Natural Observer* **4**, 2 (1901) 37–8.
90 Manchester A.S. 'Minute Book' (ref.83), 13 January 1904, 7 September 1909; date order. Gertrude Bacon was a vigorous lecturer; her talk on the 1898 Indian eclipse, given to the Manchester A.S. (North West Branch of the B.A.A.) on what seems to have been 11 October 1899 (from the B.A.A. Report, the year is unclear, but probably 1899), was delivered 'in a racy and fluent manner': *J.B.A A.* **10**, 1 (1900) 18. Exactly how this lecture related to the one delivered on the same subject on her behalf by Mary Proctor, 12 May 1898 (ref.76), is not clear.
91 'William Donald Barbour', obituary, *Leeds A.S. Journal and Transactions* **11**, for 1903 (1904) 9–10.
92 Miss C.A. Barbour, 'Astrology', 11 April 1900, *Leeds A.S. Journal and Transactions* **8**, for 1900 (1901) 46–51. This lecture had been originally billed to be given on 25 October 1899: *Leeds A.S. Journal and Transactions* **6** (1898), syllabus for 1899, p. 8. Miss C.A. Barbour, 'Mythological Astronomy', 22 November 1904, *Leeds A.S. Journal and Transactions* **12**, for 1904 (1905) 50–7. Miss Barbour's article 'Sir William Herschel' in the *Leeds A.S. Journal and Transactions* **24**, which was probably delivered to the Society as a lecture, was greatly prized by Miss Francisca Herschel 'as containing a record of her famous ancestor'; a copy was requested by William Porthouse, Vice-President of the Manchester A.S. See letter, Porthouse to Miss Barbour, 11 June 1920, in Leeds A.S. 'Album', date order.
93 Miss Tranmar, 'Chaucer's Astrology', 28 February 1904, *Leeds A S. Journal and Transactions* **13** (1905) 14–24.

15 CONCLUSION AND POSTCRIPT: THE AMATEUR ASTRONOMER INTO THE TWENTIETH CENTURY

1 'Angelo Secchi', in *D.S.B.* Also *The Cambridge Illustrated History of Astronomy*, ed. M.A. Hoskin (C.U.P., 1997) 290–1.
2 T.E.R. Phillips, obituary, 'Thomas Henry Espinall Compton Espin', *M.N.R.A.S.* **95**, 4 (February 1935) 319–22: 319. Phillips had known Espin very well over several decades, and wrote this obituary for a fellow priest–astronomer with both knowledge and insight. W.H. Davies, obituary, 'T.H.E.C. Espin', *J.B.A.A.* **45**, 3 (January 1935) 128.
3 John Willoughby Meares, 'Astronomical Notes and Drawings, Vol. II', August

1894, p. 179: one of two MS observing books loaned to the author many years ago by a member of the Meares family, and not reclaimed. On the same page Meares records that he 'Joined British Astronomical Association as an original member in 1890 Novr. 12th. Also Astr. Society of Wales.' 'Elected Fellow of the R.A.S. May 10th 1895. Given a 9¼" With-Browning (& Calver) by Father on going to India in September '96.' I have not been able to locate an R.A.S. obituary notice for Meares, who died in 1945 or 1946, at around 74 years old. For a short obituary, see *J.B.A.A.* 56 (March 1946) **56**. This obituary fails to give the date or year of Meares' death, which one presumes took place in 1945 or early 1946.

4   G.B. Airy, *Smith's Prizes. To the Members of the Senate of the University of Cambridge*, 13 March 1879: printed circular issued by Airy. Airy had issued similar circulars, advocating a more practical content in the Smith's Prize requirement, *e.g.* 5 December 1857, 25 February 1875. I am indebted to the Airy family for the loan of these documents. They have now been generously donated to Cambridge University Library, and are in process (1997) of classification.

5   'The Endowment of Research': letter, G.B. Airy to the *English Mechanic* no. 831 (25 February 1881) 587.

6   'Espin', obituary, *J.B.A.A.* **45** (ref.2) 128, for 'public house... mental stimulant'. Both of his obituarists (ref.2) emphasise Espin's generosity to his parishioners. Tow Law was a small coal-mining town, surrounded by hills. Espin's 24-inch Calver was subsequently restored by David Sinden of the Newcastle Astronomical Society, and is now in the University of Newcastle Observatory at Close House.

7   W.H. Steavenson, obituary, 'Theodore Evelyn Reece Phillips', *M.N.R.A.S.* **103**, 2 (1943) 70–2. Also B.M.P., obituary, 'T.E.R. Phillips', *J.B.A.A.* **52**, 6 (July 1942) 203–8: 204 for instruments in Phillips' Headley observatory.

8   'Espin', obituary, *M.N.R.A.S.* **95** (ref.2) 320. Espin received the prestigious Jackson–Gwilt Medal of the R.A.S. in 1903 for his work on stellar spectra.

9   *The Stargazer of Tow Law*, joint composition of the Tow Law History Society (Tow Law History Group, Tow Law, Durham, 1992) states that Espin left £12,399 at his death in 1934. His ecclesiastical benefice in 1910 was worth £304 gross and £267 net: *Crockford's Clerical Directory* (London, 1910), by name. Steavenson, obituary, 'T.E.R. Phillips', *M.N.R.A.S.* **103**, 2 (ref.7) 72 for D.Sc.

10 'Espin', obituary, *M.N.R.A.S.* **95** (ref.2) 322.

11 Waterfield (1900–1986) was a haematologist who was renowned as a Mars and cometary observer: M.J. Hendrie, obituary, 'Reginald Lawson Waterfield', *Q.J.R.A.S.* **28**, 4 (December 1987) 544–6. Will Hay (1888–1949) was an eminent Saturn and cometary observer, but was better known to the public as a comic actor and film star (*Oh, Mr Porter* etc.): R.L. Waterfield, obituary, 'William Thompson Hay', *M.N.R.A.S.* **110**, 2 (1950) 130–1. I am indebted to Sir William McCrae, F.R.S., for information regarding the changing composition of the R.A.S., imparted to me in conversation and private correspondence.

12 Gerard Gilligan *et al.*, *The History of the Liverpool Astronomical Society* (Liverpool A.S., 1996) 4. In the summer of 1971 I had the pleasure of meeting and being hosted for an afternoon by the then elderly Alan Sanderson, a retired Liverpool bank manager. In retrospect, I can see how exactly he fell into the

tradition of Grand and Enthusiastic Amateur Astronomers discussed in this book. His large Victorian house, at Hunt's Cross, Liverpool, was a veritable Aladdin's Cave of collections: rooms full of bookcases, brass microscopes, Gregorian reflecting telescopes, instruments, antiquities, and a spectacular Bassett-Lowke model railway. He was an accomplished microscopist who said that he could, from its micro-organisms, identify a vial of water taken from any lake or pond in the vicinity of Liverpool. He also possessed the only authentic Herschel 7-foot reflector which has in recent times, to my knowledge, been in private hands. Alan Sanderson told me that he spotted the lower mahogany frame of the instrument many years before; it was then being used as an umbrella stand in the hall of a house of a gentleman where he was a dinner guest. On being told that the umbrella stand had some other 'bits' – which constituted a virtually complete Herschel telescope – he offered £5 for them, and thereby acquired the instrument.

13 Several pictures of Scriven Bolton, including one with his 18¼-inch reflector, are preserved in the 'Album' of the Leeds A.S., and are still in the possession of the Society: 1907 and 1929, date order.

14 A.D. Thackeray, obituary, 'Mary Acworth Evershed', *M.N.R.A.S.* **110**, 2 (1950) 128–9.

15 Stephen H.G. James, 'Arthur Philip Norton (1876–1955): The man and his star atlas', *J.B.A.A.* **103**, 6 (1993) 289–93. I am also indebted to the late Stephen James for his correspondence regarding Norton.

16 A reprint copy of this classic amateur's chart was purchased new by the present author when a schoolboy: *Map of the Moon, Originally drawn by T. Gwyn Elger, F.R.A.S. Notes and revision by Dr H.P. Wilkins, F.R.A.S.* (George Philip & Son, London, 1959). Single folded sheet and notes.

17 Patrick Moore's *Guide to the Moon* (1953; *Survey* 1963 and 1976), *Guide to the Planets* (1955, 1971, 1976), and *The Planet Venus* (1956, 1959, 1961) have been consistently popular, while his *The Amateur Astronomer* had passed through 11 editions by 1990.

18 In addition to 'AstroFest' which draws in thousands of people over two days in February each year, there are many regional amateur astronomy conventions in Britain, generally held on an annual or biennial basis. The Leeds Astronomical Society 'AstroMeet' is held annually in autumn, while the 'Cosmos' meeting of the Cleveland and Darlington Astronomical Society, and the Manchester, Liverpool, West Yorkshire, Cardiff, Portsmouth, Kent and other conventions are generally biennial. Many societies founded post-1950 celebrated their 20th, 25th, 40th etc. anniversaries with jamboree meetings which not infrequently developed into regular conventions. Scotland's great amateur astronomical event is the convention held in late summer in Edinburgh, Dundee or elsewhere, whilst Ireland celebrates its 'Whirlpool' Star Party each year at Birr. Several long-standing weekend courses (which often have to put a ceiling on the number of participants because of insufficient space) also exist. The September course at Horncastle College, Lincolnshire, has been running for well over 20 years, and that at Alston Hall, Preston, Lancashire, for over 30. All of these events are either entirely or substantially amateur-organised, and are usually dependent upon

voluntary speakers. Few conventions attract less than 100 people, and most well over that number. They play an enormous role in popularising astronomy in Britain. A similar situation exists in many European countries, Japan, Australasia, and the U.S.A.

19 It was through the Salford Astronomical Society that I was first introduced into the world of 'organised' amateur astronomy, and I am indebted to the late Arthur Taylor, Hector Wheddon, and Alan Whittaker for their enthusiasm and encouragement. I now have the honour of being President of the Society. Following a lecture to the Society by its then President, the late Zdenek Kopal, Professor of Astronomy at Manchester University, in 1969–70, an 18-inch equatorial reflector and its observatory and dome, which had become surplus to requirements at the University's Jodrell Bank site, were given to the Salford Astronomical Society. The instrument is still extensively used by Ken Irving and other members of the Society for CCD deep-sky research, and for extensive school and public education projects. I have also had a continuous relationship with Crayford Manor House Astronomical Society, to which I have lectured annually since 1975.

20 'Butcher reaches for the stars', *Lancaster Guardian*, 21 July 1995. I am also indebted to Denis Buczynski for information passed on via correspondence.

21 The late Jack Ells, who was Britain's most distinguished amateur photometrist, demonstrated this remarkable telescope with its 12-inch mirror to me on several occasions, and on 9 February 1989 I obtained one of the best views of Jupiter and Europa that I have ever witnessed with any instrument. Because the only direct contact which the observer had with the telescope was the light coming through the hollow trunnion into the small wooden room, it was possible for the room to be electrically heated without any risk of thermal distortion to the telescope standing outside. The room contained a desk, a computer, a comfortable chair, and a stereo music system, to make the act of observing as physically pleasant as possible, even on the coldest of nights. The instrument is still in the possession of Mr Ells' family. He also built a fully-computerised 8½-inch aperture reflector – the Automatic Photoelectric Telescope – that could conduct its own photoelectric surveys of preprogrammed celestial objects without any human presence. This telescope is now used for research by Crayford Manor House Astronomical Society, and its results and analyses are reported annually to the Variable Star Section of the B.A.A. I am indebted to Roger Pickard, of the Crayford Society, for bringing me up to date on the Ells Telescopes: 30 May, 1998. See also Roger Pickard, 'The Ells Telescope', *The Observatory* **110** (December 1990) 197; and 'The Ells APT [Automatic Photoelectric Telescope]', *The Observatory* **112** (October 1992) 235.

22 Dame Kathleen Ollerenshaw, 'Starting with a CCD Imaging Camera', *1996 Yearbook of Astronomy, ed.* Patrick Moore (Macmillan, London, 1995) 169–83.

23 In the wake of NASA's Hubble Space Telescope, several British amateurs are now seriously discussing the possibility of putting an amateur-built and controlled telescope into orbit around the Earth, appropriately named the 'Humble Space Telescope'. A telescope with a 10-inch mirror, CCDs, and on-board computers, launched by a commercial telecommunications rocket, could be built for under

£1,000,000. I am endebted for this insight to Dr Michael Martin-Smith, a physician of Hull, Yorkshire, whom I met on 16 January 1998 and with whom I exchanged subsequent correspondence.

14 NOW LADIES AS WELL AS GENTLEMEN REFERENCES 94–101

94. Miss Tranmar, 'Has Science Killed Romance?' 25 March, 1907, *Leeds Astronomical Society Journal and Transactions* **15** (1907) 17–25.
95. Miss Jagger, 'Evolution of Planetary Systems', *Leeds Astronomical Society Journal and Transactions* **29–30** (1921-2) 14–17. A. Chapman, 'The Lady Astronomers of Victorian Britain', R.A.S. *Astronomy and Geophysics* **57**, 4 (August 2016) 12–13.
96. *Leeds Astronomical Society Journal and Transactions* **29–30** (1921-2) 42–3 for current list of members.
97. Mark Hurn, 'Anne Sheepshanks: Patron, Benefactor, Sister', R.A.S. *Astronomy and Geophysics* **57**, 3 (June 2016) 11. Over 2016, to mark the admission of women to the full Fellowship of the R.A.S., *Astronomy and Geophysics* ran a series of articles on astronomical women before and after 1916. J. L. E. Dreyer and H. H. Turner, *History of the Royal Astronomical Society 1820–1920* (R.A.S. 1923, reprinted 1987), p. 92.
98. Roger Hutchins, 'Elisabeth Isis Pogson [Kent]', *Oxford Dictionary of National Biography* (O.U.P., 2004). Dreyer and Turner, *History of the R.A.S.* (ref. 97), 233–4.
99. Sara Russell, 'Agnes Mary Clerke: Stars, Systems, and Problems', R.A.S. *Astronomy and Geophysics* **57**, 3 (June 2016) 16-17. Mary Brück, *Agnes Mary Clerke and the Rise of Astrophysics* (C.U.P., 2002). Barbara Becker, 'Margaret Huggins and the Tulse Hill Observatory', R.A.S. *Astronomy and Geophysics* **57**, 1 (April 2016) 13–14. Sue Bowler, 'Annie Jump Cannon, Stellar Astronomer', R.A.S. *Astronomy and Geophysics* **57**, 3 (June 2016) 14–15.
100. Sian Prosser, 'Making a Career from Outreach' [Mary Proctor], R.A.S. *Astronomy and Geophysics* **57**, 5 (October 2016) 19-20. Sylvia Dalla and Lyndsay Fletcher, 'A Pioneer of Solar Astronomy' [Annie Maunder], R.A.S. *Astronomy and Geophysics* **57**, 5 (October 2016) 21–3.
101. 'Dorothea Klumpke Roberts', obituary by J. H. Reynolds, *M.N.R.A.S.* **104**, 2 (1944) 92–3. Jeremy Shears, 'Selenography and Variable Stars' [Mary Adela Blagg], R.A.S. *Astronomy and Geophysics* **57**, 5 (October 2016) 17–18.

ERRATA to the first edition:

p. 24 Last line. For 1785 read 1783.
p. 25 Line 1. For Astronomer Royal for Ireland read Astronomer Royal of Ireland.
p. 28 Line 22 down. For Nicholas II read Nicholas I.
p. 109 Gt Melbourne telescope. For 1874 read 1869.
p. 124 Line 13. For French Government read French Académie.
p. 139 'The Irish Photometrists', para. 2. George Minchin is more accurately described as Irish rather than Anglo-Irish.
p. 139 Final paragraph, line 3. For Iran read Oran.
p. 140 Lines 15–16. For Fitzgerald read Fitz Gerald.
p. 140 Fitz Gerald and the Daramona House Group in 1890s Ireland. I am indebted to Dr Ian Elliot for the two following corrections. (1) The April 1895 research was to establish the photoelectric brightness of the stars, not their temperatures. (2) Dr Elliot further says 'The work on the temperature of the solar photosphere was done with a coelostat on loan from the Royal Society and in collaboration with P. L. Gray of Birmingham University.' Private communication, 26 January 2007.
p. 146 Line 3: for Duke of Marlborough read Earl of Macclesfield.
Line 4: for Blenheim Palace read Shirburn Castle.
p. 249 Lines 5–7, Dom Pedro II of Brazil. Dom Pedro II (1825–1891), Emperor of Brazil, was renowned for his love of learning. He was not in fact assassinated, but was forced to abdicate following the Brazilian Revolution of 1889. He died in Paris in 1891.
p. 275 Line 7. For 1797 read 1798.
p. 313 Ref. 84, last line. For Nasmyth's Penshurst Observatory read Nasmyth's Penshurst Observing Terrace.
p. 320 Ref. 81. For *Astronomical Register* **2** read **3**.
p. 333 Ch. 5, Ref. 3. 'I.F.' In 2001, the late Peter Hingley, Archivist and Librarian of the R.A.S., suggested that these initials could also refer to the astronomer Isaac Fletcher (1827-1879).
p. 348 Ref. 79, line 2. For 1852 read 1851.
p. 353 Ref. 22. Roger Hutchins, 'A Mismatch of Ideals and Resources', cited as an Oxford D.Phil. thesis, 1997, is now published as: Roger Hutchins, *British University Observatories 1772–1939* (Ashgate Press, Aldershot, Hampshire, 2007).
p. 359 Ref. 78. 'Dorothea Klumpke Roberts' obit., M.N.R.A.S. **104**, 2: for (1904) read (1944).
p. 361 Ref 3. For Blenheim Observatory read the Shirburn Castle Observatory. Although Phelps and Bartlett worked for the Earl of Macclesfield at the Shirburn Castle Observatory, and not for the Duke of Marlborough at Blenheim Palace, George Spencer, Fourth Duke of Marlborough, did maintain a private observatory at Blenheim, which was sometimes visited by the celebrated instrument-maker Jesse Ramsden.
p. 373 Ref. 16, Comet of 1811. Hind, p. 110, gives the date of the first sighting as 26 March, but I was told by a correspondent that it was in fact 28 March.

Index

1869 Code of Regulations, 151, 364
1876 Loan Exhibition, 124
34 Tauri, 397
61 Cygni, 42

α Centauri, 42
Abbott, C.J.R., 350
Abbott, Minnie (*née* Russell(?)), 350
ABC Guide to Astronomy (Hawkins), 287
absorption spectrum, 114
achromatic lens, 237
Acland, Sir Thomas, 216
actinoscope, 332
Adams, John Couch (1819–1892), 17, 25, 74, 147, 152, 157, 270, 296, 319, 333, 364
Adare(?), Lord, 99
Aedes Hartwellianae (Smyth), 80
Ainslie, Capt Maurice Anderson (1869–1951), 222
Airdrie Observatory, 244, 388
Airy, Enid, 307, 308, 309, 337, 338, 344, 348, 367
Airy, Sir George Biddell (1801–1892), 3, 4, 5, 10, 15, 16, 17, 18, 20, 21, 23, 28, 29, 31, 34, 37, 44, 45, 48, 57, 62, 74, 86, 87, 90, 99, 106, 116, 146, 147, 148, 149, 150, 153, 154, 156, 165, 168, 177, 189, 190, 199, 201, 207, 218, 231, 238, 240, 245, 296, 297, 305, 306, 307, 308, 309, 310, 311, 312, 315, 317, 319, 320, 333, 334, 338, 341, 344, 346, 348, 353, 360, 361, 362, 363, 364, 365, 366, 367, 368, 370, 374, 375, 376, 378, 386, 388, 404
Airy, Hubert, 338

Airy, Lady Richarda (1804–1875), 14, 15, 16, 24, 83, 87, 99, 308, 309, 337, 338, 344, 348, 368, 370
Airy, Wilfred, 15, 307, 308, 353, 367
Airy, William, 309
Albert, Prince, 236
Alcock, George E.D., 299
Almanacs (Hawkins), 287
Alonzo, Rosa, 310
Alston Hall, Preston, 405
Alter, Peter, 306
Altham, Henry, 167
Altona Observatory, 186
amateur, 7, 295, 302
amateur astronomical society, 218, 222, 241, 243–256, 265–271, 295, 300, 405
Amphitrite, 153
Anderson, R.G.W, 384
Andrews, Frank P., 357
Andromeda Nebula, 69, 134, 138, 198, 257, 358
Anning, Mary, 273, 396
Antoniadi, Eugène Michel (1870–1944), 253, 270, 395
Appleton, Henry, 40
Arbour, Ronald W., 302
Archer, Frederick Scott (1813–1857), 117
Architecture of the Heavens (Nichol), 177
Arcturus, 140
Argelander, Friedrich Wilhelm Augustus (1799–1875), 42
Ariel, 103, 347
Armagh Observatory, 25, 26, 35, 101, 105, 148, 309, 312, 313, 315, 348, 362

Armstrong, Mark W., 302
Arnold, John, 232
Ashbrook, Joseph, 314, 318, 320, 336, 392
Ashley, Mary (1843–1903), 284, 287, 401
asteroids, 48, 80, 152, 153
Astore, Capt William J., 369, 371, 387
Astraea, 19, 49, 320
AstroFest, 300, 405
astrometer, 71, 331
Astronomical and Physical Society of Toronto, 255
Astronomical Mania, 167
Astronomical Society of London (*later* R.A.S.), 29, 95, 317
Astronomical Society of the Pacific, 254
Astronomical Society of Wales, 213, 255, 256, 265, 270, 279, 290, 325, 326, 380, 392, 393, 398, 404
Astronomy Now, 300
Aubert, Alexander (1730–1805), 38
Auckland, Lord, 14, 308
Austen, Jane (1775–1817), 84
Aylesbury area, 168–170

Babbage, Charles (1792–1871), 6, 45, 54, 57, 276, 277, 306, 323, 329, 336, 398
Backhouse, Thomas William (1842–1920), 27, 154, 155, 313, 365
Bacon, Fred, 283
Bacon, Miss Gertrude, 253, 269, 283, 288, 291–292, 293, 391, 403
Bacon, Mrs Gertrude (*née* Myers), 403
Bacon, Revd John Mackenzie (1846–1904), 253, 283, 290, 298, 391, 403
Baden Powell, Professor, 87
Baden-Powell, Lord, 333
Bader, Paul, 383
Bailey, J.E., 396
Baily, Francis (1774–1844), 23, 30, 44, 57, 308, 315, 318
Baldwin, Mary, 53, 55, 71, 331
Baldwin, Sophia, 53, 55
Baldwin, Thomas, 331
Ball, Sir Robert Stawell (1840–1913), 25, 150, 153, 155, 156, 221, 222, 241, 249, 254, 265, 270, 312, 356, 357, 363, 387, 395
Ball, W. Valentine, 312, 363
Bamford, Samuel, 182, 372
Bangor Museum, 379, 380

Bangor University, 378
Barber, Dr, 168
Barbour, Miss C.A., 279, 291, 403
Barbour, William Donald (1833–1903), 247, 254, 255, 291, 389, 392, 399, 403
Barclay, Sheriff, 206, 208
Barden, Dr Rowland, 399
Barnum, Phineas T., 379
Barry Astronomical Society, 266
Bartlett, John (*b*.1721), 146, 361
Bath Philosophical Society, 243
Batten, Alan H., 310
Baum, Richard, 310, 314, 341, 349, 390
Beanlands, Arthur, 311
Beaufoy Circle, 78
Becker, Dr Barbara J., 353, 401
Bedford Astronomical Society, 337
Bedford Observatory, 76, 77
Bedford Record Office, 337
Bedford School, 337
Bedford–Aylesbury axis, 11
Beech, Martin, 387
Beer, Wilhelm (1797–1850), 19, 27, 81, 310
Bell, Gertrude, 273
Bell, Thomas, 110, 349, 350, 351
Bellamy, Frank Arthur, 270, 395
Belville, John Henry, 363, 367
Bennett, Arnold, 370
Bennett, James A., 25, 312, 313, 315, 340, 362, 384
Beresford, Lord John, 25, 312
Berlin Academy, 6
Berlin Observatory, 50
Berthon collapsible boat, 228, 383
Berthon dynamometer, 228, 383
Berthon mount, 229, 232, 262, 382
Berthon, Revd Edward Lyon (1813–1899), 227, 228, 229, 262, 263, 383
Bessel, Friedrich Wilhelm (1784–1846), 3, 18, 20, 42, 305, 310
Bessemer, Sir Henry (1813–1898), 142, 360
binary stars, 42, 58
Bird, Dr, 172
Bird, Isabella, 396
Bird, John, 14, 24
Birkbeck, Dr, 338
Birr Castle, 65, 66

Birt, William Radcliff (1804–1881), 82, 166, 336, 368
Bishop, George (1785–1861), 46, 47, 151, 152, 153, 177, 278, 320
Bishop, George, 400
Blacklock, Dr, 215, 218
Blagg, Mary Adela (1858–1944), 284, 290, 401, 402
Blenheim Observatory, 361
Blenheim Palace, 95, 146
Blomfield, Edward David, 314
Bloxam, Revd Dr Richard Rouse, 286, 401
Boas Hall, Marie, 306
Bodleian Library, 393
Boeddicker, Otto, 150
Boles, Tom, 302
Bolton, Scriven (d.1929), 233, 255, 256, 298, 385, 392, 405
Bond, George Phillips (1825–1865), 138, 257, 314, 354, 358
Bond, William Cranch (1789–1859), 28, 50, 103, 104, 118, 138, 314, 347, 354
Bonwick, James, 370
books, 176–178, 213, 371
botany, 8, 9, 181
Bowden, Alan, 351, 360, 387
Bowler, J. Peter, 307
Bozo (astronomically-minded tramp), 179
Bradbury, W., 386
Bradley, James (1693–1762), 3, 14, 35, 308, 315
Brahe, Tycho (1546–1601), 10, 35
Brashear, John A., 141
Bredichin, Prof Theodor Alexandrovich (1831–1904), 281, 390, 399
Breen, Hugh, 148, 362
Breen, Hugh, jr., 148, 151
Breen, James, 148, 362
Breen, John William, 362
Bremiker, Karl, 49
Brewer, Dr Ebenezer Cobham, 176
Brewster, Sir David (1781–1868), 7, 68, 76, 87, 95, 102, 185, 186, 213, 244, 306, 313, 330, 334, 342, 343, 373, 374
Bridgewater Treatises, 169
Brinkley, John (1763–1835), 25, 30
Brisbane, Sir Thomas, 23
Bristol and Exeter Railway, 214

British Association for the Advancement of Science, 3, 7, 25, 30, 33, 57, 81, 82, 99, 105, 165, 166, 235, 243, 244, 296, 315, 329, 336, 342, 387, 388
British Astronomical Association, 233, 236, 247, 252, 253, 262, 270, 280, 282–290, 298, 300, 336, 382, 391, 394, 400, 401, 402, 404, 406
   Midlands Branch, 268, 270, 395
   North West Branch, 289, 391, 394, 403, 266–268
British Museum, 124
Brock, Laurie G., 141, 360
Brodetsky, Prof S., 392
Brodie, Frederick, 139
Brooke, C.F., 352
Brougham, Henry, Lord, 162, 276, 397
Brown, Elizabeth (1830–1899), 249, 251, 252, 253, 256, 261, 273, 278, 280–282, 289, 300, 390, 391, 398, 399, 400
Brown, Jemima, 281
Brown, Thomas Crowther, 280
Browne, John S., 311
Browning, John (1835–1925), 230, 232, 382, 384
Brück, Dr Mary T., 277, 311, 312, 324, 330, 335, 337, 350, 353, 360, 364, 366, 397, 398, 399, 401
Brück, Hermann A., 133, 311, 312, 335, 337, 350, 356, 360, 364, 387
Brühl, Count von, 36
Brünnow, Franz Friedrich Ernst, 152, 364
Bryden, David, 341
Buckland, William (1784–1856), 8, 306, 307
Buczynski, Denis G., 233, 262, 301, 406
Budding, Edwin, 357
Bunsen, Robert Wilhelm (1811–1899), 113, 115, 119, 171
Burgess, Revd R.B., 170
Burgoyne, J.T., 344
Burnerd, F., 218, 381
Burnett, J.E., 359
Burnett, John, 370, 377
Burnham, Robert, jr., 358
Burnham, Sherburne Wesley (1838–1921), 50, 81, 320, 390
Burns, Robert (1759–1796), 185

Index

Bush, Martha, 201
Bush, Thomas William (1839–1928), 217, 218, 246, 260, 375, 376, 377, 200–203
Buss, Mr, 394
Butler, C.J., 320, 359
Butler, Harriet, 330
Buttmann, Günther, 322, 333
Buxton, Harry Wilmot, 333, 334

Cabinet Cyclopaedia (Lardner), 177
Cahill, Revd Dr, 170
Calendars (Hawkins), 287
Calver, George (1834–1927), 136, 211, 223, 230, 231, 232, 233, 249, 252, 266, 298, 384, 385
Cambrian Natural Observer, 256, 270, 280, 291, 398
Cambridge University, 4, 14, 135, 278, 297
Cambridge University Library, 307, 337, 344, 348, 404
Cambridge University Observatory, 17, 25, 70, 148, 149, 150, 157, 270, 309, 362, 363
Camp, T. Allan, 324
Campden Hill Observatory, 44, 77, 398
Camps, Anne, 87
Canino, Prince of, 54
Cannon, Annie Jump (1863–1941), 122, 292, 355
Cardiff Astronomical Society, 265
Cardiff Central Library, 392
Cardiff City Observatory, 325, 393
Cardwell, Donald S.L., 306
Carlisle, Earl of, 245
Carpenter, James, 116, 353
Carrington, Richard Christopher (1826–1875), 40, 90, 148, 149, 278, 281, 311, 317, 318, 362
Carte du Ciel, 19, 139
Carter Observatory, 357
Cary, William, 78, 87
Cat's Eye nebula, 115
catadioptric telescopes, 234
Catechism of Astronomy (Pinnock), 214
Catholic Emancipation Act of 1829, 21, 25
Cauchoix, Robert Aglaé, 25, 49, 320, 334
Cavalier, J.M., 340
CCD, 234, 301, 302, 406
Celestial Objects for Common Telescopes (Webb), 50, 225, 226, 241, 251, 254, 299

celestial police, 18
Celestial Scenery (Dick), 176
Cellini, Benvenuto, 205
Ceres, 79
Chadwick, Owen, 311
Challis, James (1803–1862), 17, 148, 320, 362
Chalmers, Dr Thomas, 169, 265, 369
Chambers, Dick, 316
Chance Brothers, 237
Chantrey, Sir Francis Leggatt, 277
Chapman, Clarice, 309
Chapman, Dr Allan, 258, 259, 262, 307, 308, 309, 310, 312, 315, 318, 320, 324, 330, 345, 346, 347, 349, 365, 366, 371, 379, 384, 387, 391, 394, 396, 397
Cheshire Record Office, 374
Chevallier, Revd Temple (1794–1843), 21, 25, 40, 148, 149, 150, 153, 154, 155, 165, 199, 245, 298, 311, 317, 362, 367, 375
Children, Robert, 172, 173, 174, 176, 178, 217
Christian belief, 169
Christian Philosopher (Dick), 176, 210
Christie, Sir William Henry Mahoney (1845–1922), 156, 207, 354, 366
Church Commissioners, 222
Cincinatti Observatory, 28, 244, 372
cinematography, 140
circle, 35
Clark refractors, 28, 233
Clark, Alvan (1804–1887), 47, 50, 115, 121, 139, 227, 234, 236, 278, 314, 320, 352, 359, 382, 385, 398
Clark, J.W., 323
Clayton, William, 245
Clerke, Agnes Mary (1842–1907), 157, 249, 252, 270, 275, 285, 286, 287, 289, 292, 309, 310, 318, 319, 320, 322, 331, 337, 344, 352, 353, 358, 366, 378, 396, 397, 399, 401
Clerke, Ellen Mary (1840–1906), 276, 285, 287, 397, 399, 401
Cleveland and Darlington Astronomical Society, 405
Coates, Thomas, 368, 371
Coggia's Comet, 217, 296
Coghlan, William, 96, 343
Colby, Thomas Frederick (1784–1852), 30
Cole, J.F., 87

Cole, Revd Alfred Adolphus (1821–1893), 269, 395
Colebrooke, Henry Thomas (1765–1837), 30, 315
collapsible boat, 228
Collegio Romano Observatory, 116
Collins, Timothy, 401
Colthorpe, Brian, 323
Columbia University, 290
Colvin, Christina, 330
comet Arend–Roland, 299
comet Hale–Bopp, 300
comets, 133, 152, 274
Common, Andrew Ainslie (1841–1903), 132, 135–139, 142, 301, 357, 385
Comte, Auguste, 113, 352
Conder Brow Observatory, 233, 301
Connexion of the Physical Sciences (Somerville), 277
Conter, Joseph, 90, 340
Cook, Alice Grace (*d.*1958), 283, 292, 401
Cooke refractors, 108
Cooke, Mrs Hannah, 235, 385, 386
Cooke, Thomas (1807–1868), 47, 90, 115, 137, 154, 184, 206, 207, 223, 237, 340, 350, 386, 234–237
Cooke, Troughton and Simms, 329
Cooper Key, Revd Henry (*d.*1880), 228, 230, 231, 232, 233, 237, 384
Cooper, Astley, 44
Cooper, Edward Henry, 153
Cooper, Edward Joshua (1798–1863), 25, 48, 77, 147, 149, 153, 236, 320, 320
Cooper, Samuel, 380
Copeland, Dr Ralph (1837–1905), 134, 137, 150, 153, 207, 208, 218, 221, 281, 356, 365, 399
Copeland, Fanny, 153, 364
Core, Thomas H., 268, 270, 396
Cork Academy, 26
Corless, R., 338
Cornu, M., 355
Cornwall, Revd Peter, 214
Corrie, George Elwes, 311
Cortie, Fr Aloysius Laurence (1859–1925), 253, 256, 269, 355, 356
Crabtree, William (1610–1644), 395, 396
Craig, Revd Edward, 27, 130, 174, 313, 314, 370

Crawford and Balcarres, Earl of, *see* Lindsay, James Ludovic, Lord
Crawford, William Horatio, 26, 278
Crayford Manor House Astronomical Society, 301, 316, 406
Creese, Mary, 281, 398, 399, 400
crepe ring, 47, 103
Criswick, George, 149
Crommelin, Andrew Claude de la Cherois, 283
Crommelin, Loetitia (1866–1921), 283, 400
Crossley, Edward, 131, 132, 137, 142, 320, 336, 357
Crowe, Prof Michael J., 322, 324, 389
Crowe, Sister Dolores, 312
Cubitt, William, 75
Cummergen, James L., 232, 384
Cycle of Celestial Objects (Smyth), 77, 79, 80, 85, 90, 225, 226

Dalton, John (1766–1844), 243, 275
Darby, W.A., 48
Darlington Observatory, 388
Dartmouth College Observatory, 51
Darwin, Charles Robert (1809–1882), 9, 56, 307, 324
Daubeny, Charles, 96, 99, 342
Davidson, Revd Martin (1880–1968), 283
Davies, Dr Tegfan, 265
Davies, William Henry, 248, 250, 251, 254, 390, 392, 395, 403
Davis, William, 313, 398
Davy, Sir Humphry (1778–1829), 10, 297
Dawes, Anne (*née* Welsby), 46, 87
Dawes, Very Revd Richard (1793–1867), 231
Dawes, William Rutter (1799–1868), 23, 46, 48, 50, 74, 87, 100, 103, 114, 130, 139, 217, 227, 232, 235, 319, 320, 347, 359, 384
De La Rue, Warren (1815–1889), 17, 82, 84, 85, 86, 94, 102, 107, 110, 117–118, 191, 192, 201, 337, 341, 346, 349, 353, 354
De Morgan, Augustus (1806–1871), 238, 315, 329
Dearborn Observatory, 51, 81
Deeming, T.J., 330
Dell, Thomas, 76, 82, 89, 166, 171, 214, 229, 334, 339, 368, 370

Index

Demainbray, Dr Stephen Charles Triboudet (1710–1782), 20
Dennett, Frank C., 390
Denning, William Frederick (1848–1931), 233, 249, 256
Dewhirst, Dr David W., 320, 344, 355, 362
Dewhirst, Professor, 167
Dick, Dr Thomas (1774–1857), 169, 176, 182, 206, 210, 212, 244, 265, 369, 373, 387
Dickens, Charles (1812–1870), 314, 363, 367, 370
Disraeli, Benjamin (1804–1881), 221, 375
dissociation hypothesis, 124
Dixon, Stephen M. (1866–1940), 139, 359
Doberck, W., 320
Dodgson, A., 395
Dodwell, G.F., 363
Dollond refractors, 100, 234
Dollond, George, 43, 77
Dolman, Mondeford R., 311
Dom Pedro II, Emperor of Brazil, 249, 250, 390
Dominguez, Manuel Berrocosa, 310
Donati's Comet, 90, 170
Donati, Giovanni Battista (1826–1873), 116
Donkin, Bryan (1768–1855), 17
Doppler, Johann Christian (1803–1853), 115
Dorpat Observatory, 19
Dowdell, Alan P., 382
Draper, Anna Mary (née Palmer), 120, 122, 139, 355
Draper, Henry (1837–1882), 28, 50, 116, 118, 119–121, 135, 136, 137, 139, 142, 320, 354, 355, 360
Draper, John William (1811–1882), 50, 113, 118, 119, 352, 354
Dreyer, Jean Louis Emil (1852–1926), 26, 27, 150, 309, 310, 311, 312, 313, 314, 315, 317, 318, 319, 329, 353, 356, 357, 364, 365, 397
Dugden, J., 333
Duke, Philip F., 267, 268, 394
Dun Echt Observatory, 24, 133–135, 151, 152, 207, 208, 281, 365
Dunkin, Edwin (1821–1898), 147, 157, 207, 240, 273, 317, 322, 323, 324, 361, 367, 387
Dunlop, James, 23

Dunsink Observatory, 24, 35, 149, 309, 312
Durham University, 154, 366
Durham University Library, 311, 365
Durham University Observatory, 20, 40, 148, 149, 150, 152, 155, 165, 309, 310, 311, 362
Dyer, G.P., 332
dynamometer, 228, 383
Dyson, Sir Frank Watson (1868–1939), 357, 360

η Boötis, 140
East India Company, 23
eclipse expeditions, 253, 281–283
École Polytechnique, 6
Eddington, Sir Arthur Stanley (1882–1944), 355, 356, 364
Edgeworth, Maria (1767–1849), 68, 137, 276, 330
Edgeworth, Richard Lovell (1744–1817), 176, 371
Edinburgh Astronomical Institution, 244
Edinburgh Public Library, 306, 350
Edinburgh University, 83, 176
Edmonds, Albert J., 154
Edmondson, Neil McNeil (d.1864), 149, 362
Edmunds, A.J., 365
Edwards, Revd John, 95, 100, 184, 342, 345
Edwards, Revd William, 266
Elger, Thomas Gwyn Empey (1838–1897), 166, 249, 250, 252, 253, 256, 287, 299, 368, 405
Elliott, Dr Ian, 313, 320, 359, 398
Ellis, Charlotte, 44
Ellis, Grace, 209, 378, 379
Ellis, James, 373
Ellis, Joseph, 44
Ellis, William (1828–1916), 22, 148, 157, 311, 362
Ells Automatic Photoelectric Telescope, 406
Ells, Jack (1932–1990), 301, 406
Emery, Ray, 388, 389, 394
emission spectrum, 114
Encke, Johann Franz (1791–1865), 20, 58, 320
English mount, 43
Ensor, Sir Robert K., 377
Epps, James (1773–1839), 88, 145, 360, 361

Erck, Wentworth, 47, 139, 359
Ernst II of Saxe–Coburg, Duke, 18
Ertel Circle, 49
Ertel, Traugott Lebrecht (1778–1858), 19, 24, 49
Espin, Revd Thomas Henry Espinall Compton (1858–1934), 50, 67, 226, 228, 247, 252, 269, 296–298, 300, 320, 329, 381, 382, 389, 395, 403, 404
Essays and Reviews, 169
Essays in Astronomy (Proctor), 240
Evans, B.H., 330
Evans, David S., 311, 330, 341
Evans, Franklen G., 266
Evans, Dr H.N., 322
Evans, Revd John Silas, 255, 265, 370, 378, 380, 393
Evans, Lewis, 36, 265
Everett, Alice, 156, 280, 282, 285, 289, 366, 399, 401
Everitt, Revd Mr Alfred, 170
Evershed, John (1864–1956), 282
Evershed, Mary Ackworth (*née* Orr) (1867–1949), 282, 298, 391, 400, 401, 402
Exe Valley Magazine, 217

Fairbairns of Manchester, 99, 344, 345
Fairfax, Sir William (1713–1813), 276
Fallows, Fearon (1789–1831), 23, 83
Faraday, Michael (1791–1867), 10, 275, 276, 297, 307
Farrel, William, 237
Fauvel, J., 308, 310
Federation of Astronomical Societies, 252, 300
Feist, Michael, 340
Ferguson, Robert, 176
Field, Mary Wilmer, (*later* Countess of Rosse), 96, 343
Finucane, Brendan, 313, 398
Fiott, John (Dr John Lee), 333, 335
Fitz refractor, 278
Fitzgerald, George Francis (1851–1901), 139–141
Fitzgerald–Lorenz Contraction, 140
Fitzwilliam Museum, Cambridge, 135
Flammarion, Camille, 265
Flamsteed, John (1646–1720), 13, 35, 51, 275, 308, 397

Flaugergues, M., 373
Fleckney, Patrick, 375, 376, 377
Fletcher, Isaac (1827–1879), 90, 236, 340
Fletcher, J., 340
Fletcher, John, 302
Flood, R., 310
Flora, 48
Forbes, George, 134, 356, 354
Forbes, James David, 4, 306, 348
Forester, Lord, 202
Foster, Joseph, 366
Foucault test, 232
Foucault, Jean Bernard Léon, 110
Franklin–Adams, John, 367
Franks, William Sadler (1851–1935), 202, 252, 257, 367, 375, 377
Fraunhofer, Joseph (1787–1826), 19, 43, 113, 318, 352
French Academy, 6
Fysher, Greevz, 255, 392

Galilei, Galileo (1564–1642), 227
Gascoigne, S.C.B., 341, 350, 353
Gascoigne, William (1612–1644), 396
Gaskell, Elizabeth, 174, 176, 370
Gauss, Karl Friedrich (1777–1855), 20
Gavine, Dr David M., 243, 306, 312, 350, 363, 369, 371, 378, 387
Gee, Thomas, 209
Geike, Archibald, 355
General Catalogue of Double Stars within 121° of the North Pole (Burnham), 50
gentlemen, 7
Geological Society, 33
geology, 8, 9
Geometrical Optics (Airy), 238
George III, King, 20, 30
George IV, King, 23, 244
German mount, 43, 84, 236
German research chemistry, 6
Giberne, Agnes, 252, 273, 286, 287, 401
Gill, James, 249
Gill, Lady, 157
Gill, Sir David (1843–1914), 134, 135, 137, 151, 153, 157, 221, 230, 285, 286, 356, 364
Gilligan, Gerard, 374, 390, 391, 392, 404
Gingerich, Owen, 322

Gladstone, William Ewart (1809–1898), 189, 199, 201, 217, 221, 312, 374, 375, 376
Glaisher, Cecilia Louisa, 86, 338
Glaisher, James (1809–1903), 84, 86, 90, 152, 155, 157, 338, 340, 361, 364
Glaisher, James Whitbread Lee (1848–1928), 86, 338
Glasgow University, 24, 166
glass, 110, 201, 320, 320, 334, 344, 345, 348, 350, 351
Gledhill, Joseph, 320, 357
Godlee Observatory, 268, 394
Godlee, Francis, 268
Goldfarb, S., 330
Goldney, Gabriel (1849–1905), 155, 156, 362, 366
Gonzalez, Francisco José Gonzalez, 310
Goodacre, Walter (1856–1938), 253, 336
Gordon, Elizabeth, 307
Gosse, Edmund, 386
Gott, Benjamin, 245
Göttingen Observatory, 19
Gowing, Margaret, 306
Graham Hagerty, Miss E. (1815–1908), 256, 265, 279, 393, 398
Graham, Andrew, 49, 50, 147, 149, 157, 320, 361, 362
Graham, George, 35
Graham, Dr Robert, 350
Grant, Robert (1814–1892), 24, 306, 318, 347
Gray, William, 235, 236
Great Comet of 1811, 187, 373
Great Comet of 1843, 171, 369
Great Comet of 1881 (Tebbutt), 136, 202, 240
Great Exhibition of 1851, 105, 124, 174, 348
Great Southern Telescope, 107, 349
Great Western Railway Magazine, 216
Great Western Railway, 214
Greaves, John, 13, 308
Green, Nathaniel Everett (1823–1899), 233, 385
Gregg, T., 245
Gregory, David (1661–1708), 94
Gregory, Olinthus, 29
Gresham College, 13
Greville, Charles, 36

Groombridge 1830, 42, 318
Groombridge Circle, 36, 44
Groombridge, Stephen (1755–1832), 5, 30, 35, 36, 42, 316
Grubb, Sir Howard (1844–1931), 26, 94, 109, 116, 121, 135, 137, 195, 223, 236, 249, 252, 268, 313, 320, 334, 341
Grubb, Thomas (1801–1878), 25, 49, 99, 101, 105, 236, 320, 334, 345, 350, 351
Guest, Ivor, 308, 309
Guide to Scientific Knowledge (Brewer), 176
Guide to the Knowledge of the Starry Heavens (Mann), 176
Guide to the Moon (Moore), 300
Guinand, Pierre Louis (c.1744–1824), 43, 317
Gunther, Robert T., 308, 310, 341
Gurman, S.J., 316, 317

Hackney Scientific Association, 368
Hadley, John (1682–1744), 341
Hale, George Ellery (1868–1938), 142
Hall, Asaph, 390
Hall, Jeff, 390
Halley's Comet, 187, 300
Halley's Comet (Hawkins), 287
Halley, Edmond (1656–1742), 11, 14, 23, 42
Hamburg Observatory, 149
Hamilton, Sir William, 331
Hanbury, Frederick J., 367, 377
Hanley, H.A., 334, 335
Harcourt, Revd William Venables Vernon (1789–1871), 3, 4, 7, 235, 305, 306, 315, 329
Harding, Karl Ludwig (1765–1834), 18
Hardy, Thomas (1840–1928), 10, 239, 240, 288, 386, 387
Harratt, S.R., 316, 317
Harries, Thomas, 265, 393
Harris, Richard, 346
Harrison, Thomas, 172
Hart–Davis, Adam, 383
Hart–Davis, Miss, 283
Hartnup, John (1806–1885), 22, 311
Hartnup, John, jr., 23, 311
Hartwell House, 11, 61, 75, 98, 100, 151, 166, 167, 333, 334, 335, 339
Harvard College Observatory, 28, 50, 122, 314, 355

Harvard University, 284
Harwick Museum, 188
Hastings, Dr, 386
Hawkins, Hester Periam (1846–1928), 287, 288, 289, 402
Hawkins, Revd Joshua, 287, 402
Hay, William Thompson (1889–1949), 404, 298
Hearnshaw, Dr John B., 353, 355, 360, 401
Heath, Louisa Catherine, 78
Heavens at a Glance (Mee), 265
Heidelberg University, 113
helium, 124, 134
Hencke, Karl, 19, 49
Henderson, Ebenezer, 244, 387
Henderson, Thomas (1798–1844), 23, 42, 177, 178
Hendrie, Michael J., 404
Hereford Cathedral, 225, 231
Herschel, Alexander, 208, 218, 238
Herschel, Caroline Lucretia (1750–1848), 53, 54, 64, 68, 255, 261, 274, 275, 276, 280, 289, 292, 301, 322, 330, 331, 392, 396, 397, 403
Herschel (*later* Lubbock), Constance Anne, 322
Herschel, Dietrich, 54
Herschel, Francisca, 403
Herschel, Isabella, 74
Herschel, Johann Alexander, 54
Herschel, Mrs John, 397
Herschel, Sir John Frederick William (1792–1871), 4, 7, 8, 10, 11, 17, 23, 29, 30, 33, 40, 44, 46, 47, 48, 53, 62, 63, 79, 81, 82, 83, 88, 93, 103, 104, 106, 119, 133, 165, 166, 177, 182, 186, 196, 197, 198, 200, 208, 210, 218, 234, 245, 246, 255, 264, 276, 286, 296, 305, 306, 307, 309, 310, 315, 319, 320, 322–324, 329–333, 338, 339, 340, 341, 343, 343, 344, 347, 364, 368, 371, 375, 388, 389, 398
Herschel, Lady Margaret (*née* Brodie Stewart) (1810–1884), 15, 55, 71, 83, 99, 309, 323, 331, 332, 333, 337, 338, 344, 348
Herschel, Lady Mary (*née* Pitt) (*c*.1750–1832), 53, 54, 55, 323

Herschel, Sir William (1738–1822), 29, 30, 42, 43, 44, 53, 63, 67, 93, 94, 95, 96, 109, 115, 196, 234, 243, 274, 275, 276, 277, 285, 307, 308, 318, 322, 323, 324, 329, 330, 331, 340, 341, 342, 343, 373, 397, 401, 403, 405
Hertzsprung–Russell diagram, 139, 140
Hetherington, Barry, 388
Hick, T., 245
Higgs, George (1841–1914), 141, 142, 360
Hind, John Russell (1823–1895), 23, 48, 131, 151, 152, 153, 155, 157, 177, 200, 273, 320, 332, 339, 347, 364, 369, 372, 373, 375
Hind, T. Almond, 364
Hingley, Peter D., 286, 322, 333, 342, 401
Hird, Horace, 343
Historia Coelestis Britannica (Flamsteed), 275
Hodgson, Richard, 108, 318, 350
Holden, Edward S., 51, 314, 346, 354, 392
Holehouse, Mary, 266
Home Words, 217
Hooker, John D., 142
Horncastle College, 405
Horne and Thornthwaite, 190, 218, 374
Hornsby, Thomas (1733–1810), 14, 17, 308
Horrocks, Jeremiah (1619–1641), 395, 396
Horsefield, John, 182, 372
Horton, Samuel, 85, 89, 90, 339, 340
Hoskin, Michael A., 45, 318, 319, 320, 330, 336, 343, 344, 355, 397
Hove Observatory, 340
Howarth, Helen E., 352
Howe, G. Melvyn, 323
Howell, Charles, 90, 340
Howell, James, 167, 168
Howse, H. Derek, 308, 351
Hoyle, Mr, 268, 394
Hubble Space Telescope, 406
Huggins, Lady Margaret Lindsay (*née* Murray) (1848–1916), 116, 122, 139, 252, 284, 285, 286, 292, 345, 352, 353, 397, 401
Huggins, Sir William (1824–1910), 71, 120, 121, 123, 133, 137, 139, 192, 236, 252, 266, 268, 284, 286, 345, 352, 376, 401, 114–117

Hughes, Dr David W., 310, 320
Hughes, T.M., 323
Humble Space Telescope, 406
Hunt, John L., 338, 364
Hunter, A., 401
Hussey, Thomas, 20
Hutchins, Dr Roger, 82, 157, 309, 320, 336, 353, 361, 363, 365, 366, 385
Huxley, Thomas H., 231
Hyperion, 103, 347

Imperial College, 123, 367, 398
Imperial Observatory, St Petersburg, 38
Inkster, Ian, 175, 306, 329, 368, 370, 371
International Astronomical Union, 301
Introduction to Astronomy (Hind), 177
Introduction to Practical Astronomy (Pearson), 39
Inwards, Jabez, 171, 369, 370
Irene, 372
Iris, 48
Irving, Ken, 406

Jackson, Revd Mr, 214
Jackson–Stops, Gervase, 333, 335
Jacob, Capt W.S., 90, 340
Jagger, Miss, 291
James, Stephen H.G., 358, 405
Janssen, Pierre Jules César (1824–1907), 123, 124, 355
Jenkins, Roy, 374, 375
Jennings, Revd Nathaniel, 322, 373
Jepson, Les, 64
Jeremiah Horrocks Observatory, Preston, 244
Jex, J.A., 373
Jex, Johnson (1778–1852), 183–185, 188, 209, 217, 372, 373
Jodrell Bank Radio Telescope, 143, 299, 301
Johns Hopkins University, 141
Johnson, Kevin, 313, 356
Johnson, Manuel J. (1805–1859), 17, 149, 309, 362, 364
Johnston, Joanna, 396
Jones Tybrith, G.F., 265
Jones, Dan, 325, 393
Jones, Edgar, 311
Jones, Fanny (1823–1906), 379
Jones, Dr Gwawr, 378, 379, 380

Jones, Henry, 392
Jones, John (1818–1898), 206, 209–213, 214, 215, 218, 226, 230, 234, 237, 248, 256, 260, 265, 378, 379, 380, 384, 393
Jones, Thomas (1775–1852), 25, 76, 88, 310
Joyce, Revd Jeremiah, 176
Jumbo, 211, 260, 379
Jupiter, 93

Keeler, James E. (1857–1900), 360
Keer, Norman C., 318
Keith, Thomas, 177
Kelly, Howard L., 391, 400
Ken, Bishop Thomas, 369
Kenyon College, Ohio, 384
Kersaris, Paul, 323
Kew Observatory, 20, 117, 309, 310
Kidwell, Peggy A., 280, 398, 399
Kilburn, Kevin, 267, 394, 397
Kilvert, Revd Francis, 226, 382
Kinder, Anthony J., 400, 401
King's College, London, 115
King, Alfred, 345
King, Henry C., 316, 318, 319, 320, 329, 330, 351, 386
King, Maria, 100
King, Samuel, 84
King–Hele, Desmond, 323, 324
Kingsley, Mary, 273, 396
Kirchhoff, Gustav Robert (1824–1887), 113, 114, 115, 119, 171, 194, 270, 352
Kitchener, Dr William, 95, 174, 177, 342, 370
Knobel, Edward Ball, 81, 365, 400
Knock, William, 173, 174, 176, 178, 217
Knott, George (1835–1894), 384
Königsberg Observatory, 18, 19
Kopal, Zdenek, 406
Krisciunas, Kevin, 310, 311, 314, 353

Lacaille, Nicholas Louis de (1713–1762), 23, 311
Lambert, H. Clifton, 216, 380
Lancaster University, 301
Lancaster, Samuel (*b.*1853), 184, 206, 237–239, 386
Langdon, Anne, 214
Langdon, Ellen, 215, 380
Langdon, Roger (1825–1894), 213–217, 218, 230, 237, 248, 259, 380, 383

Lankford, John, 314, 320, 355, 360, 398
Laplace, Pierre (1749–1827), 138, 276
Lardner, Dionysius, 167, 177, 371
Lassell Telescope Project, 349
Lassell, Maria, 87
Lassell, William (1799–1880), 5, 7, 22, 23, 46, 47, 72, 74, 87, 95, 99, 100–104, 106, 107, 108, 109, 110, 113, 114, 125, 126, 127, 138, 141, 153, 154, 184, 211, 217, 232, 243, 249, 278, 301, 320, 337, 341, 344, 345, 346, 347, 348, 349, 350, 351, 365, 378, 384
Lattey, Norman, 255
Laurie, Stephen, 302
Law Society, 222
Lawson, Henry, 200, 244
Le Verrier, Urbain Jean Joseph (1811–1877), 19, 28, 74, 333
Leach, Frances, 199
Leach, John (*b*.1839, *fl*. 1870–), 188–190, 199, 200, 201, 217, 218, 238, 246, 374
Lee Circle, 36, 59
Lee, Cecelia (*née* Rutter), 78, 87, 335
Lee, Dr John (1783–1866), 11, 27, 30, 36, 61, 62, 75, 76, 77, 78, 81, 82, 84, 85, 86, 87, 88, 89, 90, 145, 151, 169, 171, 229, 245, 298, 310, 314, 324, 333, 334, 335, 336, 337, 338, 339, 340, 344, 360, 368, 369, 370, 372, 375, 387, 388
Leeds Astronomical Society, 57, 166, 245–247, 251, 254, 255, 256, 262, 264, 270, 274, 279, 289, 291, 296, 298, 327, 328, 368, 378, 385, 388, 389, 392, 394, 395, 396, 399, 402, 403, 405, 405
Leeds Philosophical and Literary Society, 57, 166, 245, 368, 388
Leeds University, 399
Leisure Hour, 207
Lemmon, G.F., 230, 384
Les Femmes dans las Science (Rebière), 284
Lescarbault, Edmond Modeste, 28
Leviathan of Parsonstown, 65, 66, 98
Lewis, Dr H.N., 373
Lewis, Samuel, 311
Lick Observatory, 51, 132, 137, 142, 360
Lick, James, 360
Liebig, Justus von, 353
Lindop, Sir Norman, 317

Lindsay, James Ludovic, Lord, (Fourth Earl of Crawford and Balcarres) (1847–1913), 24, 151, 207, 312, 356, 133–135
Linné, 336
Linnean Society, 33
Literary and Philosophical Societies, 243
Liverpool Astronomical Society, 247–251, 252, 254, 268, 279, 281, 298, 325, 360, 368, 374, 390, 391, 392, 398, 404
    Pernambuco Branch, 248, 390, 398
Liverpool Dock and Harbour Board, 22, 311
Liverpool Dock Museum, 311
Liverpool Literary and Philosophical Society, 243, 247
Liverpool Museum, 106, 178, 311, 346, 351, 360, 372
Liverpool Nautical College, 254
Liverpool Observatory, 20, 2, 309, 311, 387
Llewelyn, Dillwyn, 340
Lloyd, J. Alun, 265, 326
Lloyd, Mr, 388
Llyfni Valley Astronomical Society, 266
Locke, Richard Adams, 56, 324
Lockyer, Sir Joseph Norman (1836–1920), 116, 122–124, 133, 192, 221, 232, 241, 286, 345, 352, 353, 355, 356, 381, 398
Lockyer, Mary T., 355
Lockyer, William J.S., 356
Lockyer, Winifred L., 355
Lodge, Sir Oliver Joseph (1851–1940), 270, 395
Lohrmann, Wilhelm Gotthelf (1796–1840), 27, 314
Lohse, J. Gerhardt, 356
Longbottom, Gertrude (1876–1957), 287, 290, 292, 402
Lord, Christopher J.R., 347
Lovelace, Countess of, 277
Lovell, Sir Bernard, 301
Lowe (*later* Pearson), Frances, 38, 316
Lowe, Gavin (1744–1815), 36, 38, 76, 316, 317
Lowe, H.E., 384
Lowell, Percival (1855–1916), 255, 392
Lubbock, Constance Anne (*née* Herschel), 322
Lunan, Duncan, 388

lunar geology, 105
Lyell, Sir Charles (1797–1875), 9, 307

M13, 228
M15, 227
M32, 257
M51, 72, 98, 178, 197, 198, 344
M97, 69
Mackintosh, David, 168
Maclear, Sir Thomas (1794–1879), 22, 23, 76, 82, 83, 87, 151, 230, 287, 331, 337, 383
MacLeod, Roy M., 306, 329
Mädler, Johann Heinrich (1794–1874), 19, 27, 81, 310
Madras Observatory, 23, 151, 292, 311
Main, Revd Robert (1808–1878), 20, 147, 150, 312, 361, 362
Malby, Messrs, 190, 374
Malden, B.J., 269, 395
Manchester Astronomical Society, 267, 268, 290, 291, 301, 393, 394, 395, 396, 397, 402, 403
Manchester City Library, 394
Manchester Free Trade Hall, 166
Manchester Literary and Philosophical Society, 243, 394
Manchester University, 406
Mann, Dr, 176
Manning, Brian G.W., 302
Markree Catalogue (Cooper), 49, 320
Markree Observatory, 25, 101, 149, 153, 157, 320, 334
Markwick, Ernest Elliot, 256, 393
Marlborough, Duke of, 95, 146, 341, 342
Marriott, Robert A., 231, 232, 320, 347, 352, 384, 385, 400
Mars, 82, 93, 105
Marshall, Edward G., 311
Marth, Albert (1828–1927), 48, 87, 103, 104, 153, 154, 311, 362, 365
Martin, A.R., 316
Martin, Benjamin, 176, 182, 371
Martin-Smith, Dr Michael, 407
Maskelyne, Nevil (1732–1811), 14, 308, 316, 341
Matthews, Revd Mr, 233, 385
Maudslay and Field, 107
Maudslay, Henry, 102

Maunder, Annie Scott Dill (*née* Russell) (1868–1947), 156, 252, 253, 269, 280, 282, 283, 285, 288, 289, 290, 366, 391, 399, 400, 401, 402
Maunder, Edward Walter (1851–1928), 156, 251, 253, 256, 266, 270, 282, 367, 400
Maury, Antonia, 278
Mauvais' Comet, 73, 89, 332, 339
Maw, William Henry (1838–1924), 252
Maxwell, James Clerk (1831–1879), 296
Mayhew, Henry, 173, 175, 176, 177, 205, 314, 370, 371, 372
McClean, Ellen (*née* Greg), 135
McClean, Frank, (1837–1904), 135–139, 356, 357
McClean, John Robinson, 135
McConnell, Dr Anita, 236, 329, 385, 386
McCrae, Sir William, 404
McCutcheon, Robert A., 314
McDonald Observatory, 360
McElderry, John H., 393, 399
McKenna–Lawlor, Susan, 320
McKeown, Mary, 402
McKie, J.N., 373, 385
McKim, Dr Richard J., 320
McLarin, John B., 85, 145
Meadows, A.J., 307, 355
Meares, John Willoughby (1871–1945(6?)), 253, 256, 296, 326, 327, 391, 403, 404
Mécanique Céleste (Laplace), 276
Mechanics' Institutions, 165, 166, 171, 176, 243, 371
Mechanism of the Heavens (Olmsted), 210
Mechanism of the Heavens (Somerville), 276
Mee, Arthur, 213, 255, 265, 378, 380, 382, 392, 393
Melbourne Telescope, 109, 341, 349, 350, 351, 353, 365
Melhaps' Comet, 89, 339
Melpomene, 372
Men of Invention and Industry (Smiles), 205
Merz refractors, 103, 233, 234
Merz, Georg, 19, 28
meteoritic material, 133
meteorological observers, 338
Meteorological Society, 85, 338, 340
meteorology, 85, 152, 338
meteors, 133
Metis, 49

Index   421

Meudon Observatory, 19
Michell, Ormsby McKnight, 177, 372
Microscope Demonstrator, 173, 174, 176, 177, 206, 217
middle classes, 6
Midlands Observatory, 244
Millburn, John, 371
Miller, David P., 315
Miller, William Allen (1817–1870), 115, 352
Mills Observatory, Dundee, 244, 388
Mills, Charles E., 352
Mills, Edward (1803–1865), 172, 370
Milward, Henry, 201, 376
Minchin, George M. (1845–1914), 139–141, 359
Minto, Lord, 186
mirror support system, 101, 105, 201
Mitchell, Maria (1818–1889), 261, 278, 398
Mollan, Charles, 313, 398
Monck, William Henry Stanley (1839–1915), 47, 139–141, 251, 359, 391
Moon, 19, 81, 82, 105, 113, 118, 119, 296, 299, 336
Moore, Patrick, 299, 300, 310, 336, 346, 363, 397, 405, 406
Morando, Bruno, 310
Morrell, Jack Bowes, 305, 306, 315, 329, 387
Morris, Rhys, 370, 378, 393
Morrison–Low, A.D., 359
Moscow Observatory, 19, 39
Moseley, Maboth, 398
Mount Wilson Observatory, 360
Mozans, H.J., 396, 398
Mull, Mr, 174
Müller, Friedrich Max, 30, 315
Multhauf, Lettie S., 309, 310
Munich Observatory, 19
Munnings, Revd T., 183
Murray, Andrew, 316
Murray, John, 285
Muschamp Perry, Revd Jevon John (b.c.1845), 233, 385
Museum of the History of Science, Oxford, 13, 54, 90, 95, 178, 191, 317, 322, 349, 353, 357, 361, 373, 385

Nacional de Ciencia y Tecnologia, Madrid, 310

Nairne, Edward, 308
Nasmyth focus, 104, 128, 348
Nasmyth, Alexander, 102
Nasmyth, James (1808–1890), 5, 10, 24, 27, 72, 82, 85, 94, 99, 102, 104–109, 110, 113, 114, 117, 122, 123, 126–129, 215, 218, 235, 301, 306, 312, 313, 333, 337, 338, 341, 346, 347, 348, 349, 350, 353, 386
National Academy of Sciences, 320
National Maritime Museum, 366
National Railway Museum, York, 215, 259
Natural Theology (Paley), 169
nebulae, 96, 98, 103, 104, 115, 137, 178, 274, 330, 343, 365
nebulium, 115
Neptune, 19, 48, 49, 74, 103, 106, 110, 152, 270, 310, 319, 320, 347, 349, 364
Newall, Robert Stirling, 153. 194, 235, 236, 249, 254
Newbury Amateur Astronomical Society, 302
Newby, Eric, 396
Newcastle University, 208
Newcastle University Observatory, 404
Newcastle–upon–Tyne Astronomical Society, 269, 298, 395, 404
Newcomb, Simon (1835–1909), 354
Newiss, Joan, 365
Newman, Cardinal John Henry (1801–1890), 17, 309
Newton, Sir Isaac (1642–1727), 94
Newtonian gravitation, 43
Nichol, John Pringle (1804–1859), 99, 166, 177, 178, 188, 344, 374
Nicholls, G.W., 248
Noble, Capt William Henry (1828–1904), 249, 252, 256, 300, 336, 390, 391
Norman Lockyer Observatory, 133, 193, 356
Norman Lockyer Observatory Society, 192, 356
Norman, Ann, 123
Northop, David K., 260
Northumberland Refractor, 21
Norton, Arthur Philip (1876–1955), 299, 405
Notre Dame University, 389
Nottingham Literary and Philosophical Society, 202, 377
Nottingham Observatory, 375

Nottingham University, 260, 377
Nottingham University Library, 375, 376, 377
Nottinghamshire Archive Office, 376

O'Brien Stafford, A., 362
O'Callaghan, P., 246
O'Sullivan, Dr Colm, 398
Oates, Michael, 125
Oberon, 93
object glasses, 234
observatory directors, 16
Observing Astronomical Society, 390
Oceans of the Air (Giberne), 287
Okell, Samuel, 266, 269
Olbers, Heinrich Wilhelm Matthias (1758–1840), 4, 18, 27, 310
Ollerenshaw, Dame Kathleen Mary, 301, 406
Olmsted, Denison, 177, 210, 371, 372
Orbs Around Us (Proctor), 296
Orchard Dr, 395
Orion Nebula, 69, 103, 104, 134, 138, 196, 197
Orwell, George (1903–1950), 179, 372
Osterbrock, Donald W., 306, 360
Outlines of Astronomy (Herschel), 177, 210
Owain, Steffan ab, 379
Owen's College, Manchester, 278
Owens, Captain, 210
Oxford University, 8, 13, 17, 155, 278
Oxford University Museum of Natural History, 347
Oxford University Observatory, 17, 118, 191, 270, 309, 336
Oxmantown, Lord, *see* Rosse

Paisley Observatory, 244
Palermo Observatory, 35
Paley, Revd William, 169
Palmer, Courtlandt, 120
Palomar Observatory, 360
Pannekoek, A., 370
parallaxes, 42, 140
Paramatta Observatory, 23
Paris Observatory, 292
Parry Jenkins, G., 265, 379, 381, 382
Parsons, William, *see* Rosse, Third Earl of
Paton, Charles (1849–1938), 154, 155, 261, 366

Paton, Emily, 154
Patterson, Mary C., 397
Pattinson, Hugh, 236
Peacock, William, 398
Pearce, Professor, 155, 366
Pearson, Revd William (1767–1847), 29, 30, 37, 57, 76, 166, 176, 298, 301, 312, 313, 315, 316, 317, 318, 322, 329, 334, 340, 341, 342
Peel, Sir Robert, 332
Pennington, Philip, 390
Perceval, Spencer, 14
Perigal, Henry, 338
Pernambuco Astronomical Society, 248, 390
Perry, Fr Stephen Joseph (1833–1889), 281, 399, 400
Phelps, Thomas (*b*.1694), 146, 361
Phillips, John, 17, 82, 118, 235, 243, 309, 336, 347, 350, 354
Phillips, Revd Theodore Evelyn Reece (1864–1942), 228, 233, 288, 297, 299, 300, 301, 301, 320, 329, 336, 386, 395, 403, 404
photoelectric photometry, 139
Photographic Atlas of the normal Solar Spectrum (Higgs), 141
Photographic Society of Ireland, 343
Photographs of Stars, Star Clusters and Nebulae (Roberts), 138
photography, 113
photoheliograph, 117
photometry, 71, 117, 121, 139
Physical Geography (Somerville), 277
Piazzi, Guiseppe (1746–1826), 79
Pickard, Roger D., 406
Pickering Fellowship for Women, 284
Pickering, Edward Charles (1846–1919), 122
Pike, W.S., 337
Pilkington Glassworks, 211, 237
Pinion, F.B., 386, 387
Pinnock, William, 380
Pizor, Faith K., 324
planetary nebulae, 71
plate–measuring, 122
Plea for Reflectors (Browning), 230
Pleiades, 138, 358, 359
Plotkin, Howard N., 355
Plummer, John I., 311
Pogson (*later* Kent), Elizabeth Isis, 151, 292

Pogson, Norman Robert (1829–1891), 8, 48, 61, 86, 89, 148, 151, 152, 153, 155, 200, 218, 292, 307, 311, 338, 362, 364, 364
Pole, William, 306, 349
political reform, 221
Pond, John (1767–1836), 29, 37, 44, 118, 123, 312, 316
Pontécoulant, Phillipe Gustave Doulcet de, 189
Popular Astronomy (Airy), 177
Popular History of Astronomy in the Nineteenth Century (Clerke), 270, 285
Porthouse, William, 403
post–war boom, 300
Potsdam Astrophysical Observatory, 117, 156
Practical Astronomy (Dick), 176, 206
Practical Observations on Telescopes (Kitchener), 176
Pritchard, Ada, 309, 354
Pritchard, Revd Charles (1808–1893), 17, 116, 118, 123, 235, 287, 296, 299, 309, 354
Problems in Astrophysics (Clerke), 286
Proctor, Mary (1862–1957), 283, 288, 289, 290, 292, 300, 400, 402, 403
Proctor, Richard Anthony (1837–1888), 142, 150, 177, 222, 240, 241, 254, 265, 277, 288, 289, 292, 296, 360, 363, 372, 387, 397, 398, 402
Proctor-Smyth, Sallie Duffield (1856–1941), 269, 289, 290, 399, 402
professional, 122, 295
proper motions, 42, 140
Prosper, Henry, 138
Prosper, Paul, 138
Public Libraries Act of 1850, 221
Pulkowa Observatory, 18, 19, 24, 28, 133, 249, 314
Purdy, R.L., 386

Queen's College, Cork, Observatory, 26, 264, 309, 313

Radcliffe Observatory, 14, 16, 20, 35, 89, 149, 151, 309, 317, 341, 342, 362
Radcliffe Science Library, 381, 393, 398
Radcliffe Trust, 16
Radcliffe, Dr John, 16, 308, 309

Radcliffe, R.D., 392
radial velocities, 115
Radiant Suns (Giberne), 287
radio observatories, 143
railways, 87
Ramage, John, 68, 95, 99, 188, 330, 341, 342, 374
Rambaut, Arthur Alcock (1859–1923), 359
Ramsden, Jesse (1735–1800), 25, 27, 35, 59, 232, 310, 384
Reade, Revd G.D., 333, 339
Reade, Revd Joseph Bancroft (1801–1870), 76, 89, 166, 298, 339
Rebière, Adolphus, 284, 401
Redhill Catalogue (Carrington), 40, 317, 362
Redhill Observatory, 40, 149
redshift, 115
Rees, Abraham, 312, 313, 316, 317, 322, 334, 340
Rees, W., 168
Reform Act of 1840, 22
Reform Act of 1867, 375
Regulus, 140
Reichenbach, George Friedrich von, 19
Reingold, Nathan, 314, 398
Renshaw, S. Alex, 201, 375
Repsold, Adolf, 19
Results of Astronomical Observations at the Cape of Good Hope (Herschel), 73, 332
reversing layer, 114
Reynolds, E.E., 333
Reynolds, J.H., 359
Rigaud, Stephen Peter, 17
Ring, Francis J., 323
Roberts, Dorothea (*née* Klumpke) (1861–1942), 138, 139, 292, 359
Roberts, Dr Isaac (1829–1904), 135–139, 195, 236, 250, 251, 257, 268, 292, 301, 357, 358, 359, 367, 377, 390
Roberts, Eleazar (1825–1912), 210
Roberts, Paul W., 125
Roberts, Robert (1777–1836), 212, 380
Roberts, Silas, 213
Robertson, Abram (1751–1826), 15, 17
Robertson, John, 206–209, 211, 213, 214, 216, 218, 226, 235, 378, 379, 386
Robinson, Leif J., 314, 336
Robinson, Mr, 189, 199, 374, 378

424 Index

Robinson, Thomas Romney (1792–1882), 21, 25, 26, 40, 45, 98, 99, 101, 105, 148, 149, 285, 298, 312, 313, 317, 319, 344, 345, 346, 350, 362
Robinson, William Henry (1847–1926), 268, 269, 395
Rochester, Prof G.D., 310, 311, 362
Roden, Bridgid, 343
Rodway, Rosa Helen, 41
Rodway, William, 41
Rogerson, William, 147
Romsey observatory, 229, 383
Roscoe, Henry E., 352
Ross and Company, 245, 264
Rosse, Fourth Earl of (1840–1908), 99, 150, 153, 365
Rosse, Countess of, (*née* Mary Wilmer Field), 96, 343
Rosse, Third Earl of (1800–1867), 25, 27, 64, 72, 74, 96–100, 102, 104, 105, 106, 109, 113, 114, 119, 126, 133, 145, 149, 150, 178, 197, 198, 211, 217, 230, 232, 277, 285, 312, 313, 319, 331, 336, 342, 344, 345, 346, 347, 348, 349, 350, 351, 362, 363, 384, 398
Routh, E.J., 309, 368
Rowland, Henry Augustus (1848–1901), 141
Royal Astronomical Society, 5, 7, 11, 29, 33, 34, 39, 40, 45, 46, 57, 81, 135, 136, 243, 244, 249, 275, 287, 288, 292, 295, 297, 298, 305, 312, 315, 317, 318, 322, 324, 337, 357, 361, 397, 404
Royal Cork Institution, 398
Royal Institution, 10, 38, 276, 348
Royal Institution, Cornwall, 361
Royal Insurance Lassell Telescope Project, 125, 349
Royal Meteorological Society, 86, 282, 338
Royal Mint, 73, 332, 333
Royal Naval Observatory, Cadiz, 310
Royal Observatory, Cape of Good Hope, 20, 23, 83, 135, 151, 157, 285
Royal Observatory, Edinburgh, 23, 134, 150, 153, 244, 309, 312, 356, 363, 364, 378
Royal Observatory, Greenwich, 3, 5, 13, 14, 34, 35, 37, 51, 85, 86, 95, 116, 146, 147, 148, 149, 152, 155, 156, 189, 231, 240, 251, 280, 282, 285, 309, 312, 319, 320, 335, 338, 351, 361, 362, 363, 365, 366, 367, 387
Royal Observatory, Madrid, 19, 282
Royal Scottish Museum, Edinburgh, 186
Royal Society, 6, 11, 13, 13, 27, 29, 33, 34, 44, 57, 98, 99, 102, 106, 107, 108, 116, 243, 249, 275, 276, 277, 316, 323, 324, 329, 332, 338, 349, 355, 360, 389, 390
Ruat, William, 24, 312
Rugby School, 26, 28, 233, 296
Rümker, George, 87, 149, 311, 362
Rupke, Nicholaas A., 306
Russell(?), Flossie (mistress of James Nasmyth), 350
Russell, Hester Dill, 400
Rutherfurd, Lewis Morris (1816–1892), 50, 116, 118, 122, 194, 290, 320, 354
Ryan, Tony, 313
Ryan, W.F., 384
Rylands, Thomas Glazebrook, 254, 392
Ryves, P.M. (*d*.1956), 384, 401

Sabine, Sir Edward (1788–1883), 40
Sadler, Herbert (1856–1898), 81, 254, 336
St Andrew's University, 348, 350
Salford Astronomical Society, 301, 406
Sallit, George, 302
Salt, Titus, 245
Sanderson, Alan, 404
Satterthwaite, Gilbert, 367
Saturn, 47, 93, 103, 110, 227, 328
Savary, Félix, 46, 58, 319, 329
Saxony, King of, 332
Schaaf, Larry J., 324, 332
Schiaparelli, Giovanni, 249
Schrader, Johann, 94
Schröter, Johann Hieronymus (1745–1816), 4, 18, 27, 94, 310, 341
Schumacher, Heinrich Christian (1790–1850), 186
Schuster, J.A., 315
Schwabe, Heinrich (1789–1875), 19, 40, 41, 117, 227
Science Museum, London, 61, 105, 109, 124, 129, 274, 313, 350, 351, 390, 397
Scientific Dialogues (Joyce), 176
Scientific Instrument Society, 387
Scientific Society of Boston, 253
Scott, Alexander, 186

Scott, Benjamin, 314
Scott, Gideon (1765–1833), 188, 373
Scott, Sir Walter (1771–1832), 185, 186, 187, 373
Sears, David, 28, 314
Sebastián, Dr Amparo, 310
Secchi, Angelo (1818–1878), 116, 117, 121, 122, 295, 353, 403
Secord, Dr Anne, 181, 307, 372
Sedgwick, Adam (1785–1873), 67, 306, 323, 329
segmented mirrors, 97
Self–Help (Smiles), 205
Seryddiaeth a Seryddwr [*Astronomy and Astronomers*] (Evans), 265
Shankland, Revd Thomas, 379
Shapkley, Revd Mr, 235
Shapley, Harlow, 352
Sharman, Percy, 376, 377
Sharp, Irwin, 154
Shea, Charles, 90, 340
Sheehan, William, 310, 314, 341
Sheeky, Michael, 363
Sheepshanks, Anne (1789–1876), 292
Sheepshanks, Richard (1794–1855), 16, 45, 77, 80, 103, 292, 316, 318, 335, 346
Sheffield Literary and Philosophical Society, 348
Sheffield Mechanics' Institution, 166
Sheldon, Dr Steele, 266, 269
Sheldon, Mrs, 279
Shelton, John, 24
Shimeld, J.E., 376, 377
Short, James, 24, 94, 95, 341, 342
Short, Thomas, 342
Shuckburgh Equatorial, 59
Shuckburgh, Sir George, 26, 36, 313
Sidgreaves, Fr Walter, 253, 266, 269, 394
Sidmouth Amateur Radio Society, 356
Sidmouth and District Astronomical Society, 356
silver–on–glass mirrors, 110, 119, 230
Silverton station, 214, 259, 380
Simcock, Anthony, 353
Simmonds, George Harvey, 40, 149, 362
Simms refractors, 234
Simms, William, 43, 44, 45, 84, 337, 347
Simpson, Don, 154, 261, 365, 366, 385
Simpson, R., 312

Sinden, David, 404
Sirius, 115
Sisson, John and Jeremiah, 24
skeleton tube, 107, 201
Slater, F., 130
Slavings, Ricky L., 320
Slawinski, Peter, 29
Slipher, Vesto M., 358
Smeaton, John (1724–1792), 38, 340
Smiles, Samuel, 181, 205, 206, 209, 210, 211, 212, 235, 237, 239, 313, 320, 333, 338, 340, 341, 350, 370, 377, 378, 379, 380, 384, 385, 386
Smith, Adam, 307
Smith, Capt John, 79
Smith, Robert W., 347, 349
Smyth, Caroline Mary, 87, 168, 368, 369
Smyth, Charles Piazzi (1819–1900), 22, 23, 24, 62, 79, 83, 87, 88, 108, 123, 134, 141, 142, 150, 151, 208, 249, 311, 312, 335, 337, 350, 360
Smyth, Ellen Philadelphia, 87
Smyth, Georgiana Rosetta, 168, 368, 369
Smyth, Henrietta, 87
Smyth, Dr J.C., 289
Smyth, Mary, 364
Smyth, Michael, 364
Smyth, Admiral William Henry (1788–1865), 5, 8, 10, 23, 39, 48, 60, 62, 74, 75, 77, 78, 79, 81, 82, 83, 85, 86, 89, 90, 145, 169, 217, 225, 229, 287, 300, 307, 316, 319, 329, 333, 334, 335, 338, 339, 340, 360, 364
social reform, 221
solar cycle, 41
solar eclipses, 105, 281–283
Solar Physics Observatory, 123, 124, 133, 356
solar spectrum, 113, 114
Solar System (Religious Tract Society), 176
Solly, Revd Henry, 172, 370
Solomon, Messrs, 211, 379, 383
Somerville, Martha, 373, 397
Somerville, Mary (1780–1872), 57, 97, 185, 186, 187, 255, 274, 275, 276–278, 280, 289, 292, 324, 343, 344, 373, 392, 396, 397, 398, 402, 403
Somerville, Dr Thomas, 276
Somerville, Dr William, 276, 397, 398
Sorby, Clifton, 348

South Shields Central Hall, 165, 367
South Villa Observatory, 46, 47, 151, 152, 153, 177, 365
South, John Flint, 44
South, Sir James (1785–1867), 4, 25, 30, 44, 49, 56, 58, 60, 74, 76, 79, 80, 98, 229, 277, 310, 313, 318, 319, 324, 334, 336, 343, 344, 398
Southern Equatorial Telescope, 107, 108
southern hemisphere telescope, 106
Southern Telescope Committee, 106
Southey, Robert (1774–1843), 307
Space Race, 299
spectrograph, 120
spectroscopy, 113
speculum metal, 96, 109
Splendour of the Heavens (Phillips and Steavenson), 299
Sprat, Thomas, 308
Spring Rice, T. (1790–1866), 15, 308
Sputnik, 299
Star Atlas (Norton), 299
steam hammer, 102
steam-powered polishing machine, 98, 345
Steavenson, William Herbert (1894–1975), 297, 299, 320, 336, 367, 386, 404
Stebbins, Joel, 359
Steinheil, Karl Augustus von, 110
Stella Maitland (Hawkins), 288
stellar distribution, 140
stellar parallaxes, 42, 140
stellar proper motions, 42, 140
Stephenson, George, 205
Stevens, Catherine Octavia (1864–1959), 283, 287, 290, 400, 402
Stewart, Balfour, 355
Stewart, Mrs Duncan, 331
Stone, Edmund James, 353, 357, 363
Stoney, Bindon Blood (1829–1909), 150, 343
Stoney, George Johnstone (1826–1911), 150, 359, 363
Stonyhurst College, 26, 28, 296
Stonyhurst College Observatory, 253, 281
Strangeways, Revd F.H. Fox, 216
Stride, James, 363
Struve, Friedrich Georg Wilhem (1793–1864), 19, 20, 28, 310, 314
Struve, Otto Wilhelm (1819–1905), 19, 109, 249, 309, 310, 345, 351

Stuyvesant, Peter, 118
suffragette movement, 289
Suggett, Martin, 346, 351
Sun, 40, 71, 105, 123, 140, 171, 348
Sun, Moon and Stars (Giberne), 286, 287
sunspot cycle, 19
Supernumery Computer system, 157
Sussex, Duke of, 57, 308
Symons, James (1838–1900), 378
System of the Stars (Clerke), 286, 331
System of the World (Nichol), 177

Tanner, Lloyd, 256
Taton, René, 310
Tayler, R.J., 376
Taylor *(later* Hildred), Florence, 255, 262, 269, 270, 274, 275, 278, 279, 289, 291, 392, 396, 399, 402
Taylor, Arthur, 406
Taylor, Henry, 37
Taylor, William Arthur, 301
Teague, E.T.H., 317
Teasdale, Washington, 254, 328, 392, 402
Tebbutt's Comet (Great Comet of 1881), 136, 202, 240
Tebbutt, John, 27, 253, 314, 392
Telstar, 299
Tempel, Ernst Wilhelm (1821–1889), 138
temperance movement, 289
Templar, Benjamin, 361
Temple Observatory, Rugby School, 233, 263, 352
Tennant, James Francis, 232
Test Act of 1871, 155
Thackeray, A.D., 309, 400, 405
Thackray, Arnold, 305, 306, 315, 329, 387
The Amateur Astronomer (Moore), 300
The Sky at Night, 299
The Telescope: a familiar sketch (Ward), 285
Théorie analytique du système du monde (Pontécoulant), 189
Thom, Alexander, 313
Thomas, Dillwyn, 90
Thompson, Flora, 369
Thompson, Revd Robert A., 311
Thorp, Mrs, 279
Thorp, Thomas (1850–1914), 394
Thorpe, Archdeacon, 311
Titania, 93

Tiverton Museum, 259
Todd, Alice Gilliam (*née* Bell), 151, 364
Todd, Sir Charles (1826–1910), 151, 246, 363
Todd, David, 373
Todhunter, Isaac, 189, 238, 374
Tow Law History Society, 404
Townsend, H.J., 254, 327, 328
Tranmar, Miss, 291, 389, 403
transit of Venus, 134, 240, 381, 395, 396
Trant (*later* Scott), Mrs, 389
Trant, William, 246, 255, 388, 389, 392
Trapezium, 103
Treatise on Astronomy (Herschel), 79, 177, 178, 210
Treatise on Optics (Brewster), 188
Treatise on the Globes (Keith), 177
Treatise on the Solar System (Dick), 206, 210
Treatise on the Telescope (Herschel), 200
Tregent, Mr (*b.*1808), 174, 175, 177, 178, 182, 206, 248, 258, 314, 370, 371
Trinity College Dublin, 140
Triton, 103, 110, 347
Trollope, Anthony (1815–1882), 84
Trollope, Fanny, 396
Trollope, Frances, 273
Troughton and Simms, 27, 40, 84, 129, 240
Troughton, Edward, 30, 35, 36, 38, 43, 44, 45, 59
Troughton, John, 35
Tulley refractors, 100, 234
Tulley, Charles, 39, 43, 77, 89, 95, 317, 334
Tulley, Thomas, 225
Tulley, William, 225
Turbridy, Michael, 344, 351
Turnbull, Dr David, 395
Turnbull, Joseph, 89, 339
Turner, Herbert Hall (1861–1930), 255, 270, 284, 307, 309, 312, 315, 317, 319, 329, 334, 356, 357, 364, 392, 397
Two on a Tower (Hardy), 239, 288
Tyndall, John, 231

Ulster Astronomical Society, 266, 280, 393, 399
Umbriel, 103, 347
Under an English Heaven (Kay Williams, biography of G.E.D. Alcock), 299
University of Manchester Institute of Science and Technology, 268

University of Texas, 323, 324, 389, 370, 378
Unsöld, Albrecht, 352
Urania's Mirror, 286
Uranian Society, 90, 244, 251, 340, 387
Uranus, 35, 54, 93, 275, 340, 341, 347, 397, 401
Ushaw College, 311

Vassar College, 278
Vassar College Observatory, 28, 261, 278, 398
Vassar, Matthew, 278
Vega, 120
Veitch, Betty, 187
Veitch, James (1771–1838), 95, 185–188, 208, 209, 213, 217, 258, 276, 373
Veitch, Revd Dr James, 188
Victoria, Queen, 201, 396
Vogel, Hermann Carl (1842–1907), 117

Wales, Prince of, 201
Wales, William, 37
Walker, Anne (*b.c.*1864), 157
Walker, J.M., 337
Wall, John, 301
Wallace, Alexander, 150, 363
Wallace, W. Stewart, 389
Walsall Atronomical Society, 395
Walsall Public Libraries, 395
Walsall Science and Art Institute, 269
Walton, Revd W., 169
Ward, Mary (1827–1869), 273, 285, 286, 401
Ward, Robert, 182, 227
Warington, Anarella, 79
Warner, Brian, 323, 331, 397
Warner, Deborah Jean, 314, 320, 352, 354, 359, 382, 385, 398
Waterfield, Dr Reginald Lawson (1900–1986), 298, 404
Watson, G., 373
Watson, J., 165
Watt Institution, 244
Watt, Michael J., 337
Watt, Louisa, 337
Waugh, James, 256
Wayman, Patrick A., 312, 345

Weale, John, 26, 30, 85, 89, 145, 181, 225, 313, 315, 318, 319, 338, 339, 340, 360, 372
Webb, Henrietta (*née* Montagu Wyatt), 226
Webb, Sir Henry, 226
Webb, Revd John, 226
Webb, Revd Thomas William (1806–1885), 50, 218, 225–228, 229, 232, 241, 247, 249, 262, 263, 281, 297, 300, 301, 336, 381, 382, 383, 393, 399
Weir, Thomas, 266, 267, 268, 394, 402
Weitzenhoffer, Kenneth, 359
Welsh books, 213
Welsh chapels, 169
Wesley, William H., 359
West Hendon House Observatory, 27, 154, 155, 313
West, Algernon, 146, 361
Westbury Circle, 37, 44
Whatton, A.B., 396
Wheddon, Hector, 406
Whewell, William (1794–1866), 9, 33 57, 275, 276, 306, 307, 315, 329, 398
Whitbread, Samuel Charles (1796–1879), 27, 76, 84, 85, 86, 117, 129, 145, 287, 337
White, W., 244, 340
White, W.H., 170
Whitmell, Charles, 289, 402
Whitney, Charles A., 360
Whitney, Mary Watson, 278
Whittaker, Alan, 406
Wicks, Catherine, 148, 362
Wickstead, Revd Charles, 209
Wigmore, Robert H., 390
Wilde, Mr, 394
Wilding, R., 266
Wilkins, George A., 356
Wilkins, H. Percy (1896–1960), 299, 336, 405
Wilkins, John, 9
William, T.R., 320
Williams, Revd Cadwalladr, 209, 210
Williams, Isaac, 348
Williams, John, 105, 348
Williams, L. Pearce, 307

Williams, Stuart, 395
Wilson, Curtis, 310
Wilson, Fiametta (1864–1920), 283–284, 292, 401
Wilson, James M., 320, 357
Wilson, R., 308, 310
Wilson, William Edward (1851–1908), 139, 359
Winchester School, 296
With, George Henry (1827–1904), 223, 227, 228, 230, 231–234, 238, 262, 265, 301, 382, 384, 385
Wollaston, William Hyde (1766–1828), 276
women astronomers, 273–293
Wood, Sir Charles, 15, 309, 312, 315, 361, 362
Woodham-Smith, Cecil, 369
Woods, Thomas, 314, 344
Wordsworth, William (1770–1850), 10, 175, 307, 371
Working Men's Exhibition of 1870, 200, 260, 376
World's Foundations (Giberne), 287
Wray refractors, 284
Wren, Sir Christopher, 13
Wrottesley, John, Lord (1798–1867), 22, 23, 39, 133, 245, 317

X-rays, 267

Y Dorspath heulawg [The Solar System] (Dick), 210
Yeates & Sons, 140, 359
Yeo, R.R., 315
Yeomans, Donald K., 339, 369
Yerkes Observatory, 51, 360
York Minster, 3, 235
Yorkshire Philosophical Society, 243
Young, Arthur, 184, 373
Young, Thomas (1773–1829), 276
Yr anianydd Christionogol [The Christian Philosopher] (Dick), 210

Zach, Franz Xaver von (1754–1832), 18, 309
Zodiacal Light, 70

Printed in October 2022
by Rotomail Italia S.p.A., Vignate (MI) - Italy